Design of Experiments

Statistical Principles of Research Design and Analysis

www.duxbury.com

Valuable resources @ no additional charge

Design of Experiments

Statistical Principles of
Research Design and Analysis

Second Edition

Robert O. Kuehl
The University of Arizona

Pacific Grove • Albany • Belmont • Bonn • Boston • Cincinnati • Detroit • Johannesburg • London • Madrid
Melbourne • Mexico City • New York • Paris • Singapore • Tokyo • Toronto • Washington

Sponsoring Editor: *Carolyn Crockett*
Marketing Team: *Karin Sandberg, Laura Hubrich*
Editorial Assistant: *Kimberly Raburn*
Production Coordinator: *Keith Faivre*
Production Service: *Greg Hubit Bookworks*
Manuscript Editor: *Lura Harrison*

Permissions Editor: *Connie Dowcett*
Interior Design and Illustration: *Susan Rogin*
Cover Design: *Laurie Albrecht*
Print Buyer: *Vena Dyer*
Typesetting: *Bob Lande*
Printing and Binding: *R.R. Donnelley & Sons, Crawfordsville*

For more information, contact Duxbury Press at Brooks/Cole Publishing Company:
BROOKS/COLE
511 Forest Lodge Road
Pacific Grove, CA 93950 USA
www.brookscole.com

Printed in the United States of America.

10 9 8 7 6 5 4 3 2 1

Library of Congress Cataloging-in-Publication Data

Kuehl, R. O.
    Design of experiments: statistical principles of research design
and analysis / Robert O. Kuehl.—2nd ed.
        p.   cm.
    Rev. ed. of: Statistical principles of research design and
analysis. c1994.
    Includes bibliographical references and index.
    ISBN 0–534–36834–4
    1. Science—Experiments—Statistical methods.   2. Research—
Statistical methods.   3. Statistics.   4. Numerical analysis.
I. Kuehl, R. O.   Statistical principles of research design and
analysis.   II. Title.
Q182.3.K84   2000
001.4′22—dc21   99-23308

To Mom and Dad,

Schleswig,

Uncle Henry,

and the rabbit sandwich

Contents

Preface

Objectives

My objective for this text is to present the principles of statistical design and analysis for comparative scientific studies to graduate students in the experimental sciences and applied statistics. For this second edition, the primary title *Design of Experiments* was added to make the text more identifiable while leaving the original title to portray the text's objective. *Statistical Principles of Research Design and Analysis* describes the philosophy that successful comparative experiments have clearly defined objectives that can be addressed by appropriate choices of treatment designs. In addition, the experiment design chosen to accommodate the treatment design should provide the most efficient design in the context of the experiment and resources at the scientist's disposal. I call this process *research design*—the total effort in a study that includes development of the research hypothesis, the choice of treatment design to address the research hypothesis, and the experiment design choice to facilitate efficient data collection.

Orientation and Background

The text is applications oriented, using the results of established theory, and does not include theoretical development. It has a classical design orientation that is intended to introduce to the student the fundamental principles and designs. These fundamentals constitute the necessary ingredients upon which statistical researchers and practitioners continually innovate, enhance, and expand upon in the development of useful and efficient design strategies for research. The text presumes reasonable facility with college algebra and a prerequisite introduction to statistical methods, including the basic ideas of random sampling, basic probability laws, confidence intervals, t tests, analysis of variance, regression, and F tests.

Examples and Exercises

The examples and exercises used throughout the text are based on actual research studies whenever possible. Approximately one-half of the examples and exercises with data sets involve applications to various life and agricultural sciences, while the remaining half involve various applications to engineering, industrial, and chemistry research.

Structure and Coverage

The first chapter emphasizes the role statistical science plays in the research study planning stages. Chapters 2 through 7 emphasize the association between research hypotheses and treatment designs and their analysis in the simplest setting of completely randomized designs. Chapters 8 through 17 present developments of experiment designs and their analyses from the simplest blocking structures in the earlier chapters to more complex and specialized blocking designs and analyses in the later chapters. Care has been taken to preserve the research design paradigm with examples drawn from a variety of sciences to illustrate the major topics in chapters. The reader is provided with a consistent reminder of the connection between the research objectives and the statistical design and analysis to meet those objectives. Interpretations are provided for the analyses.

The book purposely contains more material than is necessary for a one-term course, primarily because not all instructors have preferences for the same material coverage. Courses emphasizing experiment designs, analysis of treatment designs, or combinations thereof can be developed with appropriate choice of chapters.

New to the Second Edition

Chapter 1 is the keystone for everything that follows in the text since it establishes the relationship between research hypotheses and treatment designs, replication, randomization, and local control practices to set the stage for later chapters. In this edition, proper credits are given to Sir Ronald A. Fisher and his seminal paper in 1926, which consolidated his unified concepts for valid inference from designed experiments. A case study has been added that illustrates the application of design principles discussed in the previous sections of this chapter.

New or upgraded discussions, clarification statements (hopefully), as well as updated references to many topics have been added in various and sundry places throughout this edition. Some cosmetic changes were made to improve the overall appearance, including a generic analysis of variance table format in place of computer output. Also removed were illustrations with data for hand calculator computation of the basic analysis of variance sums of squares.

Computer programs constantly evolve in terms of what they do, the user interface, and how they display output. An abundance of commercial computer programs are available to help design experiments for complete and incomplete blocks, fractional factorials, mixtures, response surfaces, and so forth, including the randomization schemes. Programs for analysis are even more numerous. The attachment to a particular software package seems to be very personal with users and it is an interface that is better left alone in a textbook of this nature.

Pointers to advanced methodology for analysis choices have been added, methods often based on maximum likelihood, such as those for variance component estimation in Chapter 7, mixed model software for split plots in Chapter 14, and choices for covariance matrices for repeated measure in Chapter 15. A snippet on the generalized linear model was added in Chapter 4 to let the student know that we have models and method that do not require the normal distribution assumption. It includes reference to several of the good generalized model analysis books specifically designed for users of the generalized model.

Simultaneous statistical inference has been upgraded in Chapter 3 with discussion on different strengths of inference available to the experimenter for post analysis of variance methods. Confidence interval level of inference with simultaneous confidence intervals is the preferred choice of inference for multiple comparisons followed by the confident inequalities option that was used in the first edition. Some references are provided in Chapter 3 to other tests that require software (now being implemented in some packages) to compute critical test values but which provide much better alternatives to some of the old standbys.

Spread-location plots were added to Chapter 4 to aid evaluation of homogeneous variance assumptions, and the residual-fitted spread plots were added to evaluate adequacy of the fitted model. The unrestricted mixed model is now the model choice for the designs in Chapter 7, if for no other reason than that it extends naturally from balanced to unbalanced designs. It is perhaps a matter of trivial concern to some, but it is a subject that is fairly evenly divided between the "restricted" and the "unrestricted" model camps.

The newer Latinized resolvable incomplete block designs have been added to the cyclical design section in Chapter 10. They offer another dimension of blocking control for the α-designs while they preserve the orthogonality for that blocking criterion. Several pages on the general Taguchi method philosophy have been added to Chapter 12. Although the Taguchi method is very specialized to the manufacturing research community, it does have some design and analysis components that should be acknowledged. An example with heterogeneous slopes among treatment groups for the covariate has been added in Chapter 17 to expand the discussion on analysis of covariance assumptions.

Acknowledgments

I am indebted to many individuals who directly and indirectly contributed to the development of this book. My thanks to mentors at Iowa State and North Carolina State Universities, colleagues, and friends who have in one way or another positively influenced me through the years. In particular George, Lowell, Russ, Arnel, Charley, John, Clark, Dave, and Ted deserve a special thanks. I am grateful to colleagues and friends at the University of Arizona and elsewhere who have presented me with stimulating consulting problems over the years that influence the teaching and organization of this material. In particular I thank those (acknowledged in the text) for their generous contribution of data for examples and exercises. My special thanks to Elnora Fairbank and Helen Ferris for their dedicated assistance in various stages of manuscript preparation. I'm very grateful to Rick Axelson for his statistical programming support throughout the preparation of the original manuscript and to Harika Basaran for originally preparing the exercise and example data disks. I am indebted to John Kimmel for his confidence and support in the formative stages of this book, and special appreciation is extended to Alex Kugushev for his dedication to excellence. I wish to express my gratitude to Carolyn Crockett, who had the confidence in the viability of the book to promote a second edition. A note of thanks to Kimberly Raburn for her invaluable guidance of all things important during the revision process. Finally, I want to acknowledge the critical reviews and many helpful suggestions by the many reviewers for both editions, including Richard Alldredge, Daniel C. Coster, Shu Geng, Robert Heckard, Hui-Kuang Hsieh, David Jowett, Larry J. Ringer, Oliver Schabenberger, and G. Morris Southward.

Robert O. Kuehl

1   Research Design Principles

All of our activities associated with planning and performing research studies have statistical implications. The principles we encounter in Chapter 1 form a basis for the structure of a research study; the structure, in turn, defines the study's function. If the structure is sound the study will function properly and produce the information it was designed to provide. If the structure is faulty the study will not function properly, and it will produce either incomplete or misleading information. The statistical principles are those associated with the collection of observations to gain maximum information for a research study in an efficient manner. They include treatment design, local control of variability, replication, randomization, and the efficiency of experiments.

1.1   The Legacy of Sir Ronald A. Fisher

No one person had as much impact on the statistical principles of experimental design in his time as Ronald A. Fisher. In October 1919, Fisher joined the staff at Rothamsted Experimental Station near Harpenden, England. He had been asked to come for a period of six months to one year to apply a thorough statistical analysis to agricultural research data that had been accumulated by the staff.

It was during his tenure at Rothamsted, where he remained until 1933, that he developed and consolidated basic principles of design and analysis that we today take as necessary practices for valid research results. From 1919 to 1925, he studied and analyzed wheat experiments on Rothamstead Experimental Station that had been conducted since 1843. From his statistical investigations of these experiments and others, Fisher developed the analysis of variance and unified his ideas on basic principles of experimental design.

In 1926 he published the first full account of his ideas in a paper, "The Arrangement of Field Experiments" (Fisher, 1926). In that seminal paper he outlined and advanced three fundamental components for experiments in agricultural field trials: **local control** of field conditions for the reduction of experimental error, **replication** as a means to estimate experimental error variance, and **randomization** for valid estimation of experimental error variance. Although replication and local control were practiced at the time, his justifications with regards to experimental error variance were relatively new concepts. Randomization was a radical new concept met with skepticism and resistance by his contemporaries who, for the most part, did not understand its statistical implications. Each of these concepts is discussed in more detail in succeeding sections of this chapter.

Two books that grew out of Fisher's experiences at Rothamsted became standard references for researchers on the design and analysis experimental studies. *Statistical Principles for Research Workers* was first published in 1925 (Fisher, 1925) with 13 subsequent editions, and *The Design of Experiments* was published ten years later (Fisher, 1935) with 7 subsequent editions. His contributions to experimental design were but a few of his contributions to the science of statistics. A biography on the life and times of Fisher, written by one of his daughters (Box, 1978), pays homage to him as the consummate scientist.

1.2   Planning for Research

A **research program** is an organized effort on the part of a scientist to acquire knowledge about a natural or manufactured process. The total program may require many individual studies, each with specific objectives. The individual studies usually answer related questions and provide related pieces of information, which in concert meet the goals of the program. The design and analysis of the individual research study are the object of our attention in this book.

Good planning helps the scientist to organize the required tasks for a research study. The individual study requires the scientist to make a number of critical decisions. Consider the nutrition scientist who wanted to improve the standard method for evaluating nutritional quality of different protein sources. Although the acceptable standard procedure for evaluation was fairly rigidly defined by peer scientists, he hypothesized that the substitution of mice for rats as the standard test animal was more time- and cost-efficient. A test of his research hypothesis required a study to determine whether the mice were more efficient. Among the critical decisions he had to make were the number of mice and rats to use; the amount of protein to use in the diets; the various sources of protein necessary to validate the new protocol; the length of time to run the study; and the number of replications for the full experiment.

Documented Plans Prevent Oversights

The importance of developing a written plan cannot be overemphasized. Frequent reference to an existing document prevents serious oversights. The document also will be useful for subsequent insertions of notes and any alterations relating to the specific items in the original plan.

The astute investigator develops a checklist of specific considerations at the beginning of a study. Some typical items that a checklist can address are

- the specific objectives of the experiment
- identification of influential factors and which of those factors to vary and which to hold constant
- the characteristics to be measured
- the specific procedures for conducting tests or measuring the characteristics
- the number of repetitions of the basic experiment to conduct
- available resources and materials

Bicking (1954) presents a detailed checklist for planning a research study that can be consulted as a guide to develop a written plan.

Simple Questions to Focus Activities

Simple, but challenging, questions aid the design process, even though we may have a well-defined research hypothesis as an impetus for the research study.

Questions that focus our attention throughout the design process include "What is my objective?" "What do I want to know?" and "Why do I want to know it?" Productive follow-up questions for each activity in the process—such as "How am I going to perform this task?" and "Why am I doing this task?"—direct our attention to define the role of each activity in the research study.

Components of the research study are discussed separately in the following sections, but they are interconnected and an investigator must integrate those separate parts into an effective research study. We begin by establishing a small vocabulary to communicate our ideas.

1.3 Experiments, Treatments, and Experimental Units

Accurate communication requires that both parties respond to a common vocabulary with a common meaning. This section establishes the interpretation of some common terms and concepts as they are applied to scientific research studies.

For our purposes, an **experiment** shall be confined to investigations that establish a particular set of circumstances under a specified protocol to observe and evaluate implications of the resulting observations. The investigator establishes and controls the protocols in an experiment to evaluate and test something that for the most part is unknown up to that time.

The **comparative experiment** is the type of experiment familiar to investigators in the fields of biology, medicine, agriculture, engineering, psychology, and other experimental sciences. The adjective *comparative* implies the establishment of more than one set of circumstances in the experiment, and that responses resulting from the differing circumstances will be compared with one another.

Treatments are the set of circumstances created for the experiment in response to research hypotheses, and they are the focus of the investigation. Examples of treatments are animal diets, cultivars of a crop species, temperatures, soil types, and amounts of a nutrient. Two or more treatments are used in a comparative study, and they are compared with one another for their effects on the subjects of the study.

The **experimental unit** is the physical entity or subject exposed to the treatment independently of other units. The experimental unit, upon exposure to the treatment, constitutes a single replication of the treatment.

Experimental error describes the variation among identically and independently treated experimental units. The various origins of experimental error include (1) the natural variation among experimental units; (2) variability in measurement of the response; (3) inability to reproduce the treatment conditions exactly from one unit to another; (4) interaction of treatments and experimental units; and (5) any other extraneous factors that influence the measured characteristics.

A beef cattle feeding trial provides an example of natural variation among experimental units. Two cattle of the same breed and herd receive the same amount of a diet; yet one steer gains 2.0 pounds per day, and the other gains 2.3 pounds per day over a one-month period.

The inability to reproduce treatment conditions exactly occurs when replicate test tubes are prepared independently, containing the same mixture of compounds, and the resultant chemical product is weighed and found to differ in each of the tubes by 0.1 μg. Weighing or pipetting processes are not exact; therefore, a small amount of variation is introduced at the treatment preparation stage.

A major objective in statistical calculation is the attainment of an estimate for the *variance of experimental error*. In its simplest form, the experimental error variance is the variance of observations on experimental units for which the differences among the observations can be attributed only to experimental error. Many of the statistical procedures we use require an estimate of this variance. Examples include confidence interval estimates for a mean and the two-sample Student t tests for the hypothesis of no differences between the means of two treatment populations.

Comparative observational studies are those studies for which we would like to conduct an experiment but cannot do so for practical or ethical reasons. The investigator has in mind conditions or treatments that have causal effects on subjects for which experiments cannot be conducted to elicit responses.

Investigators in the social sciences, ecology, wildlife, fisheries, and other natural resource sciences often must conduct observational studies in lieu of direct experimentation. The basic unit of study in the investigation may be human subjects, individual animals, habitats, or other microcosms; they have the same role as the experimental unit in the designed experimental study.

The subjects are either self-selected into identifiable groups or they simply exist in their particular circumstances. The groups or circumstances are used as treatment classifications in the observational study. By contrast, the investigator assigns treatments to the experimental units in a designed experiment. For example, to study nitrification in soils from pure and mixed stands of mature pine and oak trees, a soil scientist selects existing pure and mixed stands of the two species and collects the necessary observations from the selected sites. A true experiment is quite impractical in this case because it requires an extraordinary expenditure of time to establish mature stands of the trees.

Ethical considerations sometimes prevent the use of experiments in lieu of observational studies. Consider a study to compare the severity of automobile accident injuries with and without the use of seat belts. It would be clearly unethical to assign anyone randomly to a "seat belt" or a "no seat belt" treatment and then collide the automobile into a concrete wall nor would anyone agree to this. Rather, investigators would rely on injury data collected from accidents and compare the "seat belt" data with the "no seat belt" data.

The nature of the scientific inference is the primary difference between the designed experiment and the observational study. With the designed experiment it is often possible to assign causal relationships between the responses and the treatments. Observational studies are limited to association relationships between the responses and treatment conditions.

1.4 Research Hypotheses Generate Treatment Designs

The research hypothesis establishes a set of circumstances and the consequences that follow from those circumstances. The treatments are a creation of the circumstances for the experiment. Thus, it is important to identify the treatments relative to the role they each have in the evaluation of the research hypothesis. A failure to clearly delineate the research hypothesis and objective of the study can lead to difficulties in the choice of treatments and to unsuccessful experiments.

The Relationship Between Treatments and Hypotheses

When treatments are chosen properly in response to a research hypothesis the underlying mechanisms may be better understood, whether they be physical, chemical, biological, or social. In some cases the objective may be to "pick the winner" to find one treatment that provides the desired response. In other cases, the experiment is used to elucidate underlying mechanisms associated with the treatments as they affect measured response variables. In the latter case sound research hypotheses motivate the selection of treatments.

It is incumbent on the investigator to ensure that the choice of treatments is consistent with the research hypothesis. It may be sufficient and less difficult to design a study solely to discover the best treatment. However, with a little extra effort even more fundamental information may be derived from the experiment in response to research hypotheses.

The following illustrate treatments used in actual research settings generated by research hypotheses:

- The drinking kinetics of honeybees was studied under different ambient temperatures to address the hypothesis that the energy required by honeybees to retrieve food for the colony was dependent on the ambient air temperature.

- The survival of Euphorbia seedlings under attack by a soil pathogen was determined for different types of fungicide treatment to address the hypothesis that not all fungicides were equally effective in controlling the soil pathogen.

- In traffic engineering, several methods of measuring traffic delay at an intersection were evaluated under different types of traffic signal configurations to address the hypothesis that the method of measuring delay was dependent on the type of configuration used for signaling traffic at an intersection.

- The development of social competence in young children was measured for its relationship to (1) parent education, (2) parent income, (3) family structure, and (4) age of child to address a complex research hypothesis that certain family demographics favorably affect the development of a child.

Note that in some research settings the treatments are conditions imposed by the investigator, such as those involving the honeybees and the survival of seedlings. On the other hand, in the traffic engineering and child development studies the treatments were those of existing conditions. Whether the investigator is performing a designed experiment or an observational study, he or she has the task of selecting the proper treatments to address the research hypotheses.

Frequently, additional treatments are required to fully evaluate the consequences of the hypotheses. An important component of many treatment designs is the *control treatment* discussed in the following section.

Control Treatments Are Benchmarks

The **control treatment** is a necessary benchmark treatment to evaluate the effectiveness of experimental treatments. There are several circumstances in which a control treatment is useful and necessary.

Conditions of the experiment may disallow the effectiveness of the experimental treatments that are known generally to be effective. A control of *no treatment* will reveal the conditions under which the experiment was conducted. For example, nitrogenous fertilizers are generally effective but will fail to produce responses in fields with high fertility. A control of *no nitrogenous fertilizer* will reveal the base fertility conditions for the experiment.

Sometimes treatments require manipulating the experimental units or subjects where the manipulation alone can produce a response. *Placebo* controls establish a basis for treatment effectiveness in these cases. The placebo unit or subject is processed just as the treatment units, but the active treatment is not included in their protocol. One of the most famous health experiments, the 1954 field trial of the Salk poliomyelitis vaccine, utilized placebo controls in approximately one-half of the test areas in the United States. The placebo was prepared to look just like the vaccine, but without the antigenic activity of the poliomyelitis vaccine. The placebo subjects were inoculated in the same fashion as were the subjects receiving the vaccine (Tanur et al., 1978).

Finally, the control may represent a *standard* practice to which the experimental method may be compared. In some situations it is necessary to include two distinct types of controls. For example, the no treatment and the placebo treatment can reveal the effect of manipulating the experimental unit in the absence of any treatment.

Multiple-Factor Treatment Designs Expand Inferences

In "The Arrangement of Field Experiments," Fisher (1926) noted that no proverb in connection with field experiments at that time was more repeated than that which said,

> We must ask Nature few questions, and ideally, one at a time.

He was convinced this was a mistaken view. In this regard, he wrote, "Nature ... will best respond to a logical and carefully thought out questionnaire; indeed, if we ask her a single question, she will often refuse to answer until some other topic has been discussed."

Fisher understood that in natural systems we don't know whether one treatment's influence is independent of others or its influence is related to variation in the other treatments. Consequently, the conditions under which treatments are compared can be an important aspect of the design.

For example, in a study on nitrogen production by *Rhizobium* bacteria in soil a comparison of interest was the amount of nitrogen produced in normal, saline, and sodic soils. However, the nitrogen production was known to be affected by the temperature and moisture conditions in the soil. In fact, the optimum conditions of temperature and moisture for nitrogen production may well have been different for each soil type. Consequently, the experiment was set up to test the nitrogen production at several temperatures in combination with several moisture conditions for each of the soils. Treatment designs of this type are known as **factorial treatment designs,** wherein one set of treatments, say soils, is tested over one or more other sets of treatments, such as moisture and temperature.

A **factor** is a particular group of treatments. Temperature, moisture, and soil type are each considered a factor. The several categories of each factor are termed *levels* of the factor. Temperature levels are 20°C, 30°C, and 40°C; while soil type levels are normal, saline, and sodic. A *quantitative factor* has levels associated with ordered points on some metrical scale such as temperature. The levels of a *qualitative factor* represent distinct categories or classifications, such as soil type, that cannot be arranged in order of magnitude.

The factorial arrangement consists of all possible combinations of the levels of the treatment factors. For example, the factorial arrangement for three levels of temperature with three levels of soil type consists of $3 \times 3 = 9$ factorial treatment combinations. The nine combinations are

(20°C, normal)	(30°C, normal)	(40°C, normal)
(20°C, saline)	(30°C, saline)	(40°C, saline)
(20°C, sodic)	(30°C, sodic)	(40°C, sodic)

With this arrangement the investigator was able to evaluate the nitrogen production in each of the soil types and also to determine the influence of temperature conditions on the relative production by soil type. In addition, it was also possible to evaluate separately the influence of temperature on nitrogen production. The statistical analysis of factorial arrangements is discussed in Chapters 6 and 7.

1.5 Local Control of Experimental Errors

Precise and accurate comparisons among treatments over an appropriate range of conditions are the primary objectives of most experiments. These objectives require precise estimates of means and powerful statistical tests. Reduced experimental error variance increases the possibility of achieving these objectives. **Local control** describes the actions an investigator employs to reduce or control experimental error, increase accuracy of observations, and establish the inference base of a study.

The investigator controls (1) technique, (2) selection of the experimental units, (3) blocking or ensuring parity of information on all treatments, (4) choice of experiment design, and (5) measurement of covariates. Each of these is discussed in this section in more detail.

Technique Affects Variation and Bias

If experimental tasks are performed in a slipshod manner the observations will exhibit an increase in variation. *Technique* includes simple tasks such as accurate measurement, preparation of media, pipetting of solutions, or calibration of instruments. On a more complex level the investigator may have several alternative laboratory methods or instruments for measuring chemical or physical properties. The methods may vary in their accuracy, precision, and range of application. The researcher must choose the method or instrument that provides the most accurate and precise set of observations within the budgeted resources.

When technique adversely affects precision the estimated experimental error variances are unnecessarily inflated. Outlying observations caused by recording errors or extraordinary environmental conditions can increase variation. Whatever the cause the investigator has to decide whether to include these observations in the analysis.

On the other hand, no discernible pattern may exist in the observations other than a general increase in their variability. This type of increased variation could point to faulty technique somewhere in the course of the experiment. The investigator would need to check out the selection of experimental units for the study, treatment protocol, measurement techniques, and personnel for sources of increased error and then attempt to make adjustments wherever necessary.

Poor techniques can affect the accuracy of observations, introducing bias into the results. Further, the variation introduced into the observations by poor technique is not necessarily random and, therefore, not subject to the same probability laws we associate with statistical inference.

Uniform application of treatments throughout the experiment increases the likelihood of unbiased measurement of the treatment effects. For instance, uniform amounts of food intake are required for accurate measures of differences between diets for animals. Uniform amounts of fertilizers applied to the replicate plots are required for accurate measures of crop yield differences in fertilizer trials.

Selection of Uniform Experimental Units

Heterogeneous experimental units produce large values for the experimental error variance. Precise comparisons among the treatments require the selection of uniform experimental units to reduce experimental error. However, an excessively stringent selection of experimental units can produce artificially uniform conditions. The narrow set of conditions restricts the inference base of the study. Therefore, to have reasonable confidence in the conclusions secured from the experiment, it is desirable to have units represent a sufficient range of conditions without unnecessarily increasing the heterogeneity of the experimental units.

The nature of the experiment dictates the balance between range of conditions and uniformity of units. For example, plant selections in a plant-breeding study should be tested over the range of conditions in which future varieties are expected to be cultivated; hence the range of conditions may be quite wide. If the selections are tested in several widely separated locations, then uniformity within a location becomes important. Selecting a uniform set of plots controls the variation within a location. Uniformity of units in an experiment with dairy cows requires selection of cows from the same breed, at the same stage of lactation, and with similar numbers of previous lactations.

Blocking to Reduce Experimental Error Variation

Fisher (1926) argued that no advantage had to be given up to have a valid estimate of error but two things were necessary:

> (a) that a sharp distinction should be drawn between those components of error which are to be eliminated in the field, and those which are not to be eliminated... (b) that

the statistical process of the estimation of error shall be modified so as to take account of the field arrangement, and so that the components of error actually eliminated in the field shall equally be eliminated in the statistical laboratory.

Blocking provides local control of the environment to reduce experimental error. The experimental units are grouped such that the variability of units within the groups is less than that among all units prior to grouping. The practice of blocking or grouping experimental units into homogeneous sets goes hand in hand with experimental unit selection for uniformity. Treatments are compared with one another within the groups of units in a more uniform environment, and the differences between the treatments are not confused with large differences between experimental units. The variability associated with environmental differences among groups of units can be separated from experimental error in the statistical analysis.

Experimental units are blocked into groups of similar units on the basis of a factor or factors that are expected or known to have some relationship with the response variable or the measurement that is hypothesized to respond differentially to the several treatments.

Four Major Criteria for Blocking

Four major criteria frequently used to block experimental units are (1) proximity (neighboring field plots), (2) physical characteristics (age or weight), (3) time, and (4) management of tasks in the experiment.

The classical blocking practice had its origins in agricultural field experiments when contiguous plots were placed into one group, and each of the treatments was assigned to a plot in that group. Then a second set of contiguous plots was used in the same fashion, and so forth, leading to a *complete block* design.

The rationale for this type of grouping is that plots near one another are more alike than plots separated by greater distance in cultivated fields. Blocking patterns in agricultural trials may not necessarily be nice rectangular configurations. Existing variability patterns can require rather different blocking arrangements to reduce the experimental error.

Another natural blocking unit is the animal litter. The size of litters in some animal species allows several treatments to be accommodated by one litter. Body weight is used to advantage as a blocking factor in animal experiments if variation in body weight amplifies variation in the response variable.

Industrial experiments require homogeneous batches of raw materials. A replicated experiment may require more raw material than that provided by a single batch, and variation from batch to batch may increase experimental error. A single batch sufficiently large for one replication of all treatments can serve as a blocking unit.

Blocking is used to divide the experiment into reasonably sized units for uniform management of time or tasks. Days serve as convenient blocking units if only one replication of treatments can be harvested in the field or processed in the laboratory in a single day.

Individual technicians can serve as blocking units to avoid confusion of observer or technician variability with that of the treatments. Each person can be assigned to one replicate of each treatment when several people are available to record data or perform the laboratory analyses.

A Demonstration of Variance Reduction by Blocking

A **uniformity trial** best illustrates the potential effectiveness blocking may have for variance reduction in a research study. The uniformity trial is essentially an experiment in which the experimental units are measured but have not been subjected to any treatment. For example, the classic uniformity trial in agriculture may have been a field of wheat all of the same variety divided into plots all of the same dimension. The wheat yields were then measured on each plot. Because variation in agricultural fields can generally occur on gradients, it was possible to determine which groupings of adjacent field plots led to groups of plots with the smallest variance within the groups. In following years' experiments the treatments could be allocated within the groups of similar plots based on the results from the uniformity trial.

Similarly, baseline observation prior to treatment application on the measurement of interest, or some variable that is known to have a strong relationship to the measurement of interest, is equivalent to a uniformity trial for blocking purposes. For example, measurements such as weights, age, chemical composition, or cholesterol levels prior to treatment administration may suitably be used as blocking criteria if they have strong relationships to the measurement of interest or are themselves the measurement of interest.

Suppose we have observations on ten units from either a uniformity trial or measurements prior to treatment administration: 43, 72, 46, 66, 49, 68, 50, 76, 42, and 69. The mean and variance for the observations on these ten units are $\bar{x} = 58$ and $s^2 = 175$. Grouping the units into two blocks based on the size of those measurements, we have

$$\text{Block 1:} \quad 43, 46, 49, 50, 42 \quad \bar{x} = 46 \quad s^2 = 12.5$$
$$\text{Block 2:} \quad 72, 66, 68, 76, 69 \quad \bar{x} = 70 \quad s^2 = 15.2$$

Whereas the total variance among the ten units was equal to 175, the variances within the separate blocks have been reduced to 12.5 and 15.2, respectively. The component of error eliminated by blocking will be reflected in the variance between the two block means, 46 and 70. The variability within each of the blocks is now assumed to represent the natural variation that exists among the relatively uniform experimental units unencumbered by controllable environmental differences. Likewise, the comparisons among treatments within those blocks will not be influenced by those same controllable environmental differences.

Matching Strategies to Group Units

Grouping of units often utilizes strategies to match similar units. Subjects or units are chosen for treatment groups on the basis of equality with respect to all influential factors. Any variables thought to have an influence on the value of the observed characteristics of the subjects are candidates for controlling variables. Matched subjects have common values for the controlling variables with the exception of the treatment.

The matching strategies at the study design stage attempt to achieve comparability of subjects on all factors that could unduly bias the comparison of the treatments. *Pair matching* and *nonpair matching* are two general strategies employed in the selection of units or subjects.

With pair matching, a subject from each treatment is paired with subjects having the same controlling variable values in each of the other treatment groups. The values of the controlling variables may be chosen to be (1) *exact value* matches or (2) *caliper* value matches. Exact matches for human subjects are possible with variables such as gender, profession, use of seat belt, and education level. Exact matching of roadway segments in a study of traffic performance is possible with variables such as number of lanes, lane width, presence or absence of a median, and speed limit.

The caliper match allows a certain tolerance in the value of the matching variables. A study on manipulation of a forest ecosystem may require study sites to be matched by tree species composition, slope, and aspect. Exact site matches on composition, slope, and aspect are quite difficult to attain. However, it may be possible to obtain sites with similar values for any or all of the matching variables to the extent that the variation will not have a serious effect on treatment comparisons.

Nonpair matching may be accomplished through (1) a *frequency* or (2) a *mean* strategy. The frequency method stratifies the units into groups on the basis of the controlling variables. Suppose age is a potential influential variable in a study involving human subjects. The subjects can be stratified from a frequency distribution of their ages such that there is a sufficient number of subjects in each age stratum to accommodate all treatment groups.

The mean-based nonpair strategy groups subjects or units in such a way that the average values of the controlling variables are the same in each of the treatment groups. Experimental animals can be grouped such that their average weight is the same for all treatment groups.

The nature of the research study dictates the most effective matching strategy and whether matching is a desirable protocol. Details on matching methods, including advantages and disadvantages, for comparative observational studies can be found in Cochran (1983), Kish (1987), and Fleiss (1981).

Experiment Designs Accommodate Treatment Designs

The **experiment design** is the arrangement of experimental units used to control experimental error and, at the same time, accommodate the treatment design in an experiment. A wide variety of experiment design arrangements to control experimental error exists in the published literature, and there is a natural tendency to design an experiment that conforms to the existing designs. But, a more appropriate attitude is to develop an experiment design that satisfies the demands of the current experiment.

The attainment of maximum information, precision, and accuracy in the results, along with the most efficient use of existing resources, is a guiding principle in choosing an appropriate experiment design.

The association between a treatment design and an experiment design is illustrated here under two different settings. The first illustration shows how three treatments are associated with six experimental units when each treatment is assigned to two experimental units. The second illustration shows how the three treatments are assigned to the six experimental units after the units are blocked into two sets of three homogeneous units.

An Experiment Design Without Blocking

Consider an experiment to compare three gasoline additives for their effect on carbon monoxide emission. Automobile engines are used as experimental units. Each of the three gasoline additives is to be used with two engines. A schematic representation of the design is shown in Figure 1.1. The boxes represent automobile engines as experimental units. One of the three gasoline additives is administered to an engine independently of the other engines in the experiment. The three gasoline additive treatments are randomly allocated to the six engines, two units for each treatment.

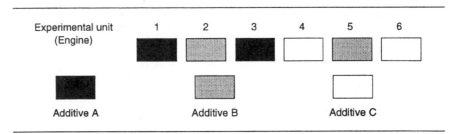

Figure 1.1 Illustration of a completely randomized design, with three treatments, each on two experimental units

This design, known as the **completely randomized design**, is the simplest of the experiment designs. The treatments are allocated to the experimental units at random. Each experimental unit has the same chance of receiving any treatment. The function of randomization in experiment designs is discussed in Section 1.8.

An Experiment with One Blocking Criterion

The completely randomized design provides little control over environmental variation. Several general classes of designs use blocking and grouping of experimental units to control environmental variation. The simplest blocking design is the **randomized complete block design** with one blocking criterion. This design employs one restriction on the random assignment of treatments to experimental units; all treatments must occur an equal number of times in each block.

Consider the arrangement of units in a randomized complete block design for an experiment to study the effects of three diets on the growth of laboratory mice. Three mice of the same gender were selected from each of two litters. The litters were used as blocks for the experiment as shown in Figure 1.2. The three diets were randomly assigned to the three mice in each litter.

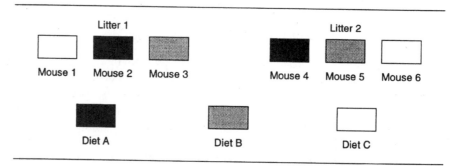

Figure 1.2   Illustration of a randomized complete block design, with three experimental diets tested in two litters of mice

The use of a litter as a block reduces experimental error variation because it isolates the variation among litters from the variation among diets. The diets can be compared under uniform conditions when each of them is tested in the same litter. A complete discussion of the randomized complete block design and its analysis is given in Chapter 8.

Covariates for Statistical Control of Variation

Covariates are variables that are related to the response variable of interest. The information on covariates is used to enact a statistical control on experimental error variance by a procedure known as the *analysis of covariance.*

It was suggested earlier that body weight can be used as a criterion for blocking. After animals are grouped according to weight no further use is made of those values. But, the actual body weights can be used effectively in the statistical model to reduce the estimates of experimental error in lieu of or in addition to using them for blocking. Body weight would be a covariate of weight gain in the experiment. Other examples of covariates might be pretest scores, field plot fertility, previous year's yield in perennial crops, or purity of raw material in a chemical process; all may vary from unit to unit. Any attribute that is measurable and thought to have a statistical relationship to the variable of primary interest is a candidate for covariate adjustment.

The prime requirement is that the covariate be unaffected by the particular treatments used in the experiment. In practice the covariates are measured before the treatments are administered or before the treatments have had time to develop a response by the unit, or we can assume the treatment never has an effect on the covariate.

The observations in the experiment consist of pairs of observations (x, y) on each experimental unit, where y is the variable of interest in the experiment and x is the covariate. Suppose there are six such pairs of observations for each of two treatment groups, and the data are plotted as shown in Figure 1.3.

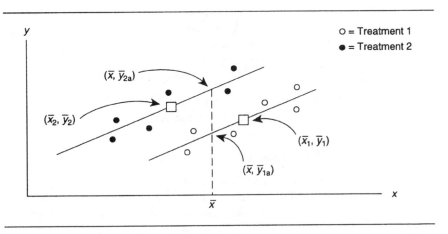

Figure 1.3 An illustration of covariate adjustment for two treatment groups

A portion of the variation in y is associated with x as well as with the treatment effects if there is a statistical relationship between the variable of interest, y, and covariate x. The analysis of covariance estimates the regression relationship between y and the covariate x to statistically reduce the experimental error variance. The average responses to each treatment, $\bar{y}_1$ and $\bar{y}_2$, are adjusted to the same value of the covariate, usually the grand mean $\bar{x}$ as shown in Figure 1.3. A comparison between the adjusted treatment means, $\bar{y}_{1a}$ and $\bar{y}_{2a}$, eliminates the influence of the covariate x on the comparison. For example, if weight gain has the covariate of initial body weight, the average weight gain response to a treatment is adjusted to remove the variation associated with initial body weight. The adjusted treatment means for weight gain represent the weight gains that would be obtained if all animals had the same initial body weight. The analysis of covariance is discussed in Chapter 17.

1.6 Replication for Valid Experiments

The scientific community regards replication of experiments to be a prime requisite for valid experimental results. **Replication** implies an independent repetition of the basic experiment. More specifically, each treatment is applied independently to each of two or more experimental units.

There are several reasons for replicating an experiment, most notably:

- *Replication* demonstrates the results to be reproducible, at least under the current experimental conditions.

- *Replication* provides a degree of insurance against aberrant results in the experiment due to unforeseen accidents.

- *Replication* provides the means to estimate experimental error variance. Even if prior experimentation provided estimates of variance, the estimate from the present experiment may be more accurate because it reflects the current behavior of observations.

- *Replication* provides the capacity to increase the precision for estimates of treatment means. Increasing replication, r, decreases $s_{\bar{y}}^2 = s^2/r$, thereby increasing the precision of $\bar{y}$.

Observational Units and Experimental Units Can Be Distinctly Different

The *observational unit* may not be equivalent to the experimental unit. The observational unit can be a sample from the experimental unit, such as individual plant samples from a field plot or serum samples from a subject.

The variance of the observations on the experimental units is the experimental error variance. It is a valid measure of the variation among the experimental units that have *independently* had the treatments administered to them.

The variance among multiple observations from the same experimental unit often is used mistakenly as a measure of experimental error for comparisons among treatments. The following examples may help to clarify the distinction between replicated experimental units and multiple observations from the same experimental unit.

Example 1.1 A simple animal diet ration study has one cage or pen of six animals assigned to ration A and a second cage or pen of six animals assigned to ration B. Weight gain or some other appropriate data are collected to test the efficacy of the rations. The necessary measurements are made on each of the animals in the pens at the end of the study. The schematic in Figure 1.4 illustrates the design.

Typically, the Student t test using the difference between the means of the two pens would be used to test a hypothesis of no difference between the rations.

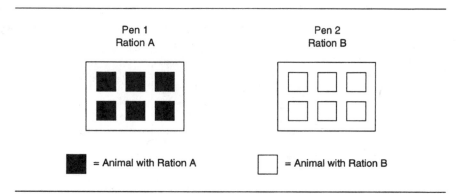

Figure 1.4 Illustration of an unreplicated experiment

However, the differences between the two pens of animals can be caused by the effect of other factors in addition to the treatments. The response to a given ration may vary with different pens of animals. This variation in response could be due to one or more of the factors contributing to experimental error.

The natural variation in the response from one pen to another, or the *pen effect*, will contribute to experimental error. Also, variation in the preparation and presentation of the treatment to the separate pens can cause a treatment–pen interaction. Therefore, any differences between the two rations in Example 1.1 cannot be attributed clearly to the rations alone. The differences may be attributable to combinations of the treatment effects, pen effects, and treatment–pen interactions.

The experiment will not have resolved clearly the question of whether the two rations differed in their effect on weight gain. The experiment has only one true replication. The *pen* is the *experimental unit* because that is the unit to which the treatment was administered independently. The *animals* in the pen are the *observational units*. The variance calculated among the observations on the animals within the pens is only an estimate of observation error within a pen of animals and not an estimate of variance among the experimental units.

> **Example 1.2** Suppose the experiment in Example 1.1 is restructured such that the animals are randomly divided into four pens of three animals each. Further, each of the two rations is randomly assigned to two pens of animals as shown in Figure 1.5. The rations are administered independently to each of the pens.

Each ration has been replicated properly in Example 1.2. The *experimental units* are the *pens* to which the rations have been administered independently (two pens per ration), and the *animals* within the pens are still the *observational units*. Therefore, the response from the experimental unit is the average pen response.

The estimate of experimental error variance, s^2, calculated as the variance among the pen means within each of the rations, is the appropriate variance for the

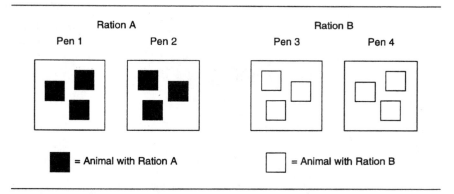

Figure 1.5   Illustration of a replicated experiment

Student t test. The variance among the animals within each pen, say s_w^2, is a measure of the variability of the observational units within the replicate pens.

Thus, two levels of variation are present in this type of study: (1) the variation among observational units *within* experimental units (s_w^2), and (2) variation *among* the experimental units (s^2).

It is important to distinguish which units of the study constitute the experimental unit and consequently which units constitute true replication of the treatment.

More details on the statistical models and analysis for the pseudoreplication and true replication exhibited by Examples 1.1 and 1.2 may be found in Addelman (1970). Examples of pseudoreplication in agronomic studies may be found in Nelson and Rawlings (1983). Hurlbert (1984) provided numerous examples of pseudoreplication in ecological field experiments.

1.7   How Many Replications?

The number of replications in a research study affects the precision of estimates for treatment means and the power of statistical tests to detect differences among the means of treatment groups. However, the cost of conducting research studies constrains the number of replications for a reasonably sized study. Thus, replication numbers are determined on the basis of practical constraints that we can assign to the problem.

Replication Numbers for Testing Hypotheses

The method for determining the number of replications is often based on a test of a hypothesis about differences among treatment group means. An elementary method for experiments with two independent samples is used here to illustrate a few attributes of the replication number problem.

The method is based on a test of a hypothesis about the difference between two treatment group means $d = m_i - m_j$, with known experimental error variance s^2, using the standard normal distribution test statistic. The method determines the number of replications required to test the difference between two independent sample means with specified Type I and Type II errors.

The required number of replications is affected primarily by four factors that are required for calculations:

- the variance (σ^2)

- the size of the difference (that has physical significance) between two means (δ)

- the significance level of the test (α), or the probability of Type I error

- the power of test $1 - \beta$, or the probability of detecting δ, where β is the probability of a Type II error

The required replication number for each treatment group, r, for two-sided alternatives is estimated with

$$r \geq 2[z_{\alpha/2} + z_\beta]^2 \left(\frac{\sigma}{\delta}\right)^2 \tag{1.1}$$

where $z_{\alpha/2}$ is the standard normal variate exceeded with probability $\alpha/2$ and z_β is exceeded with probability β. Probabilities for the standard normal variable are found in Appendix Table I.

The replication number can be estimated with knowledge of the percent coefficient of variation, %CV. The %CV is substituted for σ in Equation (1.1), where $\%CV = 100(\sigma/\mu)$. The difference (δ) must be expressed as a percentage of the overall expected mean of the experiment, $\%\delta = 100(\delta/\mu)$, in Equation (1.1).

The influence of the coefficient of variation (%CV), the percent difference (%δ), the power ($1 - \beta$), and the significance level (α) on required numbers of replications is shown in Table 1.1. Although the values in Table 1.1 only apply to two independent samples, the influences are similar in more complex experiments.

Required replication numbers generally increase if

- the variance, %CV or σ^2, increases

- the size of the difference between two means, %δ or δ, decreases

- the significance level of the test, α, decreases

- the power of test, $1 - \beta$, increases

Calculated values for required replication numbers are estimates and approximations. Often they are determined on the basis of variance estimates associated with previous studies and not the variances from the actual study that will be used

Table 1.1   Number of replications required for a given coefficient of variation ($\%CV$) and probability $(1 - \beta)$ of obtaining a significant difference of $\%\delta$ between two treatment means with a two-sided test at the α significance level

		$\alpha = .05$		$\alpha = .01$	
		$\%\delta$		$\%\delta$	
$\%CV$	$1 - \beta$	10	20	10	20
5	.80	4	1	6	2
	.95	7	2	9	3
10	.80	16	4	24	6
	.95	26	7	36	9

to compute confidence intervals and tests of hypotheses. The determination of replication numbers for analysis of variance applications will be considered in Chapters 2, 5, 6, and 7. Commercial software programs are available to determine replication numbers for many different types of experimental studies.

1.8   Randomization for Valid Inferences

> In reconciling thus the two desiderata of the reduction of error and of the valid estimation of error, . . . no principle is in the smallest degree compromised. An experiment either admits of a valid estimate of error, or it does not; whether it does so, or not, depends not on the actual arrangement of plots, but only on the way in which that arrangement was arrived at. (Fisher, 1926)

Replication of the experiment provides the data to estimate the experimental error variance. Blocking provides a means to reduce experimental error. However, replication and blocking alone do not guarantee *valid* estimates of experimental error variance or valid estimates of treatment comparisons.

Fisher (1926) was making the point that **randomization**, alone, in the experiment provides a valid estimate of error variance for justifiable statistical inference methods of estimation and tests of hypotheses. Randomization is the random assignment of treatments to experimental units.

A Rationale for Randomization

Our analysis of data from experiments assumes the observations constitute a random sample from a normally distributed population. This assumption is plausible for comparative observational studies that use random samples of the available observation units from different treatment populations. However, whether experimental units can be considered a random sample is questionable when they are carefully selected, controlled, and monitored in experiments.

Independent observations are critical for estimation and tests of hypotheses because they provide valid estimates of experimental error variance. But, the assumption of independence among the experimental units cannot be justified when relationships exist among them. For example, it is well known that field plots tend to respond more similarly when they are adjacent. Any type of proximity can produce correlated responses whether it be physical location of units or temporal performance of tasks on the units.

Fisher (1926) recognized these potential difficulties with field plot experiments and justified random assignment of treatments to experimental units as the means to obtain valid estimates of experimental error variance. In a more detailed discussion, Fisher (1935) showed that randomization provided appropriate reference populations for statistical inferences free of any assumptions about the distribution of the observations. He showed that significance tests could be based on the distribution created by randomization and that the normal theory tests provided reasonable approximations to these test results. Thus, the random allocation of treatments to the experimental units *simulates* the effect of independence and permits us to proceed as if the observations are independent and normally distributed. These **randomization tests**, illustrated in this section, form the basis for valid statistical inferences in properly randomized experiments.

Further justification for randomization (Cochran & Cox 1957; Greenberg 1951; and Ostle & Mensing 1975) is based on the need to eliminate biases in the comparison of treatments that arise through systematic assignment of treatments to experimental units. If, for example, procedure A is always performed before procedure B, any systematic variation over time will bias the resulting comparisons between A and B. Randomization over these potential systematic sources of variation ensures estimates of treatment means differ from true values only by random variation.

Randomization Tests Show Utility of Randomization

The utility of randomization can be demonstrated with a randomization test that makes no assumptions about the form of the probability distribution for the observations. Randomization creates a population of experiments that could have been performed, although only one arrangement has been chosen at random for the actual experiment.

The randomization test evaluates the test statistic for all possible arrangements of treatments on the experimental units. The *randomization distribution* is the distribution of those values that would be obtained under the null hypothesis of no treatment effects.

Example 1.3   Illustration of a Randomization Test

Consider an experiment in which two treatments are randomly assigned to seven experimental units. Four experimental units receive treatment A and three experimental units receive treatment B, with the following set of responses:

Unit:	1	2	3	4	5	6	7
Treatment:	B	B	A	A	B	A	A
Response:	14	16	19	17	15	13	17

If there is no difference between the effects of treatments A and B, then they are merely labels on the experimental units and do not affect the results. For example, if the null hypothesis is true, the response for unit 1 would be 14 regardless of the treatment applied. The same is true for each of the other units under the null hypothesis.

The A and B labels (four A's and three B's) may be allocated to the seven experimental units in $7!/4!\,3! = 35$ possible arrangements. These are the 35 experiments possible if the treatments are randomly allocated to the units.

All 35 of the possible arrangements are shown in Table 1.2, along with the difference between the group means $(\bar{y}_A - \bar{y}_B)$ based on the labels assigned to the units in each arrangement. The 35 differences $(\bar{y}_A - \bar{y}_B)$ are 35 possible differences that could occur for the 35 possible randomizations if the null hypothesis is true. They make up the randomization distribution under the null hypothesis.

Table 1.2   Thirty-five possible arrangements of four A's and three B's to seven experimental units with the mean difference $\bar{y}_A - \bar{y}_B$

| Arrange-ment | Unit: | 1 | 2 | 3 | 4 | 5 | 6 | 7 | |
	Response:	14	16	19	17	15	13	17	$\bar{y}_A - \bar{y}_B$
1		A	A	B	B	A	A	B	− 3.17
2		A	B	B	A	A	A	B	− 2.58
3		A	B	B	B	A	A	A	− 2.58
4		A	A	B	A	B	A	B	− 2.00
5		A	A	B	B	B	A	A	− 2.00
6		A	B	A	B	A	A	B	− 1.42
7		A	B	B	A	B	A	A	− 1.42
8		B	A	B	A	A	A	B	− 1.42
9		B	A	B	B	A	A	A	− 1.42
10		A	A	A	B	B	A	B	− 0.83
11		A	A	B	A	A	B	B	− 0.83
12		A	A	B	B	A	B	A	− 0.83
13		B	B	B	A	A	A	A	− 0.83
14		A	B	A	A	B	A	B	− 0.25
15		A	B	A	B	B	A	A	− 0.25

(Continued on next page)

Table 1.2 (Continued)

Arrange-ment	Unit: Response:	1 14	2 16	3 19	4 17	5 15	6 13	7 17	$\bar{y}_A - \bar{y}_B$
16		A	B	B	A	A	B	A	− 0.25
17		B	A	A	B	A	A	B	− 0.25
18		B	A	B	A	B	A	A	− 0.25
19		A	A	A	B	A	B	B	0.33
20		A	A	B	A	B	B	A	0.33
21		B	B	A	A	A	A	B	0.33
22		B	B	A	B	A	A	A	0.33
23		A	B	A	A	A	B	B	0.92
24		A	B	A	B	A	B	A	0.92
25		B	A	A	A	B	A	B	0.92
26		B	A	A	B	B	A	A	0.92
27		B	A	B	A	A	B	A	0.92
28		A	A	A	A	B	B	B	1.50
29		A	A	A	B	B	B	A	1.50
30		B	B	A	A	B	A	A	1.50
31		A	B	A	A	B	B	A	2.08
32		B	A	A	A	A	B	B	2.08
33		B	A	A	B	A	B	A	2.08
34		B	B	A	A	A	B	A	2.67
35		B	A	A	A	B	B	A	3.25

Consider an alternative hypothesis H_a: $\mu_A - \mu_B \neq 0$ to the null hypothesis H_0: $\mu_A - \mu_B = 0$. The arrangement of the actual experiment is arrangement 30 with a mean difference $(\bar{y}_A - \bar{y}_B) = 1.50$. An absolute difference of 1.50 or larger occurs with 13 arrangements. Under the null hypothesis an absolute mean difference of 1.50 or larger occurs with a frequency of $\frac{13}{35}$, yielding a significance level of 0.37. On the basis of the observed results of the experiment, arrangement 30, there is no reason to reject the null hypothesis.

Under the assumption of a true null hypothesis, H_0: $\mu_A - \mu_B = 0$, the randomization test enabled an evaluation of the test statistic from the actual experiment, $(\bar{y}_A - \bar{y}_B) = 1.50$, against the values for $\bar{y}_A - \bar{y}_B$ from all other members of the population of 35 possible experiments.

Normal Theory Tests Approximate Randomization Tests

Fisher (1935) first demonstrated that normal theory tests are good approximations to the randomization tests provided there has been a random allocation of the treatments to the experimental units and sample sizes are reasonably large.

Approximations to the randomization test by normal theory tests improve as sample size increases. Practical guides to randomization tests for various experimental situations may be found in Edgington (1987) and Manly (1991).

Rigorous treatments of the randomization models and tests of significance in experiment designs are provided by Kempthorne (1952), Scheffé (1959), Mead (1988); and Hinkelmann and Kempthorne (1994). More detailed discussions of randomization in connection with statistical inference may be found in Kempthorne (1966, 1975).

It has become common practice to describe the statistical models for experimental studies in terms of the normal theory models. The formalities of the normal theory models are more straightforward than those for the randomization models, however, the use of normal theory models for experiments can be justified only under the randomization umbrella.

Restricted Randomization for Difficult Circumstances

Randomization may result in an arrangement of treatments that is unsatisfactory from the point of view of scientific validity. Sequential arrangements, such as AAA BBB CCC or ABC ABC ABC, are possible with random assignment. However, the sequential arrays of treatments can lead to problems of bias.

Hurlbert (1984) questions the blind use of randomization in small-scale ecological studies. In small studies, there is a very high possibility of randomized layouts resulting with treatments markedly segregated from each other in space or time. Segregation could lead to spurious treatment effects in which treatment and space effects are confounded. Hurlbert (1984) argues that there must be some physical interspersion of the treatments to avoid the systematic segregation of treatments in small experiments.

Yates (1948) and Youden (1956) independently introduced restricted randomization as a solution to the problem of bad patterns from complete randomization. Restricted randomization omits certain arrangements of treatments that in the opinion of the experimenter are unacceptable for the particular study under consideration.

Industrial experiments may require elaborate dismantling of experimental apparatuses between certain types of treatments. Youden (1972) gave examples of industrial and laboratory experiments in which the cost of switching from one treatment to another might outweigh the advantages of complete randomization.

Bailey (1986, 1987) discussed some history and recent research on the topics of restricted randomization. Schemes for restricted or constrained allocation of treatments have been developed that permit the usual analysis of variance procedures (Bailey, 1986; Youden, 1972).

1.9   Relative Efficiency of Experiment Designs

Relative efficiency measures the effectiveness of blocking in experiment designs to reduce experimental error variance. In practice, relative efficiency is measured to determine the efficiency of the design *actually* used relative to another simpler design that *could have been* used but was not. For example, the efficiency of a randomized complete block design is determined relative to a completely randomized design.

The variance of a treatment mean $\sigma_{\bar{y}}^2 = \sigma^2/r$ is a measure of the precision of the estimated treatment means in an experiment. The precision of an estimate for the treatment mean is controlled by the magnitude of σ^2 and the number of replications, r, both of which are to some extent under the control of the investigator. The investigator may increase the number of replications to decrease $\sigma_{\bar{y}}^2$ and increase the precision of the mean estimate. The investigator may also attempt to reduce σ^2 through various local control activities (such as blocking), thereby increasing the precision of the experiment. A method for measuring the effectiveness of blocking is discussed in this section.

The use of $\sigma_{\bar{y}}^2$ as a measure of precision provides a means for comparing the relative precision of two experiment designs. Suppose one design has a true experimental error variance of $\sigma_1^2 = 1$, and a second design has a true experimental error variance of $\sigma_2^2 = 2$. The value of $\sigma_{\bar{y}}^2 = \sigma^2/r$ will be the same for the two designs if the second design has twice as many replications as the first design. That is, the variance of a treatment mean in each of the designs is

$$\text{Design 1:} \quad \sigma_{\bar{y}_1}^2 = \frac{\sigma_1^2}{r_1} = \frac{1}{r_1}$$

and

$$\text{Design 2:} \quad \sigma_{\bar{y}_2}^2 = \frac{\sigma_2^2}{r_2} = \frac{2}{r_2}$$

The variances $\sigma_{\bar{y}_1}^2$ and $\sigma_{\bar{y}_2}^2$ will be the same only if $r_2 = 2r_1$, or Design 2 has twice the replication as that of Design 1. Therefore, Design 1 is more efficient than Design 2 with respect to the number of replications required to have the same precision for an estimate of the treatment mean.

In practice, σ^2 is unknown for each of the designs and must be estimated from the data. Also, the degrees of freedom for the estimate of the variance changes with the designs. Under these circumstances the relative precision of the two designs is determined on the basis of the concept of *information* (Fisher, 1960). Fisher calculated the amount of information that the estimated difference between two means provides about the true difference between the population means. Information calculated from this concept is

$$I = \frac{(f+1)}{(f+3)} \frac{1}{s^2} \tag{1.2}$$

where s^2 is the estimated experimental error variance with f degrees of freedom. If σ^2 is known, then $I = 1/\sigma^2$ and the coefficient $(f+1)/(f+3)$ is replaced by unity. For any reduction in variability, s^2, there is a concurrent increase in the information one has about the mean difference of the populations. Precision and information both increase as the variability decreases.

The *relative efficiency* of two experiment designs is defined as the ratio of information in the two designs. Suppose

$$I_1 = \frac{(f_1+1)}{(f_1+3)} \frac{1}{s_1^2} \quad \text{and} \quad I_2 = \frac{(f_2+1)}{(f_2+3)} \frac{1}{s_2^2}$$

are the estimated information measures of Design 1 and Design 2, respectively. The relative efficiency of Design 1 to Design 2 is estimated as

$$RE = \frac{I_1}{I_2} = \frac{(f_1+1)(f_2+3)}{(f_1+3)(f_2+1)} \frac{s_2^2}{s_1^2} \tag{1.3}$$

If $RE = 1$, then the information in the two designs is equal, and the designs each require the same number of replications to have the same variance of treatment means $\sigma_{\bar{y}}^2$. If $RE > 1$, then Design 1 is more efficient than Design 2. For example, if $RE = 1.5$, then Design 2 requires 1.5 times as many replications as Design 1 to have the same variance of a treatment mean.

1.10   From Principles to Practice: A Case Study

The design of an experimental medical study reported by Moon et al. (1995) illustrates the process of putting principles into practice. The components of the study design provide examples of how the principles of research design, covered in this chapter, helped prepare the investigators to address their research hypothesis. The following is a brief description of the main research design elements in that study. Other details on the study can be found in the publication.

The Problem

Non-melanoma skin cancer, which includes basal cell (BCC) and squamous cell (SCC) carcinomas, is the most common type of cancer. Residents of Arizona experience a three to seven times greater incidence of these cancers than the general U.S. population. Although non-melanoma cancers are usually not life threatening, they result in substantial morbidity and treatment expense.

A history of actinic keratoses (AK), a type of skin lesion, has generally been accepted as a marker for identifying individuals at increased risk of skin cancer. In many cases AK lesions progress to non-melanoma skin cancer and are thus prema-lignant lesions. Those individuals who have a history of actinic keratoses, but few if any non-melanoma skin cancers, are considered to be at moderate risk. Also, fair-skinned, older people with a long history of sun exposure are more at risk for non-melanoma skin cancer.

Results from recent clinical and laboratory studies suggested that vitamin A and other retinoids have a preventative effect against cancer in epithelial tissues such as skin. However, these studies included a small number of subjects and did not produce reliable estimates of the vitamin A effect in the primary prevention of human skin cancer.

A five-year clinical trial was to be conducted in southern Arizona to evaluate the effectiveness of vitamin A or retinol supplementation to reduce the risk of non-melanoma skin cancer for individuals already at moderate risk.

Research Hypothesis

Retinol (vitamin A) supplementation reduces the incidence of skin cancer in mod-erate-risk individuals with a history of at least ten actinic keratoses.

Treatment Design

Important considerations for the dosage level of retinol (vitamin A) included the need to elevate the daily intake of retinol above the common intake of most adults and to avoid a dosage that could potentially induce adverse side effects known to be associated with excessive retinol intake. A placebo treatment was necessary as a comparison group for the treated group; these were subjects with the same charac-teristics as the treated subjects and on the same follow-up protocols during the course of the study. The subjects could not be told to which treatment group they were assigned in order to keep the subjects from both groups on the same regimen.

Treatment: Daily, self-administered, dietary supplement of 25,000 IU retinol in capsule form.

Placebo: Daily, self-administered, placebo capsule

Measurements of Interest

The hypothesis addressed the risk of skin cancer in its relationship to levels of reti-nol. Therefore, the measurement of interest was whether a subject developed skin cancer during the course of the study. The analysis could consider several ap-proaches to test the hypothesis. The approaches included whether a cancer developed, a binary outcome; how many cancers developed, a count variable; or how long it took for a cancer to develop, which is a time-to-event measurement used in survival analysis. Each of these approaches could be extracted by recording the time it took to develop a cancer, if one developed. Thus, the measured variables were the

- time to the first and each subsequent, if any, basal cell carcinoma (BCC)

- time to the first and each subsequent, if any, squamous cell carcinoma (SCC)

Selection of Subjects with Common Characteristics

Subjects for this study were required to be representatives of a healthy adult population with a moderate risk to non-melanoma skin cancer, willing to participate in the study, and not currently ingesting an excessive amount of vitamin A in their diet. More than 11,000 subjects were screened for the study; approximately 25% were deemed eligible. Following are some of the criteria used by the investigators to screen the subjects.

The subjects were recruited through referrals by dermatologists and media announcements. The subjects could be men or women with a history of at least ten actinic keratoses diagnosed clinically, with the most recent diagnosed in the past year. They could have had no more than two prior occurrences of SCC or BCC and no cancer diagnosis other than SCC or BCC in the previous year.

The eligible individuals had to be between 21 and 84 years of age, be ambulatory, be capable of self-care, have no diagnosis of a life-threatening disease, intend to be a resident of Arizona for the next five years, and be willing to commit to semiannual follow-up clinic visits for that period. They also had to be willing to limit non-study vitamin A supplementation to no more than 10,000 IU per day. Further, they had to be within the 95% normal range for total cholesterol, liver function, white blood cell count, hemoglobin, and platelet count.

Some Techniques to Reduce Nonrandom Error and Bias

Several precautions had to be taken to ensure that nonrandom error and bias in response was minimized in the study. This included assurance that the subjects would take the medication regularly, retain their willingness to remain in the study, return for their follow-up clinical visits for evaluation, and remain unaware of the treatment group to which they were assigned.

Conceivably, if subjects knew they were in the placebo group they could self-administer extra vitamin A on their own accord with the hope of reducing their personal risk and unknowingly bias the comparison of treatment to placebo. The investigators provided an insurance against this sort of event by not revealing to the subject or the clinician dispersing the capsules whether the subject was receiving the treatment or placebo. This type of trial is known as a *double-blind* trial.

A three-month run-in period was established to evaluate the subjects' ability and willingness to adhere to the study protocol. They received a bottle containing 100 placebo capsules, of which they were to take one per day. Subjects who had taken at least 75% of the capsules during the run-in period and were willing to continue the study were assigned to a treatment or placebo regimen, given a six-month supply of the appropriate capsules, and scheduled for a follow-up visit.

Subjects returned to the clinic every six months to receive an examination for any symptoms of BCC or SCC. As a safety measure, they were also examined for any potential side effects from elevated retinol intake. The subject's remaining medication was weighed to evaluate adherence to dosage. During the interview, subjects' questions were answered and they were motivated to adhere to the medication schedules. They were then given the next six months' supply of capsules and scheduled for the next follow-up visit. Subjects were reminded of their upcoming follow-up visit by postcard and telephone contact.

Once a year a blood specimen was collected from a random sample of subjects for analyses of retinyl palmitate levels to obtain supplementary information on group adherence to the retinol supplement. The levels of retinyl palmitate should have been greater for the treatment group than the placebo group if the treatment group had adhered to the capsule regimen.

Replication

The trial was conducted from two separate clinics, one each in Tucson and Phoenix, Arizona. The number of subjects required for the study was based on assumptions about the average annual incidence of skin cancers in the placebo and retinol treatment groups and incidences of anticipated deviation from the pre-scribed study protocols by the subjects. The required sample size was determined to be 1118 subjects in each of the treatment groups and was based on a power of .80 and a .05 two-sided Type I error rate, using techniques specific to studies on time-to-event measurements.

Blocking to Reduce Experimental Error

A subject's risk to non-melanoma skin cancer was anticipated to be related to the amount of time spent in the sun and whether the person had fair skin. Fair-skinned people are anticipated to be at greater risk for skin cancer and will sunburn more readily in a 30-minute period in southern Arizona. Anyone who spends a greater amount of time in the sun is anticipated to be at greater risk for skin cancer. Information on weekly sun exposure and anticipated skin-burning reaction after 30 minutes of sun exposure was collected from each subject during the first contact.

These were the most likely factors to interfere with risk comparisons between retinol treatment and placebo, so they were used as blocking factors prior to assignment of treatment to subjects. Subjects were categorized according to two levels of sun exposure: < 10 hours versus ≥ 10 hours per week. They were also grouped according to levels of skin reaction after 30 minutes: always or usually burns versus burns moderately, rarely, or never.

Thus, subjects were placed into one of the four block types constructed when the two factors were placed in all four combinations of their levels. For example, subjects exposed to the sun < 10 hours per week and whose skin moderately, rarely, or never burned after 30 minutes exposure to the sun would be placed in the same block. Equal numbers of subjects would be assigned to the retinol treatment and placebo in each of the blocks; thus, any potential differences between retinol treatment and placebo would not be interfered with by differential subject risks based on sun exposure and skin sensitivity to the sun.

Randomization

Subjects enter into clinical trials over a period of time as they are identified by their physicians as candidates for the study and as they respond to calls for subject volunteers. Therefore, the assignment of subjects to treatments sometimes takes place over a period of one or more years until the required number of subjects (replications) has been achieved for the study. For this study, the enrollment period required more than four years to acquire a sufficient number of subjects.

The subjects were assigned to their respective blocking type (described above) as they entered the study. They were randomly assigned to the placebo or retinol treatment by groups of four subjects with the same blocking criteria. For example, if the first two subjects entering in a block were assigned at random to the placebo the next two would automatically receive the retinol treatment. Then the randomization would start over with the next four subjects entering into any one of the blocking groups. This method of assignment in blocks of four subjects ensured an equal distribution of subjects on placebo and retinol treatment over the entry timeline of the trial, effectively removing the chance of having a greater number of subjects on one treatment for a longer period of time than on another treatment.

The randomizations were done separately for the clinics in Tucson and Phoenix. Thus, the clinics became a de facto blocking factor in the study.

Measured Covariates for Statistical Control of Experimental Error

Numerous other factors could conceivably have some relationship to risk for skin cancer. Those factors thought to have the greatest potential for affecting the comparison of risk between retinol treatment and placebo were used as blocking factors: sun exposure and skin sensitivity to sun exposure. To have included more factors in the blocking scheme might have made the study unnecessarily cumbersome, particularly if no hard evidence suggested these factors had major impacts on skin cancer risk.

Other factors the investigators considered to be potentially influential and for which measurements were recorded included age, gender, prior skin cancers (0, 1, or 2), number of moles and freckles, vitamin use, dietary vitamin A at the start of the study determined from a diet interview, and serum retinyl palmitate at the start of the study.

These factors could have been used as covariates in the data analysis to reduce experimental error for comparisons of retinol treatment to placebo. A post-randomization check after the four-year enrollment period was made to determine

Table 1.3 Cross tabulations of subjects by treatment groups and covariates in the skin cancer prevention trial

Characteristic	Placebo $n = 1140$	Retinol $n = 1157$
Age		
< 63	558	584
≥ 63	582	573
Gender		
Female	345	334
Male	795	823
Prior skin cancers		
0	932	920
1	152	177
2	45	45
> 2	11	15
Moles and freckles		
0–7	550	541
> 7	301	326
Unknown	289	290
Weekly sun exposure		
0–10 hours	452	493
> 10 hours	688	664
Skin burns		
Always/usually	490	517
Moderately/rarely/never	650	640
Clinical center		
Phoenix	429	420
Tucson	711	737
Vitamin use		
No	309	312
Sometimes	322	331
Yes	509	514
Dietary vitamin A (IU)		
1,194–6,979	337	355
6,980–10,627	324	369
10,628–41,404	365	326
Unknown	114	107
Serum retinyl palmitate (mg/ml)		
0.0–6.0	397	400
6.1–20.0	345	346
> 20.0	350	378
Unknown	48	33

whether the subjects in placebo and retinol treatment groups were equally distributed with respect to each of these factors. The cross tabulations shown in Table 1.3
indicate a relatively equal distribution of subjects on placebo and retinol treatment
for each of the major covariates considered in the study. Notice 26 subjects with
more than two previous skin cancers were enrolled in the study even though they
did not meet the selection criterion of no more than two previous skin cancers. Of
course, these could be either recording errors or an oversight by the clinicians who
could have mistakenly allowed these subjects to enroll. The reason for these types
of mistakes often remains unknown.

EXERCISES FOR CHAPTER 1

1. a. Find the definitions of *research* and *hypothesis* in a dictionary, and formulate a definition of
the term *research hypothesis*.
   b. How does a research hypothesis differ from the *statistical hypothesis* formulated for statistical
tests, such as $H_{0:} \ \mu = 0$ versus $H_a: \ \mu \neq 0$?

2. Choose a journal article in your field of study that reports the results of a comparative experiment
or observational study. Identify and briefly (one or two sentences) describe the following:
   a. Research hypothesis
   b. Treatments
   c. Experimental (observational) units
   d. Type of experiment design
   e. Criteria for any grouping, blocking, or matching done in the study
   f. Provide the article citation

3. a. Choose a practical situation from your own field of study and describe a problem whose
solution must be determined experimentally.
   b. Indicate the following for the problem described in part (a):
      (i)   a research hypothesis
      (ii) the treatments necessary to evaluate the hypothesis
      (iii) what constitutes an experimental unit

4. Choose a journal article from your special field of interest and review it for the purpose of
evaluating the application of good research design. Many aspects of research design have been discussed separately in this chapter relative to their effect on statistical and scientific inference. The
case study in Section 1.10 illustrates the elements of a reported research study that one must
identify for a validation of the reported findings from a research project.

Choose an article that reports on either an experiment conducted to compare two or more treatments or a comparative observational study conducted to compare two or more "treatment" conditions that existed a priori.

Some good questions to ask yourself for the critique are "Did they include all important elements of good research design in the study?" "If so, did they implement them properly?" "Are the elements of the work described in such a way that I could understand or duplicate what they did?"

Pay particular attention to the following items in your review:

a. Review of the literature
b. Statement of the problem
c. Research hypothesis and study objectives: discuss whether they were present and reasonable
d. Treatment design: describe how it did or did not address the hypothesis and objectives
e Experiment design or observational study design
f. Use of randomization (experiments) or random sampling (observational study) and replication
g. Statistical hypotheses and statistical analysis procedures
h. Conclusions and statistical reliability of conclusions
i. Self-evaluation of the study by the author(s) and potential for future investigations
j. Provide the article citation

Your critique should describe and evaluate the author's approach in the research itself and in the article relative to each of the important elements of the research based on the preceding list. Include a page reference from the article for your comments on each of the items.

5. A study is planned on the physiology of exercises with human subject volunteers. The two treatments in the study are two methods of aerobic exercise training (call the methods A and B). At the end of a ten-week exercise period, each subject will undergo a treadmill test for standard respiratory and cardiovascular measurements.

Individual	Sex	Age	Individual	Sex	Age
1	M	54	10	M	18
2	M	38	11	M	31
3	F	41	12	F	18
4	F	18	13	M	58
5	F	19	14	M	74
6	F	39	15	F	58
7	M	51	16	F	21
8	F	44	17	M	35
9	M	62	18	M	34
			19	F	38

Nineteen volunteers are listed in the table by sex and age. All volunteers are in good health and in the normal weight range for their age, sex, and height. Eight individuals will be tested in each of the methods (A or B), so that only 16 of the 19 volunteers will be used; a subject will participate only in one of the methods.

a. Explain how you would group the individuals prior to the assignment of treatments so that experimental error variance could be kept at a minimum.
b. Explain why you grouped as you did.
c. Show your final assignment of individuals to the treatment groups.

6. An experiment is planned to compare three treatments applied to shirts in a test of durable press fabric treatments to produce wrinkle-free fabrics. In the past formaldehyde had been used to produce wrinkle-free fabric, but it was considered an undesirable chemical treatment. This study is to consider three alternative chemicals: (a) PCA (1-2-3 propane tricarbolic acid), (b) BTCA (butane tetracarboxilic acid), and (c) CA (citric acid).

Four shirts will be used for each of the treatments. First, the treatments are applied to the shirts, which are then subjected to simulated wear and washing in a simulation machine. The chemical treatments will not contaminate one another if they are all placed in the same washing machine during the test. The machine can hold one to four shirts in a single simulation run. At the end of the simulation run each of the shirts is measured for tear and breaking strength of the fabric and how wrinkle-free they are after being subjected to the simulated wear and washing. The comparisons among the treatments can be affected by (a) the natural variation from shirt to shirt; (b) measurement errors; (c) variation in the application of the durable press treatment; and (d) variation in the run of the simulation of wear and washing by the simulation machine. Following is a brief description of three proposed methods of conducting this simple experiment.

Method I. The shirts are divided randomly into three groups of four shirts. Each group receives a durable press treatment as one batch and then each batch is processed in one run of the simulation machine. Each run of the simulation machine has four shirts that have received the same treatment. There are three runs of the simulation machine.

Method II. The shirts are divided randomly into three treatment groups of four shirts each, and the durable press treatments are applied independently to single shirts. The shirts are grouped into four sets of three, one shirt from each durable press treatment in each of the four sets, and each set of three so constructed is used in one run of the simulation machine. There are four runs of the simulation machine.

Method III. The shirts are divided randomly into three groups of four shirts. The durable press treatments are applied independently to single shirts. The simulation of wear and washing is done as in Method I.

a. Which method do you favor?
b. Why do you favor the method you have chosen?
c. Briefly, what are the disadvantages of the other two methods?

7. Explain what is meant by the term *replication* in the context of (a) an experiment in which the effectiveness of several antibiotics is tested on animals in a laboratory and (b) an observational study to determine the differences in grass species present in pure stands of mesquite and pure stands of oak in southern Arizona.

8. An experiment is planned to compare three methods of instruction. Each is tested with a single classroom of 25 students. A different instructor is to be used for each classroom and consequently each instruction method.
   a. Write a short critique of the proposed experiment.
   b. How could the experiment be improved?

9. An experiment is planned to compare the strengths of three different asphalt mixtures for road surfaces. A single batch is to be manufactured for each experimental mixture. Several asphalt specimens will be made from each of the mixtures and tested for tensile strength.
   a. Write a short critique of the proposed experiment.
   b. How could the experiment be improved?

10. Suppose you want to randomize the allocation of two treatments to 16 experimental units. How many randomizations are possible if 8 units are to be assigned to each of the treatments? How many randomizations are possible if you want to assign 6 units to one treatment and 10 units to a second treatment?

11. An experiment with four treatments and five replications of each treatment will require 20 experimental units. How many randomizations are possible for this experiment?

12. An experiment with two treatments and three replications per treatment had the following randomization of treatments to the experimental units shown along with the measured response on each unit:

Unit:	1	2	3	4	5	6
Treatment:	A	B	B	A	A	B
Response:	7	10	9	5	10	12

Conduct a randomization test of the null hypothesis, H_0: no difference in the treatment effects of A and B, versus the alternative, H_a : the effect of treatment B is greater than the effect of treatment A. Use the test statistic $\bar{y}_B - \bar{y}_A$. (*Hint*: It is only necessary to identify one-half of the randomizations directly. Each randomization has a "mirror" randomization in which the letters A and B are interchanged. For example, the "mirror" for the actual experiment randomization is the randomization B, A, A, B, B, A.)

13. In the table, the coefficients of variation and relative efficiencies (randomized complete block to completely randomized) of the same experiment conducted at four locations are given. Each trial used a randomized complete block design.

Location	Coefficient of Variation (%)	Relative Efficiency (%)
Tucson	10	100
Phoenix	10	150
Los Angeles	20	200
San Francisco	20	125

a. How many more replications of a completely randomized design would be necessary at Los Angeles to obtain the same precision as the randomized complete block design for estimating the treatment means? Explain your answer.

b. If you were asked the question in part (a) relative to San Francisco, would you require more or fewer replications in San Francisco than for Los Angeles? Explain your answer.

c. Suppose four replications are required with the randomized complete block design at Tucson to detect differences of $\delta = 20\%$ with a test at the .05 level of significance and a probability (power) of .90. How many replications would be required in Phoenix with the same criteria for a randomized complete block design? Explain your answer.

d. Would you require more or fewer replications in Los Angeles than in Tucson with the same criteria for a randomized complete block design? Explain your answer.

2 Getting Started with Completely Randomized Designs

Chapter 1 presented the principles of design in relation to the stated goals of research hypotheses—accuracy and precision of the observations and validity of the resulting analysis. Some of those principles are illustrated in this chapter, using an experiment with a completely randomized design. A statistical model is developed with parameters to describe the experiment according to the research hypothesis. The parameters are then estimated, using the least squares method. Experimental error variance is estimated and used to estimate standard errors and confidence intervals for the parameters and to test statistical hypotheses about them. The fundamental partition for the sum of squares of the observations is derived and summarized in the traditional analysis of variance table.

2.1 Assembling the Research Design

The *research hypothesis, treatment design,* and *experiment* or *observational study design* constitute the **research design** for the study. Treatments are developed to address specific research questions and hypotheses that arise in the research program. For example, a microbiologist hypothesizes that the activity of soil microbes depends upon soil moisture conditions. Treatments with different amounts of soil moisture are set up to measure the microbe activity at different levels of soil moisture to evaluate the hypothesis. A traffic engineer hypothesizes that traffic speed is related to the width of street lanes. Lanes of different width are selected by the engineer, and traffic speed is measured at each lane width to evaluate the hypothesis.

The treatment design has to be integrated into an experiment design. The investigator must decide what constitutes an experimental unit, how many replications of experimental units are required for each treatment, and which treatment to assign to each of the experimental units. The investigator must also determine whether the experimental units should be blocked into homogeneous groups to control experimental error.

Similarly, the comparative observational study on association of traffic speed with lane width requires the investigator to determine how many independent street units of each width are required for the study and how to match the street units for controlling variables.

The computational details of the statistical analysis of various designs may vary from one design to another. Many of the statistical procedures used for analysis are common to most of the designs that we encounter. This is so because the procedures themselves generally relate to specific treatment designs, each of which may appear in any number of experiment design configurations.

The intent of this chapter and Chapter 3 is to introduce useful statistical procedures for a variety of comparative studies. The procedures are extended to more complex treatment designs in the ensuing chapters, and their application in other experiment designs is demonstrated as the designs are presented. The procedures are illustrated first in this chapter with a one-way classification of treatments in a completely randomized design with equal replication of treatments.

Example 2.1   Suppression of Bacterial Growth on Stored Meats

The shelf life of stored meats is the time a prepackaged cut remains salable, safe, and nutritious. Standard packaging in ambient air atmosphere has a shelf life of about 48 hours after which the meat quality begins to deteriorate from microbial contamination, color degradation, and shrinkage. Vacuum packaging is effective in suppressing microbial growth; however, other quality losses remain a problem.

Recent studies suggested controlled gas atmospheres as possible alternatives to existing packagings. Two atmospheres which promise to combine the capability for suppressing microbial development while maintaining other meat qualities were (1) pure carbon dioxide (CO_2), and (2) mixtures of carbon monoxide (CO), oxygen (O_2), and nitrogen (N).

Research Hypothesis: Based on this new information the investigator hypothesized that some form of controlled gas atmosphere would provide a more effective packaging environment for meat storage.

Treatment Design: The treatment design developed by the investigator to address the hypothesis included packagings with (1) ambient air in a commercial plastic wrap, (2) a vacuum, (3) a mixture of gases consisting of 1% CO, 40% O_2, and 59% N, and (4) 100% CO_2. The ambient air and vacuum packagings served the role of control treatments because both were standards to which new packagings could be compared for effectiveness.

Experiment Design: A completely randomized design was used for the experiment. Three beef steaks of approximately the same size (75 g) were randomly assigned to each of the packaging conditions. The randomization method is demonstrated in Section 2.2. Each steak was packaged separately in its assigned conditions. (For the purpose of this illustration the packaging treatments are evaluated for their effectiveness in suppressing bacterial growth.) The number of psychrotrophic bacteria on the meat was measured after nine days of storage at 4°C in a standard meat storage facility. Psychrotrophic bacteria are found on the surface of the meat and are associated with spoilage of the meat product.

The results are shown in Table 2.1. Bacterial growth is expressed as the logarithm of the number of bacteria per square centimeter.

Table 2.1 Psychrotrophic bacteria [$\log(\text{count/cm}^2)$], on meat samples stored in four packaging conditions for nine days

	Psychrotrophic Bacteria		
Packaging Condition	*Log (count/cm^2)*	*Total*	*Mean*
Commercial plastic wrap	7.66, 6.98, 7.80	22.44	7.48
Vacuum packaged	5.26, 5.44, 5.80	16.50	5.50
1% CO, 40% O$_2$, 59% N	7.41, 7.33, 7.04	21.78	7.26
100% CO$_2$	3.51, 2.91, 3.66	10.08	3.36

Source: B. Nichols (1980), *Comparison of Grain-Fed and Grass-Fed Beef for Quality Changes When Packaged in Various Gas Atmospheres and Vacuum*, M.S. thesis, Department of Animal Science, University of Arizona.

2.2 How to Randomize

Randomizing Treatments in Experiment Designs

The steaks used for the experiment were relatively homogeneous experimental units, and a completely randomized design was used to avoid any subjective assignment of treatments to the steaks. The proper randomization procedure for a completely randomized design is illustrated with the meat storage study.

Step 1. Assign the sequence of numbers 1 through 12 to the experimental units, the steaks.

Step 2. Obtain a random permutation of the numbers 1 through 12 and write them down in the permutation order. A random permutation can be obtained by taking a sequence of two- or three-digit numbers from a random numbers table (Appendix Table XII) and ranking them from the smallest (1) to the largest (N). The rank numbers in the sequence constitute a random permutation. Let each number in the permutation order equal the sequence number of a steak. Suppose the permutation order is

1   6   7   12   5   3   10   9   2   8   4   11

Step 3.   Assign the first three steaks in the list (1, 6, and 7) to treatment A. Assign the next three steaks (12, 5 and 3) to treatment B and so forth. The final assignment of steaks to treatments is

Steak:	1	6	7	12	5	3	10	9	2	8	4	11
Treatment:	A	A	A	B	B	B	C	C	C	D	D	D

The random permutation of the numbers 1 through 12 will ensure that each of the possible assignments of treatments to the experimental units has an equal probability of occurrence. Many commercial statistical computing programs include routines for permutations and randomization.

An equivalent method of randomization can be used in the absence of a computer program or a random numbers table. After assigning numbers to the experimental units as in Step 1, construct corresponding numbered slips of paper or tags and draw them from a hat at random. The first r numbers drawn are the experimental units assigned to the first treatment. The second r numbers drawn are the experimental units assigned to the second treatment, and so forth. A discussion of proper and improper methods of treatment assignment to units in a random fashion is discussed by Hader (1973).

Selecting the Units for Comparative Observational Studies

The traffic engineering study on street lane widths described at the beginning of the chapter is a comparative observational study. The investigator is unable to randomly assign a unit to a treatment group in the comparative observational study. Depending on the type of research program, the basic units of the investigation are either self-selected into the treatment groups or they exist in their characteristic groups.

A probability sample of units should be selected from available members of each treatment population. Units are selected from within each population such that each unit has an equal chance of entering the sample. Note that each population represents a separate treatment classification, and random sampling is maintained only *within* the population.

The first step requires an identification of the populations that represent the conditions or treatments of interest in the observational study. A list is constructed of all available units in each population. In an ecological study, for example, the two treatments may be grassland and oak woodland plant communities. The populations consist of all grassland sites and all oak woodland sites in a particular study area. Each of the sites in both populations is assigned a unique identification code. For example, assign the numbers 0 through 99 to the 100 sites available in each classification.

Suppose the investigator wants to establish plots in ten grassland sites and ten oak woodland sites. The method for sampling the ten sites in each of the study areas begins with the selection of ten 2-digit numbers from a random numbers table (Appendix Table XII) for each of the sites. Suppose the first set of 2-digit numbers chosen is 12, 63, 34, 05, 97, 72, 42, 44, 82, 51. The sites so numbered in the study area are used to establish measurement plots. The same procedure is followed for each treatment group. If the same random numbers are drawn more than once from the table, additional 2-digit numbers may be drawn at random from the table to complete the sample. Many commercial statistical computing programs include routines for random sampling from a list.

2.3 Preparation of Data Files for the Analysis

A fairly large amount of data are collected in most studies. The data first must be organized for use by computer programs. The execution of statistical programs on computers requires a *data file*. The file can be either typed into the terminal when the program is being run or it can exist in a file that previously has been put on a computer storage device with a separate data entry program.

Each observation is identified clearly in the data file with a particular experimental unit and treatment classification in the study. The data file then has a convenient format for analysis and scrutiny of the data for any irregularities that may have occurred in either the procurement or recording of the observations. The data file for Example 2.1 might contain a sequence number identifying each beef steak, the treatment group to which the steak was assigned, and the observed bacterial count as shown in Display 2.1.

Display 2.1	Data for Meat Storage Experiment	
Steak	*Treatment*	*Log* $(count/cm^2)$
1	Commercial	7.66
6	Commercial	6.98
7	Commercial	7.80
12	Vacuum	5.26
5	Vacuum	5.44
3	Vacuum	5.80
10	Mixed Gas	7.41
9	Mixed Gas	7.33
2	Mixed Gas	7.04
8	CO_2	3.51
4	CO_2	2.91
11	CO_2	3.66

The entire data file consists of 12 lines (the titles are not part of the data file), each of which contains requisite information the computer program can utilize in record-keeping and performing calculations.

Each line of the file is a *case* or an *observation* in programs. The meat storage data file has 12 cases or observations. Each column is a *variable* in the program.

The variables in this data file are *steak*, *treatment*, and *log (count/cm²)*. The actual values listed in each line of the file are often referred to as the *data values*.

The only variables ordinarily required by a statistical program to perform calculations for this example are the *treatment* variable and the measured variable *log (count/cm²)*.

The *steak* variable is primarily a housekeeping variable used to identify the case in the data file with an experimental unit in the actual study. Housekeeping variables in a data file may include code numbers for the experimental unit, treatment names, dates, and experiment numbers. They are useful for management of large data files and for memory recall if the data file must be used long after its creation.

2.4   A Statistical Model for the Experiment

A statistical analysis is based on an underlying formal statistical model. Proper interpretation of the analysis requires an understanding of the model. In comparative studies the characteristic of the units or subjects measured upon observation is the *response variable,* identified by the variable y. The bacterial count is the response variable in the meat storage experiment.

The statistical model for comparative studies assumes there is a *reference population* of subjects or experimental units. In most cases the population is conceptual although it is possible to imagine a population of automobile engines, retail stores, field plots, pens of animals, or packaged meat. Each individual unit in the population has a value for the response variable y, and the variable y has a mean μ and variance σ^2.

A reference population is assumed for each treatment condition in the study, and the units in an experiment are assumed to be randomly selected representatives of the reference population as a result of randomization. For observational studies, we assume the observational units are randomly selected from the treatment populations.

The statistical model is illustrated in Figure 2.1 with four hypothetical treatment populations. Each population has a normal distribution for the response variable, and each has a different mean value. Such a situation would exist if the four methods of packaging meat had differing capacities for the inhibition of bacterial growth.

The population variance σ^2 is assumed to be the same for each of the populations and unaffected by the treatments as shown in Figure 2.1. That is, the variances of the treatment populations are assumed to be homogeneous.

Use of the Cell Means Model to Describe Observations

Observations are expressed as a sum of the treatment population means and the experimental errors with the *cell means* model

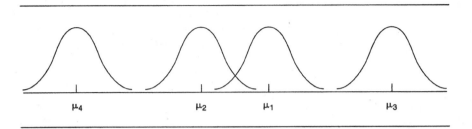

Figure 2.1   Illustration of the treatment populations

$$y_{ij} = \mu_i + e_{ij} \tag{2.1}$$

$$i = 1, 2, \ldots, t \qquad j = 1, 2, \ldots, r$$

where y_{ij} denotes the jth observation from the ith treatment group; μ_i is the mean of the ith treatment population; and e_{ij} is the experimental error. This is a **linear statistical model** for the one-way treatment classification in a completely randomized experiment design.

The model makes an allowance for the variation among the observations from a given treatment group. Because of experimental error each observation deviates from the population mean μ_i by some amount e_{ij}. The experimental error variance σ^2 is the variance of the e_{ij}, and it is assumed to be the same for all treatment populations.

The observations for the meat storage experiment are shown in Table 2.2, with $t = 4$ treatments and $r = 3$ replications with their identification y_{ij} and statistical model representation.

Use of Alternative Models to Describe Alternative Statistical Hypotheses

The statistical model for the experiment reflects our beliefs about the relationship between the treatments and the observations. The cell means model is a *full model,* $y_{ij} = \mu_i + e_{ij}$, that includes a separate mean for each of the treatment populations. A model with a reduced set of parameters is used if there are no differences among the treatment population means. The *reduced model,* $y_{ij} = \mu + e_{ij}$, states that the observations all belong to the same population with a mean μ.

The two models represent the alternative statistical hypotheses appropriate for the experiment. The reduced model represents the null hypothesis condition with no differences among the treatment means, stated as $H_0: \mu_1 = \mu_2 = \cdots = \mu_t$. The full model represents the alternate hypothesis condition when there are some differences among the treatment means, stated as $H_a: \mu_i \neq \mu_k$ for some $i \neq k$.

The investigator for the meat storage experiment must determine whether the bacterial growth differs among various packaging methods or whether one

Table 2.2 Identification of observed values from the meat storage experiment and their representation with the linear statistical model

Steak	Treatment	Observation	Log $(count/cm^2)$	y_{ij}	Model
1	1	1	7.66	y_{11}	$\mu_1 + e_{11}$
6	1	2	6.98	y_{12}	$\mu_1 + e_{12}$
7	1	3	7.80	y_{13}	$\mu_1 + e_{13}$
12	2	1	5.26	y_{21}	$\mu_2 + e_{21}$
5	2	2	5.44	y_{22}	$\mu_2 + e_{22}$
3	2	3	5.80	y_{23}	$\mu_2 + e_{23}$
10	3	1	7.41	y_{31}	$\mu_3 + e_{31}$
9	3	2	7.33	y_{32}	$\mu_3 + e_{32}$
2	3	3	7.04	y_{33}	$\mu_3 + e_{33}$
8	4	1	3.51	y_{41}	$\mu_4 + e_{41}$
4	4	2	2.91	y_{42}	$\mu_4 + e_{42}$
11	4	3	3.66	y_{43}	$\mu_4 + e_{43}$

packaging method is no better than any other in suppressing bacterial growth. From the point of view of the statistical model the investigator must determine which of the two, the full model or the reduced model, best characterizes the data in the experiment. The research questions are translated into questions about statistical populations modeled for the experiment in Figure 2.2.

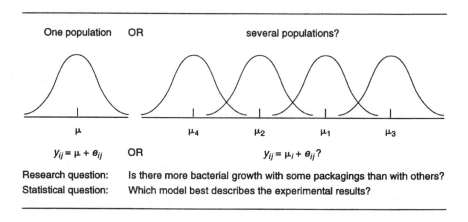

Research question: Is there more bacterial growth with some packagings than with others?
Statistical question: Which model best describes the experimental results?

Figure 2.2 The research questions, statistical questions, and alternative models for the meat storage experiment

To make some decision about the treatments the investigator requires a statistical method to estimate the parameters of the two models and, on the basis of an objective criterion, determine which of the statistical hypotheses or models best fits the data from the experiment.

The General Linear Statistical Model

The cell means model is a special case of the **general linear model**. The more general model describes relationships between two types of variables as a function linear in a set of parameters. One type of variable is the response variable y considered to be dependent on the second type of variable consisting of the design variables $x_1, x_2, \ldots, x_k$. The x_i can be variables fixed by the treatment design, such as the level of a temperature treatment, or they can be measured covariates, such as the age of the subjects. The x_i can also represent categorical treatments, such as those in the meat storage experiment.

The statistical model relates y to the x_i through a set of parameters, $\beta_0, \beta_1, \beta_2, \ldots, \beta_k$, such that it is linear in the set of parameters or

$$y = \beta_0 + \beta_1 x_1 + \beta_2 x_2 + \cdots + \beta_k x_k + e \tag{2.2}$$

Some simple examples will help clarify how the investigator can develop a model unique to a particular study.

An experiment measures the rate of a chemical reaction, y, as a response to the temperature, T, in the reaction chamber. The investigator may hypothesize that the rate increases linearly with temperature. Letting $x_1 = T$, the simple straight line equation describes the linear relationship as

$$y = \beta_0 + \beta_1 x_1 + e$$

which is the familiar simple linear regression model. If the investigator thinks the rate-to-temperature relationship may be quadratic, then let $x_1 = T$ and $x_2 = T^2$ and the model becomes

$$y = \beta_0 + \beta_1 x_1 + \beta_2 x_2 + e$$

How does the model work if the treatments are categories that cannot be represented by metrical values for the x_i? In this case the investigator can let the x_i be **indicator variables**. The indicator variable does exactly what the name implies; it indicates the treatment group to which the observation belongs. One scheme for indicator variables lets $x = 1$ indicate that the observation belongs in a particular group and lets $x = 0$ indicate that it does not belong in a particular treatment group.

The meat storage experiment has four treatments and the model can have four indicator variables, $x_1, x_2, x_3,$ and x_4. They take the values 1 and 0, as follows

$$x_1 = \begin{cases} 1 \text{ if commercial wrap} \\ 0 \text{ otherwise} \end{cases}$$

$$x_2 = \begin{cases} 1 \text{ if vacuum} \\ 0 \text{ otherwise} \end{cases}$$

$$x_3 = \begin{cases} 1 \text{ if mixed gas} \\ 0 \text{ otherwise} \end{cases}$$

$$x_4 = \begin{cases} 1 \text{ if pure } CO_2 \\ 0 \text{ otherwise} \end{cases}$$

The model may be written as

$$y = \beta_1 x_1 + \beta_2 x_2 + \beta_3 x_3 + \beta_4 x_4 + e$$

If the observation came from the commercial wrap treatment, with $x_1 = 1$ and $x_2 = x_3 = x_4 = 0$

$$y = \beta_1 + e$$

If the observation came from the vacuum treatment, with $x_2 = 1$ and $x_1 = x_3 = x_4 = 0$

$$y = \beta_2 + e$$

and so forth. Notice $\beta_0 = 0$ since it is not necessary for a description of the observations in this case.

The model $y = \beta_1 + e$ then describes the observation that has a value $\beta_1 + e$ if it comes from the commercial wrap treatment. Thus $\beta_1 = \mu_1$, the mean of the hypothesized commercial wrap population as illustrated in Figure 2.2.

The means model in Equation (2.1) comes directly from this representation by letting $\beta_1 = \mu_1, \beta_2 = \mu_2, \beta_3 = \mu_3,$ and $\beta_4 = \mu_4$. Given that the x_i take only the values 0 or 1, and if we let $\beta_1 = \mu_1$, then an observation from the commercial wrap treatment is modeled as

$$y = \mu_1 + e$$

By adding subscripts to identify specific observations

$$y_{1j} = \mu_1 + e_{1j} \qquad j = 1, 2, \ldots, r$$

is the model representing observations from the commercial wrap treatment. Generalizing the model to include all treatment groups results in the model shown in Equation (2.1), or

$$y_{ij} = \mu_i + e_{ij}$$

$$i = 1, 2, \ldots, t \quad j = 1, 2, \ldots, r$$

This model expression for experiments with treatment groups is more traditional and provides a more specific description for the experiment.

The flexibility of the general linear model in Equation (2.2) allows the investigator to include measured covariates in the model along with the treatment group variables. For example, suppose in the meat storage experiment the investigator was concerned that fluctuations in the moisture content among the meat samples could alter the rate of bacterial growth. Prior to packaging the meat, measurements are made on the moisture content of each of the meat experimental units.

If we let $x =$ moisture content (%), then a term representing a linear relationship between bacterial counts and moisture content can be added to the model as

$$y_{ij} = \mu_i + \beta x_{ij} + e_{ij}$$

where x_{ij} is the moisture content of the jth steak in the ith treatment group and β is the linear regression coefficient. This is the traditional **analysis of covariance** model which assumes the regression coefficient is the same for all treatment groups. The analysis of covariance is discussed in Chapter 17.

2.5 Estimation of the Model Parameters with Least Squares

The **least squares method** is perhaps the method most frequently used to estimate the parameters for the linear model. The least squares estimates are the estimates of μ_i that result in the smallest sum of squared experimental errors. If the experimental errors are independent with a mean of zero and have homogeneous variances the least squares estimators are unbiased with minimum variance. Random sampling for observational studies and randomization in experiments provides insurance for the independence assumption as discussed in Chapter 1. Methods to evaluate the homogeneous variance assumption are found in Chapter 4.

Estimators for the Full Model

The experimental errors for the cell means model are the differences between the observation and the population means $e_{ij} = y_{ij} - \mu_i$, where the observations y_{ij} are the only known quantities. If we denote the least squares estimators of the μ_i for the full model as $\widehat{\mu}_i$, then the estimators of the experimental errors are $\widehat{e}_{ij} = y_{ij} - \widehat{\mu}_i$. The minimum sum of squares is

$$SSE = \sum_{i=1}^{t} \sum_{j=1}^{r} \widehat{e}_{ij}^2 = \sum_{i=1}^{t} \sum_{j=1}^{r} (y_{ij} - \widehat{\mu}_i)^2 \qquad (2.3)$$

SSE is the estimated sum of squares for experimental error, and it is a measure of how well the model fits the data.

A method of differential calculus is used to determine the estimators $\widehat{\mu}_i$ that minimize the sum of squares

$$Q = \Sigma e_{ij}^2 = \Sigma (y_{ij} - \mu_i)^2$$

The method produces a set of equations that must be solved for the estimators. The conventional name for the equations is the *normal equations*; however, the designation has nothing to do with the normal probability distribution.

For t treatment groups with r replications per treatment there are t normal equations, one for each treatment mean. The normal equations are found by first differentiating

$$\sum_{i=1}^{t} \sum_{j=1}^{r} e_{ij}^2 = \sum_{i=1}^{t} \sum_{j=1}^{r} (y_{ij} - \mu_i)^2$$

with respect to each of the μ_i and setting the result equal to zero. The partial derivative for a typical equation is

$$\frac{\partial}{\partial \mu_i} \sum_{j=1}^{r} (y_{ij} - \mu_i)^2 = -2 \sum_{j=1}^{r} (y_{ij} - \mu_i) = 0$$

Simplifying the equation and substituting $\widehat{\mu}_i$ for μ_i produces

$$\sum_{j=1}^{r} \widehat{\mu}_i = \sum_{j=1}^{r} y_{ij}$$

or

$$r\widehat{\mu}_i = y_{i.}$$

where $y_{i.}$ is the total of the observations for the ith treatment.[1]

The solution for the least squares estimator of a treatment mean μ_i is

$$\widehat{\mu}_i = \frac{y_{i.}}{r} = \bar{y}_{i.} \qquad i = 1, 2, \ldots, t \tag{2.4}$$

It turns out that the estimators of the treatment population means based on the least squares criteria are the observed treatment group means.

The normal equations for the meat storage study are

$$3\widehat{\mu}_1 = 22.44$$
$$3\widehat{\mu}_2 = 16.50$$
$$3\widehat{\mu}_3 = 21.78$$
$$3\widehat{\mu}_4 = 10.08$$

---

[1] The dot notation is used to simplify the presentation of sums. The total of the observations for the ith treatment is denoted by $y_{i.}$ with the dot indicating that all observations in the ith treatment group have been summed to give this total or

$$y_{i.} = \sum_{j=1}^{r} y_{ij} = y_{i1} + y_{i2} + \cdots + y_{ir}$$

Also the total of all the observations is denoted by $y_{..}$ with the two dots indicating that summation has been completed for both subscripts or

$$y_{..} = \sum_{i=1}^{t} \sum_{j=1}^{r} y_{ij}$$

The least squares estimates for the meat storage experiment are

$$\widehat{\mu}_1 = \frac{22.44}{3} = 7.48$$

$$\widehat{\mu}_2 = \frac{16.50}{3} = 5.50$$

$$\widehat{\mu}_3 = \frac{21.78}{3} = 7.26$$

$$\widehat{\mu}_4 = \frac{10.08}{3} = 3.36$$

The sum of squares for experimental error for the model $y_{ij} = \mu_i + e_{ij}$ is

$$SSE = \sum_{i=1}^{t} \sum_{j=1}^{r} (y_{ij} - \widehat{\mu}_i)^2 = \sum_{i=1}^{t} \sum_{j=1}^{r} (y_{ij} - \bar{y}_{i.})^2 \tag{2.5}$$

Notice that SSE is the pooled sum of squares from within each of the treatment groups. The sample variance for the ith treatment group is

$$s_i^2 = \sum_{j=1}^{r} \frac{(y_{ij} - \bar{y}_{i.})^2}{(r-1)}$$

and it is an estimate of σ^2 from the data in the ith treatment group.

If we can make the assumption that σ^2 is homogeneous—that is, the same for all treatment groups—then

$$s^2 = \frac{\sum_{i=1}^{t} \left[\sum_{j=1}^{r} (y_{ij} - \bar{y}_{i.})^2 \right]}{t(r-1)} = \frac{SSE}{t(r-1)} \tag{2.6}$$

is a pooled estimate of σ^2 from all the data in the experiment. The computed sums of squares and variance for the meat storage experiment are shown in Table 2.3.

Estimators for the Reduced Model

If there are no differences among the treatment population means, then the simpler or reduced model $y_{ij} = \mu + e_{ij}$ is used to describe the data. The least squares estimator of μ for the reduced model is the grand mean of all the observations in the experiment,

$$\widehat{\mu} = \bar{y}_{..} = \frac{y_{..}}{N} \tag{2.7}$$

where $N = rt$.

The minimum sum of squares for experimental error from the reduced model under the null hypothesis is

Table 2.3 Observations, means, and within groups sums of squares for the meat storage experiment

	Commercial	Vacuum	CO, O₂, N	CO₂
	7.66	5.26	7.41	3.51
	6.98	5.44	7.33	2.91
	7.80	5.80	7.04	3.66
$\widehat{\mu}_i = \bar{y}_{i.}$	7.48	5.50	7.26	3.36
$\sum_{j=1}^{r} (y_{ij} - \bar{y}_{i.})^2$	0.3848	0.1512	0.0758	0.3150

$$SSE = 0.3848 + 0.1512 + 0.0758 + 0.3150 = 0.9268$$

$$s^2 = \frac{SSE}{t(r-1)} = \frac{0.9268}{4(2)} = 0.11585$$

$$SSE_r = \sum_{i=1}^{t} \sum_{j=1}^{r} (y_{ij} - \widehat{\mu})^2 = \sum_{i=1}^{t} \sum_{j=1}^{r} (y_{ij} - \bar{y}_{..})^2 \tag{2.8}$$

SSE_r is the total sum of squares of all the observations expressed as deviations from the grand mean.

The estimate of the grand mean for the meat storage experiment is

$$\widehat{\mu} = \bar{y}_{..} = \frac{70.80}{12} = 5.90$$

and

$$SSE_r = (7.66 - 5.90)^2 + \cdots + (3.66 - 5.90)^2 = 33.7996$$

2.6 Sums of Squares to Identify Important Sources of Variation

We can use the differences in the experimental error sums of squares for the two models to partition the total variation in the experiment. These partitions will help clarify and explain the results of the experiment. The computed experimental error sums of squares for the two models of the meat packaging data were quite different. $SSE_f = 0.9268$ for the full model with four treatment population means ($y_{ij} = \mu_i + e_{ij}$), and $SSE_r = 33.7996$ for the reduced model with only one population mean ($y_{ij} = \mu + e_{ij}$).

The smaller sum of squares for the full model indicates that the estimated experimental errors from the full model ($\widehat{e}_{ij} = y_{ij} - \widehat{\mu}_i$), in general, will be smaller values than their counterparts from the reduced model. The difference between the observations and their separate group means $\widehat{\mu}_{i.}$, shown in Table 2.4, are with one

Table 2.4 Observed values, estimates, and deviations of observed values from estimates with the reduced model and the full model

| | | Reduced Model $y_{ij} = \mu + e_{ij}$ | | Full Model $y_{ij} = \mu_i + e_{ij}$ | |
| | Observed | Estimate | Difference | Estimate | Difference |
Treatment	y	$\widehat{\mu}$	$(y_{ij} - \widehat{\mu})$	$\widehat{\mu}_i$	$(y_{ij} - \widehat{\mu}_i)$
Commercial	7.66	5.90	1.76	7.48	0.18
	6.98	5.90	1.08	7.48	-0.50
	7.80	5.90	1.90	7.48	0.32
Vacuum	5.26	5.90	-0.64	5.50	-0.24
	5.44	5.90	-0.46	5.50	-0.06
	5.80	5.90	-0.10	5.50	0.30
CO, O_2, N	7.41	5.90	1.51	7.26	0.15
	7.33	5.90	1.43	7.26	0.07
	7.04	5.90	1.14	7.26	-0.22
100% CO_2	3.51	5.90	-2.39	3.36	0.15
	2.91	5.90	-2.99	3.36	-0.45
	3.66	5.90	-2.24	3.36	0.30
		$SSE_r = 33.7996$		$SSE_f = 0.9268$	

exception in the vacuum treatment, less than the differences between the observations and the grand mean, $\widehat{\mu} = 5.90$, estimated from the reduced model. The two experimental error sums of squares and their difference follow.

Reduced Model:
$$SSE_r = \sum_{i=1}^{t} \sum_{j=1}^{r} (y_{ij} - \overline{y}_{..})^2 = 33.7996$$

Full Model:
$$SSE_f = \sum_{i=1}^{t} \sum_{j=1}^{r} (y_{ij} - \overline{y}_{i.})^2 = 0.9268$$

Difference:
$$SSE_r - SSE_f = \sum_{i=1}^{t} \sum_{j=1}^{r} (y_{ij} - \overline{y}_{..})^2 - \sum_{i=1}^{t} \sum_{j=1}^{r} (y_{ij} - \overline{y}_{i.})^2$$
$$= \sum_{i=1}^{t} \sum_{j=1}^{r} (\overline{y}_{i.} - \overline{y}_{..})^2 = r \sum_{i=1}^{t} (\overline{y}_{i.} - \overline{y}_{..})^2$$
$$= 32.8728$$

The **difference** sum of squares is the sum of squared differences between the treatment group means $\overline{y}_{i.}$ and the grand mean $\overline{y}_{..}$. The difference sum of squares, known as the **treatment sum of squares**, represents a reduction in SSE after including treatments in the model; thus, it often is referred to as the *sum of squares reduction due to treatments*. The **total sum of squares** for the experiment is SSE_r, because it is the sum of squared differences between all observations and the grand mean $\bar{y}_{..}$.

The Fundamental Partition: The total sum of squares SSE_r, from the reduced model, is the sum of the treatment sum of squares and the experimental error sum of squares SSE_f, from the full model. Therefore, we have the relationship

$$\sum_{i=1}^{t}\sum_{j=1}^{r}(y_{ij}-\bar{y}_{..})^2 = \sum_{i=1}^{t}\sum_{j=1}^{r}(\bar{y}_{i.}-\bar{y}_{..})^2 + \sum_{i=1}^{t}\sum_{j=1}^{r}(y_{ij}-\bar{y}_{i.})^2 \qquad \textbf{(2.9)}$$

or

$$SS \text{ Total} = SS \text{ Treatment} + SS \text{ Error}$$

The total sum of squares has been partitioned into two parts:

- *SS* Treatment—the sum of squared differences between the treatment group means and the grand mean

- *SS* Error—the sum of squared differences between the observations within the group and the group mean

The sum of squares formulae may be derived from an identity for the deviation of any observation from the grand mean. The equation

$$(y_{ij}-\bar{y}_{..}) = (\bar{y}_{i.}-\bar{y}_{..}) + (y_{ij}-\bar{y}_{i.}) \qquad \textbf{(2.10)}$$

partitions the deviation of any observation from the grand mean into two parts. It is a sum of (1) the deviation of the group mean from the grand mean $(\bar{y}_{i.}-\bar{y}_{..})$, and (2) the deviation of the observation from the group mean $(y_{ij}-\bar{y}_{i.})$, the latter being a measure of the experimental error associated with the observation. Squaring and summing both sides of the expression in Equation (2.10) results in

$$\sum_{i=1}^{t}\sum_{j=1}^{r}(y_{ij}-\bar{y}_{..})^2 = \sum_{i=1}^{t}\sum_{j=1}^{r}(\bar{y}_{i.}-\bar{y}_{..})^2 + \sum_{i=1}^{t}\sum_{j=1}^{r}(y_{ij}-\bar{y}_{i.})^2$$
$$+ 2\sum_{i=1}^{t}\sum_{j=1}^{r}(\bar{y}_{i.}-\bar{y}_{..})(y_{ij}-\bar{y}_{i.})$$

However, the cross-product term sums to zero, so that the resulting expression

$$\sum_{i=1}^{t}\sum_{j=1}^{r}(y_{ij}-\bar{y}_{..})^2 = \sum_{i=1}^{t}\sum_{j=1}^{r}(\bar{y}_{i.}-\bar{y}_{..})^2 + \sum_{i=1}^{t}\sum_{j=1}^{r}(y_{ij}-\bar{y}_{i.})^2$$

is identical to the sum of squares partition shown in Equation (2.9).

A summary of the formulae for the sums of squares equivalent to the definition formulae of Equation (2.9) is

$$SS\ \text{Total} = \sum_{i=1}^{t} \sum_{j=1}^{r} (y_{ij} - \overline{y}_{..})^2$$

$$SS\ \text{Treatment} = r \sum_{i=1}^{t} (\overline{y}_{i.} - \overline{y}_{..})^2$$

$$SS\ \text{Error} = SS\ \text{Total} - SS\ \text{Treatment}$$

2.7 A Treatment Effects Model

A **treatment effect** tells us something about how much the treatment changes a measurement on an experimental unit. The cell means model can be expressed differently to reflect the effects of treatments on the experimental units. Equation (2.10) may be rearranged slightly to express the observation as a function of the grand mean and the two deviations as

$$y_{ij} = \overline{y}_{..} + (\overline{y}_{i.} - \overline{y}_{..}) + (y_{ij} - \overline{y}_{i.}) \qquad \textbf{(2.11)}$$

An equivalent population model may be written as

$$y_{ij} = \overline{\mu}_{.} + (\mu_i - \overline{\mu}_{.}) + (y_{ij} - \mu_i) \qquad \textbf{(2.12)}$$

where $\overline{\mu}_{.} = \Sigma_i^t \mu_i / t$ is the average of the population means for the cell means model $y_{ij} = \mu_i + e_{ij}$.

The deviation of the group means from the grand mean $(\mu_i - \overline{\mu}_{.})$ is known as the *treatment effect,* and the model in Equation (2.12) is often written as

$$y_{ij} = \mu + \tau_i + e_{ij} \qquad \textbf{(2.13)}$$

where $\mu = \overline{\mu}_{.}$, $\tau_i = (\mu_i - \overline{\mu}_{.})$, and $e_{ij} = (y_{ij} - \mu_i)$. The several expressions for the treatment effects are shown in Display 2.2. A graphical representation of treatment effects for the meat storage experiment is shown in Figure 2.3.

	Display 2.2	Treatment Effects		
	Treatment 1	*Treatment 2*	$\cdots$	*Treatment t*
Sample	$(\overline{y}_{1.} - \overline{y}_{..})$	$(\overline{y}_{2.} - \overline{y}_{..})$	$\cdots$	$(\overline{y}_{t.} - \overline{y}_{..})$
Population	$(\mu_i - \overline{\mu}_{.})$	$(\mu_2 - \overline{\mu}_{.})$	$\cdots$	$(\mu_t - \overline{\mu}_{.})$
Effect	τ_1	τ_2	$\cdots$	τ_t

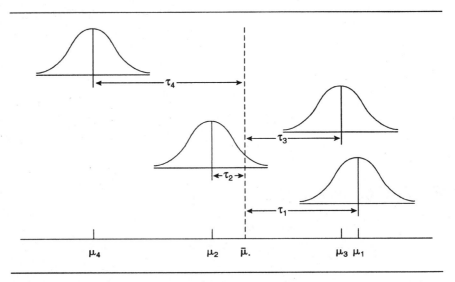

Figure 2.3   A graphical representation of treatment effects

The model in Equation (2.13) has $(t + 1)$ population parameters, which are $\mu, \tau_1, \tau_2, \cdots, \tau_t$. As a consequence of the definitions for treatment effects their sum is equal to zero:

$$\sum_{i=1}^{t} \tau_i = \sum_{i=1}^{t} (\mu_i - \overline{\mu}.) = 0 \qquad (2.14)$$

2.8   Degrees of Freedom

The **degrees of freedom** may be thought of as the number of statistically independent elements in the sums of squares. The value of the degrees of freedom represents the number of independent pieces of information in the sums of squares. The sum of squares for all the observations Σy_{ij}^2 has N statistically independent elements and, thus, has N degrees of freedom.

After the parameter μ is estimated by $\overline{y}..$ from the data, the error sum of squares for the reduced model is $SSE_r = SS \text{ Total} = \Sigma (y_{ij} - \overline{y}..)^2$. The $(y_{ij} - \overline{y}..)$ in SSE_r are not N statistically independent elements because they sum to zero and any one of the $(y_{ij} - \overline{y}..)$ is the negative of the sum of the other $(N - 1)$ values. This linear restriction on the observations is a consequence of estimating one parameter, μ, in the reduced model. In general, the degrees of freedom for SSE after fitting any model is the number of observations minus the number of parameters estimated from the data.

There were t parameters $(\mu_1, \mu_2, \cdots, \mu_t)$ estimated for the full model. Consequently, the number of statistically independent elements in the error sum of squares for the full model is $(N - t)$, so that SSE_f has $(N - t)$ degrees of freedom.

The treatment sum of squares is determined from a difference between the SSEs for two models as

$$SS\,\text{Treatment} = SSE_r - SSE_f$$

The degrees of freedom for these differences can be determined as the difference between the degrees of freedom for SSE_r and SSE_f

$$(N - 1) - (N - t) = (t - 1)$$

Thus, $(t - 1)$ degrees of freedom are associated with the sum of squares reduction due to treatments.

2.9 Summaries in the Analysis of Variance Table

The **analysis of variance table** summarizes our knowledge about variability in the observations from the experiment. The total sum of squares has been partitioned into two parts, one representing variation among the treatment means and the other representing experimental error.

The experimental error variance σ^2 is estimated by $s^2 = SSE/(N - t)$, where s^2 is referred to as the mean square for error (MSE). The other mean square of importance is the mean square for treatments (MST), computed as

$$MST = \frac{SS\,\text{Treatment}}{(t - 1)}$$

The sum of squares partitions, degrees of freedom, and mean squares are summarized in an analysis of variance table like that shown in Table 2.5.

Table 2.5 Analysis of variance table for a one-way treatment classification in a completely randomized design

Source of Variation	Degrees of Freedom	Sum of Squares	Mean Square
Total	$N - 1$	SS Total	
Treatments	$t - 1$	SS Treatment	$MST = \dfrac{SST}{t - 1}$
Error	$N - t$	SS Error	$MSE = \dfrac{SSE}{N - t}$

The MSE is an unbiased estimate of the experimental error variance σ^2; that is, the expected value of MSE is equal to σ^2, or

$$E(MSE) = \sigma^2 \qquad (2.15)$$

The expected value of MST is

$$E(MST) = \sigma^2 + r\theta_t^2 \qquad (2.16)$$

where

$$\theta_t^2 = \frac{\sum_{i=1}^{t} (\mu_i - \overline{\mu}_.)^2}{(t-1)}$$

is the variance among the treatment means. Therefore, MST estimates a combination of the experimental error variance and the variance among the treatment means in the hypothesized linear model $y_{ij} = \mu_i + e_{ij}$. The algebraic derivations of $E(MSE)$ and $E(MST)$ are shown in Appendix 2A.2 of this chapter.

2.10   Tests of Hypotheses About Linear Models

The complete analysis of variance table for the meat storage experiment is shown in Table 2.6. The analysis of variance table summarizes the magnitudes of sources of variation in the experiment. Whether the variation due to treatments is significantly greater than random experimental error requires a test of hypothesis.

Table 2.6   Analysis of variance for $\log(\text{count/cm}^2)$ of psychrotrophic bacteria from the meat storage experiment

Source of Variation	Degrees of Freedom	Sum of Squares	Mean Square	F	Pr > F
Total	11	33.800			
Wrap	3	32.873	10.958	94.58	.000
Error	8	0.927	0.116		

We could use a randomization test based on the analysis of variance information. However, as explained in Section 1.8, statistical tests based on normal distribution theory are equally good tests provided the normal distribution assumptions are valid. Consequently, we assume the observations y_{ij} are independent and also have a normal distribution with mean μ_i and variance σ^2 as shown in Figure 2.1. Methods to evaluate the assumption of normally distributed observations are found in Chapter 4.

A Sums of Squares Difference to Compare Two Models

SSE_f is a measure of how well the full model fits the observed data, and SSE_r is the equivalent measure for the simpler reduced model. The difference $(SSE_r - SSE_f)$ is then a measure of the improvement of the full model over the reduced model. Consequently, the ratio provides a means of assessing the relative improvement and forms the essential part of the familiar F test criterion:

$$\frac{(SSE_r - SSE_f)}{SSE_f}$$

The F Statistic to Test a Model Hypothesis

From statistical theory, it is known that the sums of squares of normally distributed random variables are associated with the chi-square distribution. It can be shown that SSE_f/σ^2 is distributed as the chi-square variable with $(N - t)$ degrees of freedom. The difference $SST = SSE_r - SSE_f$ with $(t - 1)$ degrees of freedom represents the reduction in the total sum of squares due to differences among the treatment means. When the treatment means are equal, $\mu_1 = \mu_2 = \cdots = \mu_t$, it can be shown that this criterion, $(SSE_r - SSE_f)/\sigma^2 = SST/\sigma^2$, also has a chi-square distribution with $(t - 1)$ degrees of freedom and that it is independent of the distribution of SSE_f/σ^2.

The ratio

$$F = \frac{(SSE_r - SSE_f)/(t - 1)}{SSE_f/(N - t)} \tag{2.17}$$

is the ratio of two chi-square distributions each divided by their respective degrees of freedom. Under the null hypothesis of no difference between treatments the ratio has the F distribution with $(t - 1)$ and $(N - t)$ degrees of freedom, respectively, in the numerator and denominator.

The test statistic computed from the analysis of variance table to test the null hypothesis $H_0: \mu_1 = \mu_2 = \cdots = \mu_t$ is

$$F_0 = \frac{MST}{MSE} \tag{2.18}$$

which has the F distribution when the null hypothesis is true.

The expected mean squares in Equations (2.15) and (2.16) show that MSE is an unbiased estimator of σ^2 under either model hypothesis. But, MST is an unbiased estimator of σ^2 only under the null hypothesis model or the reduced model; that is, if $\mu_1 = \mu_2 = \cdots = \mu_t$, then $\theta_t^2 = 0$ and $E(MST) = \sigma^2$. Under the alternate hypothesis the expected value of MST is greater than σ^2, as seen in Equation (2.16), and consequently, the expected value of the numerator of the F_0 statistic will be greater than that of the denominator. Large values of F_0 would suggest rejection of the null hypothesis in favor of the alternate.

A one-sided, upper-tail critical region is implied for the hypothesis test. The null hypothesis H_0 is rejected for a Type I error probability of α if

$$F_0 > F_{\alpha,(t-1),(N-t)}$$

where $F_0 = MST/MSE$ and $F_{\alpha,(t-1),(N-t)}$ is the value of the F distribution that is exceeded with probability α. Critical values of the F distribution are found in Appendix Table IV.

A test of the hypothesis of no differences among the four meat packaging treatments with respect to growth of bacteria in the meat storage experiment is illustrated in Display 2.3. From the analysis of variance (Table 2.6) the required mean squares are $MST = 10.958$ with 3 degrees of freedom and $MSE = 0.116$ with 8 degrees of freedom.

Display 2.3   Hypothesis Test for the Meat Storage Experiment

$$H_0 : \mu_1 = \mu_2 = \mu_3 = \mu_4$$
$$H_a : \mu_i \neq \mu_k \text{ for at least one } i \neq k$$

$\alpha = .05$ Critical Region: $F_0 > F_{.05,3,8} = 4.07$

$$F_0 = \frac{MST}{MSE} = \frac{10.958}{0.116} = 94.58$$

Since $F_0 = 94.58$ falls in the critical region, $F_0 > 4.07$, we reject the null hypothesis and conclude that the treatments differ with respect to the number of psychrotrophic bacteria observed on the meat stored under the different conditions.

2.11   Significance Testing and Tests of Hypotheses

A common practice in hypothesis testing is to report the significance level of a test statistic. The significance level of a test statistic is the probability of exceeding the value of the test statistic under the null hypothesis condition. Note the column labeled $Pr > F$ in Table 2.6. The value .000 in that column is the probability that the F statistic with 3 and 8 degrees of freedom will exceed $F_0 = 94.58$ and often is referred to as the "P-value." Since the reported value is .000, we know the probability of exceeding $F_0 = 94.58$ is less than .0001, or $P < .0001$. The magnitude of the P-value is used by many investigators to decide the statistical significance of the F test in the analysis of variance. The value is frequently reported in a discussion of results. For example, the F test for the present example may be reported as "significant at the $P < .0001$ level of significance." If the probability value is less than the traditional significance levels of .01 or .05 the null hypothesis will be rejected because the observed F_0 statistic is in the critical region.

Most analysis of variance computer programs include this probability value for the F statistic in the printed output. A technical discussion of the P-value and its relationships to the number of replications and to the alternative hypothesis can be found in Hung et al. (1997) and references cited therein.

2.12 Standard Errors and Confidence Intervals for Treatment Means

In Section 2.5 it was determined that the least squares estimators of the population means were the observed means of the treatment groups $\bar{y}_{i.}$. The observed means are averages of r independent observations, so that the variance of a treatment group mean is $\sigma_{\bar{y}_{i.}}^2 = \sigma^2/r$. The estimator of the variance is

$$s_{\bar{y}_{i.}}^2 = \frac{s^2}{r} \qquad (2.19)$$

where $s^2 = MSE$ from the analysis of variance. The estimator for the standard error of the mean is

$$s_{\bar{y}_{i.}} = \sqrt{\frac{s^2}{r}} \qquad (2.20)$$

A $100(1 - \alpha)\%$ confidence interval (CI) estimate is constructed for each of the treatment group means with upper and lower limits, respectively, where $t_{\alpha/2,(N-t)}$ is the Student t statistic exceeded with probability $\alpha/2$ and $(N - t)$ are the degrees of freedom for MSE:

$$\bar{y}_{i.} + t_{\alpha/2,(N-t)}(s_{\bar{y}_{i.}}) \quad \text{and} \quad \bar{y}_{i.} - t_{\alpha/2,(N-t)}(s_{\bar{y}_{i.}}) \qquad (2.21)$$

The standard error of the mean for any treatment group in the meat storage study is

$$s_{\bar{y}_{i.}} = \sqrt{\frac{0.116}{3}} = 0.197$$

and $t_{.025,8} = 2.306$. The treatment group means are shown in Display 2.4 with the 95% CI estimates.

Display 2.4	Means, Standard Errors, and 95% Confidence Intervals for the Meat Storage Experiment			
Treatment	_Mean_	_Standard Error_	_95% CI_	(Lower, Upper)
Commercial	7.48	0.197	7.48 ± 0.454	$(7.03, 7.93)$
Vaccum	5.50	0.197	5.50 ± 0.454	$(5.05, 5.95)$
CO, O_2, N	7.26	0.197	7.26 ± 0.454	$(6.81, 3.81)$
100% CO_2	3.36	0.197	3.36 ± 0.454	$(2.91, 3.81)$

For any group mean the 95% CI is

$$\bar{y}_{i.} \pm (2.306)(0.197) \quad \text{or} \quad \bar{y}_{i.} \pm 0.454$$

2.13 Unequal Replication of the Treatments

Unequal replication may occur because some experimental units were lost during the study; there were insufficient numbers of subjects available for all study groups; or collected data were lost, destroyed, or invalid. The most visible effects of unequal replication appear in the computations required for the analysis. A consequence of greater concern is the ensuing unequal information on all treatment groups. The loss of observations from any treatment group results in a proportional loss in precision on the estimates of the treatment group mean relative to those treatment groups with complete data.

Example 2.2 Detection of Phlebitis in Amiodarone Therapy

Phlebitis is an inflammation of a vein that can occur when intravenously administering drugs. The active drug was thought to be the main contributing factor to inflammation, although the vehicle solution used to carry the drug through intravenous administration could be a possible contributor.

Research Hypothesis: Of particular importance to the investigators was the problem of detecting the onset of phlebitis. This particular study was to explore mechanisms for early detection of phlebitis during Amiodarone therapy. They hypothesized that tissue temperature changes near the intravenous administration would be an early signal for the impending inflammation.

Treatment Design: Three intravenous treatments were administered to test animals. They were

- Amiodarone with a vehicle solution to carry the drug

- a vehicle solution only

- a saline solution

The saline solution served as a placebo control treatment to determine whether the act of administration alone affected inflammation. The vehicle solution served as a control to separate any effects of the vehicle from the effects of the drug.

Experiment Design: Rabbits used as the test animals were randomly assigned to the three treatment groups in a completely randomized design. An intravenous needle was inserted in a vein of one ear of the rabbits using one of the three treatments.

An increase in the temperature of the treated ear was considered as a possible early indicator of phlebitis. The difference in the temperatures of the two ears (treated minus untreated) was used as the response variable. Complications with the experimental protocol resulted in a different number of rabbits for each of the treatment groups. The observed temperature differences at 4.5 hours for each of the rabbits remaining in the study are given in Table 2.7.

Table 2.7   Ear temperature (°C) differences, treated minus untreated, of rabbits 4.5 hours after treatment

	Amiodarone	Vehicle	Saline	
	2.2	0.3	0.1	
	1.6	0.0	0.1	
	0.8	0.6	0.2	
	1.8	0.0	− 0.4	
	1.4	− 0.3	0.3	
	0.4	0.2	0.1	
	0.6		0.1	
	1.5		− 0.5	
	0.5			
r_i	9	6	8	$N = \sum_{i=1}^{t} r_i = 23$
total $(y_{i.})$	10.80	0.80	0.00	$y_{..} = 11.60$
mean $(\bar{y}_{i.})$	1.20	0.13	0.00	$\bar{y}_{..} = 0.50$

Source: G. Ward, Department of Pharmaceutical Sciences, University of Arizona.

The Linear Model for Unequal Replications

The cell means model for the one-way treatment classification in the completely randomized design with unequal replication is

$$y_{ij} = \mu_i + e_{ij} \tag{2.22}$$

$$i = 1, 2, \dots, t \qquad j = 1, 2, \dots, r_i$$

where r_i is the number of replications for the ith treatment group. The total number of observations is $N = \sum_{i}^{t} r_i$. The interpretation and assumptions are the same as those for the equal replication model.

The least squares estimator for treatment means determined by the methods outlined in Section 2.5 are

$$\hat{\mu}_i = \frac{y_{i.}}{r_i} \qquad i = 1, 2, \dots, t \tag{2.23}$$

or the observed treatment group mean, as for equal replication.

The Analysis of Variance for Unequal Replications

The sums of squares partitions for the analysis of variance are shown in Table 2.8. Each element of the sum of squares for treatments is a weighted square of the deviation of a treatment mean from the grand mean, $r_i(\bar{y}_{i.} - \bar{y}_{..})^2$. The weight, r_i, is the number of replications for the treatment group. The weights reflect the amount of information available for the estimation of the treatment means. The sum of squares for experimental error is the pooled sum of squares within the treatment groups. The expected mean square for treatments includes a weighted sum of squares of the treatment effects $\tau_i = (\mu_i - \bar{\mu}_.)$, where $\bar{\mu}_. = \Sigma_i^t r_i\mu_i/N$. With unequal replication, our definition of the treatment effects means their weighted sum is equal to zero, $\Sigma_i^t r_i(\mu_i - \bar{\mu}_.) = 0$.

Table 2.8 Analysis of variance for a completely randomized design with unequal replication of treatments

Source of Variation	Degrees of Freedom	Sum of Squares	Mean Square	Expected Mean Square
Total	$N - 1$	$\sum_{i=1}^{t} \sum_{j=1}^{r_i} (y_{ij} - \bar{y}_{..})^2$		
Treatments	$t - 1$	$\sum_{i=1}^{t} r_i(\bar{y}_{i.} - \bar{y}_{..})^2$	MST	$\sigma^2 + \theta_t^2$
Error	$N - t$	$\sum_{i=1}^{t} \sum_{j=1}^{r_i} (y_{ij} - \bar{y}_{i.})^2$	MSE	σ^2

$$\theta_t^2 = \frac{1}{(t-1)} \sum_{i=1}^{t} r_i(\mu_i - \bar{\mu}_.)^2$$

The analysis of variance for the Amiodarone study is shown in Table 2.9. The null hypothesis of no differences among the treatment means is tested with the statistic $F_0 = MST/MSE = 3.6081/0.2177 = 16.58$. The F_0 statistic is found in the column labeled "F" in Table 2.9. The null hypothesis is rejected at the .05 level of significance because $F_0 > F_{.05,2,20} = 3.49$ or notice that $Pr > F = .000$ in Table 2.9.

Table 2.9 Analysis of variance for ear temperature differences from the Amiodarone study

Source of Variation	Degrees of Freedom	Sum of Squares	Mean Square	F	Pr > F
Total	22	11.5696			
Treatment	2	7.2162	3.6081	16.58	.000
Error	20	4.3533	0.2177		

Standard Errors for Treatment Means and Unequal Precision

The standard error of a treatment group mean is estimated by

$$s_{\bar{y}_{i.}} = \sqrt{\frac{MSE}{r_i}} \tag{2.24}$$

The estimated standard errors of the three intravenous treatment means are

$$s_{\bar{y}_{1.}} = \sqrt{\frac{0.2177}{9}} = 0.156$$

$$s_{\bar{y}_{2.}} = \sqrt{\frac{0.2177}{6}} = 0.190$$

and

$$s_{\bar{y}_{3.}} = \sqrt{\frac{0.2177}{8}} = 0.165$$

The differences among the standard errors of the three treatment group means illustrate the decreased precision when data are lost from an experiment. For example, the vehicle treatment group with six observations has a standard error 22% greater than the Amiodarone treatment group with nine observations.

2.14   How Many Replications for the *F* Test?

In Chapter 1 the number of replications required to detect some predetermined difference between two treatment means was ascertained with the normal z statistic in conjunction with knowledge of the variance, significance level, and power of the test. The required number of replications is determined in this section by a method based on the F statistic.

The power of a test of hypothesis is the probability of rejecting a false null hypothesis. The statistic $F_0 = MST/MSE$ is used to test the null hypothesis H_0: $\tau_i = 0$. The power of the test is $1 - \beta = P(F > F_{\alpha,\nu_1,\nu_2}|H_0 \text{ false})$, where ν_1 and ν_2 are the numerator and denominator degrees of freedom, respectively. When H_0 is false, F_0 has the *non-central* F distribution with ν_1 and ν_2 degrees freedom and *non-centrality* parameter $\lambda = r \sum_i^t \tau_i^2 / \sigma^2$. If the null hypothesis is true, then the non-centrality parameter has the value $\lambda = 0$, since all $\tau_i = 0$ and F_0 has the *central F* distribution.

Tables have been computed that tabulate the power of the F test for given values of the significance level α; power $1 - \beta$; degrees of freedom ν_1 and ν_2; and Φ, a function of the non-centrality parameter, which is

$$\Phi = \sqrt{\frac{\lambda}{t}} = \sqrt{\frac{r\sum\limits_{i=1}^{t}\tau_i^2}{t\sigma^2}} \tag{2.25}$$

Charts of these power curves are found in Appendix Table IX for selected parameter values of the non-central F distribution.

The charts are used to estimate the required number of replications for given values of α, $1-\beta$, σ^2, ν_1, ν_2, and Φ. A value for Φ requires a value for σ^2 and specific values of the treatment means that lead to rejection of the null hypothesis. Given $\mu_1, \mu_2, \ldots, \mu_t$, the $\tau_i = (\mu_i - \overline{\mu}_{.})$ are evaluated for Φ in Equation (2.25).

Example 2.3 Replication Number for the Amiodarone Study

Suppose for future experiments the investigator with the Amiodarone study in Example 2.2 was interested in rejecting the null hypothesis with a power of at least .90 at the .05 level of significance if the ear temperature difference for the drug treatment group was 0.8°C while the vehicle and saline means were 0.1°C and 0°C, respectively. The average of the treatment means $\overline{\mu}_{.}$ is 0.3°C, and the treatment effects are

$$\tau_1 = \mu_1 - \overline{\mu}_{.} = 0.8 - 0.3 = 0.5$$
$$\tau_2 = \mu_2 - \overline{\mu}_{.} = 0.1 - 0.3 = -0.2$$
$$\tau_3 = \mu_3 - \overline{\mu}_{.} = 0.0 - 0.3 = -0.3$$

Therefore, $\sum_i^t \tau_i^2 = 0.38$, and we can use $MSE = 0.22$ from the analysis of variance as an estimate for σ^2. From Equation (2.25) evaluate

$$\Phi^2 = \frac{r\sum\limits_{i=1}^{t}\tau_i^2}{t\sigma^2} = \frac{r(0.38)}{3(0.22)} = r(0.58)$$

The required values for the power curve are $\nu_1 = (t-1) = 2$, $\nu_2 = t(r-1) = 3(r-1)$, and $\alpha = .05$. Using $r = 5$ as a first trial value yields $\Phi = \sqrt{2.9} = 1.7$ and $\nu_2 = 12$. From Appendix Table IX the power of the test is approximately .65, which is less than the required .90. Increasing the replication number to $r = 9$ yields $\Phi = \sqrt{5.22} = 2.3$ and $\nu_2 = 24$, with a resulting power in excess of .90. It appears that nine rabbits will be required in each treatment group to yield a power of at least .90.

It is difficult to specify desired effects for a complete set of treatments. It may be easier to specify the difference between any two treatment means that would be physically or biologically significant. Suppose it is desired to detect significance at a difference of $D = \mu_i - \mu_j$. The minimum value of Φ^2 in that case is

$$\Phi^2 = \frac{rD^2}{2t\sigma^2} \qquad\qquad\qquad \textbf{(2.26)}$$

Comments: Commercial computer programs are available to estimate replication numbers with considerable ease, although the investigator must still have available estimates of σ^2, desired power, significance levels, and magnitude of effects to declare significant.

Estimates of σ^2 ordinarily are difficult to declare. If results are available from previous experiments similar to the one contemplated, then variance estimates from those experiments may be pooled to attain a reasonable value.

When the entire experiment cannot be performed at one time or location with the required number of replications, the experiment can be repeated several times or at several locations. For example, if eight replications are required and they cannot be performed at one time, then perform the experiment twice with four replications each time.

EXERCISES FOR CHAPTER 2

---

1. A traffic engineering study on traffic delay was conducted at intersections with signals on urban streets. Three types of traffic signals were utilized in the study: (1) pretimed, (2) semi-actuated, and (3) fully actuated. Five intersections were used for each type of signal. The measure of traffic delay used in the study was the average stopped time per vehicle at each of the intersections (seconds/vehicle). The data follow.

Pretimed	Semi-actuated	Fully actuated
36.6	17.5	15.0
39.2	20.6	10.4
30.4	18.7	18.9
37.1	25.7	10.5
34.1	22.0	15.2

   Source: W. Reilly, C. Gardner, and J. Kell (1976). A technique for measurement of delay at intersections. *Technical Report* FHWA-RD-76-135, Federal Highway Administration, Office of R & D, Washington, D.C.

   a. Write the linear statistical model for this study, and explain the model components.
   b. State the assumptions necessary for an analysis of variance of the data.
   c. Compute the analysis of variance for the data.
   d. Compute the least squares means of the traffic delay and their standard errors for each signal type.
   e. Compute the 95% confidence interval estimates of the signal type means.

    f.   Test the hypothesis of no difference among the mean traffic delays of the signal types with the F test at the .05 level of significance.

    g.   Write the normal equations for the data.

2.   An experiment was conducted to test the effects of nitrogen fertilizer on lettuce production. Five rates of ammonium nitrate were applied to four replicate plots in a completely randomized design. The data are the number of heads of lettuce harvested from the plot.

Treatment (lb N/acre)	Heads of lettuce/plot
0	104, 114, 90, 140
50	134, 130, 144, 174
100	146, 142, 152, 156
150	147, 160, 160, 163
200	131, 148, 154, 163

Source: Dr. B. Gardner, Department of Soil and Water Science, University of Arizona.

    a.   Write the linear statistical model for this study, and explain the model components.

    b.   State the assumptions necessary for an analysis of variance of the data.

    c.   Compute the analysis of variance for the data.

    d.   Compute the least squares means and their standard errors for each nitrogen level.

    e.   Compute the 95% confidence interval estimates of the nitrogen level means.

    f.   Test the hypothesis of no difference among the means of the nitrogen levels with the F test at the .05 level of significance.

    g.   Write the normal equations for the data.

    h.   This experiment was conducted in a completely randomized design with the field plots in a 4×5 rectangular array of plots. Show a randomization of the five nitrogen treatments to the 20 plots using a random permutation of the numbers 1 through 20.

3.   An animal physiologist studied the pituitary function of hens put through a standard forced molt regimen used by egg producers to bring the hens back into egg production. Twenty-five hens were used for the study. Five hens were used for measurements at the premolt stage prior to the forced molt regimen and at the end of each of four stages of the forced molt regimen. The five stages of the regimen were (1) premolt (control), (2) fasting for 8 days, (3) 60 grams of bran per day for 10 days, (4) 80 grams of bran per day for 10 days, and (5) laying mash for 42 days. The objective was to follow various physiological responses associated with the pituitary function of the hens during the regimen to aid in explaining why the hens will come back into production after the forced molt. One of the compounds measured was serum T3 concentration. The data in the table are the serum T3 measurements for each of the five hens sacrificed at the end of each stage of the regimen.

Treatment	Serum T3, (ng/dl) $\times 10^{-1}$				
Premolt	94.09,	90.45,	99.38,	73.56,	74.39
Fasting	98.81,	103.55,	115.23,	129.06,	117.61
60 g bran	197.18,	207.31,	177.50,	226.05,	222.74
80 g bran	102.93,	117.51,	119.92,	112.01,	101.10
Laying mash	83.14,	89.59,	87.76,	96.43,	82.94

Source: Dr. R. Chiasson and K. Krown, Department of Veterinary Science, University of Arizona.

a. Write the linear statistical model for this study, and explain the model components.
b. State the assumptions necessary for an analysis of variance of the data.
c. Compute the analysis of variance for the data.
d. Compute the least squares means and their standard errors for each treatment.
e. Compute the 95% confidence interval estimates of the treatment means.
f. Test the hypothesis of no differences among means of the five treatments with the F test at the .05 level of significance.
g. Write the normal equations for the data.
h. This experiment was conducted in a completely randomized design with one hen in each of 25 pens. Show a random allocation of the five treatments to the 25 pens with a random permutation of the numbers 1 through 25.

4. Data were collected on student teachers relative to their use of certain teaching strategies that had been presented to them in preservice education. There were 28 student teachers who had learned to use the strategies (9 in 1979, 9 in 1980, and 10 in 1981). In 1978 there were 6 teachers who did not learn to use the strategies, and they were used as a control group. The investigator recorded the average number of strategies used per week by each of the student teachers during their student teaching assignments. The investigator wanted to know whether the number of strategies used by the student teachers was different among the years.

Average Number of Different Strategies Used			
Control 1978	1979	1980	1981
6.88	7.25	10.85	7.29
5.40	10.50	7.43	14.38
16.00	8.43	6.71	6.00
9.80	8.63	7.60	5.00
7.63	8.63	7.60	5.38
5.00	7.00	5.57	14.14
	11.13	8.71	9.25
	7.25	5.86	5.71
	10.38	7.20	7.35
			10.75

Source: Dr. A. Knorr, Family and Consumer Resources, University of Arizona.

a. Write the linear statistical model for this study, and explain the model components.
b. State the assumptions necessary for an analysis of variance of the data.
c. Compute the analysis of variance for the data.
d. Compute the least squares means and their standard errors for each treatment.
e. Compute the 95% confidence interval estimates of the treatment means.
f. Test the hypothesis of no differences among means of the four treatments with the F test at the .05 level of significance.
g. Write the normal equations for the data.

5. In a particular calibration study on atomic absorption spectroscopy the response measurements were the absorbance units on the instrument in response to the amount of copper in a dilute acid solution. Five levels of copper were used in the study with four replications of the zero level and two replications of the other four levels. The spectroscopy data for each of the copper levels are given in the table as micrograms copper/milliliter of solution.

Copper (mg/ml)				
0.00	0.05	0.10	0.20	0.50
0.045	0.084	0.115	0.183	0.395
0.047	0.087	0.116	0.191	0.399
0.051				
0.054				

Source: R. J. Carroll, C. H. Spiegelman, and J. Sacks (1988), A quick and easy multiple-use calibration-curve procedure, *Technometrics* 30, 137–141.

a. Write the linear statistical model for this study, and explain the model components.
b. State the assumptions necessary for an analysis of variance of the data.
c. Compute the analysis of variance for the data.
d. Compute the least squares means and their standard errors for each treatment.
e. Compute the 95% confidence interval estimates of the treatment means.
f. Test the hypothesis of no differences among means of the five treatments with the F test at the .05 level of significance.
g. Write the normal equations for the data.
h. Each of the dilute acid solutions had to be prepared individually by one technician. To prevent any systematic errors from preparation of the first solution to the twelfth solution, she prepared them in random order. Show a random preparation order of the 12 solutions using a random permutation of the numbers 1 through 12.

6. Consider the experiment in Exercise 3. Suppose some of the chickens were lost during the course of the experiment, resulting in the following set of observations.

Treatment	Serum T3, (ng/dl) $\times 10^{-1}$				
Premolt	94.09,	90.45,	99.38,	73.56,	
Fasting	98.81,	103.55,	115.23,	129.06,	117.61
60 g bran	197.18,	207.31,	177.50,		
80 g bran	102.93,	117.51,	119.92,	112.01,	101.10
Laying mash	82.94,	83.14,	89.59,	87.76,	

a. Write the linear statistical model for this study, and explain the model components.
b. State the assumptions necessary for an analysis of variance of the data.
c. Compute the analysis of variance for the data.
d. Compute the least squares means and their standard errors for each treatment. How has the loss of some chickens from the experiment affected the estimates of the means?
e. Compute the 95% confidence interval estimates of the treatment means.
f. Test the hypothesis of no differences among means of the five treatments with the F test at the .05 level of significance.
g. Write the normal equations for the data.

7. Use the data from Exercise 3 to determine how many chickens the biologist would need for each treatment to reject the null hypothesis at the .05 level of significance with a power of .90 if the difference between the control treatment and any new treatment was 30 units of T3 concentration.

8. Use the data from Exercise 1 to determine how many intersections the traffic engineer would need for each type of traffic signal to reject the null hypothesis at the .01 level of significance with a power of .90 if mean delays at the three traffic signal types were 20, 18, and 16 seconds, respectively.

9. The following is a small exercise to help you understand how the least squares principle works to provide a minimum sum of squares for experimental error.
   a. Use Equation (2.3) and the data from Exercise 2.1 to compute SSE for the following cases:
      (i) Use the smallest observation in each treatment group for $\hat{\mu}_i$ in Equation (2.3) to compute SSE.
      (ii) Use the largest observation in each treatment group for $\hat{\mu}_i$ in Equation (2.3) to compute SSE.
      (iii) Use the mean of the observations in each treatment group for $\hat{\mu}_i$ in Equation (2.3) to compute SSE.
      (iv) Substitute another value between the mean and the largest observation for $\hat{\mu}_i$ in Equation (2.3) to compute SSE.
      (v) Substitute another value between the mean and the smallest observation for $\hat{\mu}_i$ in Equation (2.3) to compute SSE.
   b. Plot SSE versus $\hat{\mu}_i$, with your computed values of SSE on the vertical axis and the $\hat{\mu}_i$ values on the horizontal axis.
   c. What is the value of SSE for each of the cases? Which case provides the smallest value of SSE?

10. One of the assumptions for the linear model we use to describe our data in the treatment groups ($y_{ij} = \mu_i + e_{ij}$), is that the observations were random, independent observations of the random variable Y.

   a. What would you have to do during the conduct of the study for the assumption of random and independent observations to be reasonable if
   (i) it was a designed experiment?
   (ii) it was a comparative observational study?

   b. Before you can call the least squares estimators minimum variance and unbiased estimators, what additional assumptions do you have to make about the model?

   c. Before you can compute confidence intervals and test hypotheses about the model, what additional assumption do you have to make?

   d. Suppose you could not make either of the last two assumptions (those in parts b. and c.). How would you be able to test the hypothesis of no difference among the treatment means?

2A.1   Appendix: Expected Values

The expected value of a random variable is its average value. If a random variable Y has a probability distribution with a mean μ and a variance σ^2, the expected value of Y is defined as $\mu = E(Y)$, where $E(Y)$ is read as "the expected value of Y."

The variance of Y is defined as $\sigma^2 = E(Y - \mu)^2$, which is the expected value of the square of the difference between Y and the mean.

If there are two random variables, Y_1 and Y_2, and $E(Y_1) = \mu_1$, and $E(Y_2) = \mu_2$, then the covariance between the two variables Y_1 and Y_2 is defined as

$$\sigma_{12} = E[(Y_1 - \mu_1)(Y_2 - \mu_2)]$$

The covariance indicates the relationship between Y_1 and Y_2. If large values of Y_1 are associated with large values of Y_2, then the covariance is positive. If values of Y_1 become smaller as values of Y_2 become larger or vice versa, then the covariance is negative. If the values of Y_1 and Y_2 are independent, then the covariance is zero.

The cell means model for the completely randomized design is

$$y_{ij} = \mu_i + e_{ij} \tag{2A.1}$$

$$i = 1, 2, \ldots, t \qquad j = 1, 2, \ldots, r$$

The experimental errors e_{ij} are assumed to be independent random variables with a mean of zero, variance σ^2, and zero covariance between any two errors.

The mean or expected value of any e_{ij} is $E(e_{ij}) = 0$, and the variance of any e_{ij} is $E(e_{ij}^2) = \sigma^2$. Since there is no covariance between the e_{ij}, the expectation of a product between any two error terms in the same or different treatment groups is $E(e_{ij} \cdot e_{mk}) = 0$, where $i \neq m$ or $j \neq k$. The μ_i are the population means and are constants with respect to the expectation operation. The expected value of a constant is the constant value itself, or $E(\mu_i) = \mu_i$.

The expected value of any observation described by the cell means model may be found by substitution of $\mu_i + e_{ij}$ for y_{ij} in the expectation:

$$E(y_{ij}) = E(\mu_i + e_{ij}) = E(\mu_i) + E(e_{ij}) = \mu_i \qquad (2A.2)$$

2A.2 Appendix: Expected Mean Squares

The expected values are required for MSE and MST in the analysis of variance. With any number of replications for each of the treatments the mean square for experimental error is estimated as $MSE = SSE/(N - t)$, where

$$SSE = \sum_{i=1}^{t} \sum_{j=1}^{r_i} (y_{ij} - \bar{y}_{i.})^2 \qquad (2A.3)$$

The expectation of SSE is found with the substitutions $y_{ij} = \mu_i + e_{ij}$ and $\bar{y}_{i.} = \mu_i + \bar{e}_{i.}$ into Equation (2A.3), where $\bar{e}_{i.} = \left(\dfrac{1}{r_i}\right) \sum_j^{r_i} e_{ij}$. The resulting expression is

$$SSE = \sum_{i=1}^{t} \sum_{j=1}^{r_i} [\mu_i + e_{ij} - (\mu_i + \bar{e}_{i.})]^2 = \sum_{i=1}^{t} \sum_{j=1}^{r_i} (e_{ij} - \bar{e}_{i.})^2$$

$$= \sum_{i=1}^{t} \sum_{j=1}^{r_i} e_{ij}^2 - \sum_{i=1}^{t} r_i \bar{e}_{i.}^2$$

Given $E(\bar{e}_{i.}^2) = \dfrac{1}{r_i^2} E(\sum_{j=1}^{r_i} e_{ij}^2) = \dfrac{1}{r_i^2} E(e_{i1}^2 + e_{i2}^2 + \cdots + e_{ir_i}^2) = \dfrac{1}{r_i} \sigma^2$, the expectation of SSE is

$$E(SSE) = \sum_{i=1}^{t} \sum_{j=1}^{r_i} E(e_{ij}^2) - \sum_{i=1}^{t} r_i E(\bar{e}_{i.}^2)$$

$$= N\sigma^2 - t\sigma^2$$

$$= (N - t)\sigma^2$$

and

$$E(MSE) = \frac{E(SSE)}{(N - t)} = \sigma^2 \qquad (2A.4)$$

The mean square for treatments is $MST = SST/(t - 1)$, where

$$SST = \sum_{i=1}^{t} r_i (\bar{y}_{i.} - \bar{y}_{..})^2 \qquad (2A.5)$$

and the expectation of SST requires the substitution of $\bar{y}_{i.} = \mu_i + \bar{e}_{i.}$ and $\bar{y}_{..} = \bar{\mu}_. + \bar{e}_{..}$ into Equation (2A.5), where $\bar{e}_{..} = \dfrac{1}{N} \sum_{i=1}^{t} \sum_{j=1}^{r_i} e_{ij}$ and $\bar{\mu}_. = \dfrac{1}{N} \sum_{i=1}^{t} r_i \mu_i$. The resulting expression is

$$SST = \sum_{i=1}^{t} r_i (\mu_i + \bar{e}_{i.} - \bar{\mu}_. - \bar{e}_{..})^2$$

$$= \sum_{i=1}^{t} r_i (\mu_i - \bar{\mu}_.)^2 + \sum_{i=1}^{t} r_i (\bar{e}_{i.} - \bar{e}_{..})^2 + 2 \sum_{i=1}^{t} r_i (\mu_i - \bar{\mu}_.)(\bar{e}_{i.} - \bar{e}_{..})$$

Expanding the second term of the last expression yields

$$\sum_{i=1}^{t} r_i (\bar{e}_{i.} - \bar{e}_{..})^2 = \sum_{i=1}^{t} r \bar{e}_{i.}^2 - N \bar{e}_{..}^2$$

Given

$$E(\bar{e}_{..}^2) = \frac{1}{N^2} E(\bar{e}_{..}^2) = \frac{1}{N^2} E(e_{11}^2 + e_{12}^2 + \cdots + e_{rt}^2 + \text{crossproducts}) = \frac{1}{N} \sigma^2$$

and $E(\bar{e}_{i.}^2) = \dfrac{1}{r_i} \sigma^2$, the expectation of SST is found as

$$E(SST) = \sum_{i=1}^{t} r_i (\mu_i - \bar{\mu}_.)^2 + \sum_{i=1}^{t} r_i E(\bar{e}_{i.}^2) - N E(\bar{e}_{..}^2)$$

$$= \sum_{i=1}^{t} r_i (\mu_i - \bar{\mu}_.)^2 + t \sigma^2 - \sigma^2$$

$$= \sum_{i=1}^{t} r_i (\mu_i - \bar{\mu}_.)^2 + (t - 1) \sigma^2$$

The expectation of MST is

$$E(MST) = \frac{E(SST)}{(t - 1)} = \sigma^2 + \frac{1}{(t - 1)} \sum_{i=1}^{t} r_i (\mu_i - \bar{\mu}_.)^2 \qquad \text{(2A.6)}$$

The expected value in Equation (2A.6) can be expressed in terms of the treatment effects by the substitution $\tau_i = (\mu_i - \bar{\mu}_.)$.

If all $r_i = r$, then let $\theta_t^2 = \sum_{i=1}^{t} (\mu_i - \bar{\mu}_.)^2 / (t - 1)$ and

$$E(MST) = \frac{E(SST)}{(t - 1)} = \sigma_t^2 + r \theta_t^2 \qquad \text{(2A.7)}$$

3 Treatment Comparisons

The analysis of variance and least squares estimates of treatment group means provide the basic information necessary for an in-depth analysis of research hypotheses using methods introduced in this chapter. The methods for an in-depth analysis of the responses to the treatment design include planned contrasts among treatment groups, regression response curves for quantitative treatment factors, selection of the best subset of treatments, comparison of treatments to the control, and all pairwise comparisons among treatment means. All of these methods involve a set of simultaneous decisions to be made by the investigator. This simultaneous statistical inference affects statistical errors of inference. Some of those effects and the control of those errors are discussed in this chapter.

3.1 Treatment Comparisons Answer Research Questions

The relationship between research objectives and treatment design requires us to identify treatments relative to their role in the evaluation of research hypotheses. When an experiment is conducted to answer specific questions, the treatments are selected such that comparisons among the treatments will answer the questions. For example, specific questions can be answered from the meat storage experiment in Chapter 2 about the effect of different storage conditions on the growth of bacteria on meat during storage. The four treatments for the meat storage experiment were (1) commercial wrap, (2) vacuum, (3) mixed gases, and (4) pure CO_2. The summary statistics from the experiment are shown in Display 3.1.

Questions that can be asked about meat storage conditions include

- Is the creation of an artificial atmosphere more effective in reducing bacterial growth than ambient air with commercial wrap?

Display 3.1   Summary Statistics from the Meat Storage Experiment of Example 2.1

	Treatment			
	Commercial	*Vacuum*	*CO, O_2, N*	*CO_2*
$\widehat{\mu}_i = \bar{y}_{i.}$	7.48	5.50	7.26	3.36
$t = 4$	$r_i = 3$	$s^2 = MSE = 0.116$ with 8 degrees of freedom		

- Are the gases more effective in reducing bacterial growth than a complete vacuum?

- Is pure CO_2 more effective than a mixture of CO, O_2, and N in reducing bacterial growth?

3.2   Planning Comparisons Among Treatments

Contrasts among treatment means can be constructed to answer specific questions formulated for the experiment. **Contrasts** are special forms of linear functions of observations (discussed in Appendix 3A). A contrast among means is defined as

$$C = \sum_{i=1}^{t} k_i \mu_i = k_1\mu_1 + k_2\mu_2 + \cdots + k_t\mu_t \tag{3.1}$$

where $\sum_{i=1}^{t} k_i = 0$.

The appropriate contrasts for the three specific questions from the meat storage experiment are

- commercial wrap versus artificial atmospheres:

$$C_1 = \mu_1 - \frac{1}{3}(\mu_2 + \mu_3 + \mu_4)$$

- vacuum versus gases:

$$C_2 = \mu_2 - \frac{1}{2}(\mu_3 + \mu_4)$$

- mixed gases versus pure CO_2:

$$C_3 = \mu_3 - \mu_4$$

The first contrast is the difference between the mean of the commercial wrap and the average of the means for the other treatments. The sum of the coefficients $(1, -\frac{1}{3}, -\frac{1}{3}, -\frac{1}{3})$ is zero, as it should be for a proper contrast. The second contrast is the difference between the mean of the vacuum wrap and the average of the means for the gas treatments with coefficients $(0, 1, -\frac{1}{2}, -\frac{1}{2})$, while the third contrast is a difference between the means of the two gas treatments with coefficients $(0, 0, 1, -1)$.

Estimates of the contrasts, standard errors, confidence interval estimates, and tests of hypotheses about the contrasts are computed from the observed data. Any contrast of the population treatment means, $C = \Sigma k_i \mu_i$, is estimated by the same contrast of the observed treatment means as

$$c = \sum_{i=1}^{t} k_i \bar{y}_{i.} = k_1 \bar{y}_{1.} + \cdots + k_t \bar{y}_{t.} \tag{3.2}$$

where $\bar{y}_{i.}$ is the mean of the ith treatment group.

The estimates for the three contrasts in the meat storage experiment are

$$c_1 = \bar{y}_{1.} - \frac{1}{3}(\bar{y}_{2.} + \bar{y}_{3.} + \bar{y}_{4.}) = 7.48 - \frac{1}{2}(5.50 + 7.26 + 3.36) = 2.11$$

$$c_2 = \bar{y}_{2.} - \frac{1}{2}(\bar{y}_{3.} + \bar{y}_{4.}) = 5.50 - \frac{1}{2}(7.26 + 3.36) = 0.19$$

$$c_3 = \bar{y}_{3.} - \bar{y}_{4.} = 7.26 - 3.36 = 3.90$$

Assessment of Contrast Estimates

Standard Errors of Contrasts

The variance of a contrast estimate $c = \Sigma k_i \bar{y}_{i.}$ is estimated with

$$s_c^2 = s^2 \left[\sum_{i=1}^{t} \frac{k_i^2}{r_i} \right] = s^2 \left[\frac{k_1^2}{r_1} + \frac{k_2^2}{r_2} + \cdots + \frac{k_t^2}{r_t} \right] \tag{3.3}$$

where $s^2 = MSE$ from the analysis of variance. If the r_i are equal, the variance estimator simplifies to

$$s_c^2 = \frac{s^2}{r} \left[k_1^2 + k_2^2 + \cdots + k_t^2 \right] \tag{3.4}$$

The estimator for the standard error of a contrast is the square root of the variance,

$$s_c = \sqrt{s_c^2} \tag{3.5}$$

For the meat storage experiment $MSE = 0.116$ and all $r_i = 3$. The variance and standard error estimates for the first contrast are

$$s_{c_1}^2 = \frac{0.116}{3}\left[1^2 + \left(-\frac{1}{3}\right)^2 + \left(-\frac{1}{3}\right)^2 + \left(-\frac{1}{3}\right)^2\right] = 0.052$$

and

$$s_{c_1} = \sqrt{0.052} = 0.228$$

The standard error estimates for the second and third contrasts are $s_{c_2} = 0.241$ and $s_{c_3} = 0.278$.

Interval Estimates for Contrasts

The $100(1 - \alpha)\%$ confidence interval estimator for a contrast is

$$c \pm t_{\alpha/2,(N-t)}(s_c) \tag{3.6}$$

The correct degrees of freedom for the Student t statistic are the degrees of freedom for the variance used to compute the standard error of the contrast. The variance estimate for the meat storage experiment is $MSE = 0.116$ with $(N - t) = 8$ degrees of freedom. The 95% confidence interval estimate for C_1 using $t_{.025,8} = 2.306$ is

$$2.11 \pm (2.306)(0.228)$$

The 95% interval estimates for the complete set of contrasts are shown in Table 3.1.

Table 3.1   The estimates, standard errors, and 95% confidence interval estimates of three contrasts for the meat storage experiment

		Standard	95% CI	
Contrast	Estimate	Error	Lower	Upper
C_1	2.11	0.228	1.58	2.64
C_2	0.19	0.241	− 0.37	0.75
C_3	3.90	0.278	3.26	4.54

A Sum of Squares Partition for the Contrast

A sum of squares is calculated for the treatment contrast to indicate how much of the variation in the data can be explained by that specific contrast. The sum of squares reduction for estimation of a contrast ($C = \Sigma k_i \mu_i$) can be computed from treatment means as

$$SSC = \frac{\left(\sum\limits_{i=1}^{t} k_i \bar{y}_{i.} \right)^2}{\sum\limits_{i=1}^{t} (k_i^2 / r_i)} \tag{3.6}$$

If all r_i are equal

$$SSC = r \frac{\left(\sum\limits_{i=1}^{t} k_i \bar{y}_{i.} \right)^2}{\sum\limits_{i=1}^{t} k_i^2} \tag{3.7}$$

There is 1 degree of freedom associated with the sum of squares for the contrast. The sum of squares partitions for the three contrasts in the meat storage experiment are shown in Table 3.2. The manual computations are illustrated in Display 3.2.

Table 3.2 Analysis of variance with contrasts for log(count/cm^2) of psychrotrophic bacteria from the meat storage experiment

Source	Degrees of Freedom	Sum of Squares	Mean Square	F	Pr > F
Total	11	33.80			
Treatments	3	32.87	10.96	94.58	.000
Error	8	0.93	0.12		

Contrast	DF	Contrast SS	Mean Square	F	Pr > F
1 vs others	1	9.99	9.99	86.19	.000
2 vs 3 and 4	1	0.07	0.07	0.62	.453
3 vs 4	1	22.82	22.82	196.94	.000

Hypotheses About Contrasts

The usual null hypothesis states that the contrast has a zero value. For example, the null hypothesis for the second contrast of the meat storage experiment would be

$$H_0: C_2 = \mu_2 - \frac{1}{2}(\mu_3 + \mu_4) = 0$$

or, equivalently,

$$H_0: C_2 = 2\mu_2 - \mu_3 - \mu_4 = 0$$

If the null hypothesis is true, $C_2 = 0$, the average bacterial growth for the two gas treatments is the same as that for the vacuum treatment.

**Display 3.2   Sums of Squares Computations for Contrasts
in the Meat Storage Experiment**

$$\bar{y}_{1.} = 7.48 \quad \bar{y}_{2.} = 5.50 \quad \bar{y}_{3.} = 7.26 \quad \bar{y}_{4.} = 3.36$$

	k_1	k_2	k_3	k_4
C_1:	3	-1	-1	-1
C_2:	0	2	-1	-1
C_3:	0	0	1	-1

$$SSC_1 = 3[3(7.48) - 5.50 - 7.26 - 3.36]^2/[3^2 + (-1)^2 + (-1)^2 + (-1)^2]$$
$$= 3(6.32)^2/12 = 9.99$$
$$SSC_2 = 3[2(5.50) - 7.26 - 3.36]^2/[2^2 + (-1)^2 + (-1)^2$$
$$= 3(0.38)^2/6 = 0.07$$
$$SSC_3 = 3[7.26 - 3.36]^2/[1^2 + (-1)^2]$$
$$= 3(3.9)^2/2 = 22.82$$

The two forms of the contrasts stated for the null hypotheses are equivalent except the coefficients of the latter differ from the former by a multiple of 2. The choice is purely a matter of personal preference.

The null hypotheses for the three contrasts in the meat storage experiment are

$$H_0: C_1 = 3\mu_1 - \mu_2 - \mu_3 - \mu_4 = 0$$
$$H_0: C_2 = 2\mu_2 - \mu_3 - \mu_4 = 0$$
$$H_0: C_3 = \mu_3 - \mu_4 = 0$$

The first of these hypotheses states that there is no difference between the commercial wrap and the wraps with an artificial atmosphere. The second states that there is no difference between a vacuum and the two gas atmosphere treatments. The third states that there is no difference between pure CO_2 and a mixture of CO, O_2, and N.

Testing Contrasts with the F Test

The alternate hypotheses are nonzero differences, or $H_a: C \neq 0$. The null hypothesis, $H_0: C = 0$, is tested with

$$F_0 = \frac{MSC}{MSE} \tag{3.9}$$

where $MSC = SSC$ is the 1 degree of freedom mean square for the contrast. Under the null hypothesis F_0 has the F distribution with 1 and $(N - t)$ degrees of freedom. The null hypothesis is rejected at the α level of significance if $F_0 > F_{\alpha,1,(N-t)}$.

The F_0 test statistic for each of the contrasts is given in the column labeled "F" in Table 3.2. The column labeled "$Pr > F$" shows the probability of exceeding the observed F_0 statistic. The critical value for F_0 at the $\alpha = .05$ level of significance is $F_{.05,1,8} = 5.32$. The null hypothesis is rejected for contrasts C_1 and C_3, $Pr > F = .000$, and not rejected for contrast C_2, $Pr > F = .453$.

Since the contrast C_1 has a positive value estimate, $c_1 = 2.11$, we can conclude that the mean number of bacteria on the meat wrapped in the conventional commercial wrap ($\bar{y}_{1.} = 7.48$) was greater than the average over the meats wrapped in the artificial atmospheres, $\frac{1}{3}(\bar{y}_{2.} + \bar{y}_{3.} + \bar{y}_{4.}) = \frac{1}{3}(5.50 + 7.26 + 3.36) = 5.37$. The contrast C_3 has a positive value estimate, $c_3 = 3.90$, and we can conclude that there are less bacteria on the meat in the CO_2 ($\bar{y}_{4.} = 3.36$) than on the meat in the mixed gas atmosphere ($\bar{y}_{3.} = 7.26$).

Testing Contrasts with the Student t Test

Tests of the treatment contrasts can be conducted with the Student t test as well as with the F test. The relationship between the Student t distribution and the F distribution is

$$t^2 = F \tag{3.10}$$

where the Student t has ν degrees of freedom and the F statistic has 1 numerator degree of freedom and ν denominator degrees of freedom. The relationship is always valid when there is 1 degree of freedom in the numerator of the F statistic. The degrees of freedom for the Student t will be the same as that for the denominator of the F statistic.

For any treatment contrast estimate c, with standard error s_c, the ratio

$$t_0 = \frac{c}{s_c} \tag{3.11}$$

has the Student t distribution under the null hypothesis $H_0: C = 0$.

For example, the estimate for the first contrast of the meat storage experiment was $c_1 = 2.11$ with a standard error of $s_{c_1} = 0.228$ (Table 3.1). The ratio

$$t_0 = \frac{2.11}{0.228} = 9.254$$

tests the null hypothesis of no difference between the commercial packaging and the gas atmosphere packagings. The null hypothesis is rejected at the .05 level of significance since $t_0 > t_{.025,8} = 2.306$.

Confidence Intervals or P-Values?

The P-value conveniently summarizes with a single value the significance or non-significance of hypotheses tests about the contrasts. However, the P-value conveys no quantitative information about the contrast. A small P-value, indicating significance, may or may not indicate a contrast is meaningfully different from the hypothesized value.

Confidence intervals, conversely, provide meaningful quantitative information about the contrasts as well as their status with respect to the hypotheses about their values. The results from the Amiodarone study in Example 2.2 can be used to illustrate the benefits of confidence intervals.

Recall the estimated increase in the Amiodarone-treated rabbit ears was 1.20°C, a clinically significant value, while those estimates for the vehicle and saline solutions were 0.13°C and 0.00°C, respectively, which were not clinically significant.

If Amiodarone increased the temperature more than the vehicle solution in which it was carried, then Amiodarone would be seen as contributing to tissue inflammation. Likewise, the comparison of the vehicle solution with the saline solution would convey similar information about the contribution of the vehicle solution to tissue inflammation.

The following table shows the two contrasts' estimates along with their standard errors, 95% confidence intervals, and P-values for the Student t test or F test. Case A shows the actual results for the study. Case B shows the situation where the P-values remain the same but the estimates, standard errors, and attendant 95% confidence interval limits have been artificially inflated threefold.

Contrast	Case	c	s_c	(L, U)	P-value
Amiodarone–Vehicle	A	1.07	0.25	(0.55, 1.59)	.0036
	B	3.21	0.75	(1.65, 4.77)	.0036
Vehicle–Saline	A	0.13	0.25	(−0.39, 0.65)	.6088
	B	0.39	0.75	(−1.17, 1.95)	.6088

The 95% confidence interval estimate for Amiodarone versus Vehicle in Case A indicates Amiodarone significantly elevated the ear temperature over the Vehicle solution with a positive lower limit and an upper limit of 1.59°C, which is clinically significant. Case B shows a threefold increase in the contrast and its standard error with an attendant threefold increase in the confidence interval upper limit to 4.77°C, which is dramatically higher than that for Case A. However, the P-value for both cases is the same at .0036, indicating only that the contrast is different from 0 with no information about the magnitude and direction of the difference.

The 95% confidence interval estimate for Vehicle versus Saline in Case A indicates no difference between the two solutions, with the limits including 0 and both limits less than clinically significant. Case B again shows the artificial threefold increase in the contrast estimate, standard error, and confidence interval limits. However, in this latter case the limits exceed clinical significance. The P-value of .6088 in both cases indicates no significant difference from 0 for the contrast and indicates no more than that. Case A did not approach clinical significance, but it is evident that the limits for the interval estimate of Case B did exceed clinical significance. Thus, with Case B the investigator may want to consider further action to achieve more precise estimates for the comparison of Vehicle and Saline treatments.

Orthogonal Contrasts Convey Independent Information

A certain class of contrasts, known as **orthogonal contrasts**, has special properties with respect to the sum of squares partitioning in the analysis of variance and with respect to their relationship to one another. *Orthogonality* implies that one contrast conveys no information about the other. Contrast c_2, a comparison between the vacuum packaging and the two gas atmospheres, conveys no information about contrast c_3, a comparison between the two gas atmospheres.

Suppose there are two contrasts c and d, where

$$c = \sum_{i=1}^{t} k_i \bar{y}_{i.} = k_1 \bar{y}_{1.} + k_2 \bar{y}_{2.} + \cdots + k_t \bar{y}_{t.}$$

and

$$d = \sum_{i=1}^{t} d_i \bar{y}_{i.} = d_1 \bar{y}_{1.} + d_2 \bar{y}_{2.} + \cdots + d_t \bar{y}_{t.}$$

The contrasts c and d are orthogonal if

$$\sum_{i=1}^{t} \frac{k_i d_i}{r_i} = \frac{k_1 d_1}{r_1} + \frac{k_2 d_2}{r_2} + \cdots + \frac{k_t d_t}{r_t} = 0 \qquad \textbf{(3.12)}$$

The weighted sum of crossproducts of the coefficients k_i and d_i must sum to 0. If all r_i are equal, then c and d are orthogonal if $\sum k_i d_i = 0$.

There are $(t-1)$ mutually orthogonal contrasts among t treatment means; each pair of contrasts will be an orthogonal pair. For example, there are three mutually orthogonal contrasts possible in one set of contrasts among four treatment means.

The coefficients for two of the contrasts of the meat storage experiment, c_2 and c_3, with equal treatment replications are shown in Display 3.3 along with the crossproducts of the coefficients. The two contrasts are orthogonal because their crossproducts sum to 0. Analogous calculations show that the contrast c_1, $(3, -1, -1, -1)$, is also orthogonal to c_2 and c_3.

| Display 3.3   Orthogonal Contrasts for the Meat Storage Experiment |

Contrast	k_1	k_2	k_3	k_4
		Coefficients		
c_2	0	2	-1	-1
c_3	0	0	1	-1

Sum of crossproducts $= (0)(0) + (2)(0) + (-1)(1) + (-1)(-1) = 0$

Another special feature of orthogonal contrasts is that the sums of squares for the $(t-1)$ orthogonal contrasts sum to the treatment sum of squares in the analysis of variance. The sums of squares for the three orthogonal contrasts of the example $(c_1, c_2, \text{ and } c_3)$ sum to the treatment sum of squares within rounding errors, $SST = 32.87$ as seen in Table 3.2.

Contrasts Among Treatments with Unequal Replication

The Amiodarone experiment in Example 2.2 had unequal replication of the treatment groups. The summary statistics for the experiment are shown in Display 3.4.

Display 3.4	Summary Statistics from the Amiodarone Experiment of Example 2.2	

	Treatment		
	Amiodarone	Vehicle	Saline
$\hat{\mu}_i = \bar{y}_{i\cdot}$	1.20	0.13	0.00
r_i	9	6	8

$t = 3$ $N = 23$ $s^2 = MSE = 0.2177$ with 20 degrees of freedom

Two comparisons of interest may be (1) the contrast of the Amiodarone treatment mean with the vehicle and saline means and (2) the contrast between the vehicle and saline means. The two contrasts, respectively, are estimated by

$$c_1 = 2\bar{y}_{1\cdot} - \bar{y}_{2\cdot} - \bar{y}_{3\cdot}$$

and

$$c_2 = \bar{y}_{2\cdot} - \bar{y}_{3\cdot}$$

The contrast coefficients are $(2, -1, -1)$ and $(0, 1, -1)$ with replication numbers $(9, 6, 8)$. Evaluating the contrasts using Equation (3.12),

$$\frac{2(0)}{9} + \frac{-1(1)}{6} + \frac{-1(-1)}{8} \neq 0$$

indicates that the two contrasts are not orthogonal.

For the contrasts to be orthogonal the values of the coefficients would have to be altered. For example, two contrasts for the Amiodarone experiment that satisfy the orthogonality criterion in Equation (3.12) are

$$c_1 = 7\bar{y}_{1\cdot} - 3\bar{y}_{2\cdot} - 4\bar{y}_{3\cdot}$$

and

$$c_2 = \bar{y}_{2\cdot} - \bar{y}_{3\cdot}$$

The coefficients $(7, -3, -4)$ and $(0, 1, -1)$ satisfy the condition of a contrast with $\Sigma k_i = 0$. Evaluation by Equation (3.12) is

$$\frac{7(0)}{9} + \frac{-3(1)}{6} + \frac{-4(-1)}{8} = 0$$

The two contrasts would now be orthogonal and convey no information about one another.

However, consider the first contrast, $c_1 = 7\bar{y}_{1.} - 3\bar{y}_{2.} - 4\bar{y}_{3.}$. The coefficients imply a weighted comparison among the population means. Unless the treatment populations occur in the proportions $7:3:4$ the comparison does not make sense. There is unequal information in each of the treatment groups within the experiment; however, it is not necessarily so that the treatment populations would be unequally represented in nature. In the case of the Amiodarone experiment the contrast with equal weights for the treatment means is a more sensible comparison.

Comments on Orthogonality

To present a nicely ordered analysis of variance table, the choice of contrasts for a study must not be dictated by their orthogonality. Rather, the contrasts should be constructed to answer specific research questions. For example, it may be of interest in the meat storage experiment to contrast the control treatment separately with each of the other treatments, resulting in a non-orthogonal set of contrasts, $\mu_1 - \mu_2$, $\mu_1 - \mu_3$, and $\mu_1 - \mu_4$. Orthogonal contrasts purposely were chosen for the meat storage experiment to simplify the presentation. However, they did in the meantime answer some specific meaningful questions, which is after all the purpose of a research study. The research hypotheses and the treatment design should dictate the construction of the contrasts.

3.3 Response Curves for Quantitative Treatment Factors

Many studies are conducted to determine the quantitative trend relationship between two variables. In experimental studies, one of the variables is usually under the *control* of the investigator while the other is the *observed response* variable. The factorial treatment designs discussed in Section 1.4 are an example of structured treatment designs in which a quantitative factor may have several graded levels.

For example, an experiment can be designed to measure the growth response of animals to increasing amounts of nutrient in the diet. The treatment design consists of one quantitative factor—the amount of nutrient in the diet with three levels, say, 0, 500, and 1000 parts per million (ppm). The objective of the study will be to characterize animal growth as a function of the amount of nutrient in the diet.

Polynomial Response Functions for Response Curves

The polynomial model often used to describe trend relationships between a measured response y and quantitative levels of a factor x is

$$y = \beta_0 + \beta_1 x + \beta_2 x^2 + \cdots + \beta_p x^p + e \qquad \textbf{(3.13)}$$

Several examples of response curves are shown in Figure 3.1. The linear, quadratic, and cubic polynomial response curves provide good approximations to many relationships common to the biological and physical sciences.

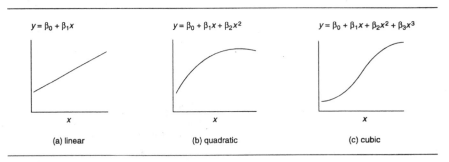

Figure 3.1   Examples of polynomial response curves: (a) linear, (b) quadratic, and (c) cubic

Example 3.1   Grain Production and Plant Density

The growth and development of plants, as with any living organism, is dependent on the availability of sufficient nutrition. The objective for most cultivated grain crops is to maximize seed or grain production under a given cultivation and soil fertility regimen. If the grain producer plants too few seeds per unit of area, maximum grain production will not be realized simply because each plant has genetically determined maximum potential. Thus, increased grain production requires more plants per unit area. On the other hand, for an entirely different reason, maximum production will not be realized if an excessive number of plants are grown per unit of area. Available nutrition per plant becomes limiting, and the plant growth and development are reduced with concomitant reduction in grain production.

The objective for a crop scientist under these circumstances is to characterize the relationship between plants per unit of area and grain production under a given cultivation and fertility regimen. An experiment was conducted to estimate a polynomial response curve to characterize the relationship.

The treatment design consisted of five plant densities (10, 20, 30, 40, and 50). Each of the five treatments was assigned randomly to three field plots in a completely randomized experiment design. The resulting grain yields are shown in Table 3.3.

Table 3.3 Amount of grain produced per plot for five plant densities

	Plant Density (x)				
	10	20	30	40	50
	12.2	16.0	18.6	17.6	18.0
	11.4	15.5	20.2	19.3	16.4
	12.4	16.5	18.2	17.1	16.6
Means $(\bar{y}_{i.})$	12.0	16.0	19.0	18.0	17.0

A graph of the observations in Figure 3.2 suggests a quadratic relationship between grain yield and plant density. The mean responses of grain yields $(\bar{y}_{i.})$ to plant density (x) shown in Table 3.3 also suggest a quadratic response polynomial as an approximate model to describe the biological relationship between grain yield and plant density for this study. The objective is to determine the lowest possible order polynomial equation that adequately describes the relationship. With five values of x it is possible to fit a fourth-degree equation in x; however, a simpler and less complex second-degree equation may do just as well.

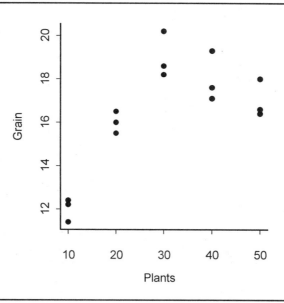

Figure 3.2 Grain yield versus plant density

Simplifying with Orthogonal Polynomials

The polynomial model can be fit to the observed data with many of the available computer regression programs. The trend analysis can be simplified by examining orthogonal contrasts among the treatment factor levels that measure the linear, quadratic, and higher level polynomial effects. These contrasts, known as **orthogonal**

polynomials, enable us to evaluate the importance of each polynomial component with a specific contrast.

In Example 3.1, there are $t = 5$ levels of the plant density factor, and $(t - 1) = 4$ orthogonal contrasts can be estimated. After transforming the polynomials in x into orthogonal polynomials, the complete orthogonal polynomial equation model for the relationship between plant density and grain yield is

$$y_{ij} = \mu + \alpha_1 P_{1i} + \alpha_2 P_{2i} + \alpha_3 P_{3i} + \alpha_4 P_{4i} + e_{ij} \tag{3.14}$$

where μ is the grand mean and P_{ci} is the cth-order orthogonal polynomial for the ith level of the treatment factor.

The transformations for the powers of x into orthogonal polynomials (P_{ci}) up to the third degree are shown in Display 3.5. The tabled values of the orthogonal polynomials for $t = 3$ to 10 are given in Appendix Table XI. The values from Display 3.5 or Appendix Table XI are valid for any distance between the values of x as long as the spacing is equal between all values of x (or factor levels) and replication numbers are the same for all treatments. The constant λ_i at the beginning of each transformation makes each of the P_i values an integer value.

Display 3.5   Transformation of the Powers of x into Orthogonal Polynomials

Mean: $P_0 = 1$

Linear: $P_1 = \lambda_1 \left[\dfrac{x - \bar{x}}{d} \right]$

Quadratic: $P_2 = \lambda_2 \left[\left(\dfrac{x - \bar{x}}{d} \right)^2 - \left(\dfrac{t^2 - 1}{12} \right) \right]$

Cubic: $P_3 = \lambda_3 \left[\left(\dfrac{x - \bar{x}}{d} \right)^3 - \left(\dfrac{x - \bar{x}}{d} \right) \left(\dfrac{3t^2 - 7}{20} \right) \right]$

t = number of levels of the factor x = value of the factor level
$\bar{x}$ = mean of the factor levels d = distance between factor levels

---

Note the term $(x - \bar{x})/d$ occurs consistently in all of the orthogonal polynomial transformations in Display 3.5. The transformation $(x - \bar{x})$ centers the x values around 0, while dividing the result by the increment between the values of x scales the values to change by one unit between levels. For example, in Example 3.1 with $\bar{x} = 30$ and $d = 10$ the resulting transformation is

x:	10	20	30	40	50
$(x - 30)$:	-20	-10	0	10	20
$(x - 30)/10$:	-2	-1	0	1	2

Some of the standard computer program packages for statistical analysis are capable of producing the orthogonal polynomials with unequal or equal spacing between the x values. Grandage (1958) gave a manual method for deriving the orthogonal polynomials with unequal spacings.

The orthogonal polynomial transformations for the plant densities are shown in Table 3.4 with $d = 10$, $t = 5$, $\bar{x} = 30$, and $x = 10, 20, 30, 40,$ or 50. Each set of coefficients P_1 through P_4 forms a contrast among the treatments since the sum of the coefficients in each of these columns is equal to 0. The contrasts are mutually orthogonal.

Table 3.4 Computations for orthogonal polynomial contrasts and sums of squares

		Orthogonal Polynomial Coefficients (P_{ci})				
Density (x)	$\bar{y}_{i.}$	*Mean*	*Linear*	*Quadratic*	*Cubic*	*Quartic*
10	12	1	-2	2	-1	1
20	16	1	-1	-1	2	-4
30	19	1	0	-2	0	6
40	18	1	1	-1	-2	-4
50	17	1	2	2	1	1
λ_c		$-$	1	1	5/6	35/12
Sum $= \Sigma P_{ci}\bar{y}_{i.}$		82	12	-14	1	7
Divisor $= \Sigma P_{ci}^2$		5	10	14	10	70
$SSP_c = r(\Sigma P_{ci}\bar{y}_{i.})^2/\Sigma P_{ci}^2$		$-$	43.2	42.0	0.3	2.1
$\widehat{\alpha}_C = \Sigma P_{ci}\bar{y}_{i.}/\Sigma P_{ci}^2$		16.4	1.2	-1.0	0.1	0.1

The treatment sum of squares can be partitioned into an additive set of 1 degree of freedom sums of squares, one sum of squares for each of the $(t-1)$ orthogonal polynomial contrasts. Consequently, it is possible to test sequentially the significance of the linear, quadratic, cubic, and so forth, terms in the model to determine the best-fitting polynomial equation.

The estimates of the α_C coefficients for the orthogonal polynomial equation in Equation (3.14) and the sum of squares for each of the orthogonal polynomial contrasts are shown in Table 3.4. The estimated orthogonal polynomial equation is found by substituting the estimates of μ, α_1, α_2, α_3, and α_4 from Table 3.4 into Equation (3.14). The estimated equation is

$$\widehat{y}_i = 16.4 + 1.2P_{1i} - 1.0P_{2i} + 0.1P_{3i} + 0.1P_{4i} \qquad (3.15)$$

The analysis of variance for this experiment is shown in the top analysis of Table 3.5. The ratio $F_0 = MST/MSE$ tests the global null hypothesis H_0: $\alpha_1 = \alpha_2 = \alpha_3 = \alpha_4 = 0$. At the .05 level of significance the critical region is $F_0 > F_{.05,4,10} = 3.48$. The null hypothesis is rejected since $F_0 = 29.28$ exceeds the critical value.

The 1 degree of freedom sum of squares partitions for each of the orthogonal polynomial contrasts are summarized in the analysis of variance at the bottom of

Table 3.5. Notice the sums of squares for the four contrasts sum to the sum of squares for Density, 87.60 with 4 degrees of freedom.

Table 3.5   Analysis of variance for the orthogonal polynomial model relationship between plant density and grain yield

Source of Variation	Degrees of Freedom	Sum of Squares	Mean Square	F	Pr > F
Density	4	87.60	21.90	29.28	.000
Error	10	7.48	0.75		
Contrast	*DF*	*Contrast SS*	*Mean Square*	*F*	*Pr > F*
Linear	1	43.20	43.20	57.75	.000
Quadratic	1	42.00	42.00	56.15	.000
Cubic	1	.30	.30	.40	.541
Quartic	1	2.10	2.10	2.81	.125

We are interested in the contribution of the separate polynomial terms of the model. One strategy to determine the best polynomial equation is to test the significance of the terms in the sequence: linear, quadratic, cubic, and so forth. Beginning with the simplest polynomial, a more complex polynomial is constructed as the data requires for adequate description.

The sequence of hypotheses is $H_0: \alpha_1 = 0$, $H_0: \alpha_2 = 0$, $H_0: \alpha_3 = 0$, and so forth. These hypotheses about the orthogonal polynomial contrasts are each tested with the F test for the respective contrasts. The ratios $F_0 = MSC/MSE$ are given in Table 3.5 for each of the polynomial contrasts estimated for the plant density study. For each sum of squares partition the null hypothesis is $H_0: \alpha_C = 0$ with critical region $F_0 > F_{.05,1,10} = 4.96$.

The null hypothesis is rejected for the linear and quadratic terms of the model $(Pr > F = .000)$ but is not rejected for the cubic $(Pr > F = .541)$ and quartic $(Pr > F = .125)$ terms. The quadratic model is sufficient for a description of the relationship between grain yield and plant density on the basis of the statistical tests.

Computing the Response Curve

The estimated quadratic response curve without the cubic and quartic terms, $\widehat{\alpha}_3 P_3$ and $\widehat{\alpha}_4 P_4$, from Equation (3.15) is

$$\widehat{y}_i = \overline{y}_{..} + \widehat{\alpha}_1 P_{1i} + \widehat{\alpha}_2 P_{2i} = 16.4 + 1.2 P_{1i} - 1.0 P_{2i} \tag{3.16}$$

The estimated value for a plant density of $x = 10$ is determined by substituting $P_1 = -2$ and $P_2 = 2$ into Equation (3.16) as $\widehat{y} = 16.4 + 1.2(-2) - 1.0(2) = 12.0$. The observed grain yields and those estimated from Equation (3.16) are shown for all plant densities in Table 3.6.

Table 3.6 Observed grain yields and those estimated from the quadratic orthogonal polynomial equation

Density	Coefficients		Estimated	Observed
x	P_1	P_2	$\widehat{y}_i$	$\overline{y}_{i.}$
10	-2	2	12.0	12
20	-1	-1	16.2	16
30	0	-2	18.4	19
40	1	-1	18.6	18
50	2	2	16.8	17

The polynomial relationship expressed as a function of y and x in actual units of the observed variables is more informative than when expressed in units of the orthogonal polynomial. A direct transformation to an equation in x requires the information in Display 3.5 and Table 3.4. The necessary quantities are $\lambda_1 = 1$, $\lambda_2 = 1$, $d = 10$, $\overline{x} = 30$, and $t = 5$. Then substituting for the P_i

$$\widehat{y} = 16.4 + 1.2P_1 - 1.0P_2$$
$$= 16.4 + 1.2(1)\left[\frac{x - 30}{10}\right] - 1.0(1)\left[\left(\frac{x - 30}{10}\right)^2 - \left(\frac{5^2 - 1}{12}\right)\right]$$

and simplifying to

$$\widehat{y} = 5.8 + 0.72x - 0.01x^2 \tag{3.17}$$

The estimated line from Equation (3.17) may be plotted as shown in Figure 3.3, along with the observed data points (squares) and treatment means $\overline{y}_{i.}$ (filled circles).

Standard Errors and Confidence Intervals for the Response Curve

The estimated response curve is composed of estimates for several parameters. The quadratic equation chosen as the best-fitting equation for the relationship between grain production and plant density has three estimates: $\overline{y}_{..} = 16.4$, $\widehat{\alpha}_1 = 1.2$, and $\widehat{\alpha}_2 = -1.0$.

The estimator for the variance of $\widehat{y} = \overline{y}_{..} + \widehat{\alpha}_1 P_1 + \widehat{\alpha}_2 P_2$ is

$$s_{\widehat{y}}^2 = s_{\overline{y}_{..}}^2 + P_1^2 s_{\widehat{\alpha}_1}^2 + P_2^2 s_{\widehat{\alpha}_2}^2 \tag{3.18}$$

The variance estimator for a polynomial contrast s_c^2 is

$$s_c^2 = \frac{s^2}{(r\Sigma P_{ci}^2)} \tag{3.19}$$

and the standard error estimator for a contrast is the square root of the variance

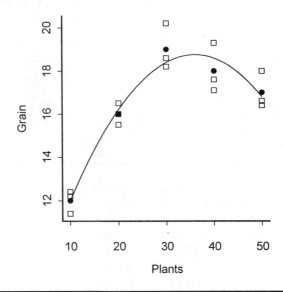

Figure 3.3   Estimated response curve, $\widehat{y} = 5.8 + 0.72x - 0.01x^2$, for grain yield as a function of plant density

$$s_c = \sqrt{s_c^2} \tag{3.20}$$

The variance estimates for the coefficients of the estimated quadratic equation are computed using $s^2 = MSE = 0.75$ from the analysis of variance in Table 3.5 and the divisors from Table 3.4. The variances are

$$s_{\bar{y}..}^2 = \frac{0.75}{(3)(5)} = 0.050$$

$$s_{\widehat{\alpha}_1}^2 = \frac{0.75}{(3)(10)} = 0.025$$

and

$$s_{\widehat{\alpha}_2}^2 = \frac{0.75}{(3)(14)} = 0.018$$

The variance of an estimated value is

$$s_{\widehat{y}}^2 = s_{\bar{y}..}^2 + s_{\widehat{\alpha}_1}^2 P_1^2 + s_{\widehat{\alpha}_2}^2 P_2^2 \tag{3.21}$$
$$= .05 + 0.025 P_1^2 + 0.018 P_2^2$$

Since the values of P_1 and P_2 differ for each value of plant density, the variance of the estimated value differs for each value of plant density. The variance of the estimated yield for plant density $x = 10$, using $P_1 = -2$ and $P_2 = 2$ in Equation (3.21), is

$$s_{\widehat{y}}^2 = 0.05 + 0.025(-2)^2 + 0.018(2)^2 = 0.222$$

and the standard error is

$$s_{\widehat{y}} = \sqrt{0.222} = 0.471$$

The $100(1 - \alpha)\%$ confidence interval for the estimated value is computed from

$$\widehat{y} \pm t_{\alpha/2,(N-t)}(s_{\widehat{y}})$$

The standard errors and 95% confidence intervals for the estimated values of grain yield for all five plant densities using $t_{.025,10} = 2.228$ are shown in Table 3.7.

Table 3.7   Observed and estimated grain yields and standard errors for estimated grain yields for each plant density

	Seed Yield		Standard	95% Confidence
Density	Observed	Estimated	Error	Interval
10	12	12.0	0.471	(10.95, 13.05)
20	16	16.2	0.305	(15.52, 16.88)
30	19	18.4	0.349	(17.62, 19.18)
40	18	18.6	0.305	(17.92, 19.28)
50	17	16.8	0.471	(15.75, 17.85)

The estimated response curve (Figure 3.3) has the advantage of portraying the relationship between y and x throughout the entire range of x values used in the experiment. With this example, it is possible to describe or estimate the grain yield for any plant density between 10 and 50 plants. The description of the relationship is not constrained, in a discussion of the results, to the five plant densities used in the study.

3.4 Multiple Comparisons Affect Error Rates

The group of contrasts exhibited in Section 3.2 are considered **multiple comparisons** because more than one comparison was made among the treatment means. Any necessary number of these comparisons can be constructed in the analysis to help answer the research questions.

However, multiple comparisons among treatment means can lead the investigator, sometimes unwittingly, into a statistical minefield. The abundant number of available procedures increases the difficulty on the part of an investigator of choosing the appropriate method for a particular situation.

Error Rates for Multiple Comparisons

The difficulties with multiple comparisons reside mainly in an understanding of the error rates associated with testing multiple hypotheses. Hypothesis tests have risks associated with decisions to reject or not reject the hypothesis. For any contrast among means the risk associated with declaring the contrast to be real when it is not is the risk of a Type I error. The risk of declaring the contrast among population means to be equal to zero when it is not is the risk of a Type II error.

The risks for Type I and Type II errors are inversely related to one another. The level of significance chosen for the hypothesis test determines the risk of Type I error. Sample size, variance, and size of the contrast for the true population means determine the Type II error rate for a given Type I error rate.

A simple test for the difference between two treatment means is the Student t test, with the statistic calculated as

$$t_0 = \frac{\overline{y}_i - \overline{y}_j}{\sqrt{s^2 \left[\dfrac{1}{r_i} + \dfrac{1}{r_j} \right]}} \tag{3.22}$$

The significance level or probability of Type I error for a single test is a *comparisonwise* error rate, α_C. It is the risk we are willing to take on a single comparison.

There are $p(p-1)/2$ pairwise comparisons among p treatment means. For example, four treatments (A, B, C, D) have $4(3)/2 = 6$ possible pairs: (A, B), (A, C), (A, D), (B, C), (B, D) and (C, D). If the six pairs of means are tested with the t_0 statistic in Equation (3.22), there is the possibility of committing 0, 1, 2, 3, 4, 5, or 6 Type I errors if the six population means are all equal.

With the possibility of up to six Type I errors for six tests we can use another form of Type I error based on the accumulated risks associated with the family of tests under consideration. The family is the set of pairwise comparisons for the example in the previous paragraph. The accumulated risks associated with a family of comparisons is often called the *experimentwise* Type I error rate, α_E. It is the risk of making at least one Type I error among the family of comparisons in the experiment.

Evaluating the Maximum Error Rate

The experimentwise Type I error rate can be evaluated for a family of independent tests. However, all pairwise tests using Equation (3.22) are not independent since the s^2 in the denominator of each of the t_0 statistics is the same, and the numerator of each test contains the same means as several of the other t_0 statistics.

Although the set of tests in the family just described are not independent, the upper limit for the value of the experimentwise Type I error rate can be evaluated by assuming independent tests. Suppose the null hypotheses are true for each of n independent tests. The probability of a Type I error for any single test is α_C (the comparisonwise rate) with $(1 - \alpha_C)$ as the probability of a correct decision. The probability of committing x Type I errors is given by the binomial distribution as

$$P(x) = \frac{n!}{x!(n-x)!}\alpha_C^x(1-\alpha_C)^{n-x} \tag{3.23}$$

for $x = 0, 1, 2, \ldots, n$ Type I errors. The probability of having no Type I errors is

$$P(x = 0) = (1 - \alpha_C)^n$$

The probability of committing *at least one* Type I error $(x = 1, 2, 3, \ldots, n)$ is $P(x \geq 1) = 1 - P(x = 0)$, or

$$\alpha_E = 1 - (1 - \alpha_C)^n \tag{3.24}$$

The probability α_E is the risk of making at least one Type I error among the n independent comparisons. It is the upper limit of the experimentwise Type I error rate for n tests among a set of treatment means.

The relationship

$$\alpha_C = 1 - (1 - \alpha_E)^{1/n} \tag{3.25}$$

expresses the comparisonwise Type I error rate as a function of the experimentwise Type I error rate.

The relationship between the two Type I error rates for a few selected values of n is shown in Table 3.8. If each of the tests is conducted with a comparisonwise error rate of $\alpha_C = .05$ the risk of at least one Type I error escalates as the number of tests increases. When $n = 1$ test is conducted both Type I errors are identical as they should be since only one Type I error can be committed. When $n = 5$ the risk probability of at least one Type I error among the five decisions has risen to $\alpha_E = .226$, and with $n = 10$ the risk rises to a probability of .401 to commit at least one Type I error.

The last column of Table 3.8 gives an indication of the comparisonwise Type I error rate required to maintain an experimentwise Type I error rate $\alpha_E = .05$. For example, when five independent tests are conducted and we want to keep the risk of committing at least one Type I error as low as 1 in 20 chances, the comparisonwise error rate for each of the $n = 5$ tests must be $\alpha_C = .01$ and for $n = 10$ tests it must be $\alpha_C = .005$.

Table 3.8   The relationship between α_C and α_E for $n = 1, 2, 3, 4, 5,$ or 10 independent tests

n	α_E when $\alpha_C = .05$	α_C when $\alpha_E = .05$
1	.050	.050
2	.098	.025
3	.143	.017
4	.185	.013
5	.226	.010
10	.401	.005

Which Experimentwise Error Rate?

Some confusion exists regarding error rates for simultaneous inference since the error rate can be defined relative to the configuration of the population means, $\mu_1, \mu_2, \ldots, \mu_t$. The *experimentwise* error rate has often been defined under the configuration $\mu_1 = \mu_2 = \cdots = \mu_t$, and under this configuration the familywise error rate is considered controlled in the *weak sense* by Hochberg and Tamhane (1987). Equality of means is unlikely to be true under most circumstances. Therefore, when inequalities do occur the weak sense experimentwise error rate offers poor protection against incorrect decisions. If no constraints are put on the relationships among the μ_i, then the familywise error rate is controlled in the *strong sense* (Hochberg & Tamhane, 1987). The latter *strong sense* definition can be interpreted as the probability of at least one incorrect decision over all parameter configurations. Hsu (1996) also presents a detailed discussion of error rates for simultaneous inference.

3.5 Simultaneous Statistical Inference

The discussion in the previous section considered multiple simultaneous inferences and the probabilities of incorrect decisions regarding those inferences. When hypotheses were tested about three contrasts in the meat storage experiment of Section 3.1, they each were tested with a comparisonwise Type I error rate of $\alpha_C = .05$. If those statements are to hold simultaneously with a Type I error rate of $\alpha_E = .05$, the three tests of hypotheses require a family error rate. According to Table 3.8, the comparisonwise error rate would have to be reduced for the three statements to hold simultaneously with a family error rate of $\alpha_E = .05$. If the contrasts were independent, then $\alpha_C = .017$ would provide an appropriate family error rate under a null hypothesis of equality among the means in the contrasts.

An investigator wants to make informed decisions from the observed data with a valid statistical method. The choice of method depends on the *type* of inference desired and the *strength* of inference desired.

The **types** of contrasts most often considered by investigators are

- planned contrasts
- orthogonal polynomial contrasts
- multiple comparisons with the best treatment
- multiple comparisons with the control treatment
- all pairwise comparisons

The first two methods were discussed in Sections 3.2 and 3.3. The other three methods are considered in later sections in this chapter.

Strength of inference refers to how much can be said about a comparison. The strongest inferences include statements about the direction and magnitude of the difference. *Simultaneous confidence interval* methods provide the strongest form of inference. The investigator is able to assert, with a given level of confidence, in which direction the effects of one treatment differ from another and also how far removed those effects are from one another.

Methods providing *confident directions* inference follow simultaneous confidence intervals in strength (Hsu, 1996). These methods, for a given contrast C_i, assert inequalities ($C_i > 0$ or $C_i < 0$) for each contrast with given levels of confidence that all statements are correct but say nothing about magnitude.

Confident inequalities declare the inequality for each contrast as $C_i \neq 0$ with a given confidence level that all declarations are correct (Hsu, 1996). These statements make no assertion about the direction or the magnitude of the inequality.

The weakest methods are *individual comparison* methods that do not consider the simultaneous nature of the inference. The weakness of these methods was demonstrated in the previous section where in Table 3.8 the probability of error for simultaneous statements, α_E, could increase considerably as the number of tests increased.

Multiple Comparison Methods, Probability Tables, and Computer Programs

Well-defined methods with explicit probability tables exist for some of the multiple comparison procedures discussed later in this chapter, given the study can be modeled as a completely randomized design with equal replication numbers for treatments.

Either approximations to the probability tables or computer programs must be employed if these methods are used with unequal replications or more complex design structures.

Fortunately, a number of statistical program suites include exact computations for these methods. However, the documentation for the programs should be checked to be certain valid methods are being used.

Approximations to the probability tables take advantage of several inequalities from probability theory. If the special computer routines are not available these approximations will provide a conservative alternative. The approximation that will be suggested for methods later in this chapter is based on the **Bonferroni inequality** because probability tables based on the inequality are readily available. This approximation can also be used for the less structured methods, such as small sets of planned contrasts and the orthogonal polynomial contrasts, to provide a measure of error protection for simultaneous inference.

Bonferroni t Statistics for Simultaneous Inference

The *Bonferroni inequality* provides a means to obtain an easy approximation to multiple comparison error rates. The inequality, translated to our context, shows that the family error rate is less than or equal to the sum of the individual comparison error rates. When n comparisons are made at the same comparison error rate, α_C, the Bonferroni inequality gives the relationship

$$\alpha_E \leq n\alpha_C$$

Equality of the relationship holds when the tests are independent. The comparisonwise error rate for the test statistic is determined by dividing the maximum desired family error rate by the number of simultaneous tests, $\alpha_C = \alpha_E/n$. For example, with three comparisons for the meat storage experiment a family error rate of $\alpha_E = .05$ requires a comparisonwise error rate of $\alpha_C = .05/3 = .017$.

Tabled values of the Student t statistic, referred to as the Bonferroni t, for selected values of α_E are given in Appendix Table V as $t_{\alpha_E/2,k,\nu}$ for two-sided tests where k is the number of comparisons and ν is the degrees of freedom. Values can also be obtained from any computer program that can compute probabilities for the Student t distribution for upper-tail probability $\alpha_E/2k$. For example, the Bonferroni t for $k = 3$ two-sided comparisons with $\nu = 8$ degrees of freedom and $\alpha_E = .05$ is $t_{.025,3,8} = 3.02$. Equivalently, the value can be found by determining the value of the Student t for $\nu = 8$ degrees of freedom exceeded with probability $\alpha_E/2k = .05/2(3) = .00833$.

Simultaneous Confidence Interval (SCI) Estimates

Confidence interval estimates for contrasts are interval estimates that hold simultaneously with confidence level $100(1 - \alpha_E)\%$. The simultaneous confidence intervals use the Bonferroni t statistic in place of the usual Student t statistic. The two-sided $100(1 - \alpha_E)\%$ confidence interval is

$$c \pm t_{\alpha_E/2,k,\nu}(s_c) \tag{3.26}$$

The three contrasts for the meat storage experiment require $t_{.025,3,8} = 3.02$ for a set of intervals that hold simultaneously with at least 95% confidence. The resulting intervals are shown in Display 3.6.

Display 3.6	Simultaneous 95% Confidence Intervals for Three Contrasts in the Meat Storage Experiment		
Contrast	*Estimate*	*Standard Error*	*95% SCI (L, U)*
$\mu_1 - \frac{1}{3}(\mu_2 + \mu_3 + \mu_4)$	2.11	0.228	(1.42, 2.80)
$\mu_2 - \frac{1}{2}(\mu_3 + \mu_4)$	0.19	0.241	(−0.54, 0.92)
$\mu_3 - \mu_4$	3.90	0.278	(3.06, 4.74)

The SCI intervals are wider than those computed as individual 95% confidence intervals in Table 3.1. There is a trade-off between the strength of the confidence statement and the interval widths. The joint confidence level is less for the shorter intervals in Table 3.1 and greater for the wider SCI intervals shown in Display 3.6.

The investigator can make joint statements with a Type I experimentwise error rate of .05 that (1) the average bacterial growth in artificial atmospheres is less than that on the meat in the commercial wrap since the lower limit of the interval is greater than 0; (2) there is no difference between the vacuum wrap and the average of the gases with respect to the bacterial growth since the interval includes 0; and (3) pure CO_2 results in fewer bacteria than the mixed gas since the lower limit of the interval is greater than 0.

The three contrasts for the meat storage experiment from Table 3.1, their estimates, standard errors, and calculated t_0 statistics are shown in Display 3.7. The calculated t_0 statistics for the first and third contrasts exceed the critical value $t_{.025,3,8} = 3.02$. The t statistics in Display 3.7 exemplify the use of *confident inequalities* inference if only their significance is evaluated. The magnitudes and signs of the contrasts and the t statistics can be used to deduce the direction and magnitude of the comparisons, although in a less direct manner than with the confidence intervals.

Display 3.7   Bonferroni t Tests for Three Contrasts in the Meat Storage Experiment

Contrast	Estimate	Standard Error	t_0
C_1 (Commercial vs. artifical)	2.11	0.228	9.25
C_2 (Vacuum vs. gases)	0.19	0.241	0.79
C_3 (CO_2 vs. mixed gas)	3.90	0.278	14.03

Scheffé's Test for Simultaneous Inference

Bonferroni t statistics can be used safely for a small number of pre-planned contrasts with preservation of the proposed experimentwise error. A method for testing *all possible* contrasts or constructing confidence intervals for *all possible* contrasts was proposed by Scheffé (1953). The method provides the prescribed experimentwise error protection for any number of contrasts. Consequently, the method is quite conservative and is generally used for unplanned contrasts or contrasts suggested by the data. The Scheffé test is shown in Display 3.8.

Simultaneous $100(1 - \alpha_E)\%$ confidence intervals for all possible contrasts are computed with the Scheffé statistic as

$$c \pm S(\alpha_E) \tag{3.27}$$

and there is a $(1 - \alpha_E)$ probability that all intervals simultaneously include the true values of the respective contrasts.

Display 3.8   The Scheffé Test

Consider any contrast, $c = \sum\limits_{i=1}^{t} k_i \bar{y}_{i.}$, among t treatment means with standard error

$$s_c = \sqrt{ s^2 \left[\sum_{i=1}^{t} \frac{k_i^2}{r_i} \right] }$$

The null hypothesis for the contrast, H_0: $C = 0$, is rejected if

$$| c | > S(\alpha_E) \qquad (3.28)$$

$S(\alpha_E)$ is the Scheffé statistic

$$S(\alpha_E) = s_c \sqrt{(t-1) F_{\alpha_E, (t-1), \nu}} \qquad (3.29)$$

where $F_{\alpha_E, (t-1), \nu}$ is the F statistic with $(t-1)$ and ν degrees of freedom exceeded with probability α_E. Also, ν is the number of degrees of freedom for experimental error variance s^2, used to estimate the standard error of the contrast, s_c.

3.6   Multiple Comparisons with the Best Treatment

In some studies, the treatments are not highly structured in their relationship to one another and structured contrasts are difficult to identify. On the other hand, treatments are related to one another because they all are under investigation for their effect on the measured response variables, and they address some specific problem of interest to the investigator. Some possible examples include testing toxins, sources of protein for diets, or mixtures of compounds for an alloy.

Under these circumstances the investigator may want to "pick the winners." The objective is to select the set of treatments or single treatment (if possible) that provides the most desirable result.

The **multiple comparisons with the best (MCB)** procedure from Hsu (1984) enables the investigator to select treatments into a subset such that the "best" population is included in the subset with a given level of confidence. The parameters of interest are

$$\mu_i - \max_{j \neq i} \mu_j, \text{ for } i = 1, 2, \ldots, t$$

where $\max\limits_{j \neq i} \mu_j$ is the maximum treatment mean not including μ_i. If $\mu_i - \max\limits_{j \neq i} \mu_j > 0$ then treatment i is the best. On the other hand, if $\mu_i - \max\limits_{j \neq i} \mu_j < 0$, then treatment i is not the best.

MCB simultaneous confidence intervals (SCI) for $\mu_i - \max_{j \neq i} \mu_j$ are constrained to include 0 with the view that no two treatments ever have identical long run averages (Hsu, 1996). The *constrained* MCB confidence interval asserts treatment i is one of the best if the interval for $\mu_i - \max_{j \neq i} \mu_j$ *includes* 0 or has a *lower* bound of 0. Conversely, if the *upper* interval bound for $\mu_i - \max_{j \neq i} \mu_j$ is 0, then treatment i is not one of the best. The MCB procedure is described in Display 3.9.

Display 3.9   Multiple Comparisons with the Best Procedure

Calculate the difference, D_i, between each treatment mean, $\overline{y}_i$, and the largest treatment mean of the remaining treatments, $\max_{j \neq i} (\overline{y}_j)$, as

$$D_i = \overline{y}_i - \max_{j \neq i} (\overline{y}_j), \text{ for } i = 1, 2, \ldots, t \qquad (3.30)$$

and the quantity M

$$M = d_{\alpha,k,\nu} \sqrt{\frac{2s^2}{r}} \qquad (3.31)$$

where $d_{\alpha,k,\nu}$ is the tabled statistic for one-sided comparisons in Appendix Table VI for an experimental error rate of α_e, $k = t - 1$ comparisons, and ν degrees of freedom for the experimental variance, $s^2 = MSE$.

100(1 − α)% Simultaneous Constrained Confidence Intervals

The lower confidence bound for $\mu_i - \max_{j \neq i} (\overline{y}_j)$ is

$$L = \begin{cases} D_i - M & \text{if } (D_i - M) < 0 \\ 0 & \text{otherwise} \end{cases}$$

and the upper confidence bound for $\mu_i - \max_{j \neq i} (\overline{y}_j)$ is

$$U = \begin{cases} D_i + M & \text{if } (D_i + M) > 0 \\ 0 & \text{otherwise} \end{cases}$$

Example 3.2   Flow Rates Through Filters

The MCB procedure is illustrated using an experiment conducted to evaluate filters with different filtering configurations. The filters were all designed to screen particles above a certain size. The investigators wanted to know which, if any, of the filters allowed the highest flow rate of a particle slurry under constant pressure.

They had developed 6 filter configurations for testing purposes. Four replicate filters of each configuration were constructed for the experiment. The 24 filters were tested in random order for a completely randomized experiment design. The average flow rates for the 6 filter types were

Filter:	A	B	C	D	E	F
Mean:	8.29	7.23	7.54	8.10	8.59	7.10

The estimate of experimental error variance for the experiment was $MSE = 0.08$ with 18 degrees of freedom.

MCB Simultaneous Confidence Intervals

We can calculate the 95% confidence interval for a comparison of filter B with the best of the other filters to illustrate the procedure. The mean for filter B is $\bar{y}_2 = 7.23$, and filter E has the largest mean among all the remaining filters, so that $\max_{j \neq 2} (\bar{y}_j) = \bar{y}_5 = 8.59$. Then $D_2 = 7.23 - 8.59 = -1.36$.

The value for $d_{\alpha,k,\nu}$ in Equation (3.31) is found from Appendix Table VI with $k = 5$, $\alpha_E = .05$, and $\nu = 18$ degrees of freedom for $MSE = 0.08$. The appropriate value is $d_{.05,5,18} = 2.41$, so that with $r = 4$ replications and $MSE = 0.08$

$$M = 2.41 \sqrt{\frac{2(0.08)}{4}} = 0.48$$

The required quantities are $D_2 - M = -1.36 - 0.48 = -1.84$ and $D_2 + M = -1.36 + .48 = -0.88$. Using the rules for upper and lower limits in Display 3.9, $L = -1.84$ because $D_2 - M < 0$ and $U = 0$ because $D_2 + M$ is not greater than 0. The upper and lower bounds are shown for each of the comparisons in Table 3.9.

Four of the filters (B, C, D, and F) have upper bounds of 0 and thus are not the "best" filters. Of the remaining two filters (A and E) neither is clearly the best, since both intervals include 0, which in turn implies that each is one of the best filters with 95% confidence. In addition, their lower bounds, -0.78 for A and -0.18 for E, are also close to 0 relative to the lower bounds for the other filters. A lower bound close to 0 indicates the treatment is close to the best (Hsu, 1996). Note the SCI not only provide the means to identify the best treatment(s) but also give information about how far removed each of the treatments is from the best. Based on the lower bound of the intervals in Table 3.9 it is easy to see that filters B, C, and F in particular are the most removed from the best treatment.

If a treatment is clearly the only best, then the lower bound will be 0. To illustrate suppose filter E had a mean of $\bar{y}_5 = 9.00$ rather than 8.59. The next largest mean is filter A with $\bar{y}_1 = 8.29$. The constrained confidence interval comparing filter E with filter A would be

Table 3.9   MCB procedure with flow rate means of six filter types

Filter	$\bar{y}_i$	$max\,(\bar{y}_j)$ $j \neq i$	D_i	$D_i - M$	$D_i + M$	95% SCI (L, U)	Select?*
A	8.29	8.59	-0.30	-0.78	0.18	$(-0.78, 0.18)$	Yes
B	7.23	8.59	-1.36	-1.84	-0.88	$(-1.84, 0)$	No
C	7.54	8.59	-1.05	-1.53	-0.57	$(-1.53, 0)$	No
D	8.10	8.59	-0.49	-0.97	-0.01	$(-0.97, 0)$	No
E	8.59	8.29	0.30	-0.18	0.78	$(-0.18, 0.78)$	Yes
F	7.10	8.59	-1.49	-1.97	-1.01	$(-1.97, 0)$	No

Select as "best" when $D_i + M > 0$.

$$D_5 - M = 9.00 - 8.29 - 0.48 = 0.23$$

$$D_5 + M = 9.00 - 8.29 + 0.48 = 1.19$$

so that with $D_5 - M > 0$ and $D_5 + M > 0$ the constrained interval bounds are $(0, 1.19)$, and the interval with a lower bound of 0 indicates filter E is the best. Likewise, filter A would not be one of the best under these circumstances since

$$D_1 - M = 8.29 - 9.00 - 0.48 = -1.19$$

$$D_1 + M = 8.29 - 9.00 + 0.48 = -0.23$$

so that with $D_1 - M < 0$ and $D_1 + M < 0$ the constrained interval bounds are $(-1.19, 0)$. Thus, with an upper bound of 0 filter A would not be the best treatment or among the set of best treatments.

Selecting the Subset with the Largest Mean

If the only interest is which treatment or treatments constitute the "best" without regard to how much their effects may differ from the others, then a simple selection rule can be used (Hsu, 1984). The MCB rule selects a treatment into the best subset with a probability of correct selection, $P(\text{CS}) = 1 - \alpha$, if

$$D_i + M > 0 \tag{3.32}$$

The MCB procedure to select the best filter types with $P(\text{CS}) = 0.95$ uses the $D_i + M$ column information given in Table 3.9. The treatments in the best subset are those for which $D_i + M > 0$. Only filters A and E have values of $D_i + M > 0, 0.18$ and 0.78 respectively. The best subset includes these two filter types with a probability $P(\text{CS}) = 0.95$ that the best filter is in the subset A and E.

The conclusions from the subset selection procedure do not differ from the SCI results. Both filter types A and E were selected to the best subset. However, the SCI are more informative because they indicated how close to 0 the lower bounds were for each of the filter types, which in turn gives the investigator more information regarding flow rate performance for each of the filters relative to that for the filters with the highest flow rates. The subset selection procedure limits the information to whether a filter is in the best subset and indicates no more than that.

Multiple Comparisons with the Smallest Mean

In some studies the treatment with the smallest mean is the "best," such as in the meat storage experiment in which a packaging would be best if it had fewer bacteria on the surface. Multiple comparisons with the smallest mean can be made with simple modifications of the rule for selecting the largest mean in Display 3.9.

Calculate the difference, D_i, between each treatment mean, $\bar{y}_i$, and the smallest mean of the remaining treatments, $\min_{j \neq i} (\bar{y}_j)$, as

$$D_i = \bar{y}_i - \min_{j \neq i} (\bar{y}_j) \text{ for } i = 1, 2, \ldots, t \qquad (3.33)$$

The $100(1 - \alpha)\%$ simultaneous confidence interval lower (L) and upper (U) limits are found as follows: The lower confidence bound for $\mu_i - \min_{j \neq i} \mu_j$ is

$$L = \begin{cases} D_i - M & \text{if } (D_i - M) < 0 \\ 0 & \text{otherwise} \end{cases}$$

and the upper confidence bound for $\mu_i - \min_{j \neq i} \mu_i$ is

$$U = \begin{cases} D_i + M & \text{if } (D_i + M) > 0 \\ 0 & \text{otherwise} \end{cases}$$

The constrained MCB confidence intervals for comparisons with the smallest treatment mean are interpreted just opposite of those for comparisons with the largest treatment mean. If the interval for $\mu_i - \min_{j \neq i} \mu_j$ *includes* 0 or has an *upper* bound of 0, then treatment i is the best treatment. Conversely, if the *lower* bound is 0, then treatment i is not the best treatment.

Multiple comparisons with the best as the treatment with the smallest mean is illustrated with the meat storage study and an objective to select the packaging material with the least amount of bacterial growth. The experimental error variance for the experiment was $MSE = 0.116$ with 8 degrees of freedom, and there were $r = 3$ replications. The required statistics for the critical value of M in Equation (3.31) are $d_{.05,3,8} = 2.42$ and $\sqrt{2MSE/r} = 0.278$, so that

$$M = d_{.05,3,8} \sqrt{\frac{2MSE}{r}} = 2.42(0.278) = 0.67$$

The treatment means and 95% simultaneous confidence interval estimates for comparisons with the smallest mean are shown in Table 3.10. The procedure is illustrated for the commercial wrap treatment with mean $\bar{y}_1 = 7.48$ and the

Table 3.10 Selection of the Treatment Subset with the Smallest Means in the Meat Storage Experiment

Treatment	$\bar{y}_i$	$min\,(\bar{y}_j)$ $j \neq i$	D_i	$D_i - M$	$D_i + M$	95% SCI (L, U)	Select?*
Commercial	7.48	3.36	4.12	3.45	4.79	(0, 4.79)	No
Vacuum	5.50	3.36	2.14	1.47	2.81	(0, 2.81)	No
Mixed	7.26	3.36	3.90	3.23	4.57	(0, 4.57)	No
Pure CO_2	3.36	5.50	-2.14	-2.81	-1.47	$(-2.81, 0)$	Yes

* Select as "best" when $D_i - M < 0$.

treatment with the minimum mean, pure CO_2, so that $\min_{j \neq 1} (\bar{y}_j) = \bar{y}_4 = 3.36$ to give $D_1 = 7.48 - 3.36 = 4.12$. The lower limit is $L = 0$ since $D_1 - M = 4.12 - 0.67 = 3.45$ is not less than 0, and the upper limit is $U = 4.79$ since $D_1 + M = 4.12 + 0.67 = 4.79$ is greater than 0. With a lower bound of 0 we can assert that the commercial wrap is not the best treatment. The best treatment is pure CO_2 since it is the only treatment in Table 3.10 with an upper confidence interval bound of 0. The upper bounds for all other treatments are considerably removed from 0 and clearly cannot be considered close to the best.

Selecting the Subset with the Smallest Mean

Again, if the only interest is which treatment or treatments constitute the best without regard to how much their effects may differ from the others, then a simple selection rule can be used (Hsu, 1984). The MCB rule selects a treatment into the best subset with a probability of correct selection, $P(\text{CS}) = 1 - \alpha$, if

$$D_i - M < 0 \tag{3.34}$$

The MCB procedure to select the best meat packaging with $P(\text{CS}) = 0.95$ uses the $D_i - M$ column information given in Table 3.10. The treatments in the best subset are those for which $D_i - M < 0$. The only treatment mean with $D_i - M < 0$ is that for pure CO_2. Pure CO_2 is selected as the treatment with the lowest bacterial growth with a probability of correct selection $P(\text{CS}) = .95$.

Unequal Replications and Complex Models

Hsu (1996) indicates that critical values must be computed separately for each comparison with unequal replication numbers in the completely randomized design. Some computer programs have incorporated routines to compute the simultaneous constrained confidence intervals with unequal replication numbers. No approximations based on probability inequalities were given by Hsu (1996).

If a more complex blocking or treatment design is used and all differences, $\mu_i - \mu_j$, have the same variance—that is, the design is *variance balanced*—then the standard MCB procedure in this section may be used.

If the variances for differences, $\mu_i - \mu_j$, are not all the same in more complex designs, then Hsu (1996) recommends several approximations based on probability inequalities. The approximation based on the *Bonferroni inequality* uses the Bonferroni t for k comparisons with ν degrees of freedom at the appropriate level of α in place of $d_{\alpha,k,\nu}$. The least squares estimate, $\widehat{\mu}_i - \widehat{\mu}_j$, and its standard error should be used; they can be obtained from most computer programs.

3.7 Comparison of All Treatments with a Control

In many studies one of the treatments acts as a control treatment for some or all of the remaining treatments. (Different types of control treatments were discussed in Section 1.4.) Determining whether the mean responses for the treatments differ from that for the control treatment is sometimes of interest. Dunnett (1955) introduced a procedure for this purpose based on an experimentwise error rate.

Simultaneous Confidence Intervals

The tabled statistic to compute $100(1 - \alpha)\%$ simultaneous confidence intervals for differences between the individual treatment means and the control mean $\mu_i - \mu_c$ using the Dunnett procedure is based on the same statistic used for the multiple comparisons with the best procedure in Section 3.6. The Dunnett test to compare each treatment mean, $\overline{y}_i$, with the control treatment mean $\overline{y}_c$ is described in Display 3.10.

Suppose filter F in Example 3.2 serves as a control. The five 95% SCI comparisons for control versus treatment means are shown in Table 3.11. The mean differences between F and the $k = 5$ other filters appear in the third column. The critical value of the Dunnett statistic for a two-sided test with an error rate of $\alpha_E = .05$ is $d_{.05,5,18} = 2.76$. The standard error of the difference is $\sqrt{2MSE/r} = \sqrt{2(0.08)/4} = 0.2$. The Dunnett criterion is

$$D(5, .05) = 2.76(0.2) = 0.55$$

An example calculation is illustrated with filter A. The lower limit is

$$L = \overline{y}_1 - \overline{y}_c - D(5, .05) = 8.29 - 7.10 - 0.55 = 0.64$$

and the upper limit is

$$U = \overline{y}_1 - \overline{y}_c + D(5, .05) = 8.29 - 7.10 + 0.55 = 1.74$$

Filter A is superior to the control filter, F, since the lower bound of the interval is greater than 0. Based on the interval estimates filters A, D, and E are superior to filter F. Also, note that filter E is the most removed from filter F since its lower bound is greater than the lower bound of either filter A or D. Filters B and C do not differ from the control since the intervals include 0.

Display 3.10   The Dunnett Method for a Comparison of All Treatments with a Control

$100(1 - \alpha)$% Simultaneous Confidence Intervals for $\mu_i - \mu_c$

The Dunnett criterion to compare k treatments to the control is

$$D(k, \alpha_E) = d_{\alpha,k,\nu}\sqrt{\frac{2s^2}{r}} \qquad (3.35)$$

Simultaneous two-sided confidence interval estimates for the differences between the individual treatment means and the control means $\mu_i - \mu_c$ are

$$\bar{y}_i - \bar{y}_c \pm D(k, \alpha_E) \qquad (3.36)$$

One-sided interval lower bounds if superiority is manifested by a treatment mean *greater* than the control mean are

$$\bar{y}_i - \bar{y}_c - D(k, \alpha_E) \qquad (3.37)$$

One-sided interval upper bounds if superiority is manifested by a treatment mean *less* than the control mean are

$$\bar{y}_i - \bar{y}_c + D(k, \alpha_E) \qquad (3.38)$$

The values of $d_{\alpha,k,\nu}$ for the two-sided or one-sided Dunnett method are found in Appendix Table VI for k treatments, an experimentwise Type I error of α_E, and ν degrees of freedom for the estimate of experimental error variance.

Table 3.11   Results of the Dunnett test comparing the mean of the control filter, F, with that of all other filters

Filter	Mean	$\bar{y}_i - \bar{y}_c$	95% SCI (L, U)	$\|\bar{y}_i - \bar{y}_c\|$	Different from control?*
F	$\bar{y}_c = 7.10$	—	–	—	—
A	8.29	1.19	(0.64, 1.74)	1.19	Yes
B	7.23	0.13	(−0.42, 0.68)	0.13	No
C	7.54	0.44	(−0.11, 0.99)	0.44	No
D	8.10	1.00	(0.45, 1.55)	1.00	Yes
E	8.59	1.49	(0.94, 2.04)	1.49	Yes

*If $|\bar{y}_i - \bar{y}_c|$ exceeds $D(5, .05) = 0.55$, then the filter mean is different from that of filter F (control).

Testing Hypotheses About $\mu_i - \mu_c$

If the investigator only wants to know if a treatment mean is significantly different from the control mean with significance level α, then the following two-sided or one-sided tests can be conducted: For the two-sided alternative with H_0: $\mu_i = \mu_c$ versus H_a: $\mu_i \neq \mu_c$, reject the null hypothesis if

$$|\,\bar{y}_i - \bar{y}_c\,| > D(k, \alpha_E) \tag{3.39}$$

For the one-sided alternative with H_0: $\mu_i \leq \mu_c$ versus H_a: $\mu_i > \mu_c$, reject the null hypothesis if

$$(\bar{y}_i - \bar{y}_c) > D(k, \alpha_E) \tag{3.40}$$

For the one-sided alternative with H_0: $\mu_i \geq \mu_c$ versus H_a: $\mu_i < \mu_c$; reject the null hypothesis if

$$(\bar{y}_i - \bar{y}_c) < -D(k, \alpha_E) \tag{3.41}$$

Results of the test with the two-sided alternatives are shown in Table 3.11. The differences for filters A, D, and E exceed $D(5, .05) = 0.55$, and their flow rates differ significantly from that for filter F. The two-sided tests exemplify a confident inequalities inference. We only state with error rate .05 that filters A, D, and E differ from filter F, and filters B and C do not differ from filter F. One can only deduce indirectly from the size and sign which differences were positive and which were the largest as opposed to the SCI, which directly give us information about size and direction.

Unequal Replications and Complex Models

Hsu (1996) indicates that critical values must be computed separately for each comparison with unequal replication numbers in the completely randomized design. Some computer programs have incorporated routines to compute the simultaneous constrained confidence intervals with unequal replication numbers. An approximation based on the Bonferroni inequality can be used that substitutes the Bonferroni t for k comparisons with ν degrees of freedom at the appropriate level of α in place of $d_{\alpha,k,\nu}$ and uses $\sqrt{2s^2[1/r_i + 1/r_c]}$ for the standard error of the difference.

If a more complex blocking or treatment design is used and all differences, $\mu_i - \mu_j$, have the same variance—that is, the design is *variance balanced*—then the standard procedure in this section may be used.

If the variances for differences, $\mu_i - \mu_c$, are not all the same in more complex designs, then Hsu (1996) presents several approximations, some of which are based on probability inequalities. The approximation based on the Bonferroni inequality uses the Bonferroni t for k comparisons with ν degrees of freedom at the appropriate level of α in place of $d_{\alpha,k,\nu}$. The least squares estimate, $\hat{\mu}_i - \hat{\mu}_c$, and its standard error should be used; they can be obtained from most computer programs.

Dunnett (1964) provided adjustments to the critical values in Appendix Table VI for unequal replication numbers. A conservative upper bound given by Fleiss (1986) for values in Appendix Table VI is $md_{\alpha,k,\nu}$, where

$$m \leq 1 + 0.07 \left(1 - \frac{r_i}{r_c}\right) \tag{3.42}$$

Dunnett has shown the optimal ratio of replication numbers is r/r_c where r is the common replication number for each treatment and r_c is the replication number for the control treatment.

3.8 Pairwise Comparison of All Treatments

Some investigators compare each treatment mean with each of the other treatment means using **pairwise comparisons**. The parameters of interest are all pairwise differences among the treatment means, $\mu_i - \mu_j$ for all $i \neq j$, resulting in $t(t-1)/2$ comparisons. Most frequently, applications of these methods have an objective to detect significant inequalities, $\mu_i \neq \mu_j$ for all $i \neq j$.

The indiscriminate use of pairwise comparison procedures in this manner for the analysis of experimental results can lead to the tendency to place a reliance on statistical significance alone to drive the inferential procedure in data analysis. It is therefore possible to lose sight of the research objectives, and the investigator's focus may diverge from the pursuit of biological or physical understanding to that of statistical significance.

Ideally, investigators would want to make the comparisons with an experimentwise error rate in the strong sense as they were made with multiple comparisons with the best (Section 3.6) and multiple comparisons with the control (Section 3.7). Several methods for pairwise comparisons will be presented in this section with some discussion of their properties.

The Tukey Method

A procedure providing an experimentwise rate in the strong sense was developed by Tukey (1949a) for pairwise comparison of all treatment means and is used to obtain $100(1 - \alpha)\%$ simultaneous confidence intervals. The test has been called by various names, including the Honestly Significant Difference. The Tukey method is described in Display 3.11.

The Tukey method is based on the Studentized range statistic

$$q = \frac{\overline{y}(\text{largest}) - \overline{y}(\text{smallest})}{\sqrt{\dfrac{s^2}{r}}} \tag{3.43}$$

where $\overline{y}(\text{largest})$ is the largest mean in an ordered group of means in an experiment and $\overline{y}(\text{smallest})$ is the smallest of the means. The difference or range is divided by the standard error of a treatment mean, from which the statistic derives the name of Studentized range statistic.

Display 3.11   The Tukey Method for All Pairwise Comparisons

For a group of k treatment means compute the Honestly Significant Difference as

$$\text{HSD}(k, \alpha_E) = q_{\alpha, k, \nu} \sqrt{\frac{s^2}{r}} \qquad (3.44)$$

where $q_{\alpha, k, \nu}$ is the Studentized range statistic for a range of k treatment means in an ordered array. Critical values for an experimentwise error rate, α_E, and ν degrees of freedom can be found in Appendix Table VII.

$100(1 - \alpha)$% Simultaneous Confidence Intervals

Simultaneous two-sided interval estimates for the absolute value of all pairwise differences, $\mu_i - \mu_j$ for all $i < j$, are

$$|\bar{y}_i - \bar{y}_j| \pm \text{HSD}(k, \alpha_E) \qquad (3.45)$$

$100(1 - \alpha)$% Confident Inequalities Test

Two treatment means are declared not equal, $\mu_i - \mu_j \neq 0$, if

$$|\bar{y}_i - \bar{y}_j| > \text{HSD}(k, \alpha_E) \qquad (3.46)$$

The absolute difference $|\bar{y}_i - \bar{y}_j|$ is given in Display 3.11 for the confidence intervals because the location of the two means in the calculated difference, $\bar{y}_i - \bar{y}_j$, is arbitrary, with the sign of the difference depending on whether one calculates $\bar{y}_i - \bar{y}_j$ or $\bar{y}_j - \bar{y}_i$. Thus, the absolute difference is equivalent to always subtracting the smaller mean from the larger. If the direction of a particular difference is necessary, then calculate the interval with the sign of the difference considered. The particulars of the study will dictate whether absolute or signed differences are best used for specific comparisons.

Example 3.3   Strength of Welds

Pairwise comparison tests are illustrated using an experiment conducted to compare the strength of welds produced by four different welding techniques. Each welding technique was used to weld five pairs of metal plates in a completely randomized design. The average strengths for the five welds of each technique were

Technique:	A	B	C	D
Mean:	69	83	75	71

> The estimate of experimental error variance for the experiment was $MSE = 15$ with 16 degrees of freedom.

Applying the Tukey method to the data of Example 3.3, the Studentized range statistic with an experimentwise error rate of $\alpha_E = .05$ is found from Appendix Table VII as $q_{.05,4,16} = 4.05$, where there are $k = 4$ treatment means in the ordered array and $\nu = 16$ degrees of freedom for $MSE = s^2$. The standard error is $\sqrt{MSE/r} = \sqrt{15/5} = 1.73$. The computed HSD is HSD(4, .05) = 4.05(1.73) = 7.0. The 95% SCI and results of the confident inequalities test are shown in Table 3.12.

Table 3.12 Results of the Tukey method for differences between weld strength means for four welding methods

Comparison	$\|\bar{y}_i - \bar{y}_j\|$	95% SCI (L, U)	Different from 0?*
A vs. B	14	(7, 21)	Yes
A vs. C	6	(−1, 13)	No
A vs. D	2	(−5, 9)	No
B vs. C	8	(1, 15)	Yes
B vs. D	12	(5, 19)	Yes
C vs. D	4	(−3, 11)	No

* The absolute difference exceeds HSD(.05) = 7.0.

Two treatment means are different with 95% confidence if the 95% SCI interval does not include 0. Method B is significantly different from all other methods, but no other treatments differ from one another. The amount method B differs from the other three methods can be assessed by the magnitude of the lower bound. Method B differs most from method A and least from method C.

Inference on the basis of confidence inequalities declares two treatment means different if the absolute difference, $\|\bar{y}_i - \bar{y}_j\|$, exceeds HSD(4, .05) = 7.0. The HSD test judges the weld strengths of method B to be different from the weld strengths of all other methods, but no other differences are significant. However, inference about the direction and magnitude of the inequalities is not possible.

Unequal Replications and Complex Models with the Tukey Method

For unequal replications in the completely randomized design, Tukey in 1953 (see Tukey, 1994) and Kramer (1956) proposed approximate simultaneous confidence intervals, which substitute

$$\sqrt{\frac{s^2}{2}\left(\frac{1}{r_i} + \frac{1}{r_j}\right)} \qquad (3.47)$$

for $\sqrt{s^2/r}$ in Equation (3.43). The approximation has become known as the Tukey–Kramer approximation, which Hayter (1984) showed to be conservative.

If the variances for differences, $\mu_i - \mu_j$, are not all the same in more complex designs, then the least squares estimates, $\widehat{\mu}_i - \widehat{\mu}_j$, and their variances, $s^2_{(\widehat{\mu}_i - \widehat{\mu}_j)}$, are used for an approximation to the exact form. The simultaneous intervals are

$$|\widehat{\mu}_i - \widehat{\mu}_j| \pm q_{\alpha,k,\nu} \sqrt{\frac{1}{2} s^2_{(\widehat{\mu}_i - \widehat{\mu}_k)}} \tag{3.48}$$

If a more complex blocking or treatment design is used and all differences, $\mu_i - \mu_j$, have the same variance—that is, the design is *variance balanced*—then the standard procedure in this section may be used, using the least squares estimates, $\widehat{\mu}_i - \widehat{\mu}_j$, and their common variance of the difference, $s^2_{(\widehat{\mu}_i - \widehat{\mu}_j)}$ in Equation (3.48). The strong sense experimentwise error rate will be exact in this case.

Tests of Homogeneity, $\mu_1 = \mu_2 = \cdots = \mu_t$

Many of the popular pairwise multiple comparison tests have been used with experimentwise error rates evaluated under a restricted assumption that all treatment means are equal, or $\mu_1 = \mu_2 = \cdots = \mu_t$, thus providing experimentwise error rates in the *weak* sense. Some of them use a single criterion for declaration of significance. Other tests are referred to as *multiple-range tests* since they use multiple criteria for declaration of significance, where the value of a criterion for one comparison depends upon how far apart the two means are in the ordered array of all the treatment means. The *Least Significant Difference, LSD(α)*, is the most common test of homogeneity that uses a single criterion. The *Student-Newman-Keuls, SNK(k,α)*, is an example of a multiple-range test that is a test of homogeneity. Both tests are illustrated below.

The Least Significant Difference (LSD)

Each hypothesis $H_0: \mu_i = \mu_j$ versus $H_a: \mu_i \neq \mu_j$ can be tested with the Student t statistic:

$$t_0 = \frac{\bar{y}_i - \bar{y}_j}{\sqrt{s^2 \left[\dfrac{1}{r_i} + \dfrac{1}{r_j} \right]}} \quad \text{for all } i \neq j \tag{3.49}$$

When the Type I error probability is set at some value α and the variance s^2 has ν degrees of freedom, the null hypothesis is rejected for any observed value of $|\bar{y}_i - \bar{y}_j|$ such that $|t_0| > t_{\alpha/2,\nu}$. The **Least Significant Difference, LSD(α)**, is an abbreviated method of conducting all possible pairwise t tests as shown in Display 3.12.

Display 3.12   Least Significant Difference for All Pairwise Comparisons

For any pair of observed treatment means, $\bar{y}_i$ and $\bar{y}_j$, the Least Significant Difference is

$$\text{LSD}(\alpha) = t_{\alpha/2,\nu}\sqrt{s^2\left[\frac{1}{r_i} + \frac{1}{r_j}\right]} \qquad (3.50)$$

The null hypothesis H_0: $\mu_i = \mu_j$ is rejected if

$$|\bar{y}_i - \bar{y}_j| > \text{LSD}(\alpha)$$

A modification by Fisher (1960) controls the weak sense experimentwise error rate. The LSD is used to test pairwise comparisons only if the null hypothesis is rejected in the analysis of variance F test. If the null hypothesis is not rejected on the basis of the F test, then all treatment means are assumed to be the same and no further testing is done. The procedure often is referred to as the protected LSD. Carmer and Swanson (1973) provided empirical demonstration that the experimentwise error rate with the protected LSD was almost the same as the significance level of the F test used as the determinate when α was set at the .05 significance level for the LSD.

Investigations by Finner (1990) and Hayter (1986) showed the LSD test was a confident inequalities method if the number of treatments was less than 3, but that for $t > 3$ it was not a confident inequalities method.

The LSD test is illustrated with the weld data from Example 3.3. The computation of the LSD requires the critical value for the Student t test, $t_{.025,16} = 2.12$, and the standard error of the difference between two treatment means, $\sqrt{2MSE/r} = \sqrt{2(15)/5} = 2.45$. The computed LSD is $\text{LSD}(.05) = 2.12(2.45) = 5.2$.

The null hypothesis H_0: $\mu_i = \mu_j$ is rejected if

$$|\bar{y}_i - \bar{y}_j| > \text{LSD}(.05) = 5.2$$

A convenient method for testing is to form a table of differences with the means ordered from smallest to largest as shown in Table 3.13. The differences in the table are each computed as a difference between a mean of the column head and a mean of lesser value in the leftmost column. An asterisk indicates the differences that exceeded $\text{LSD}(.05) = 5.2$.

The result of all pairwise comparisons in Table 3.13 indicates that the weld strength of method B exceeds that of all other methods, and the weld strength of method C exceeds that of method A.

The Student-Newman-Keuls (SNK) Multiple-Range Test

The SNK test is one of many multiple-range tests. It is based on the Studentized range statistic in Equation (3.43), but in contrast to the Tukey method it results in a

Table 3.13   Results of the LSD test on differences between the weld strength means for four welding methods

		Method			
		A	D	C	B
Method	Mean	69	71	75	83
A	69	—	2	6*	14*
D	71		—	4	12*
C	75			—	8*
B	83				—

* The difference exceeds LSD(.05) = 5.2.

homogeneity test with experimentwise error rates in the weak sense (Hsu, 1996). The critical value for the Studentized range with the SNK test is based on the range of the particular pair of means being tested within the entire set of ordered means. The test was developed independently by Newman (1939) and Keuls (1952) and is categorized as a multiple-range test since two or more ranges among the means are used for the test criteria. The SNK test is described in Display 3.13.

Display 3.13   Student-Newman-Keuls Multiple Range Test

The SNK criterion is

$$\text{SNK}(k, \alpha_E) = q_{\alpha,k,\nu} \sqrt{\frac{s^2}{r}} \quad \text{for } k = 2, 3, \dots, \tag{3.51}$$

where $q_{\alpha,k,\nu}$ is the Studentized range statistic, k is the number of means in the range, ν is the number of degrees of freedom for the estimate of experimental error variance s^2, and α_E is the experimentwise error rate for a range of k means.

For the largest and smallest means in a range of k means, say $\bar{y}_i$ and $\bar{y}_j$, the null hypothesis $H_0: \mu_i = \mu_j$ is rejected if

$$| \bar{y}_i - \bar{y}_j | > \text{SNK}(k, \alpha_E)$$

The test is not conducted if there is a range of means of size greater than k containing $\bar{y}_i$ and $\bar{y}_j$ that is not significant by the SNK criteria.

The SNK test is presented here to demonstrate the methods for multiple-range tests and also because tables of critical values for the Studentized range statistic for a constant α are readily available. A multiple-range test that provides a much stronger inference with confident inequalities is presented after the SNK test. Although the other test also uses the Studentized range statistic it has not become popular, probably because its critical values are not as easily accessible as those for the SNK test.

The treatment means are ordered from smallest to largest

$$\bar{y}_{[1]} \leq \bar{y}_{[2]} \leq \bar{y}_{[3]} \leq \cdots \leq \bar{y}_{[t]}$$

where $\bar{y}_{[1]}$ is the treatment mean with the smallest value and $\bar{y}_{[t]}$ is the treatment mean with the largest value. The critical value for each pair of means depends on the number of means in the range of the particular pair of means under test.

The SNK test of the means in Example 3.3 with an experimentwise Type I error rate of $\alpha_E = .05$ requires three critical values of the Studentized range statistic from Appendix Table VII, with $\alpha_E = .05$, $\nu = 16$, and $k = 2, 3,$ and 4. With standard error $\sqrt{MSE/r} = \sqrt{15/5} = 1.73$, the SNK statistic is computed from

$$\text{SNK}(k, .05) = q_{.05,k,16}(1.73) \qquad \text{for } k = 2, 3, 4$$

The three critical values of the Studentized range statistic and SNK($k, .05$) for Example 3.3 are

k:	2	3	4
$q_{.05,k,16}$:	3.00	3.65	4.05
SNK($k, .05$):	5.2	6.3	7.0

The critical values for the SNK test increase as the number of means in a range increases. As the distance between two means in the ordered array increases, a larger difference between the means is required to declare the means different from one another. For the minimum range of $k = 2$ the SNK(2, .05) is equal to the LSD(.05) = 5.2, and for the maximum range of $k = 4$ the SNK(4, .05) is equal to the HSD(4, .05) = 7.0. The SNK test is more conservative than the LSD but less conservative than the HSD in the required differences for rejection of the null hypothesis. The results of the SNK($k, .05$) test for Example 3.3 are shown in Table 3.14.

Table 3.14 Results of the SNK($k, .05$) on differences between the weld strength means for four welding methods

		Method					
		A	D	C	B		
Method	Mean	69	71	75	83	k	SNK($k, .05$)
A	69	–	2	6	14*	– – 4 – –	7.0
D	71		–	4	12*	– – 3 – –	6.3
C	75			–	8*	– – 2 – –	5.2
B	83						

*Differences shown are a larger mean minus a smaller mean. The difference exceeds SNK($k, .05$).

The differences between treatment means for the same range, k, are found on a diagonal running from the upper left to the lower right in Table 3.14. The SNK test commences with a comparison between the minimum and maximum means, $k = 4$. If the maximum difference does not exceed the critical value no further testing is conducted. If the maximum difference is significant the differences among the means with range $(k - 1)$ are tested.

If any pair of means with range $(k - 1)$ are not significant no further testing is conducted for any other pairs of means between that specific pair of means. By definition, no subgroup of means contained in a nonsignificant group of means can be significant. For example, if there is no significant difference between $\bar{y}_{[1]}$ and $\bar{y}_{[3]}$, then no test should be performed for $\bar{y}_{[1]}$ versus $\bar{y}_{[2]}$ or $\bar{y}_{[2]}$ versus $\bar{y}_{[3]}$. The test proceeds in this manner until there are no further significant ranges.

The maximum difference, $k = 4$, is 14 for B versus A. The difference exceeds $SNK(4, .05) = 7.0$, so the test proceeds with comparisons for a range of $k = 3$, B versus D and C versus A. The difference for C versus A does not exceed $SNK(3, .05) = 6.3$. Therefore, no more tests are conducted between pairs of means in the range C to A. They are D versus A and C. The difference for B versus D exceeds $SNK(3, .05) = 6.3$, so testing continues within the groups of means between B and D. The difference for B versus C exceeds $SNK(2, .05) = 5.2$ and is the only test necessary with a range of $k = 2$ means. The SNK test judges method B to be different from all other methods with no differences among the other methods.

With unequal replication numbers for the HSD and SNK tests the harmonic mean of the replication numbers from all treatment groups, r_h, is often used in the standard error estimate for all comparisons to simplify the calculations. However, Hsu (1996) showed its use led to invalid statistical inference in general by reducing the confidence levels considerably. The harmonic mean r_h is

$$r_h = \left[\frac{1}{t} \sum_{i=1}^{t} \left(\frac{1}{r_i} \right) \right]^{-1} \tag{3.52}$$

Multiple Range Tests for Confident Inequalities

Several multiple-range tests have been developed that provide confident inequalities inference, which is a stronger inference than that provided by the SNK test. Einot and Gabriel (1975) proposed a choice of α_k to test the difference between two means with a range of k in the set of t treatment means. A modification suggested by a number of authors has led to common use as

$$\alpha_k = \begin{cases} 1 - (1 - \alpha_E)^{\frac{k}{t}} & \text{if } k = 2, \dots, k - 2 \\ \alpha & \text{if } k = t - 1, t \end{cases} \tag{3.53}$$

for the desired experimentwise error rate α_E. The Studentized range statistic for a range of k means would be used for the critical value as $q_{\alpha_k, k, \nu}$.

The test will provide confident inequalities inference if the critical values are nondecreasing with increasing values of k for the range (Hsu, 1996). The test can be conducted fairly easily if a computer program is available to compute quantiles for the Studentized range statistic.

The values of α_k and critical values of the Studentized range statistic calculated from a computer program for the weld strength example are shown below and can be compared with those for the SNK test given earlier. The statistic for a decision regarding significance of a difference between two means is $EG(k, \alpha_k) = q_{\alpha_k,k,\nu}\sqrt{s^2/r}$. Recall there were $t = 4$ treatments, $r = 5$ replications, and a standard error of 1.73 with $\nu = 16$ degrees of freedom for the error mean square.

k:	2	3	4
α_k:	.025	.05	.05
$q_{\alpha_k,k,16}$:	3.49	3.65	4.05
$EG(k, .05)$:	6.0	6.3	7.0

Notice the values of the Studentized range statistic for $k = 3$ and 4 are the same as those for the SNK test, but the value for $k = 2$ is somewhat larger because of the different choice for α_k. The SNK test uses $\alpha_k = \alpha_E$ for all values of k. The significant comparisons using $EG(k, \alpha_k)$ are the same as those with the SNK test; however, the inference strength has increased to that of confident inequalities from strict homogeneity (which the SNK test provided). The differences between the two tests become more pronounced as the number of treatments increase.

3.9 Summary Comments on Multiple Comparisons

Research hypotheses and treatment designs are the engines that drive the methods for analysis of observed results from a study. A set of treatments structured to address certain research hypotheses lead naturally to planned comparisons, such as a set of orthogonal (or nonorthogonal) contrasts, polynomial regressions, and comparisons of all treatments with a control.

The MCB Procedure for Screening Studies

A set of unstructured treatments challenges the investigator to select an appropriate protocol for decision making. Such studies include experiments to evaluate sets of crop cultivars, industrial products, pesticides, and so forth. The MCB procedure to select a subset of treatments with the desired response is the logical choice if the objective is to screen for the best products, pesticides, or cultivars. Subsequent studies may provide the possibility for more structured research hypotheses and treatment designs with opportunities for hypothesis testing.

Multiple Contrasts Require Decisions About Error Rates

Multiple contrasts require the investigator to make some decision relevant to error rates and power of the tests. Comparisonwise error rates are applicable if individual comparisons are the conceptual units of interest. Experimentwise error rates are appropriate if a family of comparisons is the conceptual unit of interest, and the investigator wants to reduce the chance of too many incorrect decisions in the family of tests.

Probabilities for Type II errors and the power of a test can be determined on an experimentwise basis, but they are usually expressed as comparisonwise rates. The more liberal comparisonwise error rate will result in a lower Type II error rate and a more powerful test given all other conditions constant. The more conservative investigator will utilize the strong sense experimentwise error rate for a family of comparisons. The family of comparisons as a conceptual unit is important to the investigator under these circumstances. A typical family would be the comparison of all treatments with a control using the Dunnett procedure.

Most good experiments are designed to be efficient in the use of existing resources and time, and it is more efficient to design an experiment that answers multiple, related questions. A family of related comparisons will exist within any well-planned experiment, and one comparison cannot be considered in total isolation from all other comparisons; thus, tests based on experimentwise error rates are most appropriate.

Choose a Pairwise Comparison Procedure Consistent with Your Philosophy

The selection of an appropriate pairwise comparison method is difficult since they each have their advantages and disadvantages. The best alternative is to choose a test that is consistent with your philosophy and use it consistently for all of your pairwise comparison tests. A test with a single critical value for each experiment is preferable. Informative discussions on many different pairwise comparison methods can be found in Hsu (1996) and in extensive notes on multiple comparisons that were written by J. W. Tukey in 1953 (Tukey, 1994). Some discussions of their use by Jones (1984), Carmer and Walker (1985), and Saville (1990) may prove useful.

The Tukey method provides the best protection against decision errors, along with the strong inference about magnitude and direction of differences with the $100(1 - \alpha)\%$ SCI.

Hayter (1990) provided a method for one-sided simultaneous lower confidence bounds, $\mu_i - \mu_j > L$ for all $i > j$, which is very useful for directional inference in certain types of studies. The bounds are computed as

$$\widehat{\mu}_i - \widehat{\mu}_j - q_\alpha^* \sqrt{\frac{2s^2}{r}} \quad \text{for all } i > j \tag{3.54}$$

These bounds will be much sharper than those supplied by two-sided procedures such as the Tukey method. Tables of critical values for the statistic q_α^* can be found in Hayter and Liu (1996).

A General Recommendation

The general recommendation for conducting multiple comparisons among treatments is to utilize the research hypotheses and treatment design to choose appropriate multiple comparison procedures with the strength of inference desired for the study. Confidence intervals provide the strongest inference with magnitude and direction of differences followed in strength by confident directions and then confident inequalities. A confidence interval procedure that provides magnitude and direction of inference is recommended if pairwise comparisons are your only remaining alternative.

EXERCISES FOR CHAPTER 3

1. Use the data on traffic delay in Exercise 2.1.
    a. Conduct an analysis of variance for the data, and estimate the following contrasts and their standard errors:
        (i) a contrast between the pretimed and the average of the semi- and fully actuated signals
        (ii) a contrast between the semi- and fully actuated signals
    b. Compute the sum of squares of each contrast, and show that their sum is equal to the treatment sum of squares in the analysis of variance.
    c. Test the null hypothesis for each contrast, $H_0: C = 0$, with the Student t test at the .05 level of significance.
    d. Test the null hypothesis in part (c) with the F test at the .05 level of significance.
    e. What is the relationship between the two tests in parts (c) and (d)?

2. Use the data on serum T3 concentrations from the experiments with chickens in Exercise 2.3. Contrasts of interest were the serum T3 concentration differences between successive stages: (1) premolt versus fasting, (2) fasting versus 60 grams of bran, (3) 60 grams of bran versus 80 grams of bran, and (4) 80 grams of bran versus laying mash.
    a. Estimate each of the contrasts and their standard errors.
    b. Test the null hypothesis for one of the contrasts with the Student t test.
    c. Test the null hypothesis for one of the contrasts with the F test.
    d. Suppose you were to test the four contrasts each with a comparisonwise error rate of .05. Compute the maximum experimentwise error rate for this family of four tests.

3. Use the data on lettuce yields in Exercise 2.2.
    a. Compute the analysis of variance for the data.
    b. Determine the best polynomial response function that describes the relationship between lettuce yield and nitrogen fertilizer at the .05 level of significance.
    c. See Table 3.7. Construct a similar table for the results of the current problem.
    d. Plot the observed means along with the estimated equation.

4. Use the data from Exercise 2.1.
    a. Compute the 95% SCI for multiple comparisons with the best with "best" defined as the signal type with the shortest stopped time delay.
    b. Select the signal type(s) with the shortest stopped time delay with a probability of correct selection of .95.

5.   A hospital clinical laboratory measures the concentration of cholesterol in patient serum samples with a spectrophotometer. On one particular day the laboratory analyzed samples from eight patients. Two samples from each patient were prepared for analysis. The data that follow are concentrations of cholesterol (mg/dl) .

Patient	Cholesterol (mg/dl)
1	167.3, 166.7
2	186.7, 184.2
3	100.0, 107.9
4	214.5, 215.3
5	148.5, 149.5
6	171.5, 167.3
7	161.5, 159.4
8	243.6, 245.5

a.   Compute the 95% SCI for multiple comparisons with the best with "best" defined as the patient with the highest cholesterol level.

b.   Select the subset of patients that contains the patient with the highest cholesterol count with a probability of correct selection of .95.

6.   A set of comparisons of interest on the chicken experiment in Exercise 2.3 was the comparison of serum T3 for each of the other states with that for the premolt stage.

a.   Compute the 95% SCI comparisons of other stages with the premolt stage using the Dunnett method.

b.   What are your conclusions?

7.   Sections of tomato plant tissue were grown in tissue cultures with differing amounts and types of sugars in an experiment with five replications of four treatments in a completely randomized design. The tissue growth of each culture is given in the table below as mm × 10.

Control	3% Glucose	3% Fructose	3% Sucrose
45	25	28	31
39	28	31	37
40	30	24	35
45	29	28	33
42	33	27	34

a.   Compute the 95% SCI comparisons of all treatments with the control treatment using the Dunnett method.

b.   What are your conclusions?

8.   The coefficients shown to you by a colleague for a set of contrasts among treatment means follow. He wants you to check them out.

Treatment	A	B	C	D	E
C1	1	3	−1	−1	−1
C2	1	−1	0	−1	1
C3	−1	1	−1	1	−1
C4	0	0	2	−1	−1

a. Does each of the proposed set of coefficients constitute a contrast? Justify your answer.
b. Are $C1$ and $C2$ orthogonal? Justify your answer.
c. Construct a contrast orthogonal to $C4$ that is different from the others already shown.

9. Use the Scheffé test at the .05 level of significance to test the null hypotheses about the contrasts in Exercise 3.1.

10. Use the Bonferroni t test at the .05 level of significance to test the null hypothesis about the contrasts in Exercise 3.1.

11. Use the data on serum T3 in Exercise 2.3.
    a. Conduct all pairwise comparisons with the Tukey method at the .05 significance level.
    b. Conduct all pairwise comparisons with the Least Significant Difference at the .05 significance level.
    c. Conduct all pairwise comparisons with the SNK multiple-range test at the .05 significance level.
    d. How did the results differ among the three tests?
    e. Explain why the results differed.
    f. Compute the 95% SCI for all pairwise comparisons with the Tukey method.
    g. What additional information do you have after computing the 95% SCI?

12. The following is a description of an experiment on human work systems. In a human work system, such as a factory assembly line, workers are often required to move an object to a specific location with their hand.

    The specific purpose of the study was to determine the accuracy with which individuals could reach to specific target locations on a horizontal plane (for example, the top of a table) with their field of view cut off from the targets.

    Previous research led to the hypothesis that distal movements (movements away from the body) are more accurate than proximal movements (movements toward the body). It was also hypothesized that movements in the cardinal directions (straight ahead, straight toward the body, and lateral) were more accurate than movements in other, non-cardinal, directions.

    Targets were set up on the circumference of a circle with a radius of 10 inches (Figure 3.4). The subject was seated so that a movement of the right hand to the 90° target position from the starting point was a distal (away from body) movement. A movement of the right hand to the 270° target was a proximal (toward body) movement. Movements to the 0° and 180° targets were lateral movements. Distal movements to the 45° and 135° targets were non-cardinal movements as were proximal movements to 225° and 315°. The 0°, 90°, 180°, and 270° targets were cardinal directions.

Sixteen right-handed male subjects were trained for the study. In the actual trial the investigator randomly called out the target angles to the subject. The subject marked his try at the target, which was blocked from his view. The distance from the subject's mark to the target was recorded for each target. The average distance to each target location was computed from the observations on the 16 subjects. A smaller average value represented greater accuracy.

Show in a table the coefficients required for a contrast among means for each of the following comparisons of accuracy between movement directions.

$C1$:   Distal versus proximal in general
$C2$:   Cardinal versus non-cardinal
$C3$:   Lateral versus distal
$C4$:   Lateral versus proximal
$C5$:   Non-cardinal distal versus non-cardinal proximal
$C6$:   Cardinal distal versus cardinal proximal

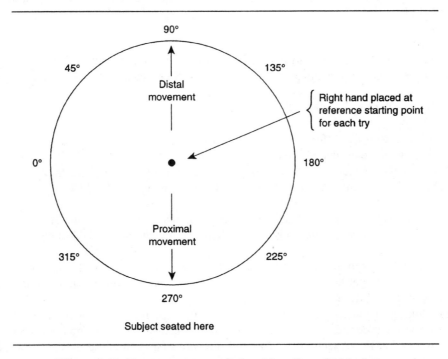

Figure 3.4   Targets set up on circle with radius of 10 inches

3A Appendix: Linear Functions of Random Variables

A linear function of the random variables $y_1, y_2, \cdots, y_n$ is defined as

$$c = \sum_{i=1}^{n} k_i y_i = k_1 y_1 + k_2 y_2 + \cdots + k_n y_n$$

If the expected value or mean of y_i is $E(y_i) = \mu_i$, then the expected value or mean of c is

$$\mu_c = E(c) = E\left(\sum_{i=1}^{n} k_i y_i\right) = \sum_{i=1}^{n} k_i E(y_i) = \sum_{i=1}^{n} k_i \mu_i$$

$$= k_1 \mu_1 + k_2 \mu_2 + \cdots + k_n \mu_n$$

The variance of a linear function $c = \Sigma\, k_i y_i$ is

$$\sigma_c^2 = \sum_{i=1}^{n} k_i^2 \sigma_i^2 + 2\sum_{i<j}^{n} k_i k_j \sigma_{ij}$$

For example, if $c = y_1 - y_2$ with $k_1 = 1$ and $k_2 = -1$, then

$$\sigma_c^2 = k_1^2 \sigma_1^2 + k_2^2 \sigma_2^2 + 2 k_1 k_2 \sigma_{12} = \sigma_1^2 + \sigma_2^2 - 2\sigma_{12}$$

If the y_i's are independent, then $\sigma_{ij} = 0$ and

$$\sigma_c^2 = \sum_{i=1}^{n} k_i^2 \sigma_i^2$$

Therefore, if y_1 and y_2 are independent in $c = y_1 - y_2$, the variance of c is $\sigma_c^2 = \sigma_1^2 + \sigma_2^2$.

The Sample Mean

The mean of a sample of r independent observations from a normal distribution with a mean μ and variance σ^2 is a linear function

$$\overline{y} = \frac{1}{r} y_1 + \frac{1}{r} y_2 + \cdots + \frac{1}{r} y_r$$

where $k_i = \dfrac{1}{r}$. The expected value of the sample mean is

$$\mu_{\overline{y}} = E(\overline{y}) = \frac{1}{r} E(y_1) + \cdots + \frac{1}{r} E(y_r) = \frac{1}{r}(r\mu) = \mu$$

Since the variances of the observations are all the same, $\sigma_1^2 = \sigma_2^2 = \cdots = \sigma_r^2 = \sigma^2$, then the variance of the mean is

$$\sigma_{\bar{y}}^2 = \frac{1}{r^2}(r\sigma^2) = \frac{\sigma^2}{r}$$

Linear Function of Sample Means

If t samples are independent and r_i is the number of observations in the ith sample, then a linear function of the sample means

$$c = k_1\bar{y}_1 + k_2\bar{y}_2 + \cdots + k_t\bar{y}_t$$

has a mean

$$\mu_c = E(c) = k_1 E(\bar{y}_1) + k_2(\bar{y}_2) + \cdots + k_t E(\bar{y}_t)$$

$$= k_1\mu_1 + k_2\mu_2 + \cdots + k_t\mu_t$$

and a variance

$$\sigma_c^2 = k_1^2\left(\frac{\sigma_1^2}{r_1}\right) + k_2^2\left(\frac{\sigma_2^2}{r_2}\right) + \cdots + k_t^2\left(\frac{\sigma_t^2}{r_t}\right)$$

If all sample variances are equal, $\sigma_1^2 = \sigma_2^2 = \cdots = \sigma_t^2 = \sigma^2$, then

$$\sigma_c^2 = \sigma^2\left(\frac{k_1^2}{r_1} + \frac{k_2^2}{r_2} + \cdots + \frac{k_t^2}{r_t}\right)$$

4 Diagnosing Agreement Between the Data and the Model

The analysis of variance can lead to erroneous inferences if certain assumptions regarding the data are not satisfied. Diagnostic methods for detecting faulty assumptions are discussed in Chapter 4 along with data transformations that can be used to address the problems. A generalization of the linear model for the analysis is suggested as an alternative to data transformations. Also, a graphical method is introduced to evaluate how well a model fits the data.

4.1 Valid Analysis Depends on Valid Assumptions

The validity of estimates and tests of hypotheses for analyses derived from the linear model rests on the merits of several key assumptions. The random experimental errors are assumed to be independent, be normally distributed with a mean of zero, and have a common variance (σ^2) for all treatment groups. Any disagreement between the data and one or more of these assumptions affects the estimates of the treatment means and tests of significance from the analysis of variance.

Summary discussions on the assumptions for the analysis of variance and effects of departures from the assumptions can be found in Eisenhart (1947) and Cochran (1947). Ito (1980) summarized research on the validity of analysis of variance test procedures under departures from assumptions.

4.2 The Effects of Departures from Assumptions

If experimental errors are positively correlated, Cochran (1947) showed that the actual precision of the treatment mean is less than the estimated precision. The usual standard error estimate is too small. Conversely, the actual precision is greater than the estimated precision if the error correlations are negative. The best insurance against excessive correlation of the observations is randomization of experimental units to the treatment groups in experiments and randomly sampling populations for observational studies.

If σ^2 differs from one group of observations to another the standard errors of treatment means will generally be greater than if σ^2 is constant over all observations. The stated significance levels for the F and t tests may be larger or smaller than the significance level actually realized. Theoretical studies by Box (1954a) produced results on the actual significance levels of the F test conducted at the .05 level of significance with unequal group variances for equal and unequal replication numbers. For a ratio of 1:3 for smallest:largest group variance the actual significance levels ranged from .056 to .074 for equal replication numbers but ranged from .013 to .14 with unequal replication numbers. With a variance ratio of 1:7 the actual significance level was .12 with equal replication numbers.

The analysis of variance F tests are quite robust against departures from the normal distribution. Ito (1980) cited the results of theoretical and empirical studies on the effects of nonnormality in which the actual significance levels ranged from .03 to .06 for tests conducted at the .05 level of significance.

Ideal conditions are seldom realized in real studies. Minor departures of the data from independence, the assumed normal distribution, and homogeneous variances generally will not cause large changes in the efficiency of estimates and significance levels of tests. Gross departures, especially excessive heterogeneity of variance or some variance heterogeneity with unequal replication numbers, can seriously affect statistical inferences. The remainder of the discussion will focus on those situations.

4.3 Residuals Are the Basis of Diagnostic Tools

The observed **residuals** form the basis for many of the primary diagnostic tools used to check the adequacy of linear model assumptions. The residuals are estimates of the experimental errors computed as the differences between the observations and the estimates of the treatment means, or

$$\widehat{e}_{ij} = y_{ij} - \widehat{\mu}_{i.} = y_{ij} - \overline{y}_{i.} \tag{4.1}$$

Examining the magnitude of the residuals and their relationship to other variables is recommended as the first step in the diagnostic process.

Residuals are used to provide visual evaluations of the analysis of variance assumptions for homogeneous variances and normal distribution of experimental errors. The homogeneous variance assumption is evaluated with a plot of the residuals versus the estimated treatment means. A normal probability plot is used to evaluate the normal distribution assumption. The techniques are demonstrated with observations from a study that do not agree satisfactorily with the linear model assumptions.

Example 4.1   Hermit Crab Counts in Coastline Habitats

A marine biologist was interested in the relationship between different coastline habitats and the populations of Hermit crabs inhabiting the site. The biologist counted Hermit crabs on 25 transects randomly located in each of six different sites of a coastline habitat. Summary statistics for the six sites, including the mean square for error, are given in Table 4.1. The data are given in Appendix 4A.

There are 150 residuals to be calculated for the data set summarized in Table 4.1. For illustration, the first five residuals for the observations in site $1, \widehat{e}_{1j} = y_{1j} - \overline{y}_{1.}$, are shown in Display 4.1.

Table 4.1   Means, standard deviations, and minimum and maximum values for Hermit crab counts from transects in six different coastline sites

Site	Mean	Median	Standard Deviation	Minimum	Maximum
1	33.80	17	50.39	0	233
2	68.72	10	125.35	0	466
3	50.64	5	107.44	0	407
4	9.24	2	17.39	0	82
5	10.00	2	19.84	0	94
6	12.64	4	23.01	0	95
		$MSE = 5170$ with 144 degrees of freedom			

Source: Department of Ecology and Evolutionary Biology, University of Arizona.

Display 4.1   Observations, Mean, and Residuals for Site 1 from Hermit Crab Study

Site	Transect	y_{1j}	$\overline{y}_{1.}$	$\widehat{e}_{1j}$
1	1	0	33.8	-33.8
1	2	0	33.8	-33.8
1	3	22	33.8	-11.8
1	4	3	33.8	-30.8
1	5	17	33.8	-16.8

A Probability Plot of the Residuals to Evaluate the Normal Distribution Assumption

The mean is considerably larger than the median at all six sites in Table 4.1, indicating a skewed and nonnormal distribution of observations. The normal probability plot of the residuals is used to evaluate the normal distribution assumption. The plot is used to visually compare the cumulative distribution of the residuals with that for the standard normal distribution. A normal probability plot arranges the residuals in increasing order and plots them against corresponding *quantiles*[1] of the standard normal distribution. The ith-ordered residual has a cumulative frequency of i/N in a sample of size N.

The quantile of a corresponding standard normal variable is determined for a cumulative proportion[2] of $f_i = (i - 0.5)/N$. The values of the five smallest and five largest residuals and the corresponding standard normal quantiles for the Hermit crab counts are shown in Table 4.2.

Table 4.2 Ordered residuals, cumulative probability (f_i), and corresponding standard normal quantiles for the Hermit crab counts

Order	Residual	f_i	Normal Quantile
1	− 68.72	0.0033	− 2.713
2	− 68.72	0.0100	− 2.326
3	− 68.72	0.0167	− 2.128
4	− 68.72	0.0233	− 1.989
5	− 67.72	0.0300	− 1.881
⋮	⋮	⋮	⋮
146	199.20	0.9700	1.881
147	263.36	0.9767	1.989
148	346.28	0.9833	2.128
149	356.36	0.9900	2.326
150	397.28	0.9967	2.713

The quantile of a normal variable is known for a given cumulative probability and the value, or equivalently the quantile, of a residual is known for its corresponding cumulative frequency in the residual data set. The corresponding quantiles of the residuals and standard normal distribution are paired and plotted as the X and Y values in a standard bivariate plot known as a *quantile-quantile* plot. If the quantiles of the residuals match the quantiles of the normal variable for the same accumulated frequency, they will plot on a straight line.

The normal probability plot of residuals for the Hermit crab counts is shown in Figure 4.1. The straight line in Figure 4.1 passes through the lower and upper quartiles (25th and 75th percentiles) of the data. The quantiles of residuals paired with their corresponding standard normal quantiles do not lie on the straight line

---

[1] The f quantile, $q(f)$, is a value such that approximately a proportion, f, of the data are less than or equal to $q(f)$.

[2] The value of f is designed to avoid a value of $f = 1$; if not there would be no finite value of the standard normal deviate. Any slight modification from $f_i = i/N$ to avoid $f = 1$ is usually adequate for these plots. The value here is used in the S-PLUS statistical programs used for the plots in this section.

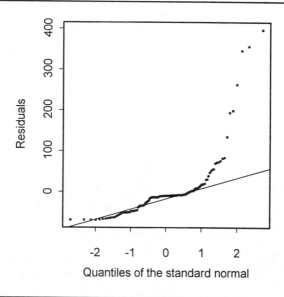

Figure 4.1 Normal probability plot of residuals from the Hermit crab study

indicator of the normal probability plot. It illustrates the probability plot for a distribution that is skewed to the right relative to the standard normal distribution. The values above the line in the upper right-hand corner of the figure are residuals with positive values larger than expected from the standard normal distribution. The series of values above the line in the lower left-hand corner are residuals with negative values smaller than expected from the standard normal distribution.

Residual Plots to Evaluate the Homogeneous Variances Assumption

Plotting the residuals against the estimated values of the treatment means provides a simple visual evaluation of the equal variances assumption for the treatment groups. If the variability of the observations around the treatment means differs from group to group the corresponding set of residuals will reflect the differences in variation. A plot of the residuals versus the estimated site means for the Hermit crab counts is shown in Figure 4.2.

The plot reflects the differences in the standard deviations among the sites in Table 4.1. The dispersion of the residuals varies considerably across the six sites. The variability of the residuals increases with the value of the estimated means. If the variances are heterogeneous the plot of residuals versus estimated values often has the funnel-shaped appearance shown in Figure 4.2. The asymmetry of the residuals around the zero value (dashed line) indicates an asymmetric distribution of the observations with a long tail to the right.

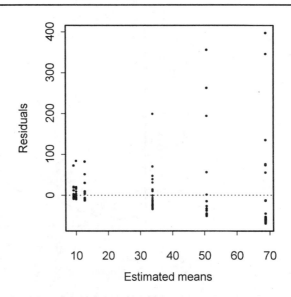

Figure 4.2   Plot of residuals versus estimated site means for the Hermit crab counts

The *spread-location* or *s-l* plot is another residual plot that can be even more revealing of heterogeneous variances. Trends in the *s-l* plot can be used to reveal relationships that may exist between the treatment group means and treatment group variances.

The square roots of the absolute values of the residuals, $\sqrt{|\widehat{e}_{ij}|}$, are used to measure *spread* of the residuals, since the size of the absolute values of the residuals will reflect the spread or variation within a treatment group. The square roots remove some of the asymmetry in the absolute residuals. The estimated treatment group means measure *location*. The *s-l* plot for the Hermit crab study is shown in Figure 4.3.

The medians of the $\sqrt{|\widehat{e}_{ij}|}$ for each of the sites are joined by straight lines in the *s-l* plot and show the increase in their magnitude with an increase in the site means. This increasing trend in the magnitude of the absolute values of the residuals reflects the increase in site variance as the site means increase.

Statistical Tests for Homogeneous Variances

Levene (Med) Test

Many formal statistical tests for homogeneity of variances exist for completely randomized designs. Conover, Johnson, and Johnson (1981) compared 56 such tests and found one of the best to be the **Levene (Med)** test.

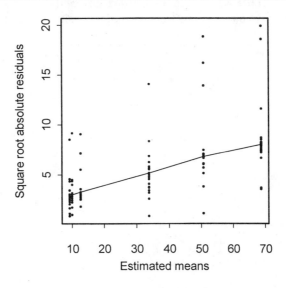

Figure 4.3   Spread-location plot for the residuals from the Hermit crab study

Let y_{ij} be the jth observation in the ith treatment group and $\tilde{y}_i$ be the median of the ith treatment group. Let $z_{ij} = |y_{ij} - \tilde{y}_i|$ be the absolute value of the difference between an observation and the median in the ith treatment group. To test for homogeneity of the variances, compute the one-way analysis of variance for z_{ij} and form the F_0 statistic

$$F_0 = \frac{MST}{MSE} = \frac{\sum\limits_{i=1}^{t} r_i(\bar{z}_{i.} - \bar{z}_{..})^2/(t-1)}{\sum\limits_{i=1}^{t}\sum\limits_{j=1}^{r_i}(z_{ij} - \bar{z}_{i.})^2/(N-t)} \tag{4.2}$$

The null hypothesis of homogeneous variances, H_0: $\sigma_1^2 = \sigma_2^2 = \cdots = \sigma_t^2$, is rejected if $F_0 > F_{\alpha,(t-1),(N-t)}$. The test statistic in Equation (4.2) is a modification of the original test introduced by Levene (1960). The modification suggested by Brown and Forsythe (1974) was a substitution of the median, $\tilde{y}_i$, for the mean, $\bar{y}_{i.}$, in the calculation of z_{ij}. The test calculations are illustrated in Table 4.3 with five observations from each of the first three sites of the Hermit crab study. The values required for the Levene (Med) test and computed from the complete data set are $MST = 14,229$; $MSE = 4,860$; and $F_0 = 14,229/4,860 = 2.93$. It can be concluded that the variances are different among the sites, since the null hypothesis is rejected with critical region $F_0 > F_{.05,5,144} = 2.28$.

Table 4.3 Illustration of the Levene (Med) test for homogeneous variances with five observations from each of three sites in the Hermit crab study

Site	y_{ij}	$\tilde{y}_i$	$z_{ij} = \|y_{ij} - \tilde{y}_i\|$	$\bar{z}_{i.}$
1	0	3	3	
	0		3	
	22		19	
	3		0	
	17		14	7.8
2	415	14	401	
	466		452	
	6		8	
	14		0	
	12		2	172.6
3	0	4	4	
	0		4	
	4		0	
	13		9	
	5		1	3.6

$$\bar{z}_{..} = 61.3$$

$$r_1 = r_2 = r_3 = r = 5, t = 3, N = 15$$

$$SST = \sum_{i=1}^{t} r_i(\bar{z}_{i.} - \bar{z}_{..})^2 = 92{,}896$$

$$SSE = \sum_{i=1}^{t} \sum_{j=1}^{r_i} (z_{ij} - \bar{z}_{i.})^2 = 216{,}539$$

$$F_0 = \frac{MST}{MSE} = \frac{92{,}896}{2} \bigg/ \frac{216{,}539}{12} = 2.57$$

F Max Test

Several formal statistical tests are valid tests for homogeneity of variances in completely randomized designs when sample sizes are equal and the observations are normally distributed. One of the simplest to compute is the *F Max* test statistic (Hartley, 1950). The null hypothesis tested with the *F Max* statistic is

$$H_0: \sigma_1^2 = \sigma_2^2 = \cdots = \sigma_t^2 \tag{4.3}$$

with the alternative hypothesis that some variances differ.

The test statistic is computed as the ratio of the largest observed variance to the smallest observed variance within the treatment groups, or

$$F_0 \; Max = \frac{\max(s_i^2)}{\min(s_i^2)} \tag{4.4}$$

where $\max(s_i^2)$ and $\min(s_i^2)$ are the largest and smallest, respectively, within treatment group variances.

The null hypothesis is rejected at the α level of significance if $F_0 \; Max > F_\alpha \; Max$, where $F_\alpha \; Max$ is the value of the $F \; Max$ variable exceeded with probability α for t treatment groups and $\nu = (r - 1)$ degrees of freedom for each s_i^2. Critical values for the $F \; Max$ statistic are found in Appendix Table VIII.

The $F \; Max$ test ordinarily would not be conducted for the crab study since the Hermit crab counts do not have a normal distribution. The test is conducted here only for illustration. From Table 4.1 the largest and smallest standard deviations are found in sites 2 and 4, respectively. The variances for these groups are $s_4^2 = 17.39^2 = 302$ and $s_2^2 = 125.35^2 = 15{,}712$. The value of the test statistic is

$$F_0 \; Max = \frac{15{,}712}{302} = 52.03$$

The critical value for $\alpha = .05$, $t = 6$, and $\nu = 24$ interpolated in Appendix Table VIII is $F_{.05} \; Max = 3.24$. The null hypothesis of homogeneous variances is rejected.

4.4 Looking for Outliers with the Residuals

Extremely large positive or negative values of the residuals will be far removed from the straight line indicator of the normal plot or far removed from the other values in the upper or lower boundaries of the residuals versus estimated means plot. The outlier potentially can affect the statistical inference because it inflates the estimate of experimental error variance and influences the estimate of a treatment mean. Outliers can be the result of errors in collecting and recording data, of mistakes in technique, or of a special combination of treatment and environment. Prior to discarding outliers it is wise to investigate the cause to avoid loss of valuable information.

Standardized Residuals

The standardized residuals are computed as

$$w_{ij} = \frac{\widehat{e}_{ij}}{\sqrt{MSE}} \tag{4.5}$$

The *standardized residual* (w_{ij}) is useful for making quick checks on the presence of outliers. The w_{ij} have approximately a standard normal distribution if the e_{ij} have a normal distribution. A residual would be considered an outlier if the standardized value fell outside the ± 3 or ± 4 limits since the probability of a standard normal value greater than 3 or 4 standard deviations from the mean of 0 is quite small.

The largest standardized residual for the Hermit crab counts from the 150th-ordered residual in Table 4.2 is

$$w = \frac{397.28}{\sqrt{5170}} = 5.53$$

The standardized residuals computed from the 148th- and 149th-ordered residuals in Table 4.2 are $4.82 = 346.28/\sqrt{5170}$ and $4.96 = 356.36/\sqrt{5170}$, respectively. The three maximum residuals are more than 4 standard deviations from a mean of 0. The normal probability plot and the residuals versus estimated values plot indicate the possibility of other outlying values. The outlying values derive from large counts of Hermit crabs. In this case, the biologist would want to ascertain the conditions on those particular transects that would cause the exceptionally high counts.

Standardized residuals are used because ordinary residuals for diagnostic purposes have some disadvantages. Estimated residuals in the same treatment group are not independent of one another, and their variances are heterogeneous from group to group with unequal replications. The variance of an estimated residual is $\sigma_{\hat{e}_i}^2 = \sigma^2(1 - 1/r_i)$ for the completely randomized design with r_i replications for the ith treatment.

The diagnostic plots with unequal replication can be influenced by the heterogeneous variances of the ordinary residuals. The plots are also affected by the correlations among the residuals regardless of the replication numbers. The correlations among the residuals tend to make the residuals exhibit more agreement with the normal distribution in the probability plot. See Cook and Weisberg (1982) for an extensive coverage of residuals beyond the scope and intent of this book.

Studentized Residuals

The heterogeneous variances of the ordinary residuals can be corrected by utilizing the *studentized residuals*. The studentized residual for the completely randomized design is the ordinary residual divided by its estimated standard deviation

$$\tilde{e}_{ij} = \frac{\hat{e}_{ij}}{\sqrt{MSE(1 - 1/r_i)}}$$

The studentized residuals have a constant variance $\sigma_{\tilde{e}_{ij}}^2 = 1$, but they are still correlated. They are recommended in place of ordinary residuals for residual plots from studies with unequal replication numbers.

The residual plots and Levene (Med) test provide good evidence that the assumptions of homogeneous variances and normal distribution are not appropriate for the Hermit crab data. Some observations are good candidates for outliers. If the departure from the assumptions, based on our judgments, is not too great, then the consequences of ignoring the lack of compliance are not severe. When there are serious departures from the assumptions, as in the Hermit crab study, some decision must be made for the next step in the analysis of the study. The topics in the next section address some possible solutions.

4.5 Variance-Stabilizing Transformations for Data with Known Distributions

Transformations are used to change the scale of observations so that they conform more closely to the assumptions of the linear model and provide more valid inferences from the analysis of variance. The probabilities of statistical inference apply only to the new scale of measurement; significance levels and averages do not apply to the original measurements. When transformations are necessary, a common practice is to conduct the analysis and make all inferences on the transformed scale but present summary means tables on the original measurement scale.

Bartlett (1947) summarized many aspects of transformations in the analysis of variance. In this section, several transformations are discussed for data that are not normally distributed but have a known probability distribution. A transformation based on an empirical relationship between the sample means and variances is discussed in Section 4.6 for data with unknown distribution.

Poisson Distribution

Observations on counts of plants in quadrats, insects on plants, bacterial colonies on plates, blemishes on a surface, and accidents per unit of time may have the Poisson distribution for which the mean is equal to the variance $\mu_y = \sigma_y^2$. The *square root* transformation is recommended to stabilize the variances for observations from the Poisson distribution. The square root transformation $x = \sqrt{y}$ will have a constant variance $\sigma_x^2 = 0.25$ for all values of μ_x. If the mean is small, say $\mu_y < 3$, then the transformation $x = \sqrt{y + \frac{3}{8}}$ is superior to $\sqrt{y}$ for stabilizing the variances (Anscombe, 1948). The correction is unnecessary if the counts are all large.

Binomial Distribution

Observations on the number of successes in n independent trials follow the binomial distribution. Examples include proportions of defective items in manufactured lots, the proportion of germinated seeds, the proportion of surviving larvae in insect studies, and the proportion of flowering plants in a transect. The estimated binomial probability is $\hat{\pi} = y/n$, where y is the number of successes in n independent trials with probability of success π. The mean and variance of the estimated binomial probability are $\mu = \pi$ and $\sigma^2 = \pi(1 - \pi)/n$, respectively, and there will be a well-defined relationship between the observed proportions and the variances in the observed data.

The *arcsin* or *angular* transformation is recommended to stabilize the variances for observations from the binomial distribution. The arcsin transformation, $x = \sin^{-1}\sqrt{\hat{\pi}}$, has a constant variance, $\sigma_x^2 = 1/4n$, for all π if the angle is expressed in radians and $\sigma_x^2 = 821/n$ if the angle is expressed in degrees. If n is small, say $n < 50$, then Anscombe (1948) recommends the substitution of $\hat{\pi}^* = (y + \frac{3}{8})/(n + \frac{3}{4})$ for $\hat{\pi} = y/n$ in the transformation. If all the observed proportions in the study are between $\hat{\pi} = 0.3$ and $\hat{\pi} = 0.7$, the binomial variance is relatively stable and the transformation is probably not necessary.

Probits and Logits

Two other transformations related to the binomial distribution are used most frequently in biological assays. The *probit* transformation is the value of the standard normal distribution that corresponds to a cumulative probability $\hat{\pi} = y/n$. The *logit* transformation is the natural logarithm of the ratio $\hat{\pi}/(1 - \hat{\pi})$ used in biological assays and the analysis of survival data. Although both of these transformations result in an amenable statistical procedure for their intended purposes the variances are not stabilized and other models must be utilized for the analysis. One such model is discussed briefly in Section 4.7. Details for the use of these transformations may be found in Cox (1970), Finney (1978), McCullagh and Nelder (1989), and Collett (1991).

Negative Binomial

Increases in the number of individuals counted can be related to the number of individuals already present; the Poisson distribution then is no longer applicable to the problem. The counted individuals tend to occur in clusters in a *contagious* distribution. Animals infected with the same disease organism, plants of the same species, or insects of the same species often occur in clusters as a result of the biological mechanisms for reproduction or disease transmission. A probability distribution frequently used for these data is the *negative binomial* with a mean μ_y and variance $\sigma_y^2 = \mu_y + \lambda^2 \mu_y^2$; the variance increases with the mean at a rate greater than with the Poisson distribution. A suggested transformation for stabilizing the variance is the inverse hyperbolic sine transformation $x = \lambda^{-1}\sinh^{-1}\sqrt{y}$. The variance of the transformed observations is $\sigma_x^2 = 0.25$. The transformation requires some knowledge of λ. A substitute transformation, $x = \log(y + 1)$, has an approximate linear relationship with the $\sinh^{-1}$ transformation (Bartlett, 1947).

4.6 Power Transformations to Stabilize Variances

The distribution of the observations cannot always be determined on the basis of sampling properties for the random variable. Under these circumstances, a transformation can be determined on the basis of an empirical relationship between the standard deviation and the mean.

Empirical Data Transformation

The power transformation alters the symmetry or asymmetry of the frequency distribution of the observations. The transformations are based on work by Box and Cox (1964) in which the standard deviation of y is supposed proportional to some power of the mean, or

$$\sigma_y \propto \mu^\beta \tag{4.6}$$

A power transformation of the observations

$$x = y^p \tag{4.7}$$

results in a standard deviation to mean proportional relationship

$$\sigma_x \propto \mu^{p+\beta-1} \tag{4.8}$$

If $p = 1 - \beta$, then the standard deviation of the transformed variable x will be constant since $p + \beta - 1 = 0$ and $\sigma_x \propto \mu^0$ in Equation (4.8).

The transformations are frequently represented as a *ladder of powers*, a phrase originating in exploratory data analysis (Tukey, 1977; Velleman & Hoaglin, 1981). Display 4.2 shows the order of the ladder of powers for some of the more useful transformations.

Display 4.2	**Transformations in the Ladder of Powers, $x = y^p$**		
p	*y^p*	*Name*	*Remarks*
2	y^2	Square	Highest usually used
1	y^1	Raw data	No transformation
$\frac{1}{2}$	$\sqrt{y}$	Square root	Poisson distribution
0	$\log(y)$	Logarithm	Holds "0" place in ladder
$-\frac{1}{2}$	$1/\sqrt{y}$	Reciprocal square root	Minus sign preserves order of observations
-1	$1/y$	Reciprocal	Reexpress time to rate

Values of p less than 1 will pull in the stretched-out upper end of the observations and stretch out the bunched-in lower end of the observations in a distribution skewed to the right. Conversely, if p is greater than 1 a left-skewed distribution becomes more symmetric by pulling in the stretched-out lower observations. The log transformation is placed at the "0" position in the ladder because its effect on the observations falls naturally in that position.

The reciprocal transformation with $p = -1$ can be useful in studies that require measurement of time to the occurrence of an event. The reciprocal of time can be roughly viewed as the rate at which a subject of the investigation arrived at the event. It is tempting to assign a value of 0 to subjects for which the event never occurs; however, care must be taken since the event was never observed. The observation may better be treated as either a member of a truncated set of observations or missing observations, depending on the circumstances.

An Empirical Estimate for the Power Transformation

An empirical estimate of p can be determined if estimates are available for the mean and standard deviation of the treatment groups. Express the relationship between the standard deviation and the mean of the ith treatment group as $\sigma_i = \alpha \mu_i^\beta$ with α as a constant of proportionality. Take the logarithm of the expression to obtain

$$\log(\sigma_i) = \log(\alpha) + \beta \log(\mu_i) \tag{4.9}$$

A plot of $\log(\sigma_i)$ versus $\log(\mu_i)$ is a straight line with intercept $\log(\alpha)$ and slope β. Estimates of the means and standard deviations can be substituted for σ_i and μ_i, and the estimate of β can be obtained from a simple linear regression analysis. The value of p for a variance-stabilizing transformation can be taken as $\widehat{p} = 1 - \widehat{\beta}$, where $\widehat{\beta}$ is the estimated slope for Equation (4.9). The determination of an empirical power transformation and the effects of the transformation are illustrated with the Hermit crab data of Example 4.1.

A plot of the logarithms of the six site standard deviations against the logarithms of the six site means from Table 4.1 is shown in Figure 4.4. The estimate of β from the regression results shown in the plot is $\widehat{\beta} = 0.99$. The estimated empirical value for p is $\widehat{p} = 1 - 0.99 = 0.01$. The value is very close to the zero position in the ladder of powers, implying a logarithmic transformation for the Hermit crab data.

When there are some zero counts among the observations, a small constant c, such that $0 < c \leq 1$, is added to the observed count y to avoid evaluation of a logarithm for 0. The values of $c = \frac{1}{2}$ or 1 are frequently used; a value of $c = \frac{1}{6}$ is suggested by Mosteller and Tukey (1977).

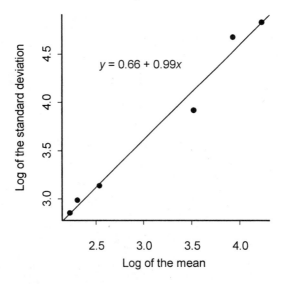

Figure 4.4 Plot and regression estimates of $\log(s_i)$ versus $\log(\bar{y}_{i.})$ from the six sites for the Hermit crab data

The Hermit crab data were transformed as $x = \log(y + \frac{1}{6})$ because there were counts of zero Hermit crabs in some transects. Summary statistics for the transformed data are shown in Table 4.4. The normal probability plot of the residuals is shown in Figure 4.5; the plot of the residuals versus the estimated treatment means is shown in Figure 4.6; and the spread-location plot is shown in Figure 4.7.

Table 4.4 Means, medians, standard deviations, minimum and maximum values, and maximum standardized residuals values, w_{ij}, for each site after the transformation, $x = \log(y + \frac{1}{6})$, for the Hermit crab data

Site	Mean	Median	Standard Deviation	Minimum	Maximum	w_{ij}
1	0.94	1.24	0.99	− 0.78	2.37	2.51
2	1.00	1.01	1.06	− 0.78	2.67	2.83
3	0.76	0.71	1.05	− 0.78	2.61	2.77
4	0.39	0.34	0.81	− 0.78	1.92	2.03
5	0.44	0.34	0.76	− 0.78	1.97	2.09
6	0.36	0.62	0.96	− 0.78	1.98	2.10
			$MSE = 0.8888$			

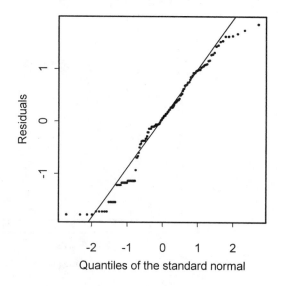

Figure 4.5   Normal probability plot of the residuals after transformation $x = \log(y + \frac{1}{6})$ for the Hermit crab data

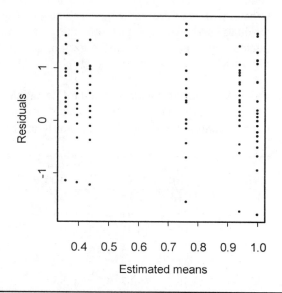

Figure 4.6   Plot of residuals versus estimated treatment means after transformation, $x = \log(y + \frac{1}{6})$, for the Hermit crab data

No outliers are evident after transformation since all of the maximum values for the standardized residuals are less than 3 standard deviations from the mean (Table 4.4). Several values diverge from the straight line of the normal probability plot in Figure 4.5, but the appearance is much improved over that of the same plot for the original data in Figure 4.1.

The dispersion of residuals in Figure 4.6 is quite similar for each of the sites, and the spread-location plot in Figure 4.7 shows no monotonic increase or decrease in the $\sqrt{|\widehat{e}_{ij}|}$, indicating relatively homogeneous variances. The maximum and minimum site variances from sites 2 and 5 are $s_2^2 = 1.06^2 = 1.1236$ and $s_5^2 = 0.76^2 = 0.5776$, respectively (Table 4.4). The computed statistics for the Levene (Med) test were $MST = 1.44$ and $MSE = 1.91$ with $F_0 = 0.75$. The null hypothesis of homogeneous variances with a critical value of $F_{.05,5,144} = 2.28$ was not rejected. The assumptions required for the analysis with the linear model appear to hold sufficiently well for the observations after transformation.

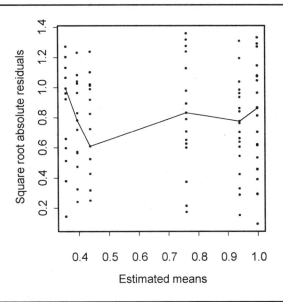

Figure 4.7 Spread-location plot for the residuals after transformation, $x = \log(y + \frac{1}{6})$, for the Hermit crab data

The Box–Cox Power Transformation for Other Designs

The regression method used in this section to estimate p for the power transformation $x = y^p$ is only effective for completely randomized designs in which group means and group standard deviations can be estimated. The estimation of p with more complex experiment designs with blocking that are encountered in later chapters requires a somewhat more rigorous approach.

The original Box–Cox transformation is $x = (y^p - 1)/p$ for $p \neq 1$ and $x = \log_e y$ for $p = 0$ (Box & Cox, 1964). The estimate of p can be found by maximizing

$$L(p) = -\frac{1}{2} \log_e [MSE(p)]$$

where $MSE(p)$ is the mean square for error from the analysis of variance using the transformation $x = (y^p - 1)/p$ for the given choice of p.

A solution can be obtained by determining $MSE(p)$ for a range of chosen values of p, plotting $MSE(p)$ against p, and reading the value where the minimum $MSE(p)$ is found. The exact value for p by this method is unlikely to be used. It is more common to use the standard transformations shown in Display 4.2, which approximate the values of the estimated p.

4.7 Generalizing the Linear Model

The linear model used in this book assumes the experimental errors have homogeneous or constant variance throughout and that their distribution is well approximated by the normal distribution. Frequently, we find in our studies that natural phenomena behave linearly over the range of interest and have errors that are homogeneous and normally distributed. If not, we often find that some suitable transformation of the response variable will provide the required linearity and error structure. This ability to use the linear model and analysis of variance so often has led to their popularity in scientific investigations. However, transformations can lead to unsatisfactory scales for interpretation (for example, arcsin of the square root).

The *generalized linear model* is a general class of linear models introduced by Nelder and Wedderburn (1972) that broadens the available variety of probabilistic models for experimental errors and forms of nonlinearity in the model. The linear model used throughout this book with normally distributed and homogeneous errors is a subset of this generalized form.

The generalized linear model introduces separate functions to allow for heterogeneous variances and nonlinearity. Rather than transforming the response variable, the generalized linear model may be better described as reexpressing the model. Dobson (1990) provides a compact introduction to the use of and applications for the generalized linear model, while a thorough coverage of the models may be found in McCullagh and Nelder (1989).

Recall linear models introduced in Section 2.4 for the observed response y, which is a realization of the random variable Y with expectation $E(Y) = \mu$, where $\mu = \beta_0 + \beta_1 x_1 + \cdots + \beta_p x_p$ is linear in the β parameters. The x variables represent variables controlled in the study or measured as covariates of the response y. The requisite assumptions are a constant σ^2 for Y and the linear relationship between μ and the x_i. However, for many types of data a change in the mean of Y introduces a change in the variance. Examples include the binary response (0 or 1) with probability of success π for the binomial distribution, for which $E(Y) = \pi$ and

$\sigma^2(Y) = \pi(1 - \pi)/n$, and count data for the Poisson distribution, for which the mean is equal to the variance.

The generalized linear model handles these issues naturally by introducing a reparametrization to allow heterogeneous variance by introducing a link function $\eta = g(\mu)$, such that $\eta = g(\mu) = \beta_0 + \beta_1 x_1 + \cdots + \beta_p x_p$ is now linear in the β parameters rather than μ. For example, the link function for binary responses with $\mu = \pi$ is the *logit* link

$$\eta = g(\mu) = \log\left(\frac{\mu}{1 - \mu}\right)$$

so that

$$\mu = \frac{e^\eta}{1 + e^\eta}$$

The range of $\mu = \pi$ is still in the interval $[0, 1]$, but $\eta = g(\mu)$ can take any real value. With the logit link function, it is the logit that is linear in the β parameters rather than $\mu = \pi$.

The natural link function for the Poisson distribution is the *log* link with $\eta = \log(\mu)$ so that $\log(\mu)$ is linearly related to the x_i. The *probit* link is another link for binary data, which is widely used for biological assays. The *identity* link $\eta = \mu$ does no reexpression and results in the usual linear model with homogeneous variances and normally distributed errors.

The emergence of readily available software has popularized the use of particular links associated with the generalized linear model, such as logistic regression with the logit link and log linear models or Poisson regression with the log link. Collett (1991) provides comprehensive coverage on the use and application of links for binary data.

Estimation and analysis with generalized linear models is based on maximum likelihood estimation methods. Coverage of the model estimation and other issues associated with generalized linear models is beyond the intent and scope of this book. Regardless, whichever model is used for statistical inference, good statistical design is fundamental to valid statistical inference.

4.8 Model Evaluation with Residual–Fitted Spread Plots

A graphic method, the *residual–fitted spread* plot, helps evaluate how well the hypothesized linear models fit the data. It has been included here as an addendum to the discussions on the evaluation of assumptions about homogeneous and normally distributed experimental errors with the residuals. Cleveland (1993) covers other useful graphic methods that can be used for exploratory and confirmatory data analysis.

The *residual–fitted spread* plot, or *r–f spread* plot, provides a graphic portrayal of the relative variation or *spread* in the experimental errors and the *fitted values*[3] of the linear model. In the case of the completely randomized design the fitted values are the estimated treatment group means. The *r–f* spread plot provides a visual companion to the ratio $(SSE_r - SSE_f)/SSE_f$, introduced in Section 2.10 as a prelude to the F test for differences among treatment group means. The ratio provides a means of assessing the relative improvement of the full model over the reduced model for the data, wherein $(SSE_r - SSE_f)$ reflects the variation attributable to the components estimated or fit in the model, and SSE_f reflects the variation in experimental error or residual variation after the model has been fit to the data.

The *r–f* spread plot includes one plot of the sorted residuals versus their cumulative frequency and one plot of the sorted fitted values *minus their mean* versus their cumulative frequency. The cumulative frequency, or the *f-value*, is the same, $f_i = (i - 0.5)/N$, $i = 1, 2, \ldots, N$, used to obtain normal quantiles for the normal probability plots introduced in Section 4.3.

The calculated ratio for the meat storage experiment in Chapter 2 with $SSE_r = 33.7996$ and $SSE_f = 0.9268$ is $(33.7996 - 0.9268)/0.9268 = 35.47$, indicating the sum of squares for the fitted-values treatment group means was 35.47 times larger than the sum of squares for the residuals. The *r–f* spread plot for the meat storage experiment is shown in Figure 4.8.

The larger spread of the fitted values relative to the spread of the residuals (Figure 4.8) reflects the large value for the ratio of their respective sums of squares. Too often statistical significance does not give a true indication of whether the treatments produce meaningful physical or biological differences. The *r–f* spread plot provides some visual evaluation of the physical differences among the treatment groups relative to the leftover residual variability, which can be used as a supplement to the formal F test in the analysis of variance. Given the differential spreads of the fitted and residual values for the meat storage experiment, one may conclude that relative to the experimental errors some of the treatments did produce meaningful biological differences among their respective means.

The transformed Hermit crab data provide a striking contrast to the meat storage study using the *r–f* spread plot. The analysis of variance for the transformed data is given in Table 4.5. An F test rejects the hypothesis of equal site means with $F_0 = 2.06/0.89 = 2.31$ and a P-value of .046. The ratio $(SSE_r - SSE_f)/SSE_f = 10.32/127.99 = 0.08$ indicates a small amount of variation among the fitted treatment means relative to the residuals. This conclusion is confirmed visually with the *r–f* spread plot in Figure 4.9. Although the site means are judged to be different by the F test, it is quite apparent that the residual variation within each of the sites is considerable relative to the differences among the averages of the transformed counts for the sites. Thus, the significance of the biological differences among the sites may be negligible.

---

[3] Fitted value stems from the notion that we "fit" a model to the data and decompose the observation y_{ij} into two parts as $y_{ij} = \widehat{\mu}_i + \widehat{e}_{ij}$ in the completely randomized design, where the fitted value is $\widehat{\mu}_i$ and the residual is $\widehat{e}_{ij}$.

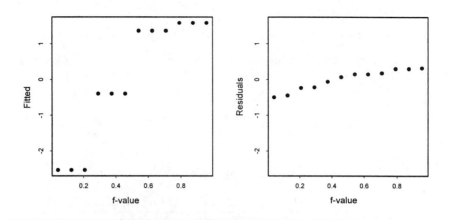

Figure 4.8 Residual–fitted spread plot to compare the spreads of the residuals and the fitted values minus their means for the meat storage experiment in Chapter 2

Table 4.5 Analysis of variance for Hermit crab data transformed by $x = \log(y + \frac{1}{6})$

Source of Variation	Degrees of Freedom	Sum of Squares	Mean Square
Total	149	138.31	
Sites	5	10.32	2.06
Error	144	127.99	0.89

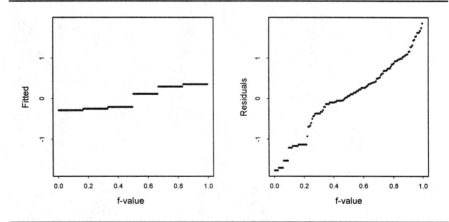

Figure 4.9 Residual–fitted spread plot to compare the spreads of the residuals and the fitted values minus their means after transformation, $x = \log(y + \frac{1}{6})$, for the Hermit crab data

EXERCISES FOR CHAPTER 4

1. A temperature-accelerated life test was performed on a type of sheathed tubular heater. Six heaters were tested at each of four temperatures: 1520°F, 1620°F, 1660°F, and 1708°F. The number of hours to failure was recorded for each of the 24 heaters in the study.

Test Temperature	Hours to Failure
1520°	1953, 2135, 2471, 4727, 6134, 6314
1620°	1190, 1286, 1550, 2125, 2557, 2845
1660°	651, 837, 848, 1038, 1361, 1543
1708°	511, 651, 651, 652, 688, 729

Source: W. Nelson (1972), A short life test for comparing a sample with previous accelerated test results. Technometrics 14, 175–185.

a. Investigate the assumptions necessary for an analysis of variance of the data. Include a normal probability plot of the residuals and a spread-location plot.
b. Determine a reasonable transformation from the ladder of powers, using the slope of the regression line based on the logs of the standard deviations and the group means.
c. Determine if your choice for the transformation resulted in data that reasonably agree with the assumptions necessary for the analysis of variance.
d. Conduct an analysis of variance of the transformed data, and partition the sum of squares for temperature into orthogonal polynomial contrasts to determine the best relationship between temperature and your response variable. Since test temperatures were unequally spaced, use the following contrast coefficients:

Temperature	1520	1620	1660	1708
Linear	− 0.773	− 0.051	0.238	0.585
Quadratic	0.382	− 0.637	− 0.328	0.583
Cubic	− 0.078	0.584	− 0.765	0.259

2. An entomologist counted the number of eggs laid by female moths on successive days in three strains of tobacco budworm (USDA, Field, and Resistant) from each of 15 matings. The data that follow are the number of eggs laid on the third day after the mating for each female in each of the strains.

Strain	Number of Eggs per Moth
USDA	448, 906, 28, 277, 634, 48, 369, 137, 29, 522, 319, 242, 261, 566, 734
Field	211, 276, 415, 787, 18, 118, 1, 151, 0, 253, 61, 0, 275, 0, 153
Resistant	0, 9, 143, 1, 26, 127, 161, 294, 0, 348, 0, 14, 21, 0, 218

Source: Dr. T. Watson and S. Kelly, Department of Entomology, University of Arizona.

a. The entomologist wants to conduct an analysis of variance on the egg counts. What probability distribution is appropriate for the data?
b. What is the suggested transformation for the probability distribution you named in (a)?
c. Determine a reasonable transformation from the ladder of powers, using the slope of the regression line based on the logs of the standard deviations and the group means. Does the transformation arrived at by this method agree with your suggestion in part (b)?
d. Transform the data with your choice of transformation, and determine if the transformed data agree with the analysis of variance assumptions.

3. A plant breeder evaluated the rooting capability of nine bermuda grass clones in a laboratory experiment. Two replications of each clone were grown in an aerated growth solution in a completely randomized design. The number of nodes that rooted on the stolons of each clone follow.

Clone	Replication 1		Replication 2	
	Rooted	Not Rooted	Rooted	Not Rooted
1	15	49	11	53
2	13	51	11	53
3	13	51	6	58
4	6	42	4	60
5	16	48	12	52
6	14	50	9	55
7	8	56	18	46
8	9	55	10	54
9	8	40	16	48

Source: Dr. W. Kneebone, Department of Plant Sciences, University of Arizona.

a. The plant breeder wants to analyze the proportion of rooted stolons or proportion of rooted nodes. What probability distribution is appropriate for the data?
b. What is the suggested transformation for the probability distribution you named in part (a)?
c. Transform the data for the proportion of rooted stolons (or nodes) with the appropriate transformation, and conduct the analysis of variance on the transformed data.
d. Use the multiple comparisons with the best procedure to select the subset of clones with the largest means and $P(CS) = 0.95$.

4. The Ames *Salmonella*/microsome assay is used to investigate the potential of environmental toxic substances for their ability to effect heritable change in genetic material. The compound 4-nitro-ortho-phenylenediamine (4NoP) was tested with strain TA98 *Salmonella*. The number of visible colonies was counted on plates dosed with 4NoP. Five dose levels of 4NoP were used in this study. The colony counts for seven of the plates at each dose level are shown.

Dose (μg/plate)	Colony Counts
0.0	11, 14, 15, 17, 18, 21, 25
0.3	39, 43, 46, 50, 52, 61, 67
1.0	88, 92, 104, 113, 119, 120, 130
3.0	222, 251, 259, 283, 299, 312, 337
10.0	562, 604, 689, 702, 710, 739, 786

Source: B. H. Margolin, B. S. Kim, and K. J. Risko (1989), The Ames *Salmonella*/microsome mutagenicity assay: Issues of inference and validation. *Journal of the American Statistical Association* 84, 651–661.

a. Since the data involve counts of bacterial colonies, can you safely assume the data have a Poisson distribution? Explain your answer.

b. The authors of the cited article suggest the negative binomial distribution as a plausible distribution. Do you agree with this conclusion? Explain your answer.

c. Determine a transformation for the data such that the analysis of variance assumptions are sufficiently satisfied by the transformed data. Conduct an analysis of variance for the transformed data.

5. Given the following random sample of $N = 15$ observations that have been ordered from smallest to largest

   14.3, 16.0, 17.3, 17.5, 17.8, 18.7, 18.8, 18.9, 20.0, 20.8, 21.4, 22.7, 23.2, 25.6, 27.8

   a. Determine the f-values and their standard normal quantiles.
   b. Plot the observations versus the standard normal quantiles.
   c. Interpret the plot relative to the form of distribution from which the observations were sampled.

6. Given the following random sample of $N = 16$ observations that have been ordered from smallest to largest

   2, 3, 4, 5, 10, 28, 34, 35, 39, 63, 87, 97, 112, 156, 188, 253

   a. Determine the f-values and their standard normal quantiles.
   b. Plot the observations versus the standard normal quantiles.
   c. Interpret the plot relative to the form of distribution from which the observations were sampled.

4A   Appendix: Data for Example 4.1

Hermit crab counts in coastline sites. A marine biologist counted Hermit crabs on 25 transects in each of six different coastline sites. The number of Hermit crabs counted on each of the transects follows.

Site	Counts
1	0, 0, 22, 3, 17, 0, 0, 7, 11, 11, 73, 33, 0, 65, 13, 44, 20, 27, 48, 104, 233, 81, 22, 9, 2
2	415, 466, 6, 14, 12, 0, 3, 1, 16, 55, 142, 10, 2, 145, 6, 4, 5, 124, 24, 204, 0, 0, 56, 0, 8
3	0, 0, 4, 13, 5, 1, 1, 4, 4, 36, 407, 0, 0, 18, 4, 14, 0, 24, 52, 314, 245, 107, 5, 6, 2
4	0, 0, 0, 4, 2, 2, 5, 4, 2, 1, 0, 12, 1, 30, 0, 3, 28, 2, 21, 8, 82, 12, 10, 2, 0
5	0, 1, 1, 2, 2, 1, 2, 29, 2, 2, 0, 13, 0, 19, 1, 3, 26, 30, 5, 4, 94, 1, 9, 3, 0
6	0, 0, 0, 2, 3, 0, 0, 4, 0, 5, 4, 22, 0, 64, 4, 4, 43, 3, 16, 19, 95, 6, 22, 0, 0

5 Experiments to Study Variances

In this chapter, the statistical model for research studies about the variances of populations is introduced. Knowledge about the assignable causes of variation is useful for improving manufacturing processes, improving the genetics of crops and livestock, enhancing quality control in the health industry, and designing research studies. The objective is to decompose the total variance into identifiable components.

5.1 Random Effects Models for Variances

The meat storage experiment from Chapters 2 and 3 included four specific treatments with no expressed interest in any other packagings for the experiment. Thus, the complete treatment population of interest consisted of the four packaging methods.

Each of the four packaging methods could be duplicated if the experiment was repeated. Under these circumstances, the statistical models used for the studies are referred to as **fixed effects** models, and the inferences are restricted to the particular set of treatments in the study.

There are other types of research studies in which we want to identify the major sources of variability in a system and estimate their variances. By nature of (1) the research objectives, (2) the treatment structure, (3) the experimental protocols, and (4) the type of inferences made from the observed results, the effects in the model are considered to be random effects, and the statistical models are referred to as **random effects** models. The following example illustrates a system in which knowledge about variability in identifiable components of an industrial process can be used to improve the process product.

Example 5.1 Castings of High Temperature Alloys

A metal alloy is produced in a high-temperature casting process. Each casting is broken down into smaller individual bars that are used in applications requiring small amounts of the alloy. The tensile strength of the alloy is critical to its intended future use.

The casting process is designed to produce bars with an average tensile strength above minimum specifications. Some variation in tensile strength among the bars is acceptable when only a small proportion of bars do not meet specifications (Figure 5.1(a)). However, excessive variation results in an unacceptable proportion of bars that do not meet specifications (Figure 5.1(b)).

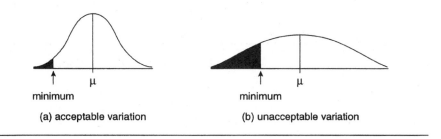

Figure 5.1   Acceptable (a) and unacceptable (b) variation in tensile strength

Two components contribute to the total variation in tensile strength of the manufactured bars: variability among fabrication castings and inconsistencies within the casting process that affect bars from the same casting. Maintaining control over the variation requires knowledge of the variability contributions by each part of the process.

An experiment was planned to isolate the variation in tensile strength due to the effects of different castings from that attributable to inconsistencies within the same casting.

High-temperature castings of the alloy were taken from three randomly selected fabrications conducted in the same facility. Each casting was broken down into individual bars. Destructive tensile strength measurements were obtained on a random sample of 10 bars from each of three castings. The tensile strength data for each of the 30 bars in pounds per square inch (psi) are given in Table 5.1.

The three castings used in the study represent a sample of the potential population of castings that could be produced in the facility. The investigators were interested in the variation in tensile strength among castings produced by the facility; thus, the concern was not with the three specific castings in the experiment.

The investigators considered the castings only a random sample of three from a population of castings produced by the facility. The effects of the castings will be random effects since they are randomly selected from a potentially infinite population of castings. The inferences will extend to the population of castings that

Table 5.1 Tensile strengths (psi) of bars from three separate castings of a high-temperature alloy

	Casting	
1	*2*	*3*
88.0	85.9	94.2
88.0	88.6	91.5
94.8	90.0	92.0
90.0	87.1	96.5
93.0	85.6	95.6
89.0	86.0	93.8
86.0	91.0	92.5
92.9	89.6	93.2
89.0	93.0	96.2
93.0	87.5	92.5

Source: G. J. Hahn and T. E. Raghunathan (1988), Combining information from various sources: A prediction problem and other industrial applications. *Technometrics* 30, 41–52.

conceivably could be produced in the facility. Likewise, the individual bars are a random sample of bars possible from a single casting, and their effects on tensile strength are random effects.

The observed tensile strength of a particular bar (y) differs from the mean of the process (μ) by some overall error, $\delta = y - \mu$. The components of the overall error are illustrated in Figure 5.2. The overall error is the sum of two components, $\delta = \delta_c + \delta_b$, where δ_c is the error component for castings and δ_b is the error component for bars in a casting.

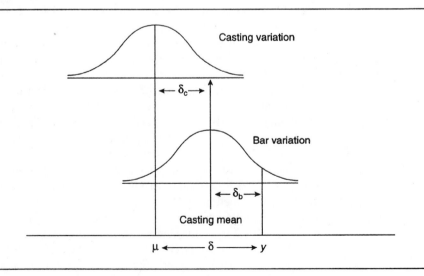

Figure 5.2 Two components of error in the metal casting process

Other Examples: Another typical study with random effects involves the inheritance of quantitative traits, such as grain yield in cultivated plant species. Many genetically distinct families of the crop are developed in a plant-breeding program. The families represent a random sample of potential families that can be developed by the plant breeder. The progeny of each family are regarded as a random sample of the progeny possible from the family. The plant breeder wants to partition the total variation into the separate contributions from the families and the progeny.

Clinical medical laboratories routinely participate in interlaboratory studies on the variability of assay results requiring random effects models for the statistical analysis. At regular intervals samples from a large homogeneous pool of serum are sent to a large number of laboratories for analysis. The participating laboratories and the samples sent to them represent a random sample of the potential populations of laboratories and serum samples. The investigators want to know if there is significant variation in assay results among the laboratories.

5.2   A Statistical Model for Variance Components

A suitable model to identify the sources of variation for the random effects in the experiment on casting high-temperature alloys is

$$y_{ij} = \mu + a_i + e_{ij} \tag{5.1}$$

$$i = 1, 2, \ldots, t \qquad j = 1, 2, \ldots, r$$

where μ is the process mean, the a_i are the random casting effects, and the e_{ij} are the random error effects for bars within castings. The effects e_{ij} and a_i are assumed to be independent of one another.

The e_{ij} error effects are assumed to be a random sample from a population with a mean of 0 and variance σ_e^2. The random effects for the groups (a_i) are assumed to be a random sample from a population with a mean of 0 and variance σ_a^2. If $\sigma_a^2 = 0$, then all group effects are equal, but if $\sigma_a^2 > 0$ there is variability among the group effects. Since the group effects in the experiment are only a sample from a larger population of effects, the differences among the specific group means, $\mu + a_i$, are of no particular interest. The variance of the distribution of group effects, σ_a^2, is the focus of interest with random effects.

The variance of an observation, σ_y^2, may be expressed as the sum of the two variances, or $\sigma_y^2 = \sigma_a^2 + \sigma_e^2$. The variances σ_a^2 and σ_e^2 are called *components of variance,* and the model in Equation (5.1) often is referred to as a **variance components** model. In the plant-breeding study the variance component among groups (σ_a^2) represents genetic variation among families, and the plant-breeder may be interested in the ratio of this genetic variation to the total variation (σ_y^2). The engineer may use the estimate of σ_y^2 to compute percentile values for the distribution of tensile strengths of the bars when they are to be used in a critical application.

An outline of the analysis of variance for the observations is shown in Table 5.2 with the expected mean squares for the Among Groups and Within Groups mean squares. The terms *Among Groups* and *Within Groups* will be used in place of *Treatments* and *Error* for the sources of variation to distinguish the random effects model from the fixed effects model. The computations for the sums of squares for Among Groups and Within Groups are the same as those for the Treatment and Error sums of squares given in Chapter 2 for the fixed effects model.

Table 5.2   Analysis of variance for the one-way classification with expected mean squares for the random effects model

Source of Variation	Degrees of Freedom	Sum of Squares	Mean Square	Expected Mean Square
Total	$N-1$	SS Total		
Among Groups	$t-1$	SSA	MSA	$\sigma_e^2 + r\sigma_a^2$
Within Groups	$N-t$	SSW	MSW	σ_e^2

5.3   Point Estimates of Variance Components

The analysis of variance method is used to estimate the variance components. The analysis of variance is computed as if the model is a fixed effects model, and the expected mean squares are derived under the assumption of the random effects model (Table 5.2). The observed mean squares are estimates of the expected mean squares, or

$$MSA = \widehat{\sigma}_e^2 + r\widehat{\sigma}_a^2$$

and   (5.2)

$$MSW = \widehat{\sigma}_e^2$$

The analysis of variance estimators of the variance components are determined by solving Equations (5.2) for the two unknowns. The solutions are

$$\widehat{\sigma}_e^2 = MSW$$

and   (5.3)

$$\widehat{\sigma}_a^2 = \frac{(MSA - MSW)}{r}$$

The estimators in Equations (5.3) are unbiased and they have the smallest variance of all estimators, which are both quadratic functions of the observations and unbiased estimators of σ_e^2 and σ_a^2.

The random effects of the model are assumed to have a normal distribution. Given the assumption of normally distributed effects, the significance of the Among Groups component of variance can be tested. The null and alternate hypotheses are H_0: $\sigma_a^2 = 0$ and H_a: $\sigma_a^2 > 0$, respectively. The test statistic is $F_0 = MSA/MSW$, and the null hypothesis is rejected at the α level of significance if $F_0 > F_{\alpha,(t-1),(N-t)}$.

The analysis of variance for the tensile strength data in Table 5.1 is shown in Table 5.3. The estimate of the components of variance for bars within castings is the mean square for Within Groups, or

$$\widehat{\sigma}_e^2 = MSW = 5.82$$

The estimate of the Among Groups component of variance is

$$\widehat{\sigma}_a^2 = \frac{(MSA - MSW)}{r} = \frac{(73.94 - 5.82)}{10} = 6.81$$

The estimated total variance of an observation on tensile strength is $\widehat{\sigma}_y^2 = \widehat{\sigma}_a^2 + \widehat{\sigma}_e^2 = 6.81 + 5.82 = 12.63$.

The F_0 ratio to test the null hypothesis H_0: $\sigma_a^2 = 0$ is $F_0 = MSA/MSW = 73.94/5.82 = 12.71$. The null hypothesis is rejected with a probability of exceeding $F_0 = 12.71$ equal to .000 (Table 5.3). The castings variation contributes significantly to the variation in the tensile strengths of the alloy.

Table 5.3 Analysis of variance for tensile strengths of bars from three castings of a high temperature alloy

Source of Variation	Degrees of Freedom	Sum of Squares	Mean Squares	F	Pr > F
Total	29	304.99			
Among Groups	2	147.88	73.94	12.71	.000
Within Groups	27	157.10	5.82		

5.4 Interval Estimates for Variance Components

Confidence interval estimates can be computed for both variance components. The exact $100(1 - \alpha)\%$ confidence interval estimator for σ_e^2 is

$$\frac{SSW}{A} < \sigma_e^2 < \frac{SSW}{B} \tag{5.4}$$

where $A = \chi_{\alpha/2,(N-t)}^2$ and $B = \chi_{(1-\alpha/2),(N-t)}^2$. A and B are values of the chi-square variable exceeded with probabilities $\alpha/2$ and $(1 - \alpha/2)$, respectively. Values of chi-square are found in Appendix Table III.

An interval with at least $100(1 - 2\alpha)\%$ confidence for σ_a^2 is

$$\frac{SSA(1 - F_u/F_0)}{rC} < \sigma_a^2 < \frac{SSA(1 - F_l/F_0)}{rD} \tag{5.5}$$

where $C = \chi^2_{\alpha/2,(t-1)}$, $D = \chi^2_{(1-\alpha/2),(t-1)}$, and $F_0 = MSA/MSW$ is the observed F_0 statistic (Williams, 1962). The quantities $F_u = F_{\alpha/2,(t-1),(N-t)}$ and $F_l = F_{(1-\alpha/2),(t-1),(N-t)}$ are values of the F variable exceeded with probabilities $\alpha/2$ and $(1 - \alpha/2)$, respectively.[1]

Given $SSW = 157.10$, $\chi^2_{.05,27} = 40.1$, and $\chi^2_{.95,27} = 16.2$, the 90% confidence interval estimate of σ_e^2 from Equation (5.4) is

$$\frac{157.10}{40.1} < \sigma_e^2 < \frac{157.10}{16.2}$$

$$3.92 < \sigma_e^2 < 9.70$$

With $SSA = 147.88$, $r = 10$, $\chi^2_{.025,2} = 7.38$, $\chi^2_{.975,2} = 0.05$, $F_0 = 12.71$, $F_{.025,2,27} = 4.24$, and $F_{.975,2,27} = 0.025$, the 90% confidence interval estimate of σ_a^2 from Equation (5.5) is

$$147.88 \left[\frac{1 - \frac{4.24}{12.71}}{10(7.38)} \right] < \sigma_a^2 < 147.88 \left[\frac{1 - \frac{0.025}{12.71}}{10(0.05)} \right]$$

$$1.34 < \sigma_a^2 < 295.18$$

The interval estimates for variance components will be quite wide when mean squares have small degrees of freedom. More groups of castings would provide a more precise interval estimate of σ_a^2.

Interpretations of the Variance Components

The mean tensile strength for the experiment was $\bar{y}_{..} = 90.9$ with standard error estimate

$$s_{\bar{y}_{..}} = \sqrt{\frac{MSA}{rt}} = \sqrt{\frac{73.94}{30}} = 1.57 \text{ psi}$$

The estimated variance of an observation on tensile strength on bars is $\hat{\sigma}_y^2 = \hat{\sigma}_a^2 + \hat{\sigma}_e^2 = 12.63$ with a standard deviation of $\hat{\sigma}_y = \sqrt{12.63} = 3.55$ psi. The variance component estimates isolated the different sources of variation in the casting process for alloy bars: The variance among castings accounted for 54% of the variation, and the variance among bars within castings accounted for 46% of

---

[1] The value of $F_{(1-\alpha)}$ cannot be read directly from Appendix Table IV, but its value can be determined from the relationship $F_{\nu_1,\nu_2,(1-\alpha)} = 1/F_{\nu_2,\nu_1,\alpha}$.

the variation. The engineer can reduce the standard deviation of $\widehat{\sigma}_y = 3.55$ psi by identifying and adjusting factors in the casting process that increase variation. The variation among castings can be caused by inconsistent alloy mixtures or temperature settings from casting to casting. The variation among bars within castings can be caused by inconsistent cooling conditions or variations in the tensile strength measurement procedure.

5.5 Courses of Action with Negative Variance Estimates

By definition, a variance component is positive. However, estimates of σ_a^2 using Equation (5.3) may be negative. There are several suggested courses of action in the case of negative estimates (Searle, 1971); Searle, Casella, & McCulloch (1992).

1. Accept the estimate as evidence of a true value of zero and use zero as the estimate, recognizing that the estimator will no longer be unbiased.

2. Retain the negative estimate, recognizing that subsequent calculations using the results may not make much sense.

3. Interpret the negative component estimate as indication of an incorrect statistical model.

4. Utilize a method different from the analysis of variance for estimating the variance components.

5. Collect more data and analyze them separately or in conjunction with the existing data and hope that increased information will yield positive estimates.

Searle (1971, Chapter 9) and Searle et al. (1992, Chapter 4) discuss several methods of estimation from the extensive literature on variance component estimation, as well as these other actions, in greater detail.

5.6 Intraclass Correlation Measures Similarity in a Group

The intraclass correlation coefficient is a measure of the similarity of observations within groups relative to that among groups. When the similarity of the observations within groups is very high, σ_e^2 will be very small. Consequently, σ_a^2 will be a larger proportion of the total variation ($\sigma_y^2 = \sigma_a^2 + \sigma_e^2$). The intraclass correlation, defined as the ratio

$$\rho_I = \frac{\sigma_a^2}{\sigma_a^2 + \sigma_e^2} \tag{5.6}$$

is used in various disciplines. Applications arise in genetics studies with various measures for the heritability of quantitative traits, in reliability studies to measure the similarity of products from the same machine or process, in medical studies to measure the repeatability of successive measurements on patients, and in survey sampling to measure the similarity of responses among people contacted by the same interviewer (Koch, 1983).

The analysis of variance was introduced by R. A. Fisher in the 1920s with an intraclass correlation model (Fisher, 1960). The model assumes all observations (y_{ij}) have the same mean (μ) and variance (σ^2), and any two members of the same group have a common correlation (ρ_I). With this model the expected mean squares for the analysis of variance in Table 5.2 are

$$E(MSA) = \sigma^2 \{1 + (r - 1)\rho_I\}$$

and   (5.7)

$$E(MSW) = \sigma^2(1 - \rho_I)$$

The estimators of ρ_I and σ^2 are found by equating the observed mean squares to the expectations shown in Equation (5.7) and solving for the unknowns. The solutions are

$$\widehat{\sigma}^2 = \frac{\{MSA + (r - 1)MSW\}}{r}$$

and   (5.8)

$$\widehat{\rho}_I = \frac{(MSA - MSW)}{\{MSA + (r - 1)MSW\}}$$

The estimate of the intraclass correlation can have a minimum value of $-1/(r - 1)$ and a maximum value of 1 (Fisher, 1960) because the expected value of MSA must be equal to or greater than zero.

The $100(1 - \alpha)\%$ confidence interval estimator for ρ_I is

$$\frac{F_0 - F_u}{F_0 + (r - 1)F_u} < \rho_I < \frac{F_0 - F_l}{F_0 + (r - 1)F_l} \tag{5.9}$$

where $F_u = F_{\alpha/2,(t-1),(N-t)}$, $F_l = F_{(1-\alpha/2),(t-1),(N-t)}$, and $F_0 = MSA/MSW$. The interval may be used for testing the hypothesis H_0: $\rho_I = 0$, where the hypothesis is not rejected if the interval includes zero (Koch, 1983).

The estimate of the intraclass correlation for castings of high-temperature alloys is

$$\widehat{\rho}_I = \frac{(73.94 - 5.82)}{\{73.94 + 9(5.82)\}} = 0.54$$

With $F_0 = 12.71$, $F_{.05,2,27} = 3.35$, and $F_{.95,2,27} = 0.051$, the 90% confidence interval estimate is

$$\frac{(12.71 - 3.35)}{\{12.71 + 9(3.35)\}} < \rho_I < \frac{(12.71 - 0.051)}{\{12.71 + 9(0.051)\}}$$

and

$$0.22 < \rho_I < 0.96$$

The interval does not include zero, and a null hypothesis of zero intraclass correlation among the bars within castings is rejected.

The interpretation of intraclass correlation can be made on the basis of the ratio in Equation (5.6). The numerator (σ_a^2) reflects the variation peculiar to the differences among groups, whereas the denominator variance $(\sigma_a^2 + \sigma_e^2)$ pertains to individuals sampled randomly from the universe of all groups without regard to group boundaries.

If the intraclass correlation is large, all the individuals in the same group are affected alike by the random effect (a_i) common to that group. Thus, the similarity among individuals within groups will be greater than that among individuals from different groups, and σ_e^2 will be small relative to σ_a^2.

On the other hand, a small intraclass correlation indicates dissimilarity among individuals within groups with σ_e^2 large relative to σ_a^2. For example, competition among plants or animals within a group for nutritional resources could lead to growth disparities within a group. This could happen if more vigorous or aggressive individuals took a greater part of the nutritional resource.

5.7 Unequal Numbers of Observations in the Groups

The random effects model for the one-way classification with unequal numbers of observations per group is

$$y_{ij} = \mu + a_i + e_{ij} \tag{5.10}$$

$$i = 1, 2, \ldots, t \quad j = 1, 2, \ldots, r_i$$

with the same assumptions and interpretations given for the random model in Equation (5.1). The analysis of variance computations are the same as those for the fixed effects model with unequal replication. The expected mean squares for Among Groups and Within Groups are, respectively,

$$E(MSA) = \sigma_e^2 + r_0\sigma_a^2$$

and

$$\tag{5.11}$$

$$E(MSE) = \sigma_e^2$$

where

$$r_0 = \frac{1}{t-1} \left[N - \sum_{i=1}^{t} \frac{r_i^2}{N} \right] \tag{5.12}$$

The analysis of variance estimators for the variance components σ_e^2 and σ_a^2 are

$$\widehat{\sigma}_e^2 = MSW$$

and $\tag{5.13}$

$$\widehat{\sigma}_a^2 = \frac{(MSA - MSW)}{r_0}$$

When the r_i are unequal the confidence interval estimator for σ_a^2 in Equation (5.5) no longer applies. The interval estimator for σ_e^2 from Equation (5.4) with $(N - t)$ degrees of freedom is valid.

5.8   How Many Observations to Study Variances?

The null hypothesis of interest in the random effects model, H_0: $\sigma_a^2 = 0$, is tested with $F = MSA/MSW$, and the power of the test is

$$1 - \beta = P(F > F_{\alpha,\nu_1,\nu_2}|H_0 \text{ false}) = P(F > F_{\alpha,\nu_1,\nu_2}|\sigma_a^2 > 0)$$

When $\sigma_a^2 > 0$, the distribution of F is the central F_{ν_1,ν_2} distribution multiplied by a constant $1/\lambda^2$, where

$$\lambda^2 = 1 + \frac{r\sigma_a^2}{\sigma_e^2} \tag{5.14}$$

The power of the test can be determined from the central F distribution as

$$1 - \beta = P\left[F > \frac{1}{\lambda^2}(F_{\alpha,\nu_1,\nu_2}) \right]$$

Given the number of groups (t), significance level (α), desired power ($1 - \beta$), and λ, the required replication numbers can be determined from charts of power curves similar to those for the fixed effects model.

A value for λ may be determined on the basis of a desired ratio for the variance components, σ_a^2/σ_e^2, or on the basis of the standard deviation of an individual observation, σ_y. Consider Example 5.1 in which the engineer manufactured several bars of a high-temperature alloy in each of several castings. If there was no variation in the strength of the bars due to castings, then the standard deviation of a bar selected at random would be $\sigma_y = \sigma_e$. The engineer may want to detect an increase in the variability among castings (σ_a^2) that causes a certain percentage increase in σ_y. Suppose P is the fixed percentage increase in σ_y that is acceptable, and beyond which the null hypothesis would be rejected. The ratio of σ_y to σ_e expressed in terms of P when the null hypothesis is rejected is

$$\frac{\sigma_y}{\sigma_e} = \frac{\sqrt{\sigma_a^2 + \sigma_e^2}}{\sigma_e} = 1 + 0.01P$$

The necessary value for the ratio σ_a^2/σ_e^2 in Equation (5.14) is

$$\frac{\sigma_a^2}{\sigma_e^2} = (1 + 0.01P)^2 - 1 \qquad (5.15)$$

Charts of power curves are given in Appendix Table X for $\alpha = .05, .01$, and selected values of ν_1 and ν_2 for the F distribution. The charts plot the power of the test, $1 - \beta$, versus λ, where λ^2 is given in Equation (5.14).

> **Example 5.2   Castings of High-Temperature Alloys Revisited**
>
> In Example 5.1 the estimates of the variance components were $\hat{\sigma}_e^2 = 5.82$ and $\hat{\sigma}_a^2 = 6.81$. The estimated standard deviations are $\hat{\sigma}_e = 2.41$ and $\hat{\sigma}_y = 3.55$. Suppose the engineer is able to run $t = 5$ castings and wants to detect an increase in the standard deviation σ_y over σ_e of $P = 35$ with a power of at least .80 at the .05 level of significance. The required value for the ratio in Equation (5.15) is
>
> $$\frac{\sigma_a^2}{\sigma_e^2} = [1 + 0.01(35)]^2 - 1 = 0.8225$$
>
> so that $\lambda = \sqrt{1 + r(0.8225)}$. If a value of $r = 10$ is chosen, then $\lambda = 3$. Entering Appendix Table X for $\nu_1 = 4$ and locating the approximate position of the line for $\nu_2 = 45$ with $\alpha = .05$ the value of $1 - \beta$ is between 0.8 and 0.9. Therefore, the engineer could measure ten bars in each of the five castings to detect an increase of 35% or more in the standard deviation due to the castings.

5.9   Random Subsamples to Procure Data for the Experiment

It is sometimes necessary or convenient to randomly sample subunits of the experimental units to procure the requisite data for a study. The observational unit in this case is a subsample taken from a larger experimental unit. Several plants may be sampled from a field plot for measurements of insect infestation. A serum sample is frequently split into two or more subsamples prior to spectrophotometric analysis. Several samples of paint are extracted from replicate batches of each paint formulation to test paint durability.

The subsamples introduce another random source of variability for the observations in addition to that among the experimental units. It is important to distinguish between the variation contributed by subsamples and that contributed by the experimental units. This distinction becomes important to estimation of standard errors for treatment means and tests of hypotheses about treatments. An introduction to this distinction was given in Examples 1.1 and 1.2 in the discussion on replication in Chapter 1.

Estimates of the variance components for experimental units and for subsamples identify the amount of variation contributed by the two sources. This information is used in Section 5.10 to determine the relative number of experimental units to minimize the standard error of the treatment means or the cost of the experiment.

Example 5.3   Pesticide Residue on Cotton Plants

Applications of pesticides are often part of insect management programs used for agronomic and horticultural crops. One of the concerns following application of pesticides is the concentration of pesticide residue that remains on the plants in the field after certain periods of time. Pesticide residues are evaluated with chemical assays in the laboratory using plants sampled from field plots treated with the pesticide.

Research Hypothesis: For one particular problem the investigators hypothesized that the ability to recover pesticide residue on cotton plant leaves differed among two standard chemistry methods that were being used on a regular basis for the residue assays.

Treatment Design: The treatments consisted of the two standard chemistry methods, methods A and B, that were used on a regular basis.

Experiment Design: Six batches of plants, each batch from a single field plot, were sampled from the field and prepared for residue analysis. Three batches were randomly allocated to each of methods A and B in a completely randomized design.

The amount of plant material in each batch sampled from the field exceeded the amount required for an assay in the laboratory by either chemistry method. Thus, two subsamples of the required quantity of plant material were taken from each batch of the prepared plant tissue and analyzed by the appropriate method. Consequently, there were two chemistry methods, three batches (replications) per method, and two subsamples from each batch. The pesticide residues determined for each of the subsamples as micrograms per unit of weight are shown in Table 5.4.

The Statistical Model with Subsamples

When there are n subsamples from each of r experimental units for t treatments the statistical model is

$$y_{ijk} = \mu + \tau_i + e_{ij} + d_{ijk} \tag{5.16}$$

$$i = 1, 2, \ldots, t \quad j = 1, 2, \ldots, r \quad k = 1, 2, \ldots, n$$

where μ is the general mean, τ_i is the fixed effect of the ith treatment, e_{ij} is the random experimental error effect for the jth experimental unit of the ith treatment,

Table 5.4 Pesticide residue (μg) found on samples of cotton plants

	Method A					Method B			
Batch	Sample	y_{ijk}	$\bar{y}_{ij.}$	$\bar{y}_{i..}$	Batch	Sample	y_{ijk}	$\bar{y}_{ij.}$	$\bar{y}_{i..}$
1	1	120			4	7	71		
	2	110	115.0			8	71	71.0	
2	3	120			5	9	70		
	4	100	110.0			10	76	73.0	
3	5	140			6	11	63		
	6	130	135.0	120.0		12	68	65.5	69.8
				$\bar{y}_{...} = 94.9$					

Source: G. Ware and B. Estesen, Department of Entomology, University of Arizona.

and d_{ijk} is the random effect for the kth subsample of the jth experimental unit of the ith treatment. It is assumed that the e_{ij} and d_{ijk} are normally distributed independent random effects with means 0 and variances σ_e^2 and σ_d^2, respectively. If treatments are random, then the fixed treatment effects (τ_i) in Equation (5.16) are replaced by the random group effects (a_i), which have a normal distribution with mean 0 and variance σ_a^2.

The Analysis of Variance with Subsamples

The observations, expressed as deviations from the grand mean, can be written as a sum of three separate deviations that represent the sources of variation in the experiment:

$$(y_{ijk} - \bar{y}_{...}) = (\bar{y}_{i..} - \bar{y}_{...}) + (\bar{y}_{ij.} - \bar{y}_{i..}) + (y_{ijk} - \bar{y}_{ij.}) \tag{5.17}$$

The deviation of any observation from the grand mean shown on the left-hand side of Equation (5.17) is the sum of three terms. They are the

- treatment deviation $(\bar{y}_{i..} - \bar{y}_{...})$ [e.g., $(\bar{y}_{1..} - \bar{y}_{...}) = 120.0 - 94.9 = 25.1$]
- experimental error $(\bar{y}_{ij.} - \bar{y}_{i..})$ [e.g., $(\bar{y}_{11.} - \bar{y}_{1..}) = 115.0 - 120.0 = -5.0$]
- sampling error $(y_{ijk} - \bar{y}_{ij.})$ [e.g., $(y_{111} - \bar{y}_{11.}) = 120.0 - 115.0 = 5.0$]

Squaring and summing both sides of Equation (5.17) results in the total sum of squares on the left-hand side expressed as a sum of the sums of squares for treatments, experimental error, and sampling, respectively. The fundamental partition of the total sum of squares is

$$SS \text{ Total} = SS \text{ Treatment} + SS \text{ Error} + SS \text{ Sampling} \tag{5.18}$$

The sums of squares partitions are summarized in the analysis of variance shown in Table 5.5 with expected mean squares for the model with fixed treatment effects.

Table 5.5   Analysis of variance for the completely randomized design with subsamples[2]

Source of Variation	Degrees of Freedom	Sum of Squares	Mean Square	Expected Mean Square
Total	$trn - 1$	SS Total		
Treatments	$t - 1$	SST	MST	$\sigma_d^2 + n\sigma_e^2 + rn\theta_t^2$
Error	$t(r - 1)$	SSE	MSE	$\sigma_d^2 + n\sigma_e^2$
Sampling	$tr(n - 1)$	SSS	MSS	σ_d^2

$$SS\,\text{Total} = \sum_{i=1}^{t} \sum_{j=1}^{r} \sum_{k=1}^{n} (y_{ijk} - \bar{y}_{...})^2$$

$$SST = SS\,\text{Treatment} = rn \sum_{i=1}^{t} (\bar{y}_{i..} - \bar{y}_{...})^2$$

$$SSE = SS\,\text{Error} = n \sum_{i=1}^{t} \sum_{j=1}^{r} (\bar{y}_{ij.} - \bar{y}_{i..})^2$$

$$SSS = SS\,\text{Sampling} = \sum_{i=1}^{t} \sum_{j=1}^{r} \sum_{k=1}^{n} (y_{ijk} - \bar{y}_{ij.})^2$$

Two sources contribute to variation in the observations that make up the estimate of a treatment mean: the variation among the replicate experimental units treated alike (σ_e^2) and the variation among the sampling units within the same experimental units (σ_d^2). Consequently, the variance of a treatment mean is

$$\sigma_{\bar{y}_{i..}}^2 = \frac{\sigma_d^2}{rn} + \frac{\sigma_e^2}{r} \qquad (5.19)$$

when there are n subsamples from each of the r replicate experimental units. The standard error of any treatment mean is estimated by

$$s_{\bar{y}_{i..}} = \sqrt{\frac{MSE}{rn}} \qquad (5.20)$$

The analysis of variance for the data in Table 5.4 is shown in Table 5.6. The null hypothesis of no differences among treatment effects, H_0: $\tau_i = 0$, is rejected if $F_0 = MST/MSE$ exceeds $F_{\alpha,(t-1),t(r-1)}$. The F_0 statistic[3] in Table 5.6 to test the null hypothesis of no difference between the means of methods A and B, $F_0 = 7550.08/190.08 = 39.72$, is exceeded with probability .003. Since the mean of method A, $\bar{y}_{1..} = 120$, exceeds that of method B, $\bar{y}_{2..} = 69.8$, it may be concluded that method A recovers more of the pesticide residue than method B.

---

[2] The sum of squares for "Error" represents the sum of squares for experimental units nested within the treatment groups. A computer program will require a term in its syntax that designates the experimental units *within* treatments.

[3] Frequently, it is necessary to specify the correct denominator for the F_0 statistic in the instructions to a computing program. By default, many programs utilize the last partition sum of squares for the denominator of the F_0 statistic for all lines of the analysis of variance table.

Table 5.6 Analysis of variance for pesticide residue from subsamples of cotton plants

Source of Variation	Degrees of Freedom	Sum of Squares	Mean Square	F	Pr > F
Total	11	8640.91			
Methods	1	7550.08	7550.08	39.72	.003
Error	4	760.33	190.08		
Sampling	6	330.50	55.08		

The standard error estimate for a method mean, Equation (5.20) is

$$s_{\bar{y}_{i.}} = \sqrt{\frac{190.08}{6}} = 5.63$$

and the standard error estimate of the difference between the two method means is

$$s_{(\bar{y}_{i..} - \bar{y}_{j..})} = \sqrt{\frac{2(190.08)}{6}} = 7.96$$

5.10   Using Variance Estimates to Allocate Sampling Efforts

The distribution of resources at the planning stage of an experiment involving subsamples requires decisions regarding the number of experimental units to use and the number of subsamples to take from each experimental unit. The objective is to have a design that results in greater precision—a smaller variance for the estimate of a treatment mean ($\sigma^2_{\bar{y}_{i.}}$) for a fixed cost. When estimates of the variance components and relative costs of the experimental and sampling units are available it is possible to provide an optimum allocation of effort between experimental units and sampling units in the experiment.

Cochran (1965) provided an optimum allocation solution based on the cost function $C = c_1 r + c_2 rn$. The value of C is the cost for a single treatment in the experiment composed of r experimental units each at a cost of c_1 and rn sampling units each at a cost of c_2. The objective may be posed as an attainment of minimum cost (C) for a fixed variance in Equation (5.19), or the attainment of a minimum variance for a fixed cost. Either way the solution for the number of sampling units (n) is

$$n = \sqrt{\frac{c_1 \sigma^2_d}{c_2 \sigma^2_e}} \qquad (5.21)$$

The value of r is found by solving the cost equation for r if the cost is fixed or by solving the variance Equation (5.19) if the variance is fixed.

Example 5.4 Pesticide Residue Revisited

An optimum allocation for plots of cotton plants and subsamples per plot is required for pesticide residue studies. The estimates of the variance components obtained from Table 5.6 are

$$\widehat{\sigma}_d^2 = MSS = 55.08$$

and

$$\widehat{\sigma}_e^2 = \frac{(MSE - MSS)}{n} = \frac{(190.08 - 55.08)}{2} = 67.50$$

Suppose the cost of one plot is $c_1 = 1.0$ relative to the cost of preparing and analyzing one subsample, $c_2 = 0.1$. The estimated number of subsamples per plot is

$$n = \sqrt{\frac{1(55.08)}{0.1(67.50)}} = 2.86$$

Three subsamples would be required from each plot. If the investigator desired a standard error for the treatment mean of $\sigma_{\bar{y}_{i..}} = 3$ or a variance of 9, the number of required plots r can be found from the substitution of the required quantities in Equation (5.19). The substitutions are

$$\widehat{\sigma}_{\bar{y}_{i..}}^2 = 9, \ \widehat{\sigma}_e^2 = 67.5, \ \widehat{\sigma}_d^2 = 55.08, \text{ and } n = 3$$

so that

$$9 = \frac{55.08}{r \cdot 3} + \frac{67.5}{r}; \ \ 9r = 85.86; \ \ r = 9.54$$

The investigator would have to use ten plots with three subsamples per plot to have a standard error of treatment means equal to 3 with relative costs of 1.0 and 0.1 for plots and subsamples, respectively.

5.11 Unequal Numbers of Replications and Subsamples

Unequal subsample and replication numbers can occur in a study. The three different possibilities are (1) unequal numbers of experimental units per treatment with unequal numbers of subsamples per experimental unit, (2) equal numbers of experimental units per treatment with unequal numbers of subsamples per experimental unit, and (3) unequal numbers of experimental units per treatment with equal numbers of subsamples per experimental unit. Any imbalance in the number of observations at the experimental unit or subsample stage affects the computations for the analysis of variance and the expected mean squares.

Example 5.5 Biology of the Tobacco Budworm

Populations of insects often develop resistance to the toxic effects of an insecticide after long-term exposure to the insecticide. When this resistance develops the insecticide is no longer effective to control the population below levels harmful to the crop.

Populations of the tobacco budworm, an insect pest harmful to the cotton plant, have developed resistance to a number of common insecticides. The insecticides are one component of the overall program of insect control in crops. Other components of the control program are also dependent on the biology of the insects in terms of their reproductive life cycle and developmental patterns.

Research Hypothesis: Entomologists hypothesized that the development of insecticide resistance could also affect other aspects of the tobacco budworm's biology. If this were true, then the changes in the insect's biology would have an effect on tobacco budworm control programs.

Treatment Design: The treatments used to address the research hypothesis included three strains of the tobacco budworm: (1) USDA, a strain very susceptible to a pyrethroid insecticide; (2) Resistant, a strain quite resistant to the insecticide; and (3) Field, a naturally occurring strain collected in a local cotton field. Both the Resistant and USDA strains were populations maintained in artificial environments to sustain their resistance characteristics. Any differences in the biology of the two strains were considered reflective of changes associated with developed insecticide resistance. The biology of the naturally occurring Field strain served as a control treatment for this supposition. One of the characteristics measured to evaluate the biology was the weight of male larvae.

Experiment Design: Six random matings between female and male moths were made from each of the strains, and the offspring from each mating were reared in separate enclosures in the laboratory. The 18 enclosures were placed randomly within the rearing facility. Unequal numbers of offspring resulted among the 18 matings as shown with the data listed in Table 5.7.

The Statistical Model and Analysis

Suppose the experiment has r_i experimental units for the ith treatment group and n_{ij} subsamples for the jth experimental unit of the ith treatment. The statistical model for unequal subsamples is

$$y_{ijk} = \mu + \tau_i + e_{ij} + d_{ijk} \qquad (5.22)$$

$$i = 1, 2, \ldots, t \quad j = 1, 2, \ldots, r_i \quad k = 1, 2, \ldots, n_{ij}$$

Table 5.7 Weight of male larvae from six matings in each of three tobacco budworm strains

Strain	Mating	Weight	n_{ij}	$n_{i.}$	$\bar{y}_{ij.}$	$\bar{y}_{i..}$
USDA	1	305, 300	2		302.5	
	2	376, 363, 389	3		376.0	
	3	282	1		282.0	
	4	309, 321	2		315.0	
	5	354, 308, 327	3		329.7	
	6	330	1	12	330.0	330.3
Field	7	280	1		280.0	
	8	311, 349, 291, 286	4		309.3	
	9	377, 342	2		359.5	
	10	346, 340, 347	3		344.3	
	11	360	1		360.0	
	12	359, 299	2	13	329.0	329.8
Resistant	13	273, 276	2		274.5	
	14	272, 253	2		262.5	
	15	315, 262, 297	3		291.3	
	16	323	1		323.0	
	17	252	1		252.0	
	18	319, 298	2	11	308.5	285.5

$$\bar{y}_{...} = 316.4$$

Source: Dr. T. Watson and S. Kelly, Department of Entomology, University of Arizona.

where μ is the general mean, τ_i is the fixed effect of the ith treatment, e_{ij} is the random experimental error effect for the jth experimental unit of the ith treatment, and d_{ijk} is the random effect for the kth subsample of the jth experimental unit of the ith treatment. We assume the e_{ij} and d_{ijk} are normally distributed independent random effects with means 0 and variances σ_e^2 and σ_d^2, respectively. If treatments are random, then the fixed treatment effects (τ_i) in Equation (5.22) are replaced by the random group effects (a_i), which have a normal distribution with mean 0 and variance σ_a^2.

The sums of squares partitions for the analysis of variance and expected mean squares for random treatment effects are shown in Table 5.8. If treatment effects are fixed, replace $c_2\sigma_a^2$ with $\sum n_{ij}(\mu_i - \bar{\mu}_.)^2/(t-1)$.

Notice that the squares of the treatment mean deviations in SST are weighted by the number of observations on the treatments, $n_{i.}$, and the squares of the experimental error deviations in SSE are weighted by the number of subsamples for the experimental units, n_{ij}. The coefficients for the variance components in the expected mean squares with random treatment effects are

Table 5.8 Analysis of variance for the completely randomized design with unequal numbers of replications and subsamples

Source of Variation	Degree of Freedom	Sum of Squares	Mean Square	Expected Mean Square
Total	$N-1$	SS Total		
Treatments	$t-1$	SST	MST	$\sigma_d^2 + c_1\sigma_e^2 + c_2\sigma_a^2$
Error	$\sum_{i=1}^{t} r_i - t$	SSE	MSE	$\sigma_d^2 + c_3\sigma_e^2$
Sampling	$N - \sum_{i=1}^{t} r_i$	SSS	MSS	σ_d^2

$$N = \sum_{i=1}^{t}\sum_{j=1}^{r_i} n_{ij} \qquad n_{i.} = \sum_{j=1}^{r_i} n_{ij}$$

$$SS\ \text{Total} = \sum_{i=1}^{t}\sum_{j=1}^{r_i}\sum_{k=1}^{n_{ij}} (y_{ijk} - \bar{y}_{...})^2$$

$$SST = \sum_{i=1}^{t} n_{i.}(\bar{y}_{i..} - \bar{y}_{...})^2$$

$$SSE = \sum_{i=1}^{t}\sum_{j=1}^{r_i} n_{ij}(\bar{y}_{ij.} - \bar{y}_{i..})^2$$

$$SSS = \sum_{i=1}^{t}\sum_{j=1}^{r_i}\sum_{k=1}^{n_{ij}} (y_{ijk} - \bar{y}_{ij.})^2$$

$$c_1 = \frac{1}{t-1}\left(A - \frac{B}{N}\right), c_2 = \frac{1}{t-1}\left(N - \frac{D}{N}\right), \text{ and } c_3 = \frac{1}{\left(\sum_{i=1}^{t} r_i - t\right)}(N - A)$$

(5.23)

where

$$A = \sum_{i=1}^{t}\sum_{j=1}^{r_i}\left(\frac{n_{ij}^2}{n_{i.}}\right), \quad B = \sum_{i=1}^{t}\sum_{j=1}^{r_i} n_{ij}^2, \text{ and } D = \sum_{i=1}^{t} n_{i.}^2$$

When the number of subsamples are equal for each of the experimental units, then $n_{ij} = n$ for all i and j; the coefficients are

$$c_1 = c_3 = n \quad \text{and} \quad c_2 = \frac{1}{t-1}\left[N - \frac{D}{N}\right] \qquad \textbf{(5.24)}$$

When $c_1 = c_3 = n$ the expected mean squares for Treatments and Error will be identical under the null hypothesis H_0: $\tau_i = 0$, and the statistic $F_0 = MST/MSE$ is used to test the hypothesis.

The analysis of variance for the data is shown in Table 5.9. The expected mean squares for analysis of variance are also shown in Table 5.9 for random treatment effects. The calculations of the coefficients for the variance components are shown in Appendix 5A.

Table 5.9   Analysis of variance for male larval weights from six matings in each of three tobacco budworm strains

Source of Variation	Degrees of Freedom	Sum of Squares	Mean Square	Expected Mean Square
Total	35	46516.75		
Strain	2	15187.05	7593.52	$\sigma_d^2 + 2.36\sigma_e^2 + 11.97\sigma_a^2$
Error	15	23082.45	1538.83	$\sigma_d^2 + 1.93\sigma_e^2$
Sampling	18	8247.25	458.18	σ_d^2

Tests of Hypotheses Require Approximate F Tests

When the subsample numbers are not equal, c_1 and c_3 can have different values. There is no exact test of the null hypothesis for treatment effects because no two mean squares have the same expected mean squares under the null hypothesis if c_1 and c_3 have different values. An approximate F_0 statistic can be calculated to test the null hypothesis of no treatment effects when $c_1 \neq c_3$. An approximate test is necessary for the tobacco budworm experiment since $c_1 = 2.36$ and $c_3 = 1.93$ in Table 5.9.

A Mean Square for Error is devised with an expectation equal to that of the Mean Square for Treatments, given a true null hypothesis with $E(MST) = \sigma_d^2 + c_1\sigma_e^2$.

The required mean square is constructed with a linear function of MSS and MSE as

$$M = a_1 MSE + a_2 MSS \qquad (5.25)$$

If $a_1 = c_1/c_3$ and $a_2 = 1 - c_1/c_3$, the expected value of M will be $\sigma_d^2 + c_1\sigma_e^2$ as required for the approximate F_0 test.

The Satterthwaite Approximation for Degrees of Freedom

Satterthwaite (1946) derived the following result for a linear function of mean squares. Given a linear function M, where

$$M = a_1 MS_1 + a_2 MS_2 + \cdots + a_k MS_k \qquad (5.26)$$

and $MS_1, MS_2, \ldots, MS_k$ are mean squares with degrees of freedom $\nu_1, \nu_2, \ldots \nu_k$, respectively, the degrees of freedom for M are approximated by

$$\nu = \frac{M^2}{\sum\limits_{i=1}^{k} \dfrac{(a_i MS_i)^2}{\nu_i}} \qquad (5.27)$$

The linear function of mean squares necessary to test the hypothesis of no difference in mean larval weights among the three tobacco budworm strains requires

$$a_1 = \frac{c_1}{c_3} = \frac{2.36}{1.93} = 1.22 \quad \text{and} \quad a_2 = 1 - \frac{c_1}{c_3} = 1 - 1.22 = -0.22$$

$$MSE = 1538.83 \text{ with } 15 \text{ d.f.} \quad \text{and} \quad MSS = 458.18 \text{ with } 18 \text{ d.f.}$$

From Equation (5.25)

$$M = a_1 MSE + a_2 MSS = 1.22(1538.83) - 0.22(458.18) = 1776.57$$

The degrees of freedom for M from Equation (5.27) are

$$\nu = \frac{1776.57^2}{\dfrac{[1.22(1538.83)]^2}{15} + \dfrac{[-0.22(458.18)]^2}{18}} = 13.4$$

The truncated value of $\nu = 13$ is used as the degrees of freedom for M. The value of the test statistic is $F_0 = MST/M = 7593.53/1776.57 = 4.27$. The critical value at the $\alpha = .05$ level of significance is $F_{.05, 2, 13} = 3.81$, and the null hypothesis is rejected. There are some differences among the mean larval weights of the three strains. The test is only approximate, and it should be noted that the approximation is degraded somewhat if some of the coefficients (a_i) in Equation (5.26) are negative.

EXERCISES FOR CHAPTER 5

1. A genetics study with beef animals consisted of several sires each mated to a separate group of dams. The matings that resulted in male progeny calves were used for an inheritance study of birth weights. The birth weights of eight male calves in each of five sire groups follow.

Sire	Birthweights
177	61, 100, 56, 113, 99, 103, 75, 62
200	75, 102, 95, 103, 98, 115, 98, 94
201	58, 60, 60, 57, 57, 59, 54, 100
202	57, 56, 67, 59, 58, 121, 101, 101
203	59, 46, 120, 115, 115, 93, 105, 75

Source: Dr. S. DeNise, Department of Animal Sciences, University of Arizona.

   a. Assume a random model for this study. Write the linear model, explain each of the terms, compute the complete analysis of variance, and show the expected mean squares.

   b. Estimate the components of variance for sires and progeny within sires, and determine the 90% confidence interval estimates.

   c. Test the null hypothesis $H_0: \sigma_a^2 = 0$ for the sires.

   d. Estimate the intraclass correlation coefficient, and give the 90% confidence interval estimate.

2. The data from Exercise 3.5 are cholesterol concentrations from laboratory analyses of 2 samples from each of 8 patients.

   a. Assume a random model for the study. Write a linear model, explain each of the terms, compute the analysis of variance, and show the expected mean squares.

   b. Estimate the components of variance for patients and samples and determine the 90% confidence interval estimates.

   c. Estimate the intraclass correlation coefficient and give the 90% confidence interval estimate.

   d. What is the interpretation of the intraclass correlation coefficient in this study?

3. Think of research problems in your field of interest for which the treatments in the study could be a random sample from a large population of treatments.

   a. Describe a particular study you could conduct.

   b. Describe how you would conduct the study.

   c. Write the linear model for your study; identify the terms; and write out the analysis of variance table showing sources of variation, degrees of freedom, and expected mean squares.

   d. Explain why it would be important to know the magnitude of the Among Group and Within Group components of variance.

   e. Describe how you would use estimates of the components of variance.

   f. What assumptions do you have to make about your study to have valid inferences from your variance component estimates?

4. A plant pathologist took four 3-pound samples from 50-ton lots of cottonseed accumulated at various cotton gins during the ginning season. The samples of seed were analyzed in the laboratory for Aflatoxin, which is a toxin produced by organisms associated with the seeds. The Aflatoxin concentrations in parts per billion for samples from eight lots of cottonseed follow.

Lot Number	Aflatoxin (ppb)
3469 – 72	39, 57, 63, 66
3849 – 52	56, 13, 25, 31
3721 – 24	64, 83, 88, 71
3477 – 80	29, 55, 21, 51
3669 – 72	38, 66, 53, 81
3873 – 76	11, 49, 34, 10
3777 – 80	23, 0, 5, 20
3461 – 64	10, 11, 23, 37

Source: Dr. T. Russell, Department of Plant Pathology, University of Arizona.

a. Assume lots and samples within lots are random effects. Write the linear model for the study, explain the terms, compute the complete analysis of variance, and show the expected mean squares.

b. Estimate the components of variance for lots and samples within lots.

c. What is the total variance estimate $(\widehat{\sigma}_y^2)$ for an individual observation?

d. What proportion of the total variation (σ_y^2) in Aflatoxin can be attributed to variation among lots and samples within lots, respectively?

e. What is the standard deviation estimate $(\widehat{\sigma}_y)$ for an individual observation?

f. Explain how the variance component estimates might be used to plan future sampling for Aflatoxin contamination.

5. Think of research problems in your field of interest that require you to take samples of the experimental (or observational) unit because the unit cannot be measured in its entirety.

a. Describe a specific study you could conduct.

b. Describe how you would conduct the study.

c. Write a linear model for your study; identify the terms; and sketch the analysis of variance showing sources of variation, degrees of freedom, and expected mean squares.

d. What would be the relative costs for experimental units (c_1) and sampling units (c_2)?

6. A study was conducted on high-energy particulate cartridge filters used with commercial respirators for protection against particulate matter. One particular test included three filters randomly selected from each of two manufacturers. Three independent replicate tests were made on each of the filters. The measurements were the percent penetration by a standard type of test aerosol.

Filter	Manufacturer 1			Manufacturer 2		
	1	2	3	4	5	6
	1.12	0.16	0.15	0.91	0.66	2.17
	1.10	0.11	0.12	0.83	0.83	1.52
	1.12	0.26	0.12	0.95	0.61	1.58

Source: R. J. Beckman and C. J. Nachtsheim (1987), Diagnostics for mixed-model analysis of variance, *Technometrics* 29, 413–426.

a.   Write a linear model for this study, explain each of the terms, compute the analysis of variance, and show the expected mean squares.

b.   Test the hypothesis that there is no difference between the average percent penetration of the
filters for the two manufacturers.

c.   Compute the means, their standard errors, and the 95% confidence interval estimates of the
means for each of the manufacturers.

d.   Suppose the relative costs, $c_1:c_2$, for the study are 200:1, where c_1 is the cost of a filter and c_2
is the cost of an independent filter test. The engineers wanted to achieve a standard error for a
mean of 0.20. How many filters and how many tests per filter would be required?

7.   A soil scientist studied the growth of barley plants under three different levels of salinity in a controlled growth medium. There were two replicate containers for each treatment in a completely randomized design and three plants were measured in each replication. The data on the dry weight of
the plants in grams follow.

Salinity	Container	Weight(g)
Control	1	11.29, 11.08, 11.10
	2	7.37, 6.55, 8.50
6 Bars	3	5.64, 5.98, 5.69
	4	4.20, 3.34, 4.21
12 Bars	5	4.83, 4.77, 5.66
	6	3.28, 2.61, 2.69

Source: Dr. T. C. Tucker, Department of Soil and Water Science,
University of Arizona.

a.   Write a linear model for an analysis of the data, explain the terms, compute the analysis of
variance, and show the expected mean squares.

b.   Test the hypothesis of no difference among the means of the salinity levels.

c.   Compute the standard error of a salinity level mean.

d.   Partition the sum of squares for salinity into two orthogonal polynomial sums of squares
(linear and quadratic), each with 1 degree of freedom, and test the null hypotheses of no linear
or quadratic regression.

e.   Suppose the relative costs, $c_1:c_2$ are 10:0.1, where c_1 is the cost of setting up and maintaining
another replicate container and c_2 is the cost of measuring the weights in a container. How
many replicate containers and plants per container would be necessary to achieve a standard
error for a treatment mean of 0.75?

8.   The porosity index is a measure used by soil scientists to assist in the prediction of water movement, storage, availability, and aeration conditions of soils. A soil scientist utilized a special sampling design to take soil samples from one of the university experiment farms to measure the
porosity index of the farm soil. The farm was partitioned into fields of approximately 4 hectares
each divided into eight sections. The sampling plan included a random selection of fields from
which sections were randomly selected. Locations for soil subsamples were randomly selected
within the sections. A special staggered sampling design from Goldsmith and Gaylor (1970) was
utilized for the study. The porosity index of each soil subsample follows.

Field	Section	Porosity	Field	Section	Porosity
1	1	3.846, 3.712	9	17	5.942
	2	5.629, 2.021		18	5.014
2	3	5.087	10	19	5.143
	4	4.621		20	4.061
3	5	4.411	11	21	3.835, 2.964
	6	3.357		22	4.584, 4.398
4	7	3.991	12	23	4.193
	8	5.766		24	4.125
5	9	5.677	13	25	3.074
	10	3.333		26	3.483
6	11	4.355, 6.292	14	27	3.867
	12	4.940, 4.810		28	4.212
7	13	2.983	15	29	6.247
	14	4.396		30	4.730
8	15	5.603			
	16	3.683			

Source: Dr. A. Warrick and M. Coelho, Department of Soil and Water Science, University of Arizona.

a. Assume all effects are random. Write a linear model for the study, explain each of the terms, compute the analysis of variance for the data, and show the expected mean squares.
b. Estimate the components of variance for fields, sections, and samples.
c. Test the null hypothesis $H_0: \sigma_a^2 = 0$ for the fields' component of variance.
d. Test the null hypothesis $H_0: \sigma_e^2 = 0$ for the sections' component of variance.

9. Use the data from Exercise 5.3 to determine how many samples the plant pathologist would have to take from each of five lots of cottonseed to detect a ratio $\sigma_a^2/\sigma_e^2 = 2$ at the .01 significance level with a power of .90.

5A   Appendix: Coefficient Calculations for Expected Mean Squares in Table 5.9

$$A = \sum_{i=1}^{t} \sum_{j=1}^{r_i} \left(\frac{n_{ij}^2}{n_{i.}} \right)$$

$$= \frac{2^2 + \cdots + 1^2}{12} + \frac{1^2 + \cdots + 2^2}{13} + \frac{2^2 + \cdots + 2^2}{11} = 7.11655$$

$$B = \sum_{i=1}^{t} \sum_{j=1}^{r_i} n_{ij}^2 = 2^2 + 3^3 + \cdots + 1^2 + 2^2 = 86$$

$$D = \sum_{i=1}^{t} n_{i.}^2 = 12^2 + 13^2 + 11^2 = 434$$

$$c_1 = \frac{1}{t-1} \left(A - \frac{B}{N} \right) = \frac{1}{2} \left(7.11655 - \frac{86}{36} \right) = 2.36$$

$$c_2 = \frac{1}{t-1} \left(N - \frac{D}{N} \right) = \frac{1}{2} \left(36 - \frac{434}{36} \right) = 11.97$$

$$c_3 = \frac{1}{\left(\sum_{i=1}^{t} r_i - t \right)} (N - A) = \frac{1}{15} (36 - 7.11655) = 1.93$$

6 Factorial Treatment Designs

The factorial treatment design was introduced in Chapter 1 as a way to investigate the relationships among several types of treatments. The basic factorial treatment design in a completely randomized experiment design and its analysis are introduced in this chapter. Planned contrasts and response curve estimation, discussed in Chapter 3, are applied to the factorial treatment design. Methods to determine the number of required replications and to analyze the factorial treatment design with one replication or unequal treatment replications are discussed as well.

6.1 Efficient Experiments with Factorial Treatment Designs

Comparisons among treatments can be affected substantially by the conditions under which they occur. Frequently, clear interpretations of effects for one treatment factor must take into account the effects of other treatment factors. A special type of treatment design, **factorial treatment design**, was developed to investigate more than one factor at a time.

Factorial treatment designs produce efficient experiments. Each observation supplies information about all of the factors, and we are able to look at responses to one factor at different levels of another factor in the same experiment. The response to any factor observed under different conditions indicates whether the factors act on the experimental units independently of one another. Interaction between factors occurs when they do not act independently of one another.

Example 6.1 Compaction Effects on Asphaltic Concrete Durability

Asphalt pavements undergo water-associated deteriorations such as cracking, potholes, and surface raveling. The weakened pavement occurs when there is a break in the adhesive bond between aggregate and the asphaltic cements that make up the pavement. Research is directed to find improved asphalt pavements more resistant to deterioration.

The ability to develop superior asphalt pavement mixes requires a reliable method to test the experimental mix for bonding strength. Several methods have been developed to compact asphaltic pavement specimens in preparation for bonding strength tests.

Two factors known to have an effect on specimen bonding strength are (1) the methods used to compact the specimen during construction and (2) the aggregate type used in the asphalt mixture. If two compaction methods produce the same relative results for strength tests with two different aggregate types, then either compaction method could be used to evaluate experimental asphalt mixes for either aggregate type. If the results are dependent on aggregate type, then one or both of the compaction methods may not be adequate for discriminating between experimental mixes of asphalt.

The factorial treatment design can be used to evaluate whether the two factors act independently on the strength of the test specimens. The factorial arrangement is illustrated in Table 6.1 for test specimens prepared by two compaction methods (static and kneading) using two types of aggregate (silicious rock and basalt) for each compaction method. For illustration, specimen bonding tensile strength values are shown in Table 6.1 for the four treatments as pressure (psi) at test failure.

Factors are types of treatments such as compaction method and aggregate type, and different categories of a factor are levels of the factors. The levels of compaction method are Static and Kneading and the levels of aggregate type are Silicious Rock and Basalt. The factors are identified by uppercase letters A and B in Table 6.1. The levels of the factors are denoted A_1, A_2, ... ; B_1, B_2, ... ; and so forth. The factorial arrangement in Table 6.1 with two factors A and B each with two levels, has $2 \times 2 = 4$ treatment combinations, A_1B_1, A_1B_2, A_2B_1, and A_2B_2.

Table 6.1   Tensile strength (psi) of asphalt specimens

Aggregate Type (A)	Compaction Method (B)		Aggregate Means
	(B_1) Static	(B_2) Kneading	
Silicious (A_1)	68	60	64.0
Basalt (A_2)	65	97	81.0
Compaction Means	66.5	78.5	

Source: A. M. Al-Marshed (1981), Compaction effects on asphaltic concrete durability. M.S. thesis, Civil Engineering, University of Arizona.

The factorial treatment design consists of all possible combinations of the levels of several factors. Experiments with factorial treatment designs often are referred to as *factorials* or *factorial experiments*.

The levels of a *quantitative* factor take metrical values, whereas the levels of a *qualitative* factor are categories of the factor. Both factors in Example 6.1 are qualitative factors; the two levels of each factor are categories. Temperature exemplifies a quantitative factor with levels of 10°C, 20°C, and 30°C, for instance.

6.2   Three Types of Treatment Factor Effects

The **effect** of a factor is a change in the measured response caused by a change in the level of that factor. Three effects of interest in a factorial experiment are *simple effects, main effects,* and *interaction effects.* These effects are illustrated with population means for the factorial treatments in Example 6.1.

The population means for a factorial experiment with two factors, A and B, can be represented with cell means μ_{ij}. The term *cell mean* is derived from a tabled display of means for each of the treatment combinations, illustrated in Table 6.2 for a 2×2 factorial.

The means of the treatment combinations, μ_{11}, μ_{12}, μ_{21}, and μ_{22}, are located in the cells of the table—hence, the designation *cell means*. The means on the margins of the table are the averages of the cell means and are referred to as the *marginal means*. The overall or grand mean is the average of the cell means, $\bar{\mu}_{..} = \frac{1}{4}(\mu_{11} + \mu_{12} + \mu_{21} + \mu_{22})$.

Table 6.2   Table of means for a 2×2 factorial experiment

A	B 1	2	Factor A Means
1	μ_{11}	μ_{12}	$\bar{\mu}_{1.} = \frac{1}{2}(\mu_{11} + \mu_{12})$
2	μ_{21}	μ_{22}	$\bar{\mu}_{2.} = \frac{1}{2}(\mu_{21} + \mu_{22})$
Factor B Means	$\bar{\mu}_{.1}$ $= \frac{1}{2}(\mu_{11} + \mu_{21})$	$\bar{\mu}_{.2}$ $= \frac{1}{2}(\mu_{12} + \mu_{22})$	

Simple Effects Are Contrasts
The **simple effects** of a factor are contrasts between levels of one factor at a single level of another factor. The simple effect (l_1) of aggregate type (A) on tensile strength with static compaction (B_1) calculated from the cell means in Table 6.1 is

$$l_1 = \mu_{21} - \mu_{11} = 65 - 68 = -3$$

It measures the difference in tensile strength between basalt and silicious rock specimens when the static compaction method is used to form the specimens. The average tensile strength of silicious rock specimens was greater than that for basalt specimens. Similarly, the simple effect

$$l_2 = \mu_{22} - \mu_{12} = 97 - 60 = 37$$

measures the difference in tensile strength between basalt and silicious rock specimens when the kneading compaction method is used. In this case, the average tensile strength of the basalt specimens was greater than that for the silicious rock specimens.

Main Effects Are Average Effects of a Factor

The **main effects** of a factor are contrasts between levels of one factor averaged over all levels of another factor. The main effect of aggregate type on tensile strength is the difference between the marginal means for aggregate type in Table 6.2:

$$l_3 = \overline{\mu}_{2.} - \overline{\mu}_{1.} = 81 - 64 = 17 \tag{6.1}$$

The difference in tensile strength between basalt and silicious rock specimens is 17 psi in favor of the basalt when averaged over both compaction methods.

Upon close inspection the main effect for aggregate type can be expressed as the average of the two simple effects. Thus, from Equation (6.1)

$$l_3 = \overline{\mu}_{2.} - \overline{\mu}_{1.} = \frac{1}{2}(\mu_{21} + \mu_{22}) - \frac{1}{2}(\mu_{11} + \mu_{12})$$

$$= \frac{1}{2}(\mu_{22} - \mu_{12}) + \frac{1}{2}(\mu_{21} - \mu_{11})$$

$$= \frac{1}{2}(l_1 + l_2)$$

or $17 = \frac{1}{2}(-3 + 37)$. The main effect contrast, $l_3 = 17$, implies the basalt specimens are stronger than the silicious rock specimens. The simple effect contrast for kneading compaction, $l_2 = 37$, supports the same conclusion. However, the simple effect contrast for static compaction, $l_1 = -3$, suggests the opposite conclusion. The difference between the two simple effects indicates aggregate type and compaction method do not act independently of one another in their influence on specimen strength.

Interaction Effects Are Differences Between Simple Effects

The **interaction effect** measures differences between the simple effects of one factor at different levels of the other factor. Consider the two simple effects of aggregate type on tensile strength shown in Display 6.1.

The difference between the two simple effects, $l_4 = l_2 - l_1$, measures interaction between aggregate type and compaction method factors as they affect the tensile strength of the specimens. The difference between basalt and silicious rock specimens was 40 units greater with kneading compaction than it was with static

---

**Display 6.1   Interaction Effect of Compaction Method
with Aggregate Type**

Compaction Method	Simple Effect of Aggregate Type
Static	$l_1 = 65 - 68 = -3$
Kneading	$l_2 = 97 - 60 = 37$
Difference	$l_4 = 37 - (-3) = 40$

---

compaction. Thus, the effect of aggregate type on tensile strength measurements of asphalt is dependent on the method employed to compact the specimen for testing.

The example illustrates the caution that must be exercised in interpreting main effects and represents a situation in which interpretations should not be based on main effects. The effect of aggregate type on tensile strength of the specimens differed considerably between the static and kneading compaction methods. Although the main effect measurement suggested the basalt specimens would be stronger, it was only true for the kneading compaction method.

The 2×2 factorial interaction contrast is derived from the difference between the two simple effects for factor A.

$$AB = (\mu_{22} - \mu_{12}) - (\mu_{21} - \mu_{11}) \qquad (6.2)$$
$$= (\mu_{22} - \mu_{21} - \mu_{12} + \mu_{11})$$

The same expression may be derived from the simple effects for B.

The presence or absence of interaction is illustrated graphically in Figures 6.1 and 6.2 for a factorial arrangement with two factors, A and B, each at two levels. The graphic measure of a simple effect for each factor is illustrated in Figure 6.1. The response to A is graphed separately for each level of B. In the absence of interaction the parallel lines show an identical response to A for both levels of B. Under these circumstances, the factors act independently, and the main effects can be used to interpret the effects of the two factors separately.

The presence of interaction is illustrated in Figure 6.2. The response lines are not parallel when there is an interaction between the two factors. Differences in the magnitude of the responses (Figure 6.2a) or in direction of the responses (Figure 6.2b) represent interaction between the factors. The factors do not act independently, and interpretations should be based on simple effect contrasts.

Example 6.2   Compaction Effects on Asphaltic Concrete Revisited

Research Objectives: The variation in tensile strength of asphaltic concrete test specimens, as discussed in Example 6.1, was known to be associated with the compaction method and aggregate type used to construct the specimens. A civil engineer conducted an experiment to evaluate differences among a set of compaction methods for their effect on the tensile strength of test specimens and to determine to what extent the aggregate type affected the comparisons among the compaction methods.

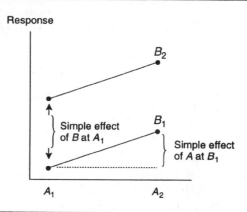

Figure 6.1   Illustration of no interaction in a factorial arrangement

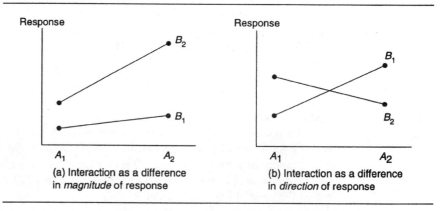

Figure 6.2   Illustration of interaction in a factorial arrangement

Treatment Design: A factorial arrangement was used with "compaction method" and "aggregate type" as factors. There were two levels of aggregate type—A_1 (basalt) and A_2 (silicious rock)—and four levels of compaction method—C_1 (static pressure), C_2 (regular kneading), C_3 (low kneading), and C_4 (very low kneading).

Experiment Design: Three replicate specimens of the asphalt concrete were constructed and tested for each of the eight treatment combinations. The 24 specimens were prepared and tested in random order for a completely randomized design.

The data for tensile strength measurements (psi) on the 24 specimens are shown in Table 6.3.

Table 6.3 Tensile strength (psi) of asphaltic concrete specimens for two aggregate types with each of four compaction methods

| Aggregate Type | | Compaction Method | | |
| | | | Kneading | |
	Static	Regular	Low	Very Low
Basalt	68	126	93	56
	63	128	101	59
	65	133	98	57
Silicious	71	107	63	40
	66	110	60	41
	66	116	59	44

Source: A. M. Al-Marshed (1981), Compaction effects on asphaltic concrete durability. M.S. thesis, Civil Engineering, University of Arizona.

6.3   The Statistical Model for Two Treatment Factors

The Cell Means Model

The observations from a factorial experiment with two factors, A with a levels and B with b levels, can be represented with the **cell means model**. The cell means model for the $a \times b$ factorial with r replications in a completely randomized design is

$$y_{ijk} = \mu_{ij} + e_{ijk} \tag{6.3}$$

$$i = 1, 2, \ldots, a \quad j = 1, 2, \ldots, b \quad k = 1, 2, \ldots, r$$

where μ_{ij} is the mean of the treatment combination $A_i B_j$ and e_{ijk} are random experimental errors with mean 0 and variance σ^2.

Least Squares Estimates of Cell Means

The cell means can be estimated by the least squares method outlined in Chapter 2. The sum of squares for experimental error is

$$SS \text{ Error} = \sum_{i=1}^{a} \sum_{j=1}^{b} \sum_{k=1}^{r} \widehat{e}_{ijk}^2 = \sum_{i=1}^{a} \sum_{j=1}^{b} \sum_{k=1}^{r} (y_{ijk} - \widehat{\mu}_{ij})^2 \tag{6.4}$$

The least squares estimators for μ_{ij} are the observed cell means of the treatment combinations.

$$\widehat{\mu}_{ij} = \frac{y_{ij.}}{r} = \bar{y}_{ij.} \qquad (6.5)$$

$$i = 1, 2, \ldots, a \qquad j = 1, 2, \ldots, b$$

The observed marginal means are unbiased estimates of the factor marginal means, so that $\widehat{\bar{\mu}}_{i.} = \bar{y}_{i..}$ and $\widehat{\bar{\mu}}_{.j} = \bar{y}_{.j.}$. The overall mean $\bar{\mu}_{..}$ is estimated with the observed grand mean, $\bar{y}_{...}$. Estimated cell and marginal means for the asphaltic concrete specimens are shown in Table 6.4.

Table 6.4 Estimated cell and marginal means for tensile strength of asphaltic concrete specimens

| Aggregate Type | Compaction Method | | | | Aggregate Means ($\bar{y}_{i..}$) |
| | Static | Kneading | | | |
		Regular	Low	Very Low	
Basalt	65.3	129.0	97.3	57.3	87.3
Silicious	67.7	111.0	60.7	41.7	70.3
Compaction means ($\bar{y}_{.j.}$)	66.5	120.0	79.0	49.5	$\bar{y}_{...} = 78.8$

Additivity and Factor Effects

The cell means μ_{ij} represent the true response for the treatment combination of level i for A and level j for B. In the absence of interaction the cell mean can be expressed as a sum of a general mean, μ, plus an effect contributed by A, say α_i, and an effect contributed by B, say β_j, so that $\mu_{ij} = \mu + \alpha_i + \beta_j$.

The effect for the ith level of factor A can be defined as $\alpha_i = \bar{\mu}_{i.} - \bar{\mu}_{..}$. The effect of A is a deviation of the marginal mean from the grand mean. The effect for the jth level of B can be defined as $\beta_j = \bar{\mu}_{.j} - \bar{\mu}_{..}$. The effects will be **fixed effects** if the levels of the factors are reproducible. In the absence of interaction, the cell mean is the sum of the grand mean and the effects of the factors for that cell:

$$\mu_{ij} = \bar{\mu}_{..} + (\bar{\mu}_{i.} - \bar{\mu}_{..}) + (\bar{\mu}_{.j} - \bar{\mu}_{..}) \qquad (6.6)$$

The effect of the ijth treatment combination, $\tau_{ij} = (\mu_{ij} - \bar{\mu}_{..})$, is the sum of the two factor effects:

$$(\mu_{ij} - \bar{\mu}_{..}) = (\bar{\mu}_{i.} - \bar{\mu}_{..}) + (\bar{\mu}_{.j} - \bar{\mu}_{..}) \qquad (6.7)$$

and the effects of the factors are *additive* in the absence of interaction.

In the presence of interaction the treatment effect will not be equal to the sum of the individual factor effects as shown in Equation (6.7). An interaction effect, denoted as $(\alpha\beta)_{ij}$, can be defined as the difference between the two sides of Equation (6.7), or

$$(\alpha\beta)_{ij} = (\mu_{ij} - \bar{\mu}_{..}) - (\bar{\mu}_{i.} - \bar{\mu}_{..}) - (\bar{\mu}_{.j} - \bar{\mu}_{..}) \tag{6.8}$$
$$= \mu_{ij} - \bar{\mu}_{i.} - \bar{\mu}_{.j} + \bar{\mu}_{..}$$

The new set of parameters, α_i, β_j, and $(\alpha\beta)_{ij}$, can be used to write the linear model for the factorial as an *effects* model

$$y_{ijk} = \mu + \alpha_i + \beta_j + (\alpha\beta)_{ij} + e_{ijk} \tag{6.9}$$

$$i = 1, 2, \ldots, a \qquad j = 1, 2, \ldots, b \qquad k = 1, 2, \ldots, r$$

where μ is the grand mean $\bar{\mu}_{..}$, α_i is the effect of the ith level of A, β_j is the effect of the jth level of B, and $(\alpha\beta)_{ij}$ is the interaction effect between the ith level of A and the jth level of B. By the nature of their definitions the sums of the effects are equal to zero. That is,

$$\sum_{i=1}^{a} \alpha_i = 0, \quad \sum_{j=1}^{b} \beta_j = 0, \quad \sum_{i=1}^{a} (\alpha\beta)_{ij} = \sum_{j=1}^{b} (\alpha\beta)_{ij} = 0 \tag{6.10}$$

6.4   The Analysis for Two Factors

Fundamental Sum of Squares Partition

The fundamental partition of the total sum of squares can be derived from the equation

$$(y_{ijk} - \bar{y}_{...}) = (\bar{y}_{ij.} - \bar{y}_{...}) + (y_{ijk} - \bar{y}_{ij.}) \tag{6.11}$$

The deviation of an observation from the grand mean $(y_{ijk} - \bar{y}_{...})$ is the sum of two parts:

- the treatment effect $(\bar{y}_{ij.} - \bar{y}_{...})$
- experimental error $(y_{ijk} - \bar{y}_{ij.})$

For example, using the observations in Table 6.3 and the means in Table 6.4, the total deviation for the first observation in Table 6.3 is

$$(y_{111} - \bar{y}_{...}) = 68 - 78.8 = -10.8.$$

The treatment deviation is

$$(\bar{y}_{11.} - \bar{y}_{...}) = (65.3 - 78.8) = -13.5$$

and the experimental error is

$$(y_{111} - \bar{y}_{11.}) = (68 - 65.3) = 2.7$$

and

$$-10.8 = -13.5 + 2.7$$

The latter deviation $(y_{ijk} - \bar{y}_{ij.}$ is a measure of experimental error for the observation in a properly replicated experiment. If both sides of Equation (6.11) are squared and summed over all observations, the result is

$$\sum_{i=1}^{a} \sum_{j=1}^{b} \sum_{k=1}^{r} (y_{ijk} - \bar{y}_{...})^2 = r \sum_{i=1}^{a} \sum_{j=1}^{b} (\bar{y}_{ij.} - \bar{y}_{...})^2 \qquad (6.12)$$

$$+ \sum_{i=1}^{a} \sum_{j=1}^{b} \sum_{k=1}^{r} (y_{ijk} - \bar{y}_{ij.})^2$$

or

$$SS \text{ Total} = SS \text{ Treatment} + SS \text{ Error}$$

Any crossproducts formed by squaring the right-hand side of Equation (6.11) sum to zero. There are a total of rab observations, so that SS Total has $(rab - 1)$ degrees of freedom. With $t = ab$, SS Treatment has $(ab - 1)$ degrees of freedom and the remaining $ab(r - 1)$ degrees of freedom are associated with SS Error.

Sums of Squares for Factorial Effects

The treatment effect $(\bar{y}_{ij.} - \bar{y}_{...})$ in Equation (6.11) can be expressed as the identity

$$(\bar{y}_{ij.} - \bar{y}_{...}) = (\bar{y}_{i..} - \bar{y}_{...}) + (\bar{y}_{.j.} - \bar{y}_{...}) + (\bar{y}_{ij.} - \bar{y}_{i..} - \bar{y}_{.j.} + \bar{y}_{...}) \qquad (6.13)$$

where the treatment effect is a sum of three effects:

- factor A main effect $(\bar{y}_{i..} - \bar{y}_{...})$
- factor B main effect $(\bar{y}_{.j.} - \bar{y}_{...})$
- interaction $(\bar{y}_{ij.} - \bar{y}_{i..} - \bar{y}_{.j.} + \bar{y}_{...})$

For example, the treatment effect for basalt with low kneading in Table 6.4 is

$$(\bar{y}_{13.} - \bar{y}_{...}) = 97.3 - 78.8 = 18.5$$

The main and interaction effects are

- Basalt main effect :

$$(\bar{y}_{1..} - \bar{y}_{...}) = 87.3 - 78.8 = 8.5$$

- Low kneading main effect:

$$(\bar{y}_{.3.} - \bar{y}_{...}) = 79.0 - 78.8 = 0.2$$

- Interaction:

$$(\bar{y}_{13.} - \bar{y}_{1..} - \bar{y}_{.3.} + \bar{y}_{...}) = 97.3 - 87.3 - 79.0 + 78.8 = 9.8$$

and

$$18.5 = 8.5 + 0.2 + 9.8$$

If both sides of Equation (6.13) are squared and summed over all observations the left-hand side is the SS Treatment. The treatment sum of squares is partitioned into three components represented by the effects on the right-hand side of Equation (6.13). Any crossproducts formed by squaring the right-hand side will sum to zero.

The sum of squares for the first component will be the sum of squares among the marginal means for A

$$SSA = rb \sum_{i=1}^{a} (\bar{y}_{i..} - \bar{y}_{...})^2 \tag{6.14}$$

and the second will be the sum of squares among the marginal means for B

$$SSB = ra \sum_{j=1}^{b} (\bar{y}_{.j.} - \bar{y}_{...})^2 \tag{6.15}$$

The sum of squares for the third component

$$SS(\boldsymbol{AB}) = r \sum_{i=1}^{a} \sum_{j=1}^{b} (\bar{y}_{ij.} - \bar{y}_{i..} - \bar{y}_{.j.} + \bar{y}_{...})^2 \tag{6.16}$$

is the sum of squares for interaction. That part of the treatment sum of squares is not explained by the sum of squares attributed to the two factors as SSA and SSB. Consequently, the additive partition of SS Treatment is

$$SS \text{ Treatment} = SSA + SSB + SS(\boldsymbol{AB})$$

The $(ab - 1)$ degrees of freedom for the treatment sum of squares are allocated to the three partitions of SS Treatment. The factors, $\boldsymbol{A}$ and $\boldsymbol{B}$, have a and b levels respectively, therefore SSA and SSB have $(a - 1)$ and $(b - 1)$ degrees of freedom. The remaining degrees of freedom allocated to the sum of squares for interaction are the degrees of freedom for treatments $(ab - 1)$ minus the degrees of freedom for the separate factor sums of squares $(a - 1)$ and $(b - 1)$, or $(ab - 1) - (a - 1) - (b - 1) = (a - 1)(b - 1)$. The degrees of freedom for interaction sums of squares in factorials is the product of the degrees of freedom for the factors included in the interaction.

The complete partition of the total sum of squares for a factorial arrangement with two factors is summarized in the analysis of variance shown in Table 6.5.

The derivation of the sum of squares partitions from solutions for the least squares normal equations of the effects model is shown in Appendix 6A. The

Table 6.5   Analysis of variance for a two-factor treatment design

Source of Variation	Degrees of Freedom	Sum of Squares	Mean Square	Expected Mean Square
Total	$rab - 1$	SS Total		
Factor A	$a - 1$	SSA	MSA	$\sigma_e^2 + rb\theta_a^2$
Factor B	$b - 1$	SSB	MSB	$\sigma_e^2 + ra\theta_b^2$
AB Interaction	$(a - 1)(b - 1)$	$SS(AB)$	$MS(AB)$	$\sigma_e^2 + r\theta_{ab}^2$
Error	$ab(r - 1)$	SSE	MSE	σ_e^2

$$\theta_a^2 = \sum_{i=1}^{a} (\bar{\mu}_{i.} - \bar{\mu}_{..})^2 / (a - 1) \qquad \theta_b^2 = \sum_{j=1}^{b} (\bar{\mu}_{.j} - \bar{\mu}_{..})^2 / (b - 1)$$

$$\theta_{ab}^2 = \sum_{i=1}^{a} \sum_{j=1}^{b} (\mu_{ij} - \bar{\mu}_{i.} - \bar{\mu}_{.j} + \bar{\mu}_{..})^2 / (a - 1)(b - 1)$$

Table 6.6   Analysis of variance for tensile strength of asphalt specimens in a 4×2 factorial arrangement

Source of Variation	Degrees of Freedom	Sum of Squares	Mean Square	F	Pr > F
Total	23	19,274.50			
Compaction (C)	3	16,243.50	5,414.50	569.95	.000
Aggregate (A)	1	1,734.00	1,734.00	182.53	.000
Interaction (AC)	3	1,145.00	381.67	40.18	.000
Error	16	152.00	9.50		

analysis of variance for the data from Table 6.3 on the asphalt concrete specimens is shown in Table 6.6.

Tests of Hypotheses About Factor Effects

Inferences about individual factor effects depend upon the presence or absence of interaction. Significance of interaction is determined before any determinations of significance for main effects of the factors.

In the absence of interaction, $(\alpha\beta)_{ij} = \mu_{ij} - \bar{\mu}_{i.} - \bar{\mu}_{.j} + \bar{\mu}_{..} = 0$ from Equation (6.8), and $\theta_{ab}^2 = 0$ in the expected mean square for AB interaction. The null hypothesis for interaction

$$H_0: (\alpha\beta)_{ij} = \mu_{ij} - \bar{\mu}_{i.} - \bar{\mu}_{.j} + \bar{\mu}_{..} = 0 \quad \text{for all } i, j$$

versus the alternative

$$H_a: (\alpha\beta)_{ij} = \mu_{ij} - \bar{\mu}_{i.} - \mu_{.j} + \bar{\mu}_{..} \neq 0 \quad \text{for some } i, j$$

is tested with

$$F_0 = \frac{MS(\boldsymbol{AB})}{MSE} \tag{6.17}$$

with critical value $F_{\alpha,(a-1)(b-1),ab(r-1)}$.

If there are no differences among the marginal means for A, then $\alpha_i = \overline{\mu}_{i.} - \overline{\mu}_{..} = 0$ and $\theta_a^2 = 0$ in the expected mean square for A. The null hypothesis of equality among the means

$$H_0: \quad \overline{\mu}_{1.} = \overline{\mu}_{2.} = \cdots = \overline{\mu}_{a.}$$

versus the alternative

$$H_a: \quad \overline{\mu}_{i.} \neq \overline{\mu}_{k.} \quad \text{for some } i, k$$

is tested with the ratio

$$F_0 = \frac{MS\boldsymbol{A}}{MSE} \tag{6.18}$$

with critical value $F_{\alpha,(a-1),ab(r-1)}$.

If there are no differences among the marginal means for B, then $\beta_j = \overline{\mu}_{.j} - \overline{\mu}_{..} = 0$ and $\theta_b^2 = 0$ in the expected mean square for B. The null hypothesis of equality among the means

$$H_0: \quad \overline{\mu}_{.1} = \overline{\mu}_{.2} = \cdots = \overline{\mu}_{.b}$$

versus the alternative

$$H_a: \quad \overline{\mu}_{.j} \neq \overline{\mu}_{.m} \quad \text{for some } j, m$$

is tested with the ratio

$$F_0 = \frac{MS\boldsymbol{B}}{MSE} \tag{6.19}$$

with critical value $F_{\alpha,(b-1),ab(r-1)}$.

F Tests for Aggregate Type and Compaction Method Effects

The F test for interaction, $F_0 = MS(\boldsymbol{AC})/MSE = 381.67/9.50 = 40.18$ in Table 6.6, indicates a significant interaction between aggregate type and compaction method since F_0 exceeds $F_{.05,3,16} = 3.24$. The marginal means for aggregate type are significantly different since $F_0 = MS\boldsymbol{A}/MSE = 1734.00/9.50 = 182.53$ exceeds $F_{.05,1,16} = 4.49$. The marginal means for compaction method are also different since $F_0 = MSC/MSE = 5414.50/9.50 = 569.95$ exceeds $F_{.05,3,16} = 3.24$. The significance level for each of the tests is listed as $Pr > F = .000$ in the rightmost column of Table 6.6.

The significant interaction can modify any inferences based on the significant differences among the marginal means of aggregate and compaction. The summary table of cell and marginal means in Table 6.4 will aid in the interpretation of the results.

Standard Errors and Interval Estimates for Means

The standard errors for the estimated marginal and cell means of the factorial experiment are

Aggregate:
$$s_{\bar{y}_{i..}} = \sqrt{\frac{MSE}{rb}} = \sqrt{\frac{9.5}{(3)(4)}} = 0.89$$

Compaction:
$$s_{\bar{y}_{.j.}} = \sqrt{\frac{MSE}{ra}} = \sqrt{\frac{9.5}{(3)(2)}} = 1.26 \qquad \textbf{(6.20)}$$

Cell means:
$$s_{\bar{y}_{ij.}} = \sqrt{\frac{MSE}{r}} = \sqrt{\frac{9.5}{3}} = 1.78$$

The Student t with $ab(r-1)$ degrees of freedom is required for interval estimates of the marginal and cell means. A 95% confidence interval estimate requires $t_{.025,16} = 2.12$. The interval estimate for a cell mean is

$$\bar{y}_{ij.} \pm t_{.025,16}(s_{\bar{y}_{ij.}}) \qquad \textbf{(6.21)}$$

For example, the 95% confidence interval estimate for the basalt aggregate with static compaction mean is

$$65.3 \pm 2.12(1.78)$$
$$(61.5, \ 69.1)$$

Interval estimates are calculated similarly for the other means upon substitution of the appropriate mean and standard error estimate from Equation (6.20).

Multiple Contrasts Assist Interpretations of Significant Interaction Effects

The significant interaction between aggregate type and compaction method indicates the simple effects of one factor differ among levels of another factor. Consequently, tests of hypotheses about the treatment factors initially should be based on simple effect contrasts among the cell means.

The specific research hypotheses for the study will dictate the contrasts among the cell means required to investigate the simple effects. One general research question for this study might ask which of the aggregate types results in the strongest specimens for each of the compaction methods. Another hypothesis might address the effect of the kneading compaction methods relative to static compaction.

Contrasts among the cell means in Table 6.4 can be used to test the aggregate type simple effects (basalt versus silicious) at each level of compaction to address the question of which aggregate type results in the strongest specimens for each compaction method. The four contrasts and their 95% simultaneous confidence intervals are shown in Display 6.2. The experimentwise error rate for the family of four tests can be controlled with the Bonferroni t statistic, $t_{.025,4,16} = 2.81$ from Appendix Table V.

Display 6.2   Estimated Contrasts for Aggregate Simple Effects

Compaction Method	Contrast (Basalt–Silicious)	95% SCI (L, U)
Static	$c_1 = \bar{y}_{11.} - \bar{y}_{21.} = 65.3 - 67.7 = -2.4$	$(-9.5, 4.7)$
Regular	$c_2 = \bar{y}_{12.} - \bar{y}_{22.} = 129.0 - 111.0 = 18.0$	$(10.9, 25.1)$
Low	$c_3 = \bar{y}_{13.} - \bar{y}_{23.} = 97.3 - 60.7 = 36.6$	$(29.5, 43.7)$
Very low	$c_4 = \bar{y}_{14.} - \bar{y}_{24.} = 57.3 - 41.7 = 15.6$	$(8.5, 22.7)$

Standard error $s_c = \sqrt{\dfrac{MSE}{r}[1^2 + (-1)^2]} = \sqrt{\dfrac{9.5}{3}(2)} = 2.52$

Summary Statements About Effects of the Factors

There is no difference in specimen tensile strength between the two aggregate types with static compaction since the interval includes 0. The specimens constructed from basalt have greater tensile strength than those constructed from the silicious rock for the kneading compaction methods since lower limits of the 3 intervals are greater than 0 and the greatest difference between the aggregate types was found with the low kneading compaction.

The comparisons to consider depend on the nature of the problem and the information required from the study. Those comparisons discussed in the preceding paragraphs were made to illustrate the interpretation of interaction. They exploited the notion that inferences regarding comparisons between aggregate types or among compaction methods depend on the other factor in the study.

A comparison among the marginal means for a factor can be informative when all of its simple effects are of a similar direction and magnitude. The comparisons among marginal means result in a more general inference about the factor, and they are more precise with smaller standard errors; see Equation (6.20). In the presence of interaction caution must be exercised in making generalizations. The difference between the marginal means for basalt and silicious aggregate types, $c = (87.3 - 70.3) = 17.0$, with standard error $s_c = \sqrt{2MSE/rb} = \sqrt{2(9.5)/12} = 1.26$, will be significant. The main effect estimate is a difference averaged over all compaction methods. The simple effects shown in Display 6.2 for aggregate type range from $c_1 = -2.4$ to $c_3 = 36.6$, depending on the compaction method. Generalizations about aggregate types based on the main effect estimate would be misleading.

Residual Analysis to Evaluate Assumptions

The assumptions for the model regarding homogeneity of variances and normal distribution can be evaluated with the residuals as discussed in Chapter 4. The residuals for the two-factor factorial are computed as the deviations of the observed values from the estimated means for each cell in the arrangement. The residual for any cell is $\widehat{e}_{ijk} = y_{ijk} - \widehat{y}_{ijk} = y_{ijk} - \bar{y}_{ij.}$. For example, the residual for the first observation in Table 6.3 is

$$\widehat{e}_{111} = y_{111} - \bar{y}_{11.} = 68 - 65.3 = 2.7$$

The plots for the square root of absolute residuals versus the estimated values and the normal probability plot of the residuals are shown in Figures 6.3a and b. The plots show no strong evidence of heterogeneous variance or nonnormality. The Levene (Med) test for homogeneity of variance is left as an exercise for the reader with reference to Chapter 4 on methods for evaluating assumptions.

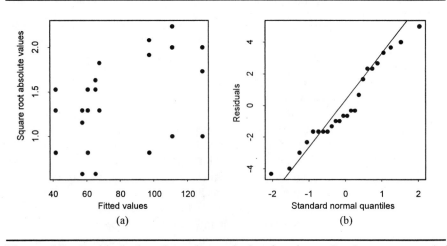

Figure 6.3 Residual plots from the analysis of variance for data on strength in Example 6.2: (a) square root of the absolute residuals vs. estimated values and (b) normal probability plot of residuals

6.5 Using Response Curves for Quantitative Treatment Factors

Response trend curves for quantitative treatment factors were estimated with orthogonal polynomials in Section 3.3. Recall that the estimated response curve has the advantage of portraying the relationship between the response variable and the treatment factor throughout the numerical range of the factor that was used in the study. The evaluation of response trend curves is extended to the two-factor experiment in this section. The analysis is discussed initially for experiments with one quantitative factor and one qualitative factor. Subsequently, responses will be analyzed for experiments with two quantitative factors.

One Quantitative Factor and One Qualitative Factor

Ascertaining whether there is interaction between the quantitative and qualitative factor is the first objective in the analysis. The response to the quantitative factor will be different across levels of the qualitative factor when the two factors interact. In that case, the response curves for the quantitative factor can be estimated separately for each level of the qualitative factor. In the absence of interaction, the response trend to the quantitative factor will be similar at all levels of the qualitative factor, and a single response curve will suffice for description of the process with respect to the quantitative factor.

Example 6.3 Heavy Metals in Sewage Sludge

Sludge is a dried product remaining from processed sewage; it contains nutrients beneficial to plant growth. It can be used for fertilizer on agricultural crops provided it does not contain toxic levels of certain elements such as heavy metals. Typically, the levels of metals in sludge are assayed by growing plants in media containing different doses of the sludge.

Research Hypothesis: A soil scientist hypothesized the concentration of certain heavy metals in sludge would differ among the metropolitan areas from which the sludge was obtained. The variation could be a result of any number of causes, including different industrial bases surrounding the areas. If this were true, then recommendations for applications on crops would have to be preceded by knowledge about the source of the material. An assay was planned to determine whether there was significant variation in heavy metal concentrations among diverse metropolitan areas.

Treatment Design: The investigator obtained sewage sludge from treatment plants located in three different metropolitan areas. Barley plants were grown in a sand medium to which the sludge was added as a fertilizer. The sludge was added to the sand at three different rates: 0.5, 1.0, and 1.5 metric tons/acre. The factorial arrangement for the treatment design consisted of one qualitative factor, "city," with three levels and one quantitative factor, "rate," with three levels.

Experiment Design: Each of the nine treatment combinations was assigned to four replicate containers in a completely randomized design. The containers were arranged completely at random in a growth chamber. At a certain stage of growth the zinc content in parts per million was determined for the barley plants grown in each of the containers. The data are shown in Table 6.7, and the analysis of variance is shown in Table 6.8. The manual calculations for linear and quadratic sums of squares partitions are shown in Table 6.9.

Table 6.7 Zinc content (ppm) of barley plants grown in media containing sludge at three rates from three metropolitan areas

City and Rate(MT/hectare)								
A			B			C		
0.5	1.0	1.5	0.5	1.0	1.5	0.5	1.0	1.5
26.4	25.2	26.0	30.1	47.7	73.8	19.4	23.2	18.9
23.5	39.2	44.6	31.0	39.1	71.1	19.3	21.3	19.8
25.4	25.5	35.5	30.8	55.3	68.4	18.7	23.2	19.6
22.9	31.9	38.6	32.8	50.7	77.1	19.0	19.9	21.9

Source: J. Budzynski, Department of Soil and Water Science, University of Arizona.

Table 6.8 Analysis of variance for zinc content in barley plants grown in media containing sewage sludge at three different rates from three metropolitan areas

Source of Variation	Degrees of Freedom	Sum of Squares	Mean Square	F	Pr > F
Total	35	9993.38			
Rate (R)	2	1945.45	972.72	50.71	.000
Rate linear	1	1944.00	1944.00	101.35	.000
Rate quadratic	1	1.45	1.45	0.08	.786
City (C)	2	5720.67	2860.34	149.13	.000
Rate × City (RC)	4	1809.40	452.35	23.58	.000
Rate linear × City (RC)	2	1760.15	880.07	45.88	.000
Rate quadratic × City	2	49.25	24.63	1.28	.293
Error	27	517.86	19.18		

The analysis of variance in Table 6.8 indicates significant interaction between Rate and City and significant main effects for both factors ($Pr > F = .000$). The 2 degrees of freedom for Rate sum of squares partition into 1 degree of freedom each for the linear and quadratic rates. The F test indicates significance for the linear regression partition ($F_0 = 101.35$) and nonsignificance ($F_0 = 0.08$) for the quadratic partition for Rate.

The 2 degrees of freedom for each of the Rate by City interaction sums of squares indicate the variability among cities in linear and quadratic regression coefficients for Rate. The interaction between Rate linear regression and City is significant ($F_0 = 45.88$), but the interaction between Rate quadratic regression and City is not significant ($F_0 = 1.28$).

Interpret Factor Effects with Regression Contrasts

The significant interactions between city and linear regression on rate of sludge application suggests that interpretations should be based on separate regression lines for each city. The estimated linear regression lines for each city are plotted in

Table 6.9 Calculation of linear and quadratic contrast sums of squares partitions for rate and rate $\times$ city interaction

City	Rate (metric tons/hectare)			Linear $\Sigma P_{1j}\bar{y}_{ij.}$	Quadratic $\Sigma P_{2j}\bar{y}_{ij.}$
	0.5	1.0	1.5		
A	24.55	30.45	36.18	11.63	− 0.17
B	31.18	48.20	72.60	41.42	7.38
C	19.10	21.90	20.05	0.95	− 4.65
Means ($\bar{y}_{.j.}$)	24.94	33.52	42.94	18.00	0.84
Linear (P_{1j})	− 1	0	1		
Quadratic (P_{2j})	1	− 2	1		

$$
\begin{aligned}
SS[R \text{ linear}] &= ra[\Sigma P_{1j}\bar{y}_{.j.}]^2/\Sigma P_{1j}^2 = 12[18]^2/2 = 1944 \\
SS[R \text{ quadratic}] &= ra[\Sigma P_{2j}\bar{y}_{.j.}]^2/\Sigma P_{2j}^2 = 12[0.84]^2/6 = 1.41 \\
SS[R \text{ linear} \times C] &= r\Sigma_i[\Sigma_j P_{1j}\bar{y}_{ij.}]^2/\Sigma P_{1j}^2 - SS[R \text{ linear}] \\
&= \frac{4[11.63^2 + 41.42^2 + 0.95^2]}{(2)} - SS[R \text{ linear}] \\
&= 1760 \\
SS[R \text{ quad} \times C] &= r\Sigma_i[\Sigma_j P_{2j}\bar{y}_{ij.}]^2/\Sigma P_{2j}^2 - SS[R \text{ quad}] \\
&= \frac{4[(-0.17)^2 + 7.38^2 + (-4.65)^2]}{(6)} - SS[R \text{ quad}] \\
&= 49
\end{aligned}
$$

Figure 6.4 along with the cell means. The plot illustrates the Rate(linear) by City interaction. The response to rate is linear for each city. The significance of the interaction between city and the linear partition for rate shows up in the plot as a different linear response of zinc to rate for each city.

The linear regression contrasts for rate are simple effects for rate computed for each of the cities. The linear regression lines can be computed for each city from the estimated effects in Table 6.9 or with a standard regression computer program. The regression line can be computed following the procedure in Section 3.3 using the cell means in Table 6.9. The linear orthogonal polynomial coefficient estimate. for city A is

$$
\begin{aligned}
a_1 &= \Sigma P_{1j}\bar{y}_{1j}/\Sigma P_{1j}^2 \\
&= \frac{-1(24.55) + 0(30.45) + 1(36.18)}{[(-1)^2 + 0^2 + 1^2]} = \frac{11.63}{2} = 5.8
\end{aligned}
$$

The mean for city A is $\bar{y}_{1..} = 30.39$. With $\lambda_1 = 1$, $\bar{R} = 1.0$, and $d = 0.5$ (Display 3.5), the transformation to an equation in terms of rate (R) is

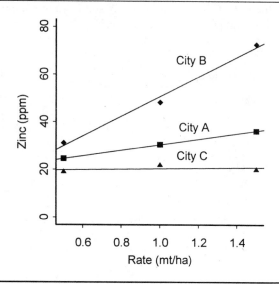

Figure 6.4   Mean zinc content versus rate of sludge application for three cities

$$\hat{y} = \bar{y}_{1..} + a_1 P_1$$
$$= 30.39 + 5.8 \frac{(R - 1.0)}{0.5} = 18.79 + 11.6(R)$$

The linear contrast for city C (0.95 in Table 6.9) is considerably smaller than that for the other two cities, 11.63 and 41.42 for cities A and B, respectively. The standard error for the rate linear contrast on the basis of cell means in Table 6.9 for any particular city is

$$s_c = \sqrt{\frac{(MSE)[(-1)^2 + 0^2 + 1^2]}{r}} = \sqrt{\frac{(19.18)[2]}{4}} = 3.1$$

The 95% simultaneous confidence intervals for the three linear contrasts require the Bonferroni $t_{.05,3,27} = 2.55$. The 95% SCI for cities A, B, and C are, respectively, (3.73, 19.54), (33.52, 49.33), and (−6.96, 8.86). The linear responses for cities A and B are significantly positive, with city B having the largest positive linear contrast. The interval for city C includes 0, and we can conclude that zinc will accumulate in barley crops fertilized with increasing amounts of the sludge by-product to the greatest extent from city B and to a lesser extent from that of city A, but that there will be no significant accumulation from that of city C.

Two Quantitative Factors

The response to two quantitative factors can be represented by a polynomial equation with two independent variables. The degree of the polynomial will depend on the number of levels for each of the factors. First-degree equations can represent a factor with two levels, second-degree equations for three levels, and so forth. The geometric representation of a polynomial equation with two independent variables is a response surface in three dimensions. For example, suppose the levels of factor A and factor B are represented by two metrical variables x_1 and x_2 in a quadratic polynomial. The second-degree polynomial equation

$$y = \beta_o + \beta_1 x_1 + \beta_2 x_1^2 + \beta_3 x_2 + \beta_4 x_2^2 + \beta_5 x_1 x_2 \tag{6.22}$$

is an empirical function commonly used for the approximation of a response surface in experimental studies. The quadratic surface can be explored for factor levels that result in the optimum response or different combinations of factor levels with equivalent responses.

The analysis of the factorial experiment with two quantitative factors consists of orthogonal polynomial partitions for the factor main effect and interaction sums of squares. The nature of the polynomial response function can be determined from these partitions. A graph of the responses can be used as an aid to interpret the role that each of the factors plays in the response.

Example 6.4   Water Uptake by Barley Plants

Deposited salts accumulate in soils irrigated for agronomic and horticultural crops. The increased soil salinity eventually suppresses plant development and crop yields.

Research Hypothesis: An investigator hypothesized that exposure of plants to high levels of salts in their media over time eventually inhibits the plant's uptake of water and nutrients from the soil, thus suppressing the growth and development of plants. An experiment was conducted with barley plants to measure the effect of growth medium salinity on water uptake by the plants.

Treatment Design: A factorial arrangement was used with "salinity of media" and "age of plant" in days as the two factors. The plants were grown in nutrient solutions with the salinity level adjusted to three different levels. The salinity levels expressed as units of osmotic pressure were 0, 6, and 12 bars. Plants were harvested at 14, 21, and 28 days.

Experiment Design: Each of the nine treatment combinations of salinity and days was assigned to two replicate containers in a completely randomized design. The containers were placed in the growth chamber in a completely randomized arrangement.

One of the measurements made at harvest was the amount of water uptake by the plants during the experiment. Water uptake is expressed as milliliters of water uptake per 100 grams of plant dry weight. The data are shown in Table 6.10, and the analysis of variance is shown in Table 6.11. Manual calculations for the sums of squares partitions are illustrated in Table 6.12.

Computational Notes: The arrangement in Table 6.12 is convenient for manual computation of the sums of squares partitions from the cell means. Main effect and

Table 6.10 Water uptake (ml/100 g) by barley plants at 14, 21, and 28 days grown in solutions with salinity levels of 0, 6, and 12 bars

Salinity	0 bars			6 bars			12 bars		
Days	14	21	28	14	21	28	14	21	28
	2.2	5.0	13.2	3.7	5.9	9.4	2.8	4.5	7.6
	3.3	5.7	12.4	4.5	7.2	11.0	3.4	5.9	8.3
Means ($\bar{y}_{ij.}$)	2.75	5.35	12.80	4.10	6.55	10.20	3.10	5.20	7.95

Source: Dr. T. C. Tucker, Department of Soil and Water Science, University of Arizona.

Table 6.11 Analysis of variance for water uptake by barley plants

Source of Variation	Degrees of Freedom	Sum of Squares	Mean Square	F	Pr > F
Total	17	184.73			
Salinity (S)	2	9.51	4.75	8.52	.008
S linear	1	7.21	7.21	12.92	.006
S quadratic	1	2.30	2.30	4.12	.073
Days (D)	2	151.99	75.99	136.24	.000
D linear	1	147.00	147.00	263.55	.000
D quadratic	1	4.99	4.99	8.94	.015
Salinity × Days (SD)	4	18.21	4.55	8.16	.005
S lin × D lin	1	13.52	13.52	24.24	.001
S lin × D quad	1	2.94	2.94	5.27	.047
S quad × D lin	1	1.21	1.21	2.18	.174
S quad × D quad	1	0.53	0.53	0.96	.353
Error	9	5.02	0.56		

interaction partitions can be computed in the same table. Main effect partitions are normally computed from marginal means. However, cell means are used in Table 6.12 to compute the main effect partitions: thus, repeated values of polynomial contrast coefficients for main effect partitions are necessary for the cell means that contribute to each of their respective marginal means. For example, S_l, the linear contrast for salinity, requires a -1 for each cell at the 0-bar level, a 0 for each cell at the 6-bar level, and a $+1$ for each cell at the 12-bar level.

The coefficients for the interaction partitions are determined as the product of the coefficients for the corresponding components of the interaction. For example, the coefficients for the interaction between salinity(linear) and day(linear) in Table 6.12 are formed as the products of the coefficients for the linear contrast of the main effects for the two factors. Each coefficient for $S_l D_l$ is a product of the corresponding coefficients for S_l and D_l. The computation is exhibited in Display 6.3.

Table 6.12 Calculation of linear and quadratic sums of squares partitions for salinity and day main effect and interaction sums of squares

rs	Days	Means ($\bar{y}_{ij.}$)	S_l	S_q	D_l	D_q	S_lD_l	S_lD_q	S_qD_l	S_qD_q
0	14	2.75	-1	1	-1	1	1	-1	-1	1
	21	5.35	-1	1	0	-2	0	2	0	-2
	28	12.80	-1	1	1	1	-1	-1	1	1
6	14	4.10	0	-2	-1	1	0	0	2	-2
	21	6.55	0	-2	0	-2	0	0	0	4
	28	10.20	0	-2	1	1	0	0	-2	-2
2	14	3.10	1	1	-1	1	-1	1	-1	1
	21	5.20	1	1	0	-2	0	-2	0	-2
	28	7.95	1	1	1	1	1	1	1	1
		$\Sigma P_{cij}\bar{y}_{ij.}$	-4.7	-4.6	21.0	6.7	-5.2	-4.2	2.7	3.1
		ΣP_{cij}^2	6.0	18.0	6.0	18.0	4.0	12.0	12.0	36.0
		SS^*	7.2	2.3	147.0	5.0	13.5	2.9	1.2	0.5
		$Effect^\dagger$	-0.78	-0.25	3.50	0.37	-1.30	-0.35	0.23	0.09

$^* = r(\Sigma P_{cij}\bar{y}_{ij.})^2/\Sigma P_{cij}^2$; $^\dagger Effect = \Sigma P_{cij}\bar{y}_{ij.}/\Sigma P_{cij}^2$

Display 6.3 **Computation of Coefficients for Orthogonal Polynomial Interaction Contrasts**								
S_l: -1	-1	-1	0	0	0	1	1	1
D_l: -1	0	1	-1	0	1	-1	0	1
S_lD_l: 1	0	-1	0	0	0	-1	0	1

Interpretations for the Regression Contrasts

The F_0 statistics in Table 6.11 indicate significant interaction of the salinity(linear) partition with the day(linear) and day(quadratic) partitions. Main effect partitions for salinity(linear), day(linear), and day(quadratic) were also significant. None of the salinity quadratic effects were significant at the .05 level of significance. The response of water uptake to salinity level was the primary focus of the investigation. A profile plot facilitates the interpretation of the results with significant interaction. A plot of the cell means and the linear regression of water uptake on salinity for each day is shown in Figure 6.5.

The linear regression lines for water uptake on salinity computed separately for each day are shown in Figure 6.5 along with the estimated treatment means. The salinity linear contrasts for each day computed from the S_l column in Table 6.12 are shown in Display 6.4 along with their 95% simultaneous confidence intervals using the Bonferroni $t_{.05,3,9} = 2.93$. The 95% SCI indicate salinity had no effect on the water uptake by the plants for the first three weeks up to day 21 since the intervals for 14 and 21 days include 0. However, by the end of the fourth week, day 28,

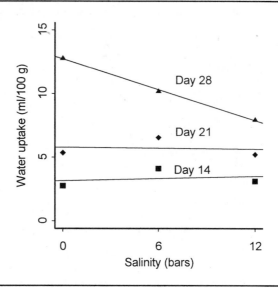

Figure 6.5   Water uptake of barley plants at three salinity levels for 14, 21, and 28 days

	Display 6.4 Linear Contrasts for Salinity for Each Day	
		95% SCI
Day	*Linear Contrast*	(*L, U*)
14	$S_l = -1(2.75) + 0(4.10) + 1(3.10) = 0.35$	(−1.85, 2.55)
21	$S_l = -1(5.35) + 0(6.55) + 1(5.20) = -0.15$	(−2.35, 2.05)
28	$S_l = -1(12.80) + 0(10.20) + 1(7.95) = -4.85$	(−7.05, −2.65)
Standard error	$s_c = \sqrt{MSE[(-1)^2 + 0^2 + 1^2]/r} = \sqrt{0.56} = 0.75$	

water uptake by the plants decreased with an increase in the salinity of the medium since the upper limit of the confidence interval for the linear contrast was -2.65.

The differences among the linear responses to salinity resulted in significant interactions between the salinity(linear) partition and the day partitions. The significant interaction between salinity(linear) and day(quadratic) effects indicates that the linear response to salinity changes in a quadratic manner over days.

From day 14 to day 28 the S_l contrast values decrease in a quadratic trend. The decrease in the contrast value from day 14 to day 21 is (−0.15) − 0.35 = −0.50, whereas the decrease in the contrast from day 21 to day 28 is (−4.85) − (−0.15) = −4.70.

Computing a Response Surface Equation

The polynomial equation relating water uptake to salinity and days will include the terms judged significant in the analysis of variance tests. The significant terms were S_l, D_l, D_q, S_lD_l, and S_lD_q. The equation in terms of S and D will be

$$y = \beta_o + \beta_1 S + \beta_2 D + \beta_3 D^2 + \beta_4 SD + \beta_5 SD^2 \tag{6.23}$$

The coefficients for Equation (6.23) may be estimated directly from the data with a multiple regression program, or they may be evaluated from the orthogonal polynomial equation as shown in Section 3.3. The polynomial coefficient contrasts in Display 3.5 are used for the transformations. Let P_{cs} and P_{cd} represent the polynomial contrasts for salinity and day, respectively. The orthogonal polynomial equation can be written as

$$y = \mu + \alpha_1 P_{1s} + \gamma_1 P_{1d} + \gamma_2 P_{2d} + (\alpha\gamma)_{11} P_{1s}P_{1d} + (\alpha\gamma)_{12} P_{1s}P_{2d} \tag{6.24}$$

where α_1 is the linear polynomial coefficient for salinity, γ_1 is the linear coefficient for day, and $(\alpha\gamma)_{12}$ is the interaction coefficient for salinity linear by day quadratic. The estimates of the coefficients in Equation (6.24) are calculated in Table 6.12 as Effect $= \Sigma P_{cij}\bar{y}_{ij.}/\Sigma P_{cij}^2$. For example, the estimate for $(\alpha\gamma)_{11}$ from the S_lD_l line in Table 6.12 is $-5.2/4 = -1.3$. The term $(\alpha\gamma)_{11}P_{1s}P_{1d}$ in Equation (6.24) with $\lambda_1 = 1$ becomes

$$-1.3\lambda_1 \left(\frac{S-6}{6}\right) \lambda_1 \left(\frac{D-21}{7}\right) = -0.0310(S-6)(D-21)$$

The remaining estimates are computed in the same manner with $\widehat{\mu} = \bar{y}_{...} = 6.44$, $\lambda_1 = 1$, and $\lambda_2 = 3$, and the resulting equation is

$$\widehat{y} = 5.70 - 0.0133(-6) + 0.50(D-21) + 0.0227(D-21)^2$$
$$-0.0310(S-6)(D-21) - 0.0036(S-6)(D-21)^2$$

These equations can be used to explore response surfaces for maxima or minima, or they may be used to determine values of the factors that result in equivalent responses. Specialized tools for these methods are discussed later in Chapter 13, Response Surface Designs.

6.6 Three Treatment Factors

The inclusion of additional factors in the treatment design increases the complexity of interaction patterns among the treatment factors. The number of treatment combinations increases rapidly as factors are added to the design. The three-factor design with a levels of A, b levels of B, and c levels of C has abc treatment combinations. A fourth factor, D, with d levels further increases the number of treatments by a multiple of d.

The two-factor design enables the investigation of the *first-order,* or two-factor, interaction AB. The two additional first-order interactions, AC and BC, in the three-factor design broaden the inferences from the study. In addition, there is a *second-order,* or three-factor, interaction, ABC, to consider. *Third-order,* or four-factor, interactions such as $ABCD$ enter the increasingly complex interaction inference structure as factors are added to the design.

Example 6.5   Shrimp Culture in Aquaria

The California brown shrimp spawn at sea and the hatched eggs undergo larval transformations while being transported toward the shore. By the time they transform to post-larval stage they enter estuaries, where they grow rapidly into subadults and migrate back offshore as they approach sexual maturity.

The shrimp encounter wide temperature and salinity variation in their life cycle as a result of their migrations during the cycle. Thus, a knowledge of how temperature and salinity affect their growth and survival is of great importance to understanding their life history and ecology.

There was at the time of this experiment great interest in commercial culture of the shrimp. From the standpoint of mariculture another important factor was stocking density in the culture tanks that affects intraspecific competition.

Research Objective: The investigators wanted to know how water temperature, water salinity, and density of shrimp populations influenced the growth rate of shrimp raised in aquaria and whether the factors acted independently on the shrimp populations.

Treatment Design: A factorial arrangement was used with three factors: "temperature" (25°C, 35°C); "salinity" of the water (10%, 25%, 40%); and "density" of shrimp in the aquarium (80 shrimp/40 liters, 160 shrimp/40 liters). The levels chosen were those considered most likely to exhibit an effect if the factor was influential on shrimp growth.

Experiment Design: The experiment design consisted of three replicate aquaria for each of the 12 treatment combinations of the $2 \times 2 \times 3$ factorial. Each of the 12 treatment combinations was randomly assigned to three aquaria for a completely randomized design. The 36 aquaria were stocked with post-larval shrimp at the beginning of the test. The weight gain of the shrimp in four weeks for each of the 36 aquaria is shown in Table 6.13 on a per-shrimp basis.

Table 6.13 Four-week weight gain of shrimp cultured in aquaria at different levels of temperature (T), density of shrimp populations (D), and water salinity (S)

T	D	S	Weight Gain (mg)
25°C	80	10%	86, 52, 73
		25%	544, 371, 482
		40%	390, 290, 397
	160	10%	53, 73, 86
		25%	393, 398, 208
		40%	249, 265, 243
35°C	80	10%	439, 436, 349
		25%	249, 245, 330
		40%	247, 277, 205
	160	10%	324, 305, 364
		25%	352, 267, 316
		40%	188, 223, 281

Source: Dr. J. Hendrickson and K. Dorsey, Department of Ecology and Evolutionary Biology, University of Arizona.

The Statistical Model for Three Factors

The cell means model for a three-factor experiment with r replications of each of the abc treatment combinations in a completely randomized design is

$$y_{ijkl} = \mu_{ijk} + e_{ijkl} \tag{6.25}$$

$$i = 1, 2, \ldots, a \quad j = 1, 2, \ldots, b \quad k = 1, 2, \ldots, c \quad l = 1, 2, \ldots, r$$

The cell mean μ_{ijk} expressed as a function of the factorial main effects and interactions is

$$\mu_{ijk} = \mu + \alpha_i + \beta_j + \gamma_k + (\alpha\beta)_{ij} + (\alpha\gamma)_{ik} + (\beta\gamma)_{jk} + (\alpha\beta\gamma)_{ijk} \tag{6.26}$$

where $\mu = \overline{\mu}_{\ldots}$ is the general mean and α_i, β_j, and γ_k are the main effects of A, B, and C. The respective two-factor interaction effects are $(\alpha\beta)_{ij}$, $(\alpha\gamma)_{ik}$, and $(\beta\gamma)_{jk}$; and the three-factor interaction effect is $(\alpha\beta\gamma)_{ijk}$. The definitions of main effects and two-factor interactions follow from the derivations given in Equations (6.6) to (6.8) for the two-factor experiment. The main effects are

$$\alpha_i = (\overline{\mu}_{i..} - \overline{\mu}_{...}), \; \beta_j = (\overline{\mu}_{.j.} - \overline{\mu}_{...}), \; \gamma_k = (\overline{\mu}_{..k} - \overline{\mu}_{...})$$

and a typical two-factor interaction is

$$(\beta\gamma)_{jk} = (\overline{\mu}_{.jk} - \overline{\mu}_{...}) - \beta_j - \gamma_k \tag{6.27}$$
$$= \overline{\mu}_{.jk} - \overline{\mu}_{.j.} - \overline{\mu}_{..k} + \overline{\mu}_{...}$$

The three-factor interaction occurs when the main effects and two-factor interactions do not satisfactorily explain the variation in the cell mean deviations $(\mu_{ijk} - \overline{\mu}_{...})$. The three-factor interaction is the difference between the cell mean deviation and the sum of the main effects and two-factor interaction effects:

$$(\alpha\beta\gamma)_{ijk} = (\mu_{ijk} - \overline{\mu}_{...}) - [\alpha_i + \beta_j + \gamma_k + (\alpha\beta)_{ij} + (\alpha\gamma)_{ik} + (\beta\gamma)_{jk}] \qquad \textbf{(6.28)}$$
$$= \mu_{ijk} - \overline{\mu}_{ij.} - \overline{\mu}_{i.k} - \overline{\mu}_{.jk} + \overline{\mu}_{i..} + \overline{\mu}_{.j.} + \overline{\mu}_{..k} - \overline{\mu}_{...}$$

The Analysis for Three Factors

The sum of squares for treatments is partitioned into main effect and interaction sums of squares as

$$SS \text{ Treatment} = SSA + SSB + SSC + SS(AB)$$
$$+ SS(AC) + SS(BC) + SS(ABC) \qquad \textbf{(6.29)}$$

Keep in mind that the degrees of freedom for main effect sums of squares are $(a - 1)$, $(b - 1)$, and $(c - 1)$, respectively, for factors A, B, and C. The degrees of freedom for two-factor interactions are the product of the main effect degrees of freedom for the included factors. Likewise, the degrees of freedom for a three-factor or higher interaction are the product of the main effect degrees of freedom for the included factors, so that $SS(ABC)$ has $(a - 1)(b - 1)(c - 1)$ degrees of freedom.

The sums of squares partitions and analysis of variance table for the three-factor shrimp growth experiment are shown in Table 6.14. The Mean Square for Error in Table 6.14 is the denominator of the F_0 statistic to test the null hypothesis for any set of factorial effects with the fixed effects model. The F_0 statistics in Table 6.14 lead to the rejection of the null hypothesis for the TS two-factor interaction, the TDS three-factor interaction, and all main effects. The cell and marginal means for all factors are shown in Table 6.15.

Table 6.14 Analysis of variance for weight gain of shrimp cultured in aquaria

Source of Variation	Degrees of Freedom	Sum of Squares	Mean Square	F	Pr > F
Total	35	537,327.01			
Temp (T)	1	15,376.00	15,376.00	5.30	.030
Salinity (S)	2	96,762.50	48,381.25	16.66	.000
Density (D)	1	21,218.78	21,218.78	7.31	.012
TS	2	300,855.17	150,427.58	51.80	.000
TD	1	8,711.11	8,711.11	3.00	.096
SD	2	674.39	337.19	0.12	.891
TDS	2	24,038.39	12,019.19	4.14	.029
Error	24	69,690.67	2,903.78		

Table 6.15 Cell and marginal means of four-week weight gain of shrimp cultured in aquaria at different levels of temperature (T), density of shrimp populations (D), and water salinity (S)

Cell Means ($\bar{y}_{ijk.}$)

		Density				
		80		160		
		Temperature		Temperature		
	Salinity	25°	35°	25°	35°	$\bar{y}_{..k.}$
	10%	70	408	71	331	220
	25%	466	275	333	312	346
	40%	359	243	252	231	271
$T \times D$ Means		298	309	219	291	

($\bar{y}_{ij..}$)

$T \times S$ Means ($\bar{y}_{i.k.}$)

T	10%	25%	40%	$\bar{y}_{i...}$
25°	71	399	306	259
35°	370	293	237	300

$D \times S$ Means ($\bar{y}_{.jk.}$)

D	10%	25%	40%	$\bar{y}_{.j..}$
80	239	370	301	303
160	201	322	242	255

Interpretations must be conditioned on some measure of statistical significance in conjunction with the biological significance of the responses. Standard errors of cell and marginal means are required for any subsequent statistical tests of specific comparisons. The standard error for any mean is $s_{\bar{y}} = \sqrt{MSE/n}$, where n is the number of observations in the mean. The standard error of the difference between any pair of means is $s(\bar{y}_i - \bar{y}_j) = \sqrt{2MSE/n}$. The estimated standard errors for the shrimp culture experiment are shown in Table 6.16.

Some Preliminary Interpretations About the Factor Effects

The significance of the three-factor interaction indicates that temperature, salinity, and density are interrelated in their effect on the shrimp growth. The significant three-factor interaction implies that the interaction between two factors is not constant over levels of the third factor. Consider the interaction between density and salinity separately at temperatures of 25°C and 35°C, as shown in the graphs of cell means in Figure 6.6.

A comparison of the simple effects of salinity at each level of density and temperature can be used to interpret the results. The simple effects of salinity are best estimated as orthogonal polynomial linear and quadratic contrasts at each combination of density and temperature. Sums of squares partitions computed for the three factor interaction TDS are $SS(T \times D \times S\ linear) = 11,051$ and $SS(T \times D \times S\ quadratic) = 12,987$ with P-values .063 and .045, respectively, indicating the salinity quadratic coefficient is dependent on the levels of temperature and density.

Table 6.16 Standard errors for cell and marginal means in a three-factor treatment design

---

Temperature: $a = 2$ levels; Density: $b = 2$ levels; Salinity: $c = 3$ levels

---

Main Factor Means

Temperature	Salinity	Density
$\sqrt{\dfrac{MSE}{rbc}} = \sqrt{\dfrac{2903.78}{18}}$	$\sqrt{\dfrac{MSE}{rab}} = \sqrt{\dfrac{2903.78}{12}}$	$\sqrt{\dfrac{MSE}{rac}} = \sqrt{\dfrac{2903.78}{18}}$
$= 12.7$	$= 15.6$	$= 12.7$

Two Factor Marginal Means

Density by Temperature		Density by Salinity
$\sqrt{\dfrac{MSE}{rc}} = \sqrt{\dfrac{2903.78}{9}}$		$\sqrt{\dfrac{MSE}{ra}} = \sqrt{\dfrac{2903.78}{6}}$
$= 18.0$		$= 22.0$

Salinity by Temperature	Cell Means
$\sqrt{\dfrac{MSE}{rb}} = \sqrt{\dfrac{2903.78}{6}}$	$\sqrt{\dfrac{MSE}{r}} = \sqrt{\dfrac{2903.78}{3}}$
$= 22.0$	$= 31.1$

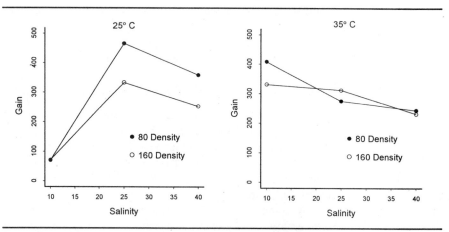

Figure 6.6 Weight gain of shrimp cultured in aquaria in a $2 \times 2 \times 3$ factorial arrangement of temperature, density, and salinity

The salinity quadratic coefficients were computed as orthogonal polynomial contrasts for the four combinations of temperature and density from the cell means in Table 6.15 following a template similar to that provided in Table 6.12. For example, the quadratic coefficient for salinity at 25°C and a density of 80 is

$$\frac{[70 - 2(466) + 359]}{[1^2 + (-2^2) + 1^2]} = -\frac{503}{6} = -83.8$$

with a standard error $\sqrt{2903.78/6(3)} = 12.7$. The 95% SCI estimates were computed for the four coefficients based on the Bonferroni $t_{.05,4,24} = 2.70$.

The 95% SCI estimates of the quadratic coefficients for salinity at 25°C are $(-118.1, -49.5)$ for a density of 80 and $(-91, -22.9)$ for a density of 160, while the estimates at 35°C are $(-17.5, 51.1)$ for a density of 80 and $(-44.6, 24.0)$ for a density of 160. Clearly the quadrature at 25°C is significant since the 95% SCI do not include 0 and not significant at 35°C since those intervals include 0.

6.7 Estimation of Error Variance with One Replication

Situations arise in research studies wherein only one observation is available in each cell of a factorial arrangement. The experimental error variance cannot be estimated with only one replication of the treatment combinations. The sums of squares partitions for factor main effects and interaction are equal to the total sum of squares for the observations.

Additivity describes the case when there is no interaction between factors. Under additivity of factors the mean square partition for interaction can be used as an estimate of experimental error. The additivity of main effects or absence of interaction is not guaranteed, and some means of evaluating the presence of interaction is required.

Error Variance Estimates with Two Quantitative Factors

The additivity of quantitative factors can be investigated with the interaction components for linear and possibly quadratic regression partitions (Section 6.5). For example, the sum of squares for linear × linear interaction can be partitioned out of the interaction sum of squares with the assumption that the remaining sum of squares for deviations from linear × linear interaction is experimental error. These sums of squares for deviations from the linear × linear interaction would include all higher orders of polynomial interaction such as linear × quadratic, and so forth. The mean square for deviations from the linear × linear interaction can be used as the mean square for error. The number of 1 degree of freedom interaction terms that are partitioned from the interaction is a matter of judgment based on the number of degrees of freedom available for a reasonably powerful test for interaction and main effects.

Error Variance Estimates with a Qualitative and a Quantitative Factor

The same approach can be used if one of the factors is qualitative and the other is quantitative. In this case, the sum of squares for the interaction between the qualitative factor and the linear effect of the quantitative factor can be partitioned out of the interaction sum of squares (Section 6.5). The remaining deviations' sum of squares can be used to estimate experimental error.

Error Variance Estimates with Two Qualitative Factors

If both factors are qualitative the problem is somewhat more difficult. Tukey (1949b) gave a method for isolating a 1 degree of freedom sum of squares to test for nonadditivity in a two-way classification with one observation per cell. The term for nonadditivity in the linear model is a simple product of the main effects, $\lambda \alpha_i \beta_j$, where the parameter λ represents the added parameter for nonadditivity. The product of main effects is a multiplicative form of interaction, and if there is nonadditivity from this specific type of interaction between the main effects, α_i and β_j, then $\lambda \neq 0$. Under this model the cell means are a sum of the general mean, the factor effects, and the product term, or

$$\mu_{ij} = \bar{\mu}_{..} + (\bar{\mu}_{i.} - \bar{\mu}_{..}) + (\bar{\mu}_{.j} - \bar{\mu}_{..}) + \lambda(\bar{\mu}_{i.} - \bar{\mu}_{..})(\bar{\mu}_{.j} - \bar{\mu}_{..})$$

The sum of squares for nonadditivity requires a computation involving the deviations of the A and B means from the grand mean, $(\bar{y}_{i.} - \bar{y}_{..})$ and $(\bar{y}_{.j} - \bar{y}_{..})$, respectively. The technique is illustrated with Example 6.6.

> ### Example 6.6 Hearing Levels in Adult Males
>
> The data in Table 6.17 are the percentage of men aged 55 to 64 with hearing levels 16 decibels above the audio metric zero. The row categories were sound levels in cycles per second (hertz), and the column categories were seven occupational categories.

The required computations include

$$P_j = \sum_{i=1}^{a} (\bar{y}_{i.} - \bar{y}_{..}) y_{ij} \tag{6.30}$$

$$P = \sum_{j=1}^{b} P_j (\bar{y}_{.j} - \bar{y}_{..}) = 17{,}416 \tag{6.31}$$

and

$$\sum_{i=1}^{a} (\bar{y}_{i.} - \bar{y}_{..})^2 = 6{,}944, \qquad \sum_{j=1}^{b} (\bar{y}_{.j} - \bar{y}_{..})^2 = 162 \tag{6.32}$$

The 1 degree of freedom sum of squares for nonadditivity is

Table 6.17 Percentage of men with hearing levels 16 decibels above the audio metric zero classified in a 7×7 factorial arrangement with one observation per cell

A	B 1	2	3	4	5	6	7	$\bar{y}_{i.}$	$(\bar{y}_{i.} - \bar{y}_{..})$
1	2.1	6.8	8.4	1.4	14.6	7.9	4.8	6.6	− 31.6
2	1.7	8.1	8.4	1.4	12.0	3.7	4.5	5.7	− 32.5
3	14.4	14.8	27.0	30.9	36.5	36.4	31.4	27.3	− 10.9
4	57.4	62.4	37.4	63.3	65.5	65.6	59.8	58.8	20.6
5	66.2	81.7	53.3	80.7	79.7	80.8	82.4	75.0	36.8
6	75.2	94.0	74.3	87.9	93.3	87.8	80.5	84.7	46.5
7	4.1	10.2	10.7	5.5	18.1	11.4	6.1	9.4	− 28.8
$\bar{y}_{.j}$	31.6	39.7	31.4	38.7	45.7	41.9	38.5	$\bar{y}_{..} = 38.2$	
$(\bar{y}_{.j} - \bar{y}_{..})$	− 6.6	1.5	− 6.8	0.5	7.5	3.7	0.3		
P_j	6719	7730	5046	7776	6850	7313	7192		

Source: C. Daniel (1978), Patterns in residuals in the two-way layout. *Technometrics* 20, 385–395. Data originally published in J. Roberts and J. Cohrssen (1968), *Hearing Levels of Adults*, Table 4, p. 36. U.S. National Center for Health Statistics Publications, Series 11, No. 31. Rockland, Md.

$$S(\text{Nonadditivity}) = \frac{P^2}{\sum_{i=1}^{a} (\bar{y}_{i.} - \bar{y}_{..})^2 \sum_{j=1}^{b} (\bar{y}_{.j} - \bar{y}_{..})^2} \qquad (6.33)$$

$$= \frac{17{,}416^2}{(6{,}944)(162)} = 269.6$$

The analysis of variance for the data is shown in Table 6.18 with the sum of squares for error partitioned into a 1 degree of freedom sum of squares for nonadditivity and a residual sum of squares.

Table 6.18 One degree of freedom partition for nonadditivity in the analysis of variance for a 7×7 factorial with one observation per cell

Source of Variation	Degrees of Freedom	Sum of Squares	Mean Square
Rows	6	48,589.1	8,098.2
Columns	6	1,141.5	190.2
Error	36	1,444.7	40.1
Nonadditivity	1	269.6	269.6
Residual	35	1,175.1	33.6

The null hypothesis of no nonadditivity is tested with the statistic F_0 = MS(Nonadditivity)/MS(Residual) = 269.6/33.6 = 8.02. The null hypothesis is rejected with a critical region $F_0 > F_{.05,1,35} = 4.12$.

Several methods have been developed to ascertain the source of nonadditivity in a two-way table. Daniel (1978) used a method based on the residuals in each of the cells, $y_{ij} - \overline{y}_{i.} - \overline{y}_{.j} + \overline{y}_{..}$. Technical discussions and examples of several other methods and models for nonadditivity can be found in Johnson and Graybill (1972), Bradu and Gabriel (1978), and Mandel (1971).

6.8 How Many Replications to Test Factor Effects?

Procedures were given in Section 2.14 to estimate replication numbers based on the test for differences among treatment means with the F_0 statistic. The values of Φ (Equation (2.25)) can be applied directly to a test for differences among cell means in the factorial arrangement with the null hypothesis H_0: $\mu_{11} = \mu_{12} = \cdots = \mu_{ab}$. In this case, the factorial structure is ignored and the cell means model y_{ijk} = $\mu_{ij} + e_{ijk}$ expressed in the effects model form is $y_{ijk} = \mu + \tau_{ij} + e_{ijk}$, where τ_{ij} is the effect of the ijth treatment combination in the factorial arrangement. Then

$$\Phi^2 = \frac{r \sum_{i=1}^{a} \sum_{j=1}^{b} \tau_{ij}^2}{ab\sigma^2} \qquad (6.34)$$

is used to estimate replication numbers from the charts based on the values of τ_{ij} required to be significant.

If replication numbers based on the factorial effects are required, the non-centrality parameters are

$$\lambda_a = br \sum_{i=1}^{a} \frac{\alpha_i^2}{\sigma^2}, \quad \lambda_b = ar \sum_{j=1}^{b} \frac{\beta_j^2}{\sigma^2}, \quad \text{and} \quad \lambda_{ab} = r \sum_{i=1}^{a} \sum_{j=1}^{b} \frac{(\alpha\beta)_{ij}^2}{\sigma^2} \qquad (6.35)$$

respectively, for A and B main effects and AB interaction. Then Φ is determined as $\Phi = \sqrt{\lambda/(\nu_1 + 1)}$, where ν_1 are the numerator degrees of freedom for the F_0 statistic.

6.9 Unequal Replication of Treatments

Missing data in research studies is inevitable. The design is no longer balanced with a complete data set, and standard computing formulae no longer apply. Before the advent of modern computing, a complete data set was most advantageous because relatively simple formulae could be used for manual computations. Much effort was put into developing methods for the analysis of variance sum of squares partitions when there were unequal numbers of observations among the cells of the factorial arrangement. General statistical routines programmed to accommodate known statistical theory have removed the computational burdens associated with the analysis of incomplete data sets.

Orthogonality Is Lost with Missing Observations

The sum of squares for one factorial effect will convey some information about other factorial effects when there are unequal numbers of observations for the treatment combinations and the sum of squares partitions are computed in the usual manner. This non-orthogonal relationship in the sum of squares partitions for the analysis of variance requires us to consider carefully the estimates we use for parameters in the model and the statistics we use to test the critical hypotheses in an analysis of the study.

Orthogonal contrasts were introduced in Chapter 3 as contrasts that do not convey information about one another. Orthogonality carries the same meaning with observations in a factorial treatment design. When there are equal numbers of observations on each treatment combination, the sums of squares in the analysis of variance constitute an orthogonal partition of the treatment sum of squares. In Section 6.4 the additive partition of SS Treatment for a balanced two-factor experiment was

$$SS \text{ Treatment} = SS\mathbf{A} + SS\mathbf{B} + SS(\mathbf{AB})$$

and the sum of squares for one factorial effect did not convey any information about other factorial effects.

The example data shown in Display 6.5 illustrate the complications introduced by unequal treatment replication in a factorial treatment arrangement. The data in the cells represent the average speed in excess of the posted speed limit traveled by automobiles involved in 20 fatal accidents, 10 occurring in rainy weather and 10 occurring in clear weather. The factors for the study are W (Weather) and R (Type of Roadway). Notice that 8 of the 10 accidents in rainy weather were on interstate highways, whereas 8 of the 10 accidents in clear weather were on two-lane highways.

Observation of the marginal means for weather indicates the average speed in excess of the speed limit for rainy weather was slightly higher than that for clear weather, $\bar{y}_{1..} = 13$ versus $\bar{y}_{2..} = 12$. However, the observed cell means indicate an entirely different result. The average speeds in fatal accidents in clear weather were greater by 5 miles per hour than in rainy weather for both interstate $(20 - 15)$ and two-lane $(10 - 5)$ highways.

The unequal treatment replications lead to contradictory results from cell means and marginal means. An excess of accidents occurred on the interstates in rainy weather and an excess of accidents occurred on the two-lane highways in clear weather. Thus, the marginal mean for clear weather is biased downward by the excess of accidents on the two-lane highways with overall slower speeds, and

| **Display 6.5** | **Unequal Treatment Replication in a 2 × 2 Factorial for Speeds in Excess of Posted Speed Limit** | | | |

	Cell Means		Sums	Means
	Interstate	Two-Lane		
Rainy	15	5	130	13
	$r_{11} = 8$	$r_{12} = 2$	$r_{1.} = 10$	
Clear	20	10	120	12
	$r_{21} = 2$	$r_{22} = 8$	$r_{2.} = 10$	
Sums	160	90	250	
	$r_{.1} = 10$	$r_{.2} = 10$	$r_{..} = 20$	$\bar{y}_{..} = 12.5$
Means	16	9		

the marginal mean for rainy weather is biased upward by the excess of accidents on the interstates with overall faster speeds.

The sum of squares for treatments with unequal replication numbers is computed correctly as

$$SS\ \text{Treatment} = 8(15 - 12.5)^2 + 2(5 - 12.5)^2 + 2(20 - 12.5)^2 + 8(10 - 12.5)^3$$

$$= 325$$

The sums of squares for main effects and interaction computed *incorrectly* with methods outlined in Section 6.4 are

$$SSW = 10(13 - 12.5)^2 + 10(12 - 12.5)^2 = 5$$

$$SSR = 10(16 - 12.5)^2 + 10(9 - 12.5)^2 = 245$$

and

$$SS(WR) = SS\ \text{Treatment} - SSW - SSR = 325 - 5 - 245 = 75$$

Inspection of the cell means and the sum of squares for interaction indicate another contradiction in the usual analysis methods. The calculated sum of squares, $SS(WR) = 75$, indicates some interaction is present in the study. However, inspection of the cell means gives no indication of interaction. The observed simple effect of weather is equal to 5 for both highway types. The sum of squares for interaction would be 0 if the sum of squares partitions were computed correctly.

The general principles for a correct analysis of factorial treatment designs with unequal treatment replication are illustrated with the two-factor experiment on durability of asphaltic concrete.

Example 6.7 Asphaltic Concrete Durability Revisited Again

The experiment on tensile strength of asphaltic concrete specimens in Example 6.2 is used to illustrate the analysis of a factorial treatment design with unequal replication of the treatments. For illustration, suppose specimens were constructed with basalt or silicious aggregate types for the three kneading compaction methods. Suppose some of the specimens were damaged prior to testing, resulting in an unequal number of specimens being available among the treatments for tensile strength tests. The data with unequal replication numbers are shown in Table 6.19.

Table 6.19 Tensile strength (psi) of asphaltic concrete specimens for two aggregate types with each of three kneading compaction methods

| Aggregate Type | Compaction Method Kneading | | | Aggregate Means ($\bar{y}_{i..}$) |
	Regular	Low	Very Low	
Basalt	106	93	56	
	108	101		
		98		
Means ($\bar{y}_{1j.}$)	107.0	97.3	56	93.7
Silicious	107	63	40	
	110	60	41	
	116		44	
Means ($\bar{y}_{2j.}$)	111.0	61.5	41.7	72.6
Compaction means ($\bar{y}_{.j.}$)	109.4	83.0	45.3	

Establish Estimators with the Cell Means Model

The cell means model can be used to establish the appropriate estimators for population parameters and hypotheses that we are required to test. The cell means model is

$$y_{ijk} = \mu_{ij} + e_{ijk} \tag{6.36}$$

$$i = 1, 2, \ldots, a \quad j = 1, 2, \ldots, b \quad k = 1, 2, \ldots, r_{ij}$$

where μ_{ij} is the cell mean for the ith level of factor A and the jth level of factor B, e_{ijk} is the normally distributed random independent experimental error with mean 0 and variance σ^2, and r_{ij} is the number of replicate observations in cell (ij). We will assume there is at least one observation in each cell of the factorial arrangement, so that $r_{ij} > 0$ for all i and j.

Least Squares Estimators for Cell and Marginal Means

The cell means can be estimated by the least squares method following the procedures outlined in Chapter 2 for the cell means model. The estimators of the cell means are the observed cell means

$$\widehat{\mu}_{ij} = \frac{1}{r_{ij}} \sum_{k=1}^{r_{ij}} y_{ijk} = \bar{y}_{ij.} \tag{6.37}$$

and the estimator for experimental error variance is

$$\widehat{\sigma}^2 = s^2 = \frac{1}{N - ab} \Sigma (y_{ijk} - \bar{y}_{ij.})^2 \tag{6.38}$$

where $N = \Sigma r_{ij.}$.

The estimates of the cell means for the tensile strength of the asphalt concrete specimens are shown in Table 6.19, and the estimate of experimental error is

$$s^2 = \frac{1}{14 - 6} [(106 - 107.0)^2 + \cdots + (44 - 41.7)^2] = \frac{89.83}{8} = 11.23$$

The unbiased least squares estimators of the marginal means are

$$\widehat{\bar{\mu}}_{i.} = \frac{1}{b} \sum_{j=1}^{b} \widehat{\mu}_{ij} \quad \text{and} \quad \widehat{\bar{\mu}}_{.j} = \frac{1}{a} \sum_{i=1}^{a} \widehat{\mu}_{ij} \tag{6.39}$$

with standard error estimators

$$s_{\widehat{\bar{\mu}}_{i.}} = \sqrt{\frac{s^2}{b^2} \sum_{j=1}^{b} \frac{1}{r_{ij}}} \quad \text{and} \quad s_{\widehat{\bar{\mu}}_{.j}} = \sqrt{\frac{s^2}{a^2} \sum_{i=1}^{a} \frac{1}{r_{ij}}} \tag{6.40}$$

The least squares estimates of the marginal means for the asphalt concrete specimens and their standard error estimates are shown in Table 6.20.

For example, the least squares estimate of the marginal mean for the basalt aggregate type is

$$\widehat{\bar{\mu}}_{1.} = \frac{1}{3}(107.0 + 97.3 + 56.0) = 86.8$$

with standard error estimate

$$s_{\widehat{\bar{\mu}}_{1.}} = \sqrt{\frac{11.23}{3^2} \left(\frac{1}{2} + \frac{1}{3} + \frac{1}{1} \right)} = 1.51$$

The observed marginal means, $\bar{y}_{i..}$ and $\bar{y}_{.j.}$, shown in Table 6.19 do not have the same value as the least squares estimates of the marginal means in Table 6.20. The observed marginal means estimate weighted functions of the population means where the weights are proportional to the number of replications in the cells. The expected values of the observed marginal means are

Table 6.20 Least squares estimates of the marginal means for tensile strength of asphalt concrete specimens and their standard errors

Aggregate	Mean $\widehat{\overline{\mu}}_{i.}$	Standard Error $s_{\widehat{\overline{\mu}}_{i}}$
Basalt	86.8	1.51
Silicious	71.4	1.21

Compaction	$\widehat{\overline{\mu}}_{.j}$	$s_{\widehat{\overline{\mu}}_{j}}$
Regular	109.0	1.53
Low	79.4	1.53
Very low	48.8	1.93

$$E(\bar{y}_{i..}) = \frac{1}{r_{i.}} \sum_{j=1}^{b} r_{ij}\mu_{ij}$$

and

$$E(\bar{y}_{.j.}) = \frac{1}{r_{.j}} \sum_{i=1}^{a} r_{ij}\mu_{ij}$$

(6.41)

If the number of observations in the treatment cells of the study is proportional to the frequency with which those treatment combinations occur in the population, then the observed marginal means provide the appropriate estimators for the marginal means in the population. The proportional relationship of observation numbers to population frequencies is common in sample surveys. However, the proportional relationship would not be expected to hold for a designed experiment or comparative observational study, and the least squares estimates in Table 6.20 should be used.

Hypotheses Unchanged by Unequal Treatment Replication

The hypotheses of interest in the factorial treatment design with unequal replication numbers are unchanged from those of interest with equal replication numbers. The initial research question in the factorial treatment design considers the existence of interaction between factors A and B. The interaction effect measures the differences between the simple effects of A at different levels of B. The difference between levels i and k of A at levels j and m of B is the general form of interaction:

$$(\mu_{ij} - \mu_{kj}) - (\mu_{im} - \mu_{km}) = \mu_{ij} - \mu_{kj} - \mu_{im} + \mu_{km}$$

The null hypothesis of no interaction can be expressed in terms of the cell means as

$$H_0: \ \mu_{ij} - \mu_{kj} - \mu_{im} + \mu_{km} = 0 \quad \text{for all } i, j, k, \text{ and } m \tag{6.42}$$

In the absence of interaction, the effect of the individual factors on the response variable can be explored separately with tests of hypotheses about the marginal means. The null hypothesis of interest for factor A is the equality of the marginal means, or

$$H_0: \ \overline{\mu}_{1.} = \overline{\mu}_{2.} = \cdots = \overline{\mu}_{a.} \tag{6.43}$$

and that for factor B is

$$H_0: \ \overline{\mu}_{.1} = \overline{\mu}_{.2} = \cdots = \overline{\mu}_{.b} \tag{6.44}$$

Weighted Squares of Means for Tests of Hypotheses

Among the many methods put forth for analyzing factorial experiments with unequal replication only the method of *weighted squares of means* proposed by Yates (1934) provides the sum of squares partitions to test all three hypotheses in Equations (6.42) through (6.44). A description of other methods and the hypotheses that can be tested with them can be found in Speed, Hocking, and Hackney (1978). The tests for equality of marginal means are of interest only in the absence of interaction.

Computing Sum of Squares for Interaction from Full and Reduced Models

The sum of squares partition for interaction is determined from the principle of full and reduced models introduced in Chapter 2. The full model expressed in terms of the factorial effects is

$$y_{ijk} = \mu + \alpha_i + \beta_j + (\alpha\beta)_{ij} + e_{ijk} \tag{6.45}$$

The solutions obtained from the least squares normal equations are used to compute the sum of squares of experimental error for the full model as

$$SSE_f = \sum_{i=1}^{a} \sum_{j=1}^{b} \sum_{k=1}^{r_{ij}} [y_{ijk} - \widehat{\mu} - \widehat{\alpha}_i - \widehat{\beta}_j - (\widehat{\alpha\beta}_{ij})]^2 \tag{6.46}$$

Under the null hypothesis of no interaction the reduced model is

$$y_{ijk} = \mu + \alpha_i + \beta_j + e_{ijk} \tag{6.47}$$

The solutions obtained from the least squares normal equations are used to compute the sum of squares of experimental error for the reduced model as

$$SSE_r = \sum_{i=1}^{a} \sum_{j=1}^{b} \sum_{k=1}^{r_{ij}} [y_{ijk} - \widehat{\mu} - \widehat{\alpha}_i - \widehat{\beta}_j]^2 \tag{6.48}$$

The sum of squares for interaction is computed as

$$SS(\boldsymbol{AB}) = SSE_r - SSE_f \tag{6.49}$$

The mean square for experimental error is $MSE = (SSE_f)/(N - ab)$, and the mean square for interaction is $MS(\boldsymbol{AB}) = SS(\boldsymbol{AB})/(a - 1)(b - 1)$. The usual F_0 statistic, $F_0 = MS(\boldsymbol{AB})/MSE$, tests the null hypothesis of no interaction in Equation (6.42). The calculations are illustrated in Appendix 6A.

Weighted Squares of Means to Test Equality of Main Effects in the Absence of Interaction

Tests of hypotheses can be conducted for the equality of marginal means of the factors if the test for interaction is not significant and it can safely be assumed there is no interaction. The correct sum of squares partitions for the weighted squares of means method to test the null hypotheses in Equations (6.43) and (6.44) are shown in Display 6.6. The analysis is based on the sums of squares of the cell means designated as the observations $x_{ij} = \overline{y}_{ij.}$.

Display 6.6   Weighted Squares of Means Sum of Squares Partitions

Factor A	Factor B
$SS\boldsymbol{A}_w = \sum_i^a w_i(\overline{x}_{i.} - \overline{x}_{[1]})^2$	$SS\boldsymbol{B}_w = \sum_j^b v_j(\overline{x}_{.j} - \overline{x}_{[2]})^2$
$w_i = \left[\dfrac{1}{b^2}\displaystyle\sum_{j=1}^b \dfrac{1}{r_{ij}}\right]^{-1}$	$v_j = \left[\dfrac{1}{a^2}\displaystyle\sum_{i=1}^a \dfrac{1}{r_{ij}}\right]^{-1}$
$\overline{x}_{[1]} = \sum_i^a w_i\overline{x}_{i.}\Big/\sum_i^a w_i$	$\overline{x}_{[2]} = \sum_j^b v_j\overline{x}_{.j}\Big/\sum_j^b v_j$
$H_0\colon \overline{\mu}_{1.} = \overline{\mu}_{2.} = \cdots = \overline{\mu}_{a.}$	$H_0\colon \overline{\mu}_{.1} = \overline{\mu}_{.2} = \cdots = \overline{\mu}_{.b}$

The sum of squares partitions required by the weighted squares of means can be computed by many statistical programs. However, the programs may have several options for the type of sum of squares partitions that are computed for the analysis. It is important that the correct options be used for the programs so that the correct sum of squares is computed by the program.[1]

---

[1] Programs used for analysis of variance will provide the correct sum of squares for the weighted squares of means. Instructions on the use of most programs will indicate if and how different types of sum of squares partitions can be obtained. The correct sum of squares options for several well known programs are

Program	Sum of Squares
SAS GLM	Type III
SPSS MANOVA	UNIQUE
MINITAB GLM	Adjusted
BMDP 2V	Default
Splus	summary.aov(...,ssType=3)

The analysis of variance for the asphaltic concrete specimens tensile strengths in Example 6.7 is shown in Table 6.21.

Table 6.21   Analysis of variance for tensile strength of asphaltic concrete specimens with unequal replication of treatments with Yates' weighted squares of means

Source of Variation	Degrees of Freedom	Sum of Squares	Mean Square	F	Pr > F
Total	13	10,963.21			
Aggregate	1	710.45	710.45	63.27	.000
Compaction	2	6,806.45	3,403.23	303.07	.000
Interaction	2	953.45	476.72	42.45	.000
Error	8	89.83	11.23		

Interpretation of the Example

The null hypothesis of no interaction between aggregate type and compaction method is

$$H_0: \; \mu_{ij} - \mu_{kj} - \mu_{im} + \mu_{km} = 0 \quad \text{for all } i, j, k, \text{ and } m \tag{6.50}$$

Interaction is significant since the statistic $F_0 = MS(\boldsymbol{AB})/MSE = 476.72/11.23 = 42.45$, in Table 6.21, is significant with $Pr > F = .000$.

With significant interaction between aggregate type and compaction method it will be necessary to look at the simple effects of one factor at each of the levels of the other factor to understand the nature of the interaction. Comparisons between the cell means for aggregate type at each level of compaction method are shown in Display 6.7.

Display 6.7   **Bonferroni t Tests for Simple Effects of Aggregate Type for Each Compaction Method**

Compaction Method	$\widehat{\mu}_{1j} - \widehat{\mu}_{2j}$	Standard Error	t_0
Regular	$107.0 - 111.0 = -4.0$	$\sqrt{11.23 \left[\dfrac{1^2}{2} + \dfrac{(-1)^2}{3} \right]} = 3.1$	-1.29
Low	$97.3 - 61.5 = 35.8$	$\sqrt{11.23 \left[\dfrac{1^2}{3} + \dfrac{(-1)^2}{2} \right]} = 3.1$	11.55
Very low	$56.0 - 41.7 = 14.3$	$\sqrt{11.23 \left[\dfrac{1^2}{1} + \dfrac{(-1)^2}{3} \right]} = 3.9$	3.67

The t_0 statistics were computed for the contrast between aggregate types for each of the compaction methods in Display 6.7. The critical value for the Bonferroni t statistic with three comparisons is $|t_0| > t_{.025,3,8} = 3.02$. There is no significant difference between tensile strengths of the basalt and silicious rock specimens with the regular kneading compaction method. The tensile strengths of the basalt specimens were significantly greater than those for the silicious rock specimens for low and very low kneading compaction methods.

Tests for Marginal Means

The tests of hypotheses for equality of the marginal means for A and B ordinarily are not considered when interaction is significant and would not be considered for the current example. However, for the sake of illustration, the procedure is illustrated for the case when no interaction exists and tests about the marginal means would be of interest. The two hypotheses to test for the asphaltic concrete example are (1) no differences among the marginal means for aggregate type

$$H_0: \ \overline{\mu}_{1.} = \overline{\mu}_{2.} \tag{6.51}$$

and (2) no differences among the marginal means for compaction method

$$H_0: \ \overline{\mu}_{.1} = \overline{\mu}_{.2} = \overline{\mu}_{.3} \tag{6.52}$$

The sum of squares for the weighted squares of means method are shown in Table 6.21. The statistic $F_0 = MSA/MSE = 710.45/11.23 = 63.26$ tests the equality of marginal means for aggregate type. The statistic for the test of equality among the marginal means of compaction method is $F_0 = MSC/MSE = 3403.23/11.23 = 303.05$. Both of the statistics are significant with $Pr > F = .000$ in Table 6.21.

Some Comments About Missing Data and Missing Cells

A method was illustrated in this section to analyze the data from a study with unequal subclass replication in a factorial treatment design. The method provides the correct estimators for population means and credible tests of hypotheses about the factors. Other methods of analysis for the unbalanced factorial treatment designs provide different tests of hypotheses about the factor effects.

Searle, Speed, and Henderson (1981) discussed five methods for calculating sums of squares in the analysis of variance that were currently used in computer programs. All of the methods give the same results for balanced data, but they can yield different results for unbalanced data. Related articles by Hocking and Speed (1975), Speed and Hocking (1976), and Speed, Hocking, and Hackney (1978) provide additional information on the computing methods and the hypotheses that are tested in the analyses by different computer programs. More extensive illustrations of some methods, including that used in this section along with some of the theoretical background, are found in Searle (1971, 1987) and Milliken and Johnson (1984).

All of the methods discussed in the context of the sum of squares partitions for main effects and interaction provide inadequate tests of hypotheses about factorial effects when entire cells of the factorial arrangement are missing. Under these circumstances an analysis based on the cell means model is recommended by Urquhart, Weeks, and Henderson (1973), Hocking and Speed (1975), Urquhart and Weeks (1978), and Searle (1987).

EXERCISES FOR CHAPTER 6

---

1. A chemical production process consists of a first reaction with an alcohol and a second reaction with a base. A 3×2 factorial experiment with three alcohols and two bases was conducted with four replicate reactions conducted in a completely randomized design. The collected data were percent yield.

| Base | Alcohol | | | | | |
	1		2		3	
1	91.3	89.9	89.3	88.1	89.5	87.6
	90.7	91.4	90.4	91.4	88.3	90.3
2	87.3	89.4	92.3	91.5	93.1	90.7
	91.5	88.3	90.6	94.7	91.5	89.8

Source: P. R. Nelson (1988), Testing for interactions using analysis of means. *Technometrics* 30, 53–61.

a. Write a linear model for this experiment, explain the terms, and compute the analysis of variance for the data.

b. Make a table of cell and marginal means, and show their respective standard errors.

c. Test the null hypotheses of no base × alcohol interaction effects. What do you conclude from the test? What do you recommend as the next step in your analysis?

d. Use multiple contrasts among cell means to help explain the interaction. For example, compare the two bases for each alcohol.

e. Conduct residual analyses with a normal plot and then with a predicted plot; also conduct a Levene (Med) test. What do you conclude?

2. A company tested two chemistry methods for the determination of serum glucose. Three pools of serum were used for the experiment. Each pool contained different levels of glucose through the addition of glucose to the base level of an existing serum pool. Three samples of serum from each pool were prepared independently for each level of glucose with each of the two chemistry methods. The concentration of glucose (mg/dl) for all samples was measured on one run of a spectrophotometer.

Glucose Level	Method 1			Method 2		
	1	*2*	*3*	*1*	*2*	*3*
	42.5	138.4	180.9	39.8	132.4	176.8
	43.3	144.4	180.5	40.3	132.4	173.6
	42.9	142.7	183.0	41.2	130.3	174.9

Source: Dr. J. Anderson, Beckman Instruments Inc.

a. Write a linear model for this experiment, explain the terms, conduct an analysis of variance for the data, and compute the residuals. Is a transformation of the data necessary? Explain.

b. If a transformation is necessary, compute the transformation for the data and the analysis of variance.

c. Test the hypotheses of no method × glucose interaction effects. What do you conclude? Should you test for main effects? Why?

d. Prepare a table of cell and marginal means with their respective standard errors.

e. Test the difference between method means for each level of glucose, and interpret the results.

3. A study of the effect of temperature on percent shrinkage in dyeing fabrics was made on two replications for each of four fabrics in a completely randomized design. The data are the percent shrinkage of two replicate fabric pieces dried at each of the four temperatures.

Fabric	Temperature			
	210° F	215° F	220° F	225°F
1	1.8, 2.1	2.0, 2.1	4.6, 5.0	7.5, 7.9
2	2.2, 2.4	4.2, 4.0	5.4, 5.6	9.8, 9.2
3	2.8, 3.2	4.4, 4.8	8.7, 8.4	13.2, 13.0
4	3.2, 3.6	3.3, 3.5	5.7, 5.8	10.9, 11.1

a. Write a linear model for the experiment, explain the terms, and compute the analysis of variance for the data.

b. Test the null hypothesis of no fabric × temperature interaction.

c. Partition the temperature main effect sum of squares into 1 degree of freedom partitions for linear and quadratic regression sum of squares, and test the null hypotheses of no linear or quadratic response to temperature.

d. Partition the temperature × fabric interaction sum of squares into temperature linear × fabric and temperature quadratic × fabric interaction sum of squares, and test the null hypotheses of no interaction for the respective partitions.

e. Prepare a profile plot of the cell means versus temperature for each fabric, and interpret the results. For example, the following questions may be asked: "How does drying temperature affect the fabric shrinkage?" "How does the relationship between shrinkage and temperature differ among the fabric types?"

4. An experiment in soil microbiology was conducted to determine the effect of nitrogen fertility on nitrogen fixation by *Rhizobium* bacteria. The experiment was conducted with four crops: alfalfa, soybeans, guar, and mungbean. Two plants were inoculated with the *Rhizobium* and grown in a Leonard jar with one of three rates of nitrogen in the media: 0, 50, or 100 ppm N. Four replications,

Leonard jars, were used for each of the 12 treatment combinations. The treatments were arranged in a completely randomized design in a growth chamber. The acetylene reduction was measured for each treatment when the plants were at the flowering stage. Acetylene reduction reflects the amount of nitrogen that is fixed by the bacteria in the symbiotic relationship with the plant.

		Crop		
Nitrogen	Alfalfa	Soybean	Guar	Mungbean
0	2.6, 1.1	6.5, 2.6	0.3, 0.1	0.8, 0.9
	0.9, 1.2	3.9, 4.3	0.4, 0.4	2.2, 1.2
50	0.0, 0.0	0.6, 0.6	0.0, 0.1	0.7, 0.4
	0.0, 0.0	0.3, 0.8	0.0, 0.2	0.3, 0.8
100	0.0, 0.0	0.0, 0.1	0.0, 0.2	0.3, 0.1
	0.0, 0.0	0.1, 0.0	0.0, 0.0	0.0, 0.1

Source : Dr. I. Pepper, Department of Soil and Water Science, University of Arizona.

a. Write a linear model for this experiment, explain the terms, and compute the analysis of variance.
b. Perform a residual analysis and determine whether a transformation of the data is necessary. Transform the data if necessary, and compute the analysis of variance for the transformed data.
c. Test the null hypotheses of no crop, nitrogen, or crop × nitrogen interaction effects.
d. Partition the nitrogen main effect and the nitrogen × crop interaction sum of squares into 1 degree of freedom partitions for linear and quadratic regression.
e. Test the null hypotheses of no nitrogen linear or quadratic main effects and the null hypotheses of no nitrogen linear or quadratic interaction with crops.
f. Make a profile plot of the cell means versus level of nitrogen for each crop, and interpret the experiment. For example, you can ask the question, "How does the addition of nitrogen to the media affect the nitrogen fixation by the *Rhizobium*?" or "Is the effect of the addition of nitrogen on nitrogen fixation the same for each crop?"
g. Note that two treatment combinations, alfalfa with 50 and 100 ppm N, have all observations with a value of zero. This phenomenon is possible if the presence of a threshold level of nitrogen in the growth medium completely inhibits *Rhizobium* activity. How does this affect the assumptions for the analysis of variance regarding homogeneity of variance? Do you have any suggestions to accommodate this situation in your analysis of the data?

5. An agronomist conducted an experiment to determine the combined effects of an herbicide and an insecticide on the growth and development of cotton plants (delta pine smoothleaf). The insecticide and the herbicide were incorporated into the soil used in the containers to grow the cotton plants. Four containers each with five cotton plants were used for each treatment combination. The containers were arranged in the greenhouse in a completely randomized design. Five levels (lb/acre) were used for both the insecticide and the herbicide to give 25 treatment combinations. The data that follow are cell means for the dry weight of the roots (grams/plant) when the plants were three weeks old.

Insecticide	Herbicide				
	0	0.5	1.0	1.5	2.0
0	122.0	72.50	52.00	36.25	29.25
20	82.75	84.75	71.50	80.50	72.00
40	65.75	68.75	79.50	65.75	82.50
60	68.00	70.00	68.75	77.25	68.25
80	57.50	60.75	63.00	69.25	73.25

Mean Square for Experimental Error = 174 with 75 degrees of freedom

Source: Dr. K. Hamilton, Department of Plant Sciences, University of Arizona.

a. Compute 1 degree of freedom regression sum of squares partitions for herbicide, insecticide, and interaction sums of squares. Compute no higher polynomial than cubic regression for herbicide or insecticide.
b. Test the null hypotheses for each of the partitions, and determine the form of the polynomial regression that adequately describes the response.
c. Transform the orthogonal polynomial equation into an equation in terms of herbicide and insecticide. Either use a standard regression program or the transformation equations in Chapter 3.
d. Interpret the results from plots of the cell means or the estimated polynomial equation.

6. An experiment was conducted on the durability of coated fabric subjected to standard abrasive tests. The $2 \times 2 \times 3$ factorial design included two different fillers (F_1, F_2) in three different proportions (25%, 50%, 75%) with or without surface treatment (S_1, S_2). Two replicate fabric specimens were tested for each of the 12 treatment combinations in a completely randomized design. The data are weight loss (mg) of the fabric specimens from the abrasion test.

	Surface and Filler Treatments			
	S_1		S_2	
Proportion of Filler	F_1	F_2	F_1	F_2
25%	194	239	155	137
	208	187	173	160
50%	233	224	198	129
	241	243	177	98
75%	265	243	235	155
	269	226	229	132

Source: G. Box (1950), Problems in the analysis of growth and wear curves. Biometrics 6, 362–389.

a. Write a linear model for the experiment, explain the terms, and compute the analysis of variance for the data.
b. Prepare a table of cell and marginal means with their respective standard errors.
c. Test the null hypothesis for all main and interaction effects.
d. Compute the 1 degree of freedom regression sum of squares partitions for proportion of filler and interaction between proportion of filler and the other factors.

e.   Test the null hypotheses for the regression partitions.

f.   Plot cell means versus proportion of filler for the four treatment combinations of surface and filler type, and interpret the results of your analysis.

7.   A soil scientist conducted an experiment to evaluate a four-electrode resistance network to compute electroconductivity (EC) of soil in specially constructed acrylic conductivity cells. The objective of the study was to evaluate the relationship between measured EC and soil water salinity at different water contents of soils. Three basic soil textures were included in the experiment since EC is specific to soil texture. The cells were constructed of acrylic tubing, 4-cm long by 8.2-cm diameter, and packed with soil. Two cells were used for each treatment combination. The three soil types were loamy sand, loam, and clay. The salinity of the soil water, three levels, was based on the EC of the water at 2, 8, and 16 dS/m (decisiemens/meter). The water content of the soil was three levels, at 0%, 5%, and 15%. The resulting experiment was a $3 \times 3 \times 3$ factorial arrangement with two replications in a completely randomized design. The EC values of the soil determined on the basis of readings from the four-electrode network follow.

Salinity	2			8			16		
Water	0	5	15	0	5	15	0	5	15
Loamy	0.60	1.69	3.47	0.05	0.11	0.06	0.07	0.08	0.22
sand	0.48	2.01	3.30	0.12	0.09	0.19	0.06	0.14	0.17
Loam	0.98	2.21	5.68	0.15	0.23	0.40	0.07	0.23	0.43
	0.93	2.48	5.11	0.26	0.35	0.75	0.21	0.35	0.35
Clay	1.37	3.31	5.74	0.72	0.78	2.10	0.40	0.72	1.95
	1.50	2.84	5.38	0.51	1.11	1.18	0.57	0.88	2.87

Source: H. Bohn and T. Tabbara, Department of Soil and Water Science, University of Arizona.

a.   Write a linear model for this experiment, explain the terms, and compute the analysis of variance for the data.

b.   Prepare a table of cell and marginal means and their respective standard errors.

c.   Test the null hypotheses for all main effects and interactions.

d.   Compute the linear and quadratic orthogonal polynomial regression sum of squares partitions for salinity and water and their interactions, including the interactions with texture. Note that the levels of salinity and water are unequally spaced; therefore, the standard orthogonal polynomial coefficients given in Appendix Table XI do not apply. Some statistical computing programs will automatically compute the orthogonal polynomial coefficients, given values for the levels of the factors (for example, MANOVA in SPSS and 2V in BMDP). The following are orthogonal polynomial coefficients that may be used to compute the orthogonal partitions:

Water linear:	-0.617	-0.154	0.772
Water quadratic:	0.535	-0.802	0.267
Salinity linear:	-0.671	-0.067	0.738
Salinity quadratic:	0.465	-0.814	0.349

8. Five nickel rods of 1-mm diameter were put in a metallic clamp in a suspension of aluminum oxide. A 100-volt tension was applied between the nickel rods and the vessel containing the suspension of aluminum oxide. The thickness of the aluminum oxide layer deposited on the nickel rods was recorded at three height positions of the five rods. The data are thickness of the deposit in microns.

Height	*Clamp Position of Nickel Rod*				
	1	*2*	*3*	*4*	*5*
1	125	130	128	134	143
2	126	150	127	124	118
3	130	155	168	159	138

Source: H. Hamaker (1955), Experimental design in industry. *Biometrics* 11, 257–286.

a. Write a linear model for this experiment, explain the terms, and state the model assumptions.
b. Do you believe the model assumptions are valid for this experiment? Explain.
c. Suppose the assumptions are reasonably valid. Compute the analysis of variance for the data.
d. Compute the 1 degree of freedom sum of squares for nonadditivity.
e. Is the additive model for clamp position and height sufficient for this data?

9. An entomologist conducted an experiment on the drinking rate energetics of honeybees to determine the effects of ambient temperature and viscosity of the liquid on energy consumption by the honeybee. The temperature levels were 20° C, 30° C, and 40° C. The viscosity of the liquid was controlled by the sucrose concentrations, which were 20, 40, and 60 percent of total dissolved solids in the drinking liquid for the bees. The entomologist recorded the energy expended by the bees as joules/second. The data given below are for three replications for each of the nine treatment combinations in a completely randomized design.

Temperature °C	*Sucrose %*		
	20	40	60
20	3.1, 3.7, 4.7	5.5, 6.7, 7.3	7.9, 9.2, 9.3
30	6.0, 6.9, 7.5	11.5, 12.9, 13.4	17.5, 15.8, 14.7
40	7.7, 8.3, 9.5	15.7, 14.3, 15.9	19.1, 18.0, 19.9

Source: Dr. S. Buckman, USDA Bee Research Lab, Tucson, Arizona.

a. Compute 1 degree of freedom regression sum of squares partitions for temperature, sucrose %, and interaction sums of squares.
b. Test the null hypotheses for each of the partitions, and determine the form of the polynomial regression that adequately describes the response.
c. Transform the orthogonal polynomial equation into an equation in terms of herbicide and insecticide. Either use a standard regression program or the transformation equations in Chapter 3.
d. Construct a profile plot such as that in Figure 6.5 with the estimated cell means and the estimated polynomial equation.
e. Interpret the results.

10. *Unequal Replication Numbers:* A biologist incubated adrenal glands of rats in vitro under stimulation by ACTH and measured steroid production of the glands. The glands were taken from the animals at four different stages of growth and were subjected to two different treatments. Glands from four animals were used for each treatment combination. However, several laboratory analysis were invalid, which resulted in unequal replication numbers for the treatments. The data given below are steroid production per 100 mg of gland per hour.

	Treatment	
Stage	*1*	*2*
1	6.98, 6.58	8.62, 9.40, 9.20
2	6.07, 7.16, 6.34	9.42, 6.67, 8.64
3	5.38, 7.31, 6.65, 7.44	4.96, 6.80, 7.61
4	7.02, 9.23, 7.32	7.17, 7.65, 6.52, 6.86

Source: Dr. R. Chaisson, Department of Veterinary Science, University of Arizona.

a. Compute the analysis of variance in order to test the global hypothesis of no Stage by Treatment interaction.
b. Compute the least squares means and their standard errors for marginal and cell means.
c. Estimate the contrast between the two treatments for levels 1, 2, 3, and 4 of the growth stages $(\widehat{\mu}_{1j.} - \widehat{\mu}_{2j.}$ for $j = 1, 2, 3, 4)$, and test the hypothesis of no difference at the .05 level of significance between the two means in each case.
d. How do the least squares means differ from the observed means?

11. *Unequal Replication Numbers*: Suppose the experiment on the chemical production process in Exercise 6.1 had unequal replications among the six treatment combinations of the two factors, Base and Alcohol. The data are given below.

	Alcohol		
Base	*1*	*2*	*3*
1	90.7, 91.4	89.3, 88.1	89.5, 87.6
		90.4	88.3, 90.3
2	87.3, 88.3	94.7	93.1, 90.7
	91.5		91.5

a. Compute the analysis of variance in order to test the global hypothesis of no Base by Alcohol interaction.
b. Compute the least squares means and their standard errors for marginal and cell means.
c. Estimate the contrast between the two bases for levels 1, 2, and 3 of Alcohol $(\widehat{\mu}_{1j.} - \widehat{\mu}_{2j.};\ j = 1, 2, 3)$, and test the hypothesis of no difference at the .05 level of significance between the two means in each case.
d. How do the least squares means differ from the observed means?

6A Appendix: Least Squares for Factorial Treatment Designs

Equal Treatment Replication

The sum of squares partitions for the data from a factorial treatment design can be derived from solutions to the least squares normal equations for a factorial effects model. The full model for a factorial treatment design with two factors will be used to illustrate the derivation.

For the sake of simplicity in notation the full model is written as

$$y_{ijk} = \mu + \alpha_i + \beta_j + \gamma_{ij} + e_{ijk} \tag{6A.1}$$

$$i = 1, 2, \ \ldots \ , a \quad j = 1, 2, \ \ldots \ , b \quad k = 1, 2, \ \ldots \ , r$$

where μ is the general mean, α_i is the effect of factor A, β_j is the effect of factor B, γ_{ij} is the interaction effect, and e_{ijk} is the random independent experimental error. The interaction term $(\alpha\beta)_{ij}$ used in the main body of this chapter has been replaced by γ_{ij} to simplify the notation for the presentation in this appendix.

The least squares estimates for the parameters in the full model are those that minimize the sum of squares for experimental error

$$Q = \sum_{i=1}^{a} \sum_{j=1}^{b} \sum_{k=1}^{r} e_{ijk}^2 = \sum_{i=1}^{a} \sum_{j=1}^{b} \sum_{k=1}^{r} (y_{ijk} - \mu - \alpha_i - \beta_j - \gamma_{ij})^2 \tag{6A.2}$$

The normal equations from the minimization include one equation for μ and one equation for each of the factorial effects, $\alpha_1, \alpha_2, \ldots , \alpha_a$; $\beta_1, \ \beta_2, \ldots , \beta_b$; and $\gamma_{11}, \gamma_{12}, \ldots , \gamma_{ab}$. The normal equations are obtained from the following set of derivatives:

$$\frac{\partial}{\partial \mu} \sum\sum\sum (y_{ijk} - \mu - \alpha_i - \beta_j - \gamma_{ij})^2 = 0$$

$$\frac{\partial}{\partial \alpha_i} \sum_j\sum_k (y_{ijk} - \mu - \alpha_i - \beta_j - \gamma_{ij})^2 = 0 \quad i = 1, 2, \ldots, a$$

$$\frac{\partial}{\partial \beta_j} \sum_i\sum_k (y_{ijk} - \mu - \alpha_i - \beta_j - \gamma_{ij})^2 = 0 \quad j = 2, 3, \ldots, b \tag{6A.3}$$

$$\frac{\partial}{\partial \gamma_{ij}} \sum_k (y_{ijk} - \mu - \alpha_i - \beta_j - \gamma_{ij})^2 = 0 \quad \begin{matrix} i = 1, 2, \ldots, a \\ j = 1, 2, \ldots, b \end{matrix}$$

Simplifying, the set of normal equations to solve is

$$\mu: \quad abr\widehat{\mu} + br\sum_j \widehat{\alpha}_i + ar\sum_j \widehat{\beta}_j + r\sum_i\sum_j \widehat{\gamma}_{ij} = y_{...}$$

$$\alpha_i: \quad br\widehat{\mu} + br\widehat{\alpha}_i + r\sum_j \widehat{\beta}_j + r\sum_j \widehat{\gamma}_{ij} = y_{i..} \quad i = 1, 2, \ldots, a$$

$$\beta_j: \quad ar\widehat{\mu} + r\sum_i \widehat{\alpha}_i + ar\widehat{\beta}_j + r\sum_i \widehat{\gamma}_{ij} = y_{.j.} \quad j = 1, 2, \ldots, b \tag{6A.4}$$

$$\gamma_{ij}: \quad r\widehat{\mu} + r\widehat{\alpha}_i + r\widehat{\beta}_j + r\widehat{\gamma}_{ij} = y_{ij.} \quad \begin{matrix} i = 1, 2, \ldots, a \\ j = 1, 2, \ldots, b \end{matrix}$$

Upon close inspection the a equations derived for the factor A effects sum to the first equation for μ; the b equations derived for the factor B effects sum to the first equation for μ as do the ab equations for interaction; the γ_{ij} equations summed over the j subscript will give the α_i equation; and the γ_{ij} equations summed over the i subscript will give the β_j equation. These linear dependencies require constraints imposed on the estimates to provide a unique solution to the equations. Any constraints that lead to a solution will suffice. One set of constraints commonly used are the sum-to-zero constraints. The sum-to-zero constraints are $\sum \widehat{\alpha}_i = 0$, $\sum \widehat{\beta}_j = 0$, $\sum_i^a \widehat{\gamma}_{ij} = 0$ ($j = 1, 2, \ldots, b$), and $\sum_j^b \widehat{\gamma}_{ij} = 0$ ($i = 1, 2, \ldots, a$).

With the constraints, the equations are

$$\mu: \quad abr\widehat{\mu} = y_{...}$$

$$\alpha_i: \quad br\widehat{\mu} + br\widehat{\alpha}_i = y_{i..} \quad i = 1, 2, \ldots, a$$

$$\beta_j: \quad ar\widehat{\mu} + ar\widehat{\beta}_j = y_{.j.} \quad j = 1, 2, \ldots, b \tag{6A.5}$$

$$\gamma_{ij}: \quad r\widehat{\mu} + r\widehat{\alpha}_i + r\widehat{\beta}_j + r\widehat{\gamma}_{ij} = y_{ij.} \quad \begin{matrix} i = 1, 2, \ldots, a \\ j = 1, 2, \ldots, b \end{matrix}$$

The solutions are

$$\widehat{\mu} = \frac{y_{...}}{abr} = \overline{y}_{...}$$

$$\widehat{\alpha}_i = \frac{y_{i..}}{br} - \widehat{\mu} = \overline{y}_{i..} - \overline{y}_{...} \quad i = 1, 2, \ldots, a$$

$$\widehat{\beta}_j = \frac{y_{.j.}}{ar} - \widehat{\mu} = \overline{y}_{.j.} - \overline{y}_{...} \quad j = 1, 2, \ldots, b \tag{6A.6}$$

$$\widehat{\gamma}_{ij} = \frac{y_{ij.}}{r} - \widehat{\mu} - \widehat{\alpha}_i - \widehat{\beta}_j = \overline{y}_{ij.} - \overline{y}_{i..} - \overline{y}_{.j.} + \overline{y}_{...} \quad \begin{matrix} i = 1, 2, \ldots, a \\ j = 1, 2, \ldots, b \end{matrix}$$

The estimate of the sum of squares for experimental error is obtained with a substitution of the estimates $\widehat{\mu}$, $\widehat{\alpha}_i, \widehat{\beta}_j$, and $\widehat{\gamma}_{ij}$ into Equation (6A.2) as

$$SSE_f = \sum_i \sum_j \sum_k (y_{ijk} - \widehat{\mu} - \widehat{\alpha}_i - \widehat{\beta}_j - \widehat{\gamma}_{ij})^2 = \sum_i \sum_j \sum_k (y_{ijk} - \bar{y}_{ij.})^2 \quad \textbf{(6A.7)}$$

The difference between the total sum of squares and SSE_f is known as the reduction in sum of squares due to fitting the model and is sometimes written as $R(\mu, \alpha, \beta, \gamma)$. With equal replication for all treatment combinations the sum of squares for each factorial effect can be derived from the computation used to compute $R(\mu, \alpha, \beta, \gamma)$. The reduction in sum of squares due to fitting the full model is

$$R(\mu, \alpha, \beta, \gamma) = \widehat{\mu}y_{...} + \sum_{i=1}^{a} \widehat{\alpha}_i y_{i..} + \sum_{j=1}^{b} \widehat{\beta}_j y_{.j.} + \sum_{i=1}^{a} \sum_{j=1}^{b} \widehat{\gamma}_{ij} y_{ij.} \quad \textbf{(6A.8)}$$

For balanced data with equal replication numbers for each treatment combination, the sum of squares partitions for the analysis of variance can be taken from the individual terms in Equation (6A.8) as

$$CF = \widehat{\mu}y_{...} = \frac{(y_{...})^2}{abr}$$

$$SSA = \sum_{i=1}^{a} \widehat{\alpha}_i y_{i..} = br \sum_{i=1}^{a} (\bar{y}_{i..} - \bar{y}_{...})^2$$

$$SSB = \sum_{j=1}^{b} \widehat{\beta}_j y_{.j.} = ar \sum_{j=1}^{b} (\bar{y}_{ij.} - \bar{y}_{...})^2 \qquad \textbf{(6A.9)}$$

$$SS(AB) = \sum_{i=1}^{a} \sum_{j=1}^{b} \widehat{\gamma}_{ij} y_{ij.} = r \sum_{i=1}^{a} \sum_{j=1}^{b} (\bar{y}_{ij.} - \bar{y}_{i..} - \bar{y}_{.j.} + \bar{y}_{...})^2$$

The sum of squares partitions shown in Equation (6A.9) are those shown in Section 6.4. They may be derived from considerations of full and reduced models. For example, the sum of squares for interaction $SS(AB)$ can be found as the difference between the experimental error sums of squares for the reduced model without the interaction terms and full model with interaction terms included. The models and sums of squares are

Full Model: $y_{ijk} = \mu + \alpha_i + \beta_j + \gamma_{ij} + e_{ijk}$ with SSE_f

Reduced Model: $y_{ijk} = \mu + \alpha_i + \beta_j + e_{ijk}$ with SSE_r

The interaction sum of squares is found as $SS(AB) = SSE_r - SSE_f$, which is $\sum_i \sum_j \widehat{\gamma}_{ij} y_{ij.}$ [the same as the last term shown in Equation (6A.8)].

The equivalence of $SSE_r - SSE_f$ to $\sum_i \sum_j \widehat{\gamma}_{ij} y_{ij.}$ can be shown by solving the normal equations for the reduced model and computing the reduction in sum of squares due to fitting the reduced model as $R(\mu, \alpha, \beta)$. The normal equations for the reduced model are obtained from those for the full model in Equation (6A.5) by removing the equations for γ_{ij} and the $\widehat{\gamma}_{ij}$ terms in the remaining equations. The solutions for $\widehat{\mu}$, $\widehat{\alpha}_i$, and $\widehat{\beta}_j$ will be those shown in Equation (6A.6). The reduction in sum of squares due to fitting the reduced model will be

$$R(\mu, \alpha, \beta) = \widehat{\mu}y_{...} + \sum_{i=1}^{a} \widehat{\alpha}_i y_{i..} + \sum_{j=1}^{b} \widehat{\beta}_j y_{.j.} \qquad \textbf{(6A.10)}$$

The difference between $R(\mu, \alpha, \beta, \gamma)$ and $R(\mu, \alpha, \beta)$ is seen to be $\Sigma_i \Sigma_j \widehat{\gamma}_{ij} y_{ij.}$, and, therefore, the differences between the sums of squares for experimental error of the two models will be equivalent to the same quantity. That is,

$$SS(AB) = SSE_r - SSE_f = R(\mu, \alpha, \beta, \gamma) - R(\mu, \alpha, \beta) \qquad \textbf{(6A.11)}$$

The sums of squares SSA and SSB can be derived in a similar fashion for balanced data, or $SSA = R(\mu, \alpha) - R(\mu)$ and $SSB = R(\mu, \beta) - R(\mu)$.

Unequal Treatment Replication

The derivation of the interaction sum of squares with unequal treatment replication in the factorial treatment design is demonstrated with the simple 2×2 factorial example shown in Table 6A.1.

Table 6A.1 Example data for a 2×2 factorial with unequal treatment replication

		B		
		1	*2*	$y_{i..}$
	1	6,5,3	2,4	20
A		$r_{11} = 3$	$r_{12} = 2$	$r_{1.} = 5$
	2	5,4	3	12
		$r_{21} = 2$	$r_{22} = 1$	$r_{2.} = 3$
	$y_{.j.}$	23	9	32
		$r_{.1} = 5$	$r_{.2} = 3$	$r_{..} = 8$

Full Model

The full model normal equations for the example data in Table 6A.1 are derived by the methods shown at the beginning of this appendix. The coefficients for the parameters in the equations will reflect the unequal replication numbers. In general, the equations will be

$$\mu: \quad N\widehat{\mu} + \sum_{i=1}^{a} r_{i.}\widehat{\alpha}_i + \sum_{j=1}^{b} r_{.j}\widehat{\beta}_j + \sum_{r=1}^{a}\sum_{j=1}^{b} r_{ij}\widehat{\gamma}_{ij} = y_{...}$$

$$\alpha_i: \quad r_{i.}\widehat{\mu} + r_{i.}\widehat{\alpha}_i + \sum_{j=1}^{b} r_{ij}\widehat{\beta}_j + \sum_{j=1}^{b} r_{ij}\widehat{\gamma}_{ij} = y_{i..}$$

$$\beta_j: \quad r_{.j}\widehat{\mu} + \sum_{i=1}^{a} r_{ij}\widehat{\alpha}_i + r_{.j}\widehat{\beta}_j + \sum_{i=1}^{a} r_{ij}\widehat{\gamma}_{ij} = y_{.j.}$$

$$\gamma_{ij}: \quad r_{ij}\widehat{\mu} + r_{ij}\widehat{\alpha}_i + r_{ij}\widehat{\beta}_j + r_{ij}\widehat{\gamma}_{ij} = y_{i..}$$

The normal equations for the data in Table 6A.1 are

$8\widehat{\mu}$ + $5\widehat{\alpha}_1$ + $3\widehat{\alpha}_2$ + $5\widehat{\beta}_1$ + $3\widehat{\beta}_2$ + $3\widehat{\gamma}_{11}$ + $2\widehat{\gamma}_{12}$ + $2\widehat{\gamma}_{21}$ + $\widehat{\gamma}_{22}$ = 32

$5\widehat{\mu}$ + $5\widehat{\alpha}_1$ + $3\widehat{\beta}_1$ + $2\widehat{\beta}_2$ + $3\widehat{\gamma}_{11}$ + $2\widehat{\gamma}_{12}$ = 20

$3\widehat{\mu}$ + $3\widehat{\alpha}_2$ + $2\widehat{\beta}_1$ + $\widehat{\beta}_2$ + $2\widehat{\gamma}_{21}$ + $\widehat{\gamma}_{22}$ = 12

$5\widehat{\mu}$ + $3\widehat{\alpha}_1$ + $2\widehat{\alpha}_2$ + $5\widehat{\beta}_1$ + $3\widehat{\gamma}_{11}$ + $2\widehat{\gamma}_{21}$ = 23

$3\widehat{\mu}$ + $2\widehat{\alpha}_1$ + $\widehat{\alpha}_2$ + $3\widehat{\beta}_2$ + $2\widehat{\gamma}_{12}$ + $\widehat{\gamma}_{22}$ = 9

$3\widehat{\mu}$ + $3\widehat{\alpha}_1$ + $3\widehat{\beta}_1$ + $3\widehat{\gamma}_{11}$ = 14

$2\widehat{\mu}$ + $2\widehat{\alpha}_1$ + $2\widehat{\beta}_2$ + $2\widehat{\gamma}_{12}$ = 6

$2\widehat{\mu}$ + $2\widehat{\alpha}_2$ + $2\widehat{\beta}_1$ + $2\widehat{\gamma}_{21}$ = 9

$\widehat{\mu}$ + $\widehat{\alpha}_2$ + $\widehat{\beta}_2$ + $\widehat{\gamma}_{22}$ = 3

Since there are linear dependencies in the equations the "sum-to-zero constraints" can be imposed on the equations to obtain a solution. The constraints are

$$\widehat{\alpha}_1 + \widehat{\alpha}_2 = 0, \ \widehat{\beta}_1 + \widehat{\beta}_2 = 0, \ \widehat{\gamma}_{11} + \widehat{\gamma}_{11} + \widehat{\gamma}_{12} = 0, \ \widehat{\gamma}_{11} + \widehat{\gamma}_{21} = 0$$

$$\widehat{\gamma}_{21} + \widehat{\gamma}_{21} = 0, \text{ and } \widehat{\gamma}_{12} + \widehat{\gamma}_{22} = 0$$

The solutions to the equations after the constraints are applied are

$$\widehat{\mu} = \frac{91}{24}, \ \widehat{\alpha}_1 = \frac{1}{24}, \ \widehat{\alpha}_2 = -\frac{1}{24}, \ \widehat{\beta}_1 = \frac{19}{24}, \ \widehat{\beta}_2 = -\frac{19}{24}$$

$$\widehat{\gamma}_{11} = \frac{1}{24}, \ \widehat{\gamma}_{12} = -\frac{1}{24}, \ \widehat{\gamma}_{21} = -\frac{1}{24}, \ \widehat{\gamma}_{22} = \frac{1}{24}$$

The sum of squares for experimental error can be determined with

$$SSE_f = \sum_i\sum_j\sum_k y_{ijk}^2 - R(\mu, \alpha, \beta, \gamma)$$

where

$$R(\mu, \alpha, \beta, \gamma) = \widehat{\mu} y_{...} + \sum_i \widehat{\alpha}_i y_{i..} + \sum_j \widehat{\beta}_j y_{.j.} + \sum_i \sum_j \widehat{\gamma}_{ij} y_{ij.}$$

The calculation for $R(\mu, \alpha, \beta, \gamma)$ is

$$R(\mu, \alpha, \beta, \gamma) = \widehat{\mu} y_{...} + \widehat{\alpha}_1 y_{1..} + \widehat{\alpha}_2 y_{2..} + \widehat{\beta}_1 y_{.1.} + \widehat{\beta}_2 y_{.2.}$$

$$+ \widehat{\gamma}_{11} y_{11.} + \widehat{\gamma}_{12} y_{12.} + \widehat{\gamma}_{21} y_{21.} + \widehat{\gamma}_{22} y_{22.}$$

$$= \frac{1}{24} [91(32) + 1(20) + (-1)(12) + 19(23) + \cdots + (-1)(9) + 1(3)]$$

$$= 132.833$$

Given $\sum y_{ijk}^2 = 140$, the sum of squares for experimental error from the full model is

$$SSE_f = 140 - 132.833 = 7.167$$

Reduced Model

The normal equations for the reduced model without the γ_{ij} interaction terms are obtained by eliminating the equations for the γ_{ij} and eliminating the $\widehat{\gamma}_{ij}$ terms from the remaining equations shown for the full model. The equations for the reduced model are

$$
\begin{array}{rcrcrcrcrcr}
8\widehat{\mu} & + & 5\widehat{\alpha}_1 & + & 3\widehat{\alpha}_2 & + & 5\widehat{\beta}_1 & + & 3\widehat{\beta}_2 & = & 32 \\
5\widehat{\mu} & + & 5\widehat{\alpha}_1 & & & + & 3\widehat{\beta}_1 & + & 2\widehat{\beta}_2 & = & 20 \\
3\widehat{\mu} & & & + & 3\widehat{\alpha}_2 & + & 2\widehat{\beta}_1 & + & \widehat{\beta}_2 & = & 12 \\
5\widehat{\mu} & + & 3\widehat{\alpha}_1 & + & 2\widehat{\alpha}_2 & + & 5\widehat{\beta}_1 & & & = & 23 \\
3\widehat{\mu} & + & 2\widehat{\alpha}_1 & + & \widehat{\alpha}_2 & & & + & 3\widehat{\beta}_2 & = & 9
\end{array}
$$

The sum-to-zero constraints are $\widehat{\alpha}_1 + \widehat{\alpha}_2 = 0$ and $\widehat{\beta}_1 + \widehat{\beta}_2 = 0$. The solutions are

$$\widehat{\mu} = \frac{212}{56}, \quad \widehat{\alpha}_1 = \frac{3}{56}, \quad \widehat{\alpha}_2 = -\frac{3}{56}, \quad \widehat{\beta}_1 = \frac{45}{56}, \quad \widehat{\beta}_2 = -\frac{45}{56}$$

The reduction in sum of squares due to fitting the reduced model is

$$R(\mu, \alpha, \beta) = \widehat{\mu} y_{...} + \widehat{\alpha}_1 y_{1..} + \widehat{\alpha}_2 y_{2..} + \widehat{\beta}_1 y_{.1.} + \widehat{\beta}_2 y_{.2.}$$

$$= \frac{1}{56} [212(32) + 3(20) + (-3)(12) + 45(23) + (-45)(9)]$$

$$= 132.821$$

The sum of squares for experimental error for the reduced model is

$$SSE_r = \sum_i \sum_j \sum_k y_{ijk}^2 - R(\mu, \alpha, \beta) = 140 - 132.821 = 7.179$$

The sum of squares for interaction is

$$SS(\boldsymbol{AB}) = R(\mu, \alpha, \beta, \gamma) - R(\mu, \alpha, \beta) = 132.833 - 132.821 = 0.012$$

$$SS(\boldsymbol{AB}) = SSE_r - SSE_f = 7.179 - 7.167 = 0.012$$

Main Effect Sums of Squares

Tests of equality for marginal means for A and B, H_0: $\bar{\mu}_{1.} = \bar{\mu}_{2.}$ and H_0: $\bar{\mu}_{.1} = \bar{\mu}_{.2}$, in the absence of interaction require the sums of squares partitions from the method of weighted squares of means (Yates, 1934). Some computer programs that compute the required sums of squares were indicated in Section 6.9. The hypothesis tested by the sum of squares partition for a main effect depends greatly on the computational technique used in the least squares estimation process. Details of the results from different techniques can be found in Hocking and Speed (1975), Speed and Hocking (1976), Speed, Hocking, and Hackney (1978), and Searle, Speed, and Henderson (1981).

7 Factorial Treatment Designs: Random and Mixed Models

The subject of variance component analyses is expanded to more complex designs in this chapter. Variance component models are developed for several variations on the factorial treatment design. Random effects models and models with mixtures of fixed and random effects for factorial arrangements are introduced in this chapter. The concept of the factorial treatment design is extended to include experiments with factors nested within other factors. Designs are discussed for experiments that have a combination of crossed and nested factors. Information for determining replication numbers is included in the discussions. Rules are given for deriving expected mean squares for a variety of balanced factorial experiments.

7.1 Random Effects for Factorial Treatment Designs

Random effects were introduced in Chapter 5 for studies in which the effects of the treatment factor were random samples from a population of treatment effects. The objective was to decompose the total variance into identifiable components. The variability caused by one source or factor can depend on the conditions under which it is evaluated. Thus, some of the total variance is associated with the interaction between two or more factors. The following example illustrates the interaction variance between two factors.

Example 7.1 Evaluating Machine Performance with Variance Components

A manufacturer was developing a new spectrophotometer for use in medical clinical laboratories. The development process was at the pilot stage of assembly after which machine performance was to be evaluated from assembly line production.

Research Question: A critical component of instrument performance is the consistency of measurements from day to day among machines. In this particular instance, the scientist who developed the instrument wanted to know if the variability of measurements among machines operated over several days was within acceptable standards for clinical applications.

Treatment Design: The scientist set up a factorial treatment design with "machines" and "days" as factors. Four machines were to be tested on four separate days in a 4 × 4 arrangement.

Experiment Design: Four machines were randomly selected from the pilot assembly production. Eight replicate serum samples were prepared each day from the same stock reagents. Two serum samples were randomly assigned to each of the four machines on each of the four days for a completely randomized design with two replications of each treatment combination. The same technician prepared the serum samples and operated the machines throughout the experiment. The observations on triglyceride levels (mg/dl) in the serum samples are shown in Table 7.1.

Table 7.1 Triglyceride levels (mg/dl) in serum samples run on four machines on each of four days

	Machine			
Day	1	2	3	4
1	142.3, 144.0	148.6, 146.9	142.9, 147.4	133.8, 133.2
2	134.9, 146.3	145.2, 146.3	125.9, 127.6	108.9, 107.5
3	148.6, 156.5	148.6, 153.1	135.5, 138.9	132.1, 149.7
4	152.0, 151.4	149.7, 152.0	142.9, 142.3	141.7, 141.2

Source: Dr. J. Anderson, Beckman Instruments, Inc.

Machines were random factors because they represented a random sample from a potential population of machines to be manufactured, and the Days were a random sample from a population of days on which the machines could be run. The factorial arrangement enabled evaluation of interaction between Machines and Days. The consistency of machine performance would be evidenced by the absence of interaction.

Statistical Model for Variances with Two Treatment Factors

Variability due to the interaction of random factors can play an important role in the inferential process. A random effects model for the two-factor experiment in a completely randomized design is

$$y_{ijk} = \mu + a_i + b_j + (ab)_{ij} + e_{ijk} \tag{7.1}$$

$$i = 1, 2, \ldots, a \quad j = 1, 2, \ldots, b \quad k = 1, 2, \ldots, r$$

The random effects a_i, b_j, and $(ab)_{ij}$ are assumed to be independent and normally distributed with means of 0 and variances σ_a^2, σ_b^2, and σ_{ab}^2, respectively. The effects are assumed to be independent of one another. The random errors e_{ijk} are assumed to be independent and normally distributed with mean 0 and variance σ^2.

The observations y_{ijk} in the random effects model have a normal distribution with mean μ and variance

$$\sigma_y^2 = \sigma^2 + \sigma_a^2 + \sigma_b^2 + \sigma_{ab}^2 \tag{7.2}$$

The components of variance in Equation (7.2) become the focus of any investigation with random effects.

The factorial analysis of variance for the measured concentration of triglycerides in the serum for the spectrophotometer tests described in Example 7.1 is shown in Table 7.2. The analysis of variance computations are those presented in Chapter 6. The expected mean squares are included in the analysis of variance table. A complete set of rules for expected mean square determination applicable to several types of factorial models, including random effects models, is given in Section 7.6.

Table 7.2 Analysis of variance for spectrophotometric readings from four machines on each of four days

Source of Variation	Degrees of Freedom	Mean Square	Expected Mean Square
Day	3	$MSD = 445$	$\sigma^2 + r\sigma_{dm}^2 + rb\sigma_d^2$
Machine	3	$MSM = 549$	$\sigma^2 + r\sigma_{dm}^2 + ra\sigma_m^2$
Interaction	9	$MS(DM) = 87$	$\sigma^2 + r\sigma_{dm}^2$
Error	16	$MSE = 18$	σ^2

Point Estimators for the Variance Components

With equal numbers of observations for each treatment combination the analysis of variance method discussed in Section 5.3 can be used to estimate the components of variance. The analysis of variance is computed for the factorial experiment just as it is for the fixed effects model. The estimates of the components are determined by equating the observed mean squares to the corresponding expected mean squares and solving for the unknown component values. The estimates for the four components of variance in Table 7.2 are

Error: $\qquad \widehat{\sigma}^2 = MSE = 18.0$

Interaction: $\qquad \widehat{\sigma}^2_{dm} = \dfrac{MS(DM) - MSE}{r} = \dfrac{87 - 18}{2} = 34.5$

Machines: $\qquad \widehat{\sigma}^2_m = \dfrac{MSM - MS(DM)}{ra} = \dfrac{549 - 87}{2(4)} = 57.8$

Days: $\qquad \widehat{\sigma}^2_d = \dfrac{MSD - MS(DM)}{rb} = \dfrac{445 - 87}{2(4)} = 44.8$

The estimate of total variation for a single observation is

$$\widehat{\sigma}^2_y = \widehat{\sigma}^2 + \widehat{\sigma}^2_d + \widehat{\sigma}^2_m + \widehat{\sigma}^2_{dm} = 18.0 + 44.8 + 57.8 + 34.5 = 155.1$$

and the estimated standard deviation is $\widehat{\sigma}_y = 12.5$.

Tests of Hypotheses About the Variance Components

The significance of the contribution by the components for Machines, Days, and their interaction can be assessed with the F test. The denominator for the F_0 statistic is the mean square with the same expectation as the numerator mean square under the null hypothesis. The test for no interaction, H_0: $\sigma^2_{dm} = 0$, requires that the mean square for error be used for the denominator since it will have the same expectation as $MS(DM)$ under the null hypothesis. The statistic is

$$F_0 = \frac{MS(DM)}{MSE} = \frac{87}{18} = 4.83$$

and the null hypothesis of no interaction is rejected with $F_0 > F_{.05,9,16} = 2.54$.

The situation is different for tests involving components of variance for main effects. The expected mean squares for main effects are equal to the expected mean square for interaction when the null hypothesis is true. Therefore, the correct F_0 statistic to test H_0: $\sigma^2_d = 0$ is

$$F_0 = \frac{MSD}{MS(DM)} = \frac{445}{87} = 5.11$$

and the null hypothesis for the Days component is rejected with $F_0 > F_{.05,3,9} = 3.86$. Likewise, the statistic to test H_0: $\sigma^2_m = 0$ is

$$F_0 = \frac{MSM}{MS(DM)} = \frac{549}{87} = 6.31$$

and the null hypothesis for the Machines component is rejected with $F_0 > F_{.05,3,9} = 3.86$.

An interval estimate for σ^2 can be calculated as shown in Section 5.4 since SSE/σ^2 is a chi-square variable with $ab(r-1)$ degrees of freedom.

Interpretations of the Variance Component Estimates

Each of the components contributes significantly to the variation of a measurement from this particular model of spectrophotometer. The Error component, $\widehat{\sigma}^2 = 18.0$, represents the variation in preparation of the serum samples. The Machines component, $\widehat{\sigma}_m^2 = 57.8$, is the variability in machine performance and contributes 37% of the variation. The Days component, $\widehat{\sigma}_d^2 = 44.8$, is the variability associated with a new start-up utilizing new reagents for the analysis of samples and other sources of variability that can be identified with day-to-day operational differences. The Interaction component, $\widehat{\sigma}_{dm}^2 = 34.5$, contributes 22% of the total variation. The significant interaction implies that the relative performance of the several machines does not vary consistently with the day-to-day changes in the operation. An inconsistency in the calibration of the machines from day to day could be one possible explanation of the interaction.

The factorial design has made it possible to identify several sources of variability in the measurements made by this model of spectrophotometer. The investigator, based on experience, will be able to decide if any of the contributing sources of variability exceeds an acceptable level and correct any deficiencies in the machine or operating conditions if necessary.

Variance Components for Three-Factor Studies

The expected mean squares for a three-factor experiment with random factors are shown in Table 7.3. Some complications arise in the construction of F_0 statistics for tests of hypotheses about the components of variance in random models with more than two factors. The mean square for error can be used to test the hypothesis of no three-factor interaction, and the mean square for three-factor interaction, $MS(ABC)$, can be used to test hypotheses for two-factor interaction components. However, upon inspection of the expected mean squares in Table 7.3 it can be seen that there is no legitimate mean square for the denominator of the F_0 statistic to test null hypotheses about the main effect variance components.

Approximate F Tests Required for Some Hypotheses

It is necessary to construct a mean square for the denominator of the F_0 statistic to test the significance of main effect variance components. The construction of approximate F_0 statistics was discussed in Chapter 5 with unequal subsample numbers. For example, to test the hypothesis $H_0: \sigma_b^2 = 0$, there are two possible F_0 statistics that approximate the F distribution.

The first approximation is constructed with MSB as the numerator and $M = MS(AB) + MS(BC) - MS(ABC)$ as the denominator. The second approximation uses $M1 = MSB + MS(ABC)$ as the numerator and

Table 7.3 Expected mean squares for a three-factor experiment with random effects

Source of Variation	Degrees of Freedom	Expected Mean Square
A	$a - 1$	$\sigma^2 + r\sigma^2_{abc} + rc\sigma^2_{ab} + rb\sigma^2_{ac} + rbc\sigma^2_a$
B	$b - 1$	$\sigma^2 + r\sigma^2_{abc} + rc\sigma^2_{ab} + ra\sigma^2_{bc} + rac\sigma^2_b$
C	$c - 1$	$\sigma^2 + r\sigma^2_{abc} + rb\sigma^2_{ac} + ra\sigma^2_{bc} + rab\sigma^2_c$
AB	$(a - 1)(b - 1)$	$\sigma^2 + r\sigma^2_{abc} + rc\sigma^2_{ab}$
AC	$(a - 1)(c - 1)$	$\sigma^2 + r\sigma^2_{abc} + rb\sigma^2_{ac}$
BC	$(b - 1)(c - 1)$	$\sigma^2 + r\sigma^2_{abc} + ra\sigma^2_{bc}$
ABC	$(a - 1)(b - 1)(c - 1)$	$\sigma^2 + r\sigma^2_{abc}$
Error	$abc(r - 1)$	σ^2

$M2 = MS(\mathbf{AB}) + MS(\mathbf{BC})$ as the denominator. The first ratio may be easier to use because approximate degrees of freedom need be computed for only one of the mean squares by the Satterthwaite procedure (Equation 5.27). However, it is possible to construct a negative mean square when some of the mean squares have negative signs in the function. Gaylor and Hopper (1969) discuss some of the problems associated with approximating the F distribution with linear combinations of mean squares. The recommended F_0 statistic would be the second with synthesized mean squares $M1$ and $M2$.

7.2 Mixed Models

Many experiments are designed to study the effects of one factor on the population mean and the effects of another on the population variance. These experiments have a mixture of fixed factors and random factors. Models for factorial arrangements that include random factors and fixed factors are called **mixed models**, because they contain a mixture of random and fixed effects.

The model and analysis for mixed effects consist of two parts because there are two types of inferences. The inferences for the random effects factor apply to the variation in a population of effects, whereas the inferences for the fixed effects factor are restricted to the specific levels used for the experiment. The experiment described in the following example included a mixture of random and fixed factors.

Example 7.2　Evaluation of Two Chemistry Methods on Four Days

New methods of chemistry frequently are developed to assay for compounds in a clinical laboratory setting. Given the choice among two or more chemistry methods, the clinical chemist must evaluate the relative performance of the methods.

Research Question: In this example, a chemist wanted to know whether the two chemistry methods consistently provided equivalent results in an assay for triglycerides in human serum. A methods comparison test was constructed by the clinical chemist to evaluate the difference in the performance of the two chemistry methods for the assay.

Treatment Design: A factorial treatment design was used with the factors "chemistry methods" and "days." The two methods each were to be tested on four days, for a 2 × 4 arrangement.

Experiment Design: Each day two replicate samples of serum were prepared with each of the two chemistry methods. The samples were analyzed in random order on the same spectrophotometer for a completely randomized experiment design with two replications. The same technician prepared the serum samples and operated the spectrophotometer for each of the tests. The observations on triglyceride levels (mg/dl) in the serum samples are shown in Table 7.4.

Table 7.4 Triglyceride levels (mg/dl) in serum samples from a factorial arrangement for two chemistry methods, fixed factor, and four days, random factor

Method	Day			
	1	2	3	4
1	142.3, 144.0	134.9, 146.3	148.6, 156.5	152.0, 151.4
2	142.9, 147.4	125.9, 127.6	135.5, 138.9	142.9, 142.3

Source: Dr. J. Anderson, Beckman Instruments, Inc.

Chemistry Method is a fixed factor because the two methods are reproducible in a repeated experiment. The inferences are restricted to a comparison between the two methods used in the experiment. Days, on the other hand, represent random effects because the four days are considered a random sample of days on which the two chemistry methods could be tested.

The advantage of the factorial arrangement in this example is the ability to evaluate the consistency of the two methods under repeated runs. In the absence of interaction the two methods would produce the same relative results from day to day. The presence of interaction would indicate one or both of the methods were inconsistent in repeated applications.

Statistical Model for One Fixed and One Random Factor

The linear model for a two-factor experiment with a fixed effect, factor A, and a random effect, factor B, is

$$y_{ijk} = \mu + \alpha_i + b_j + (ab)_{ij} + e_{ijk} \tag{7.3}$$

$$i = 1, 2, \ldots, a \quad j = 1, 2, \ldots, b \quad k = 1, 2, \ldots, r$$

where μ is the mean, α_i is the fixed effect for factor A, b_j is the random effect for factor B, $(ab)_{ij}$ is the interaction effect, and e_{ijk} is random experimental error. The random effects, b_j and e_{ijk}, are assumed independent and normally distributed with means 0 and variances σ_b^2 and σ^2, respectively.

The interaction effects $(ab)_{ij}$ are assumed to be random effects, independent and normally distributed with mean 0 and variance σ_{ab}^2. The interaction effects are assumed random when one of the factors involved is a random effect.

Analysis of Mixed-Factor Experiments

The expected mean squares for the mixed-model analysis of variance are different from those for either the completely fixed or completely random effects models. The expected mean squares and the analysis for the factorial experiment with one random and one fixed factor are illustrated for the chemistry method experiment in Example 7.2. The analysis of variance for the 16 observations is shown in Table 7.5. The expected mean squares are included in the analysis of variance table. Note the use of θ_m^2 for the variance of the fixed effect as defined in Table 6.5.

Table 7.5 Analysis of variance for a factorial experiment with one fixed effects factor, Method, and one random effects factor, Day

Source of Variation	Degrees of Freedom	Mean Square	Expected Mean Square
Method	1	$MSM = 329$	$\sigma^2 + r\sigma_{md}^2 + rb\theta_m^2$
Day	3	$MSD = 144$	$\sigma^2 + r\sigma_{md}^2 + ra\sigma_d^2$
Interaction	3	$MS(MD) = 62$	$\sigma^2 + r\sigma_{md}^2$
Error	8	$MSE = 14$	σ^2

Source: Dr. J. Anderson, Beckman Instruments, Inc.

Tests of Hypotheses About Variance Components and Means

The null hypothesis of no interaction, $H_0: \sigma_{md}^2 = 0$, is tested with

$$F_0 = \frac{MS(MD)}{MSE} = \frac{62}{14} = 4.43$$

and the null hypothesis is rejected since $F_0 > F_{.05,3,8} = 4.07$. The presence of interaction with days suggests a possibility of differences among the chemistry methods that vary with days. Comparisons can be made between chemistry methods within each day with contrasts among cell means.

A test of the null hypothesis of no difference between the marginal means of the fixed effect factor would only be appropriate in the absence of interaction. For illustrative purposes only, the F_0 statistic for Chemistry Methods is

$$F_0 = \frac{MSM}{MS(MD)} = \frac{329}{62} = 5.31$$

and the null hypothesis would not be rejected since $F_0 < F_{.05,1,3} = 10.13$. The observed means for the two chemistry methods were $\bar{y}_{1..} = 147$ and $\bar{y}_{2..} = 138$ mg/dl.

The F_0 statistic to test the significance of the variance component for Days, σ_d^2, is

$$F_0 = \frac{MSD}{MS(MD)} = \frac{144}{62} = 2.32$$

and the null hypothesis H_0: $\sigma_d^2 = 0$ would not be rejected since $F_0 < F_{.05,3,3} = 9.28$.

Standard Errors for the Fixed-Factor Means

The standard error for the difference between the two chemistry methods on a given day,

$$\sigma_{(\bar{y}_{ij.} - \bar{y}_{kj.})} = \sqrt{\frac{2(\sigma^2 + r\sigma_{md}^2)}{r}} \tag{7.4}$$

is estimated by

$$s_{(\bar{y}_{ij.} - \bar{y}_{kj.})} = \sqrt{\frac{2MS(MD)}{r}} = \sqrt{\frac{2(62)}{2}} = 7.9 \tag{7.5}$$

As a general rule the mean square used for the standard error of a difference between the means of a fixed effect factor is the denominator mean square of F_0 used to test the null hypothesis about the fixed effect. Thus, the standard error of the difference between the marginal means of the two chemistry methods is estimated by

$$s_{(\bar{y}_{i..} - \bar{y}_{k..})} = \sqrt{\frac{2MS(MD)}{rb}} = \sqrt{\frac{2(62)}{2(4)}} = 3.9 \tag{7.6}$$

Three-Factor Experiments with Random and Fixed Factors

The problem of constructing F_0 statistics for tests of hypotheses was discussed in Section 7.1 for the three-factor experiment with three random factors. Similar difficulties occur with the mixed effects model for experiments with two or more random factors. The expected mean squares are given in Table 7.6 for three-factor experiments that have either one or two random effect factors with the remaining factor(s) fixed.

Table 7.6 Expected mean squares for three-factor experiments with (1) one fixed effect factor and two random effect factors and (2) two fixed effect factors and one random effect factor

Source of Variation	Expected Mean Square	
	A Fixed, B and C Random	*A and B Fixed, C Random*
A	$\sigma^2 + r\sigma^2_{abc} + rc\sigma^2_{ab} + rb\sigma^2_{ac} + rbc\theta^2_a$	$\sigma^2 + r\sigma^2_{abc} + rb\sigma^2_{ac} + rbc\theta^2_a$
B	$\sigma^2 + r\sigma^2_{abc} + rc\sigma^2_{ab} + ra\sigma^2_{bc} + rac\sigma^2_b$	$\sigma^2 + r\sigma^2_{abc} + ra\sigma^2_{bc} + rac\theta^2_b$
C	$\sigma^2 + r\sigma^2_{abc} + rb\sigma^2_{ac} + ra\sigma^2_{bc} + rab\sigma^2_c$	$\sigma^2 + r\sigma^2_{abc} + rb\sigma^2_{ac} + ra\sigma^2_{bc} + rab\sigma^2_c$
AB	$\sigma^2 + r\sigma^2_{abc} + rc\sigma^2_{ab}$	$\sigma^2 + r\sigma^2_{abc} + rc\theta^2_{ab}$
AC	$\sigma^2 + r\sigma^2_{abc} + rb\sigma^2_{ac}$	$\sigma^2 + r\sigma^2_{abc} + rb\sigma^2_{ac}$
BC	$\sigma^2 + r\sigma^2_{abc} + ra\sigma^2_{bc}$	$\sigma^2 + r\sigma^2_{abc} + ra\sigma^2_{bc}$
ABC	$\sigma^2 + r\sigma^2_{abc}$	$\sigma^2 + r\sigma^2_{abc}$
Error	σ^2	σ^2

The difficulties that are encountered in the construction of F_0 statistics for some hypotheses are immediately apparent upon inspection of the expected mean squares. In some instances, there is no legitimate mean square available for the denominator of the F_0 statistic among the existing mean squares. The mean squares for the F_0 statistic must be synthesized as discussed in Sections 5.11 and 7.1.

Alternative Mixed Model with Restrictions on the Interaction

There are several versions of the mixed model based on the definition used for the interaction effects; see Hocking (1973, 1985) and Searle et al. (1992) for a technical discussion. The alternative model places a constraint on the interaction effects. The model is

$$y_{ijk} = \mu + \alpha_i + g_j + (ag)_{ij} + e_{ijk} \tag{7.7}$$

$$i = 1, 2, \ldots, a \quad j = 1, 2, \ldots, b \quad k = 1, 2, \ldots, r$$

where μ is the mean, α_i is the fixed effect for factor A, g_j is the random effect for factor B, $(ag)_{ij}$ is the interaction effect, and e_{ijk} is random experimental error. The random effects, g_j and e_{ijk}, are assumed independent and normally distributed with means 0 and variances σ^2_g and σ^2, respectively. The interaction effects $(ab)_{ij}$ are assumed to be random effects and normally distributed with mean 0 and variance $\frac{(a-1)}{a}\sigma^2_{ag}$.

Since one of the factors, α_i, is fixed the alternative model has the interaction effect sum to zero over levels of the fixed factor, so that

$$\sum_{i=1}^{a} (ag)_{ij} = (ag)_{.j} = 0$$

With this model, interaction effects summed over levels of the random factor $\sum_j^b (ag)_{ij} = (ag)_{i.}$ will not be equal to zero, because they represent only a random sample of the interaction effects at each level of the fixed factor. However, at any given level of the random factor there is a finite set of interaction effects equal to the number of levels for the fixed factor, and the summation, $\sum_i^a (ag)_{ij} = (ag)_{.j}$, is equal to zero. Consequently, there is a covariance between two interaction effects at the same level of the random effect and different levels of the fixed effect, which is $-\frac{1}{a}\sigma_{ag}^2$. For example, with the experiment in Example 7.2 observations made on the same day with the two chemistry methods will be correlated, but observations on the same chemistry method on two different days will not be correlated.

Thus, the primary distinction for the alternative model is the presence of correlation between the interaction effects. As a consequence, the expectations of some mean squares are different with the alternative model. For example, with two factors, A fixed and B random, the expected mean squares are

$$E(MSA) = \sigma^2 + r\sigma_{ag}^2 + rb\theta_a^2$$

$$E(MSB) = \sigma^2 + ra\sigma_g^2 \qquad (7.8)$$

$$E[MS(AB)] = \sigma^2 + r\sigma_{ag}^2$$

$$E(MSE) = \sigma^2$$

The expected mean square for the random main effect factor does not include the interaction component in Equation (7.8), whereas previously in Table 7.5 the interaction component was present. This difference in the expected mean square for the random main effect can have considerable impact on statistical inference. For example, if the restricted model was used for Example 7.2 a test of the hypothesis $H_0: \sigma_d^2 = 0$ would require the test statistic $F_0 = MSD/MSE = 144/14 = 10.29$ and the null hypothesis would be rejected with $F_0 > F_{.05,3,8} = 4.07$, which is just opposite of the conclusion with the original test for the model without restrictions on the interaction.

Hocking (1973) discusses the relationship between the two models and shows the relationship between the variance components for the two models to be

$$\sigma_g^2 = \sigma_b^2 + \frac{1}{a}\sigma_{ab}^2 \qquad (7.9)$$

$$\sigma_{ag}^2 = \sigma_{ab}^2$$

The original model without restrictions on the interaction terms assumed the random interaction effects $(ab)_{ij}$ were uncorrelated with mean 0 and variance σ_{ab}^2. Also, b_i and e_{ijk} are uncorrelated random effects with variances σ_b^2 and σ^2, respectively. There was no assumption that the sum of $(ab)_{ij}$ over levels of the fixed factor $\sum_i^a (ab)_{ij} = (ab)_{.j}$ is equal to zero.

Therefore, a reasonable choice must be made with regard to the model that is most appropriate for the experimental situation. The model without restrictions on the interactions has one major advantage, which is that the expected mean squares for unbalanced data are consistent with the unrestricted model (Hartley & Searle, 1969). The restricted model is not considered in the unbalanced case. If there is a possibility of correlation between effects of a fixed factor for a given level of the random effect, and data are balanced, then the restricted interaction model may be appropriate. If not, or if data are unbalanced, then the model discussed originally in this section with no correlation among the interaction effects is most appropriate.

7.3 Nested Factor Designs: A Variation on the Theme

The standard factorial treatment design has two prominent characteristics: Each level of every factor occurs with all levels of the other factors, and the interaction among factors can be examined.

In certain types of studies the levels of one factor, B, will not be identical across all levels of another factor, A. Each level of factor A will contain different levels of factor B.

The levels of B are said to be nested within the levels of A. The designs are referred to here as **nested factor designs.** They are also called *hierarchical designs*. The following example illustrates a design to study components of variance for nested factors.

Example 7.3   Glucose Standards in Clinical Chemistry

Clinical laboratories perform patient serum assays critical for correct medical diagnoses. The laboratories maintain quality control programs to monitor the performance of assays and ensure the physician is receiving accurate information for diagnosis.

The important sources of variation in the assays are days on which the assays are conducted, the replicate runs within days, and the replicate serum sample preparations within runs. The quality control program requires that a spectrophotometer be tested with several runs on each of several days with serum standards used in the laboratory for control runs. Replicate serum preparations are evaluated within each of the runs.

The data in Table 7.7 are observations from a design used for quality control on the analysis of glucose standards. Glucose standard serums are maintained in the laboratory specifically for quality control runs. There were $c = 3$ replications of the standard prepared for each of $b = 2$ runs on each of $a = 3$ days.

The design is a nested design with two independent and unique runs on each of three days. The nesting of runs within days occurs because a run on any one day has no relationship with a run on any other day. For example, the first run on day 1 has nothing in common with the first runs on days 2 or 3.

The runs are numbered 1 through 6 in Table 7.7 to reflect their independence from one another from day to day.

In the same manner the replicate serum preparations are nested within runs. The Days, Runs, and Replicates represent factors at the first, second, and third levels of the hierarchy.

Table 7.7 Glucose (mg/dl) in quality control standards

	Day 1		Day 2		Day 3	
	Run 1	Run 2	Run 3	Run 4	Run 5	Run 6
	42.5	42.2	48.0	42.0	41.7	40.6
	43.3	41.4	44.6	42.8	43.4	41.8
	42.9	41.8	43.7	42.8	42.5	41.8
Day mean ($\bar{y}_{i..}$)	42.4		44.0		42.0	

Source: Dr. J. Anderson, Beckman Instruments, Inc.

Other Examples with Nested Factors

Consider a genetics study involving animals wherein each sire (male parent) is mated to a random sample of dams (female parents), and each mating results in a litter of several offspring as shown in Display 7.1. The Sires, Dams, and Offspring represent factors of interest in the study. There are different dams for each sire, and the dams are nested within sires. The offspring or progeny from a dam are different from those of other dams, and the progeny are nested within dams. The factors in the nested design form a hierarchy. The hierarchy or nesting of the factors is illustrated in Display 7.1. The uppermost level of the hierarchy represents sires, followed by dams and progeny in the second and third levels, respectively.

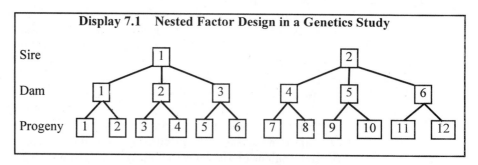

Display 7.1   Nested Factor Design in a Genetics Study

There are $b = 3$ different dams mated to each of the $a = 2$ sires for a total of $ab = 6$ dams, and there are $c = 2$ progeny per dam for a total of $abc = 12$ progeny.

Nested designs occur in education research that utilizes several elementary schools. The classrooms are nested within the schools, and the students are nested within the classrooms.

An experiment on fabric dye formulations requires that several replicate batches of each formulation be independently mixed and each batch tested on several specimens of a common fabric. The replicate batches are nested within dye formulations, and fabric specimens are nested within batches.

Statistical Model for Nested Factors

The factors in the nested design hierarchy can be fixed or random factors. The design for glucose standards in Example 7.3 would have all random factors if days, runs, and replicate serum preparations were considered random samples of their respective populations.

The linear model for a nested design with three random nested factors—A, B within A, and C within B—is

$$y_{ijk} = \mu + a_i + b_{j(i)} + c_{k(ij)} \tag{7.10}$$

$$i = 1, 2, \ldots, a \quad j = 1, 2, \ldots, b \quad k = 1, 2, \ldots, c$$

where a_i is the effect of factor A, $b_{j(i)}$ is the effect of factor B nested within A, and $c_{k(ij)}$ is the effect of factor C nested within B. The subscript $j(i)$ refers to the factor represented by the j subscript nested within the factor represented by the i subscript. The effects a_i, $b_{j(i)}$, and $c_{k(ij)}$ are assumed to be independent random effects with means 0 and variances σ_a^2, $\sigma_{b(a)}^2$, and $\sigma_{c(b)}^2$, respectively.

Factor A is a fixed effect in the genetics study of Display 7.1 if only two sires are available for the study and the investigator wants to restrict the genetic results to those two sires. The dam effects are random if they represent six dams randomly selected from a potential population of dams. If factor A effects are fixed, then the random effect a_i shown in Equation (7.10) is replaced by the fixed effect notation α_i.

Fixed effects for both factors A and B can occur in the study. Suppose a large metropolitan area fire department wants to evaluate the effect of two different fire station crew rotations on crew efficiency in its six fire districts. One rotation is tested in three of the districts chosen at random, and the other rotation is tested in the remaining three districts. A random sample of crews is tested in each district after the evaluation test period. The rotation factor, A, is fixed because only two types of crew rotations are under consideration. Likewise, the district factor, B, is fixed because only the six districts occurring in the metropolitan area are under consideration. If factors A and B are fixed, then the random effects a_i and $b_{j(i)}$ are replaced by fixed effects α_i and $\beta_{j(i)}$ in Equation (7.10).

Analysis for Random Nested Factors

The objectives of studies that utilize the random effects nested design depend on the subject matter of the study, but estimation of the components of variance and tests of hypotheses about the components are involved. The analysis of variance sums of squares partitions are used for both the estimation and testing procedures.

The computations for the analysis of variance are identical to those for subsampling as shown in Table 5.5. The expectations for the mean squares with all effects random are given in the abbreviated analysis of variance outline shown in Table 7.8.

Table 7.8 Expected mean squares for the analysis of variance of a nested design with three random factors, A, B, and C

Source of Variation	Degrees of Freedom	Mean Square	Expected Mean Square
Total	$abc - 1$		
A	$a - 1$	MSA	$\sigma^2_{c(b)} + c\sigma^2_{b(a)} + bc\sigma^2_a$
B within A	$a(b - 1)$	$MS(B/A)$	$\sigma^2_{c(b)} + c\sigma^2_{b(a)}$
C within B	$ab(c - 1)$	$MS(C/B)$	$\sigma^2_{c(b)}$

Point Estimators for Variance Components

The analysis of variance estimates of the variance components are found by equating the observed mean squares to the expected mean squares and solving for the components of variance. The three estimators for the analysis in Table 7.8 are

$$\widehat{\sigma}^2_{c(b)} = MS(C/B)$$
$$\widehat{\sigma}^2_{b(a)} = \frac{[MS(B/A) - MS(C/B)]}{c} \tag{7.11}$$

and

$$\widehat{\sigma}^2_a = \frac{[MSA - MS(B/A)]}{bc}$$

The null hypotheses of interest are $H_0: \sigma^2_a = 0$ if the a_i are random effects and $H_0: \overline{\mu}_{1..} = \overline{\mu}_{2..} = \cdots = \overline{\mu}_{t..}$ for fixed effects with the F_0 statistic $F_0 = MSA/MS(B/A)$. The hypothesis $H_0: \sigma^2_{b(a)} = 0$ is tested with $F_0 = MS(B/A)/MS(C/B)$.

Analysis of Glucose Quality Control Standards

The analysis of variance for the glucose standards in Example 7.3 is shown in Table 7.9. Many computer programs will print a table of expected mean squares and variance component estimates upon request. The coefficient for the Day component (σ^2_a) is $bc = 6$, while the coefficient for the Run/Day component is $c = 3$.

Table 7.9 Analysis of variance for glucose quality control standards

Source of Variation	Degrees of Freedom	Sum of Squares	Mean Square	F Value	Pr > F
Total	17	43.88			
Day	2	13.76	6.88	1.26	.400
Run/Day	3	16.36	5.45	4.75	.021
Rep/Run	12	13.76	1.15		

The estimates for the components of variance are

Reps: $\widehat{\sigma}^2_{c(b)} = MS(\text{Rep/Run}) = 1.15$

Run: $\widehat{\sigma}^2_{b(a)} = \left[\dfrac{MS(\text{Run/Day}) - MS(\text{Rep/Run})}{c} \right] = \dfrac{(5.45 - 1.15)}{3} = 1.43$

Day: $\widehat{\sigma}^2_a = \dfrac{[MS\text{Day} - MS(\text{Run/Day})]}{bc} = \dfrac{(6.88 - 5.45)}{6} = 0.24$

The estimate of total variance for a glucose standard analysis is

$$\widehat{\sigma}^2_y = \widehat{\sigma}^2_a + \widehat{\sigma}^2_{b(a)} + \widehat{\sigma}^2_{c(b)} = 0.24 + 1.43 + 1.15 = 2.82$$

with standard deviation $\widehat{\sigma}_y = 1.68$. Approximately 9% of the variation is attributed to day-to-day variation, $\widehat{\sigma}^2_a$; 51% to run-within-day variation, $\widehat{\sigma}^2_{b(a)}$; and 41% to replicates-within-runs variation, $\widehat{\sigma}^2_{c(b)}$. The grand mean of the study was $\overline{y}_{...} = 42.8$, and the percent coefficient of variation for the glucose standards in this set of runs was $\%CV = (100)(\widehat{\sigma}_y / \overline{y}_{...}) = (100)(1.68/42.8) = 3.9\%$.

Standard Errors for Means

The variances for the grand mean for the study $\overline{y}_{...}$ and a day mean $\overline{y}_{i..}$, are

$$\sigma^2_{\overline{y}_{...}} = \frac{\sigma^2_{c(b)}}{abc} + \frac{\sigma^2_{b(a)}}{ab} + \frac{\sigma^2_a}{a} \quad \text{and} \quad \sigma^2_{\overline{y}_{i..}} = \frac{\sigma^2_{c(b)}}{bc} + \frac{\sigma^2_{b(a)}}{b} \qquad (7.12)$$

respectively. The estimates are

$$s^2_{\overline{y}_{...}} = \frac{MS\ \text{Day}}{abc} = \frac{6.88}{(3)(2)(3)} = 0.38$$

and

$$s_{\bar{y}_i.}^2 = \frac{MS(\text{Run/Day})}{bc} = \frac{5.45}{(2)(3)} = 0.91$$

The standard error estimates are $s_{\bar{y}_{...}} = \sqrt{0.38} = 0.62$ and $s_{\bar{y}_{i..}} = \sqrt{0.91} = 0.95$.

The variances, standard deviations, standard errors, coefficient of variation, and means are all useful statistics for quality control monitoring. Cumulative records of these statistics are often maintained in a laboratory and inspected regularly. If the values deviate from some established norms the analyst is alerted to the fact that the process must be investigated for the source or cause of the deviation.

Tests of Hypotheses About Variances

The test of the null hypothesis for the day component of variance is

$$F_0 = \frac{MS\text{Day}}{MS(\text{Run/Day})} = \frac{6.88}{5.45} = 1.26$$

and it is not significant with $Pr > F = .400$ (Table 7.9). The test of the null hypothesis for the run-within-day component of variance is

$$F_0 = \frac{MS(\text{Run/Day})}{MS(\text{Rep/Run})} = \frac{5.45}{1.15} = 4.75$$

and it is significant with $Pr > F = .021$ (Table 7.9). These results indicate the analyst could most effectively increase precision by concentrating efforts to reduce the variation among runs on the same day and replicates within runs to obtain more precise estimates.

Analysis with Fixed-Factor Effects

The F_0 ratios for tests of hypotheses with fixed effects in the nested design can be determined from the expected mean squares shown in Table 7.10. The fixed effects are defined such that $\Sigma_i \alpha_i = 0$ and $\Sigma_j \beta_{j(i)} = 0$ for $i = 1, 2, \ldots, a$.

Table 7.10   Expected mean squares for the analysis of variance of a nested design with fixed and mixed effects for A and B

Mean Square	A and B Fixed	A Fixed and B Random
MSA	$\sigma_{c(b)}^2 + bc\Sigma\, \alpha_i^2/(a-1)$	$\sigma_{c(b)}^2 + c\sigma_{b(a)}^2 + bc\Sigma\, \alpha_i^2/(a-1)$
$MS(B/A)$	$\sigma_{c(b)}^2 + c\Sigma\, \beta_{j(i)}^2/a(b-1)$	$\sigma_{c(b)}^2 + c\sigma_{b(a)}^2$
$MS(C/B)$	$\sigma_{c(b)}^2$	$\sigma_{c(b)}^2$

With A fixed and B random the null hypothesis for factor A effects, H_0: $\alpha_1 = \alpha_2 = \cdots = \alpha_a$, is tested with $F_0 = MSA/MS(\boldsymbol{B/A})$. When both factors are fixed the null hypothesis for factor A effects is tested with $F_0 = MSA/MS(C/B)$. The null hypothesis for factor B effects with both factors fixed is H_0: $\beta_{1(i)} = \beta_{2(i)} = \cdots = \beta_{b(i)}$ for all i, and it is tested with $F_0 = MS(B/A)/MS(C/B)$.

Staggered Nested Designs to Equalize Information About Variances

The nested factors design contains more information on factors at lower levels in the hierarchy of the design than at higher levels. In larger studies the discrepancies in degrees of freedom among sources of variation can be considerable. *Staggered nested designs* were developed to equalize the degrees of freedom for the mean squares at each level of the hierarchy.

The staggered designs have unequal numbers of levels for factors that are nested within other factors. The levels for factor B nested within factor A vary from one level of factor A to another in such a way that the degrees of freedom for MSA and $MS(B/A)$ are more equal. Likewise, the levels for factor C nested within factor B can vary across levels of factor B to achieve degrees of freedom for $MS(C/B)$ similar to that for the other mean squares.

Anderson (1960) and Bainbridge (1963) gave some of the early results on the use of staggered designs. Smith and Beverly (1981) provide a general discussion on the use and analysis of staggered designs. Goldsmith and Gaylor (1970) enumerated 61 staggered designs for three stages of factors (A, B, and C), such that all three variance components can be estimated by the analysis of variance method. The designs enumerated by Goldsmith and Gaylor included two or three of the five possible fundamental structures in a staggered design shown in Display 7.2.

Display 7.2 Five Fundamental Structures for Three-Stage Staggered Nested Designs (Levels of B within A are b_i and the levels of C within B are c_{ij})

			Structures			
Stage	1		2	3	4	5
A						
B/A						
C/B						
b_i	2		2	2	1	1
c_{ij}	2		2, 1	1	2	1

The analysis of a three-stage staggered design follows the pattern established for the analysis with unequal subsamples and replications in Table 5.8. Leone et al. (1968) gave computational formulae for the four-stage nested designs with unequal levels at all stages. They provided formulae for the analysis of variance sums of squares, expected mean squares, and estimates of variances for the means at various stages of the design ($\overline{y}_{\ldots}, \overline{y}_{i\ldots}, \overline{y}_{ij\ldots}$, and $\overline{y}_{ijk\cdot}$). Gates and Shiue (1962) provided computational formulae for the analysis of variance sums of squares and expected mean squares for a general S-stage hierarchical classification.

Example 7.4 A Staggered Design for Soil Samples

The data in Exercise 5.8 were generated from a staggered nested design to estimate components of variance for characteristics of soil samples. The three factors in the design were fields (F), sections within fields (S), and locations within sections (L). Suppose $a = 12$ fields and $b_i = 2$ sections per field are sampled. Suppose in fields 1 through 6 that $c_{ij} = 2$ locations per section are sampled, and in fields 7 through 12 $c_{ij} = 1$ location per section is sampled. This staggered design has six replications each of structures 1 and 3 from Display 7.2 as shown in Display 7.3. The design is one of those listed by Goldsmith and Gaylor (1970).

Display 7.3 Staggered Nested Design for Soil Samples

The analysis of variance outline for this design shown in Table 7.11 follows the format shown in Table 5.8 and utilizes Equation (5.23) to compute the coefficients for the variance components in the expected mean squares. The near equal degrees of freedom for the mean squares provide near equal information on each source of variation. The variance components are estimated by the analysis of variance method as

$$\widehat{\sigma}^2_{c(b)} = MS(L/S), \quad \widehat{\sigma}^2_{b(a)} = \frac{[MS(S/F) - MS(L/S)]}{1.50}$$

and

(7.13)

$$\widehat{\sigma}^2_a = \frac{[1.50MSF - 1.48MS(S/F) - (1.50 - 1.48)MS(L/S)]}{(2.97)(1.50)}$$

Table 7.11 Analysis of variance outline with expected mean squares for the staggered nested design

Source of Variation	Degrees of Freedom	Mean Square	Expected Mean Square
Field	$a - 1 = 11$	$MS\boldsymbol{F}$	$\sigma_{c(b)}^2 + 1.48\sigma_{b(a)}^2 + 2.97\sigma_a^2$
Sections	$\sum_{i=1}^{a} b_i - a = 12$	$MS(\boldsymbol{S}/\boldsymbol{F})$	$\sigma_{c(b)}^2 + 1.50\sigma_{b(a)}^2$
Locations	$N - \sum_{i=1}^{a} b_i = 12$	$MS(\boldsymbol{L}/\boldsymbol{S})$	$\sigma_{c(b)}^2$

7.4 Nested and Crossed Factors Designs

Certain experimental conditions give rise to factorial arrangements that contain crossed and nested factors. In this case, some factors are crossed in the usual factorial arrangement of levels, whereas other factors are nested within the cells of the factorial arrangement or within levels of at least one other factor. These designs are sometimes called *nested factorial designs* (Anderson & McLean, 1974; Hicks, 1973; Smith & Beverly, 1981).

Example 7.5   Spectrophotometer Evaluation

Companies that manufacture machines and instruments have research and development departments to create new instruments or improve their current line of instruments. Performance tests are a regular part of the development stage of any new machine. The machines are tested for their mechanical or electrical functions, their accuracy in performing their designated function, and so forth.

Research Problem: In one such setting a researcher was developing a new spectrophotometer for applications in medical laboratories. A model spectrophotometer had been constructed according to the proposed design and was ready for evaluation of its spectral capabilities in a laboratory setting. It was necessary to determine whether this particular design determined spectral properties over the required range of serum glucose standards. The researcher had to determine whether the variability and consistency of results over multiple runs and days were within the required specifications.

Treatment Design: A factorial treatment design was used with "concentrations" of glucose and "days" as factors. The serum samples were enhanced with three different levels of glucose to cover the range of glucose concentrations the instrument should be able to analyze. All three concentrations were analyzed on each day, so that concentrations were crossed with days in a 3×3 factorial arrangement. Two runs of the instrument were made on each day, so that runs were nested within each day.

Experiment Design: Four replicate serum samples were prepared for each of the three concentrations of the glucose standards each day. Two samples of each concentration were assigned randomly to each run of each day. The six samples were analyzed in random order on each run. The same technician prepared the samples and operated the instrument throughout the experiment.

The design had nested and crossed factors with $a = 3$ concentrations crossed with $b = 3$ days with $c = 2$ runs nested within each day and $r = 2$ replicate serum sample preparations for each concentration in each run. The glucose concentrations observed on the spectrophotometer are shown in Table 7.12.

Table 7.12 Glucose concentrations (mg/dl) for three standard concentration samples from two runs of a spectrophotometer on each of three days

	Day 1		Day 2		Day 3	
Concen.	Run 1	Run 2	Run 3	Run 4	Run 5	Run 6
1	41.2	41.2	39.8	41.5	41.9	45.5
	42.6	41.4	40.3	43.0	42.7	44.7
2	135.7	143.0	132.4	134.4	137.4	141.1
	136.8	143.3	130.3	130.0	135.2	139.1
3	163.2	181.4	173.6	174.9	166.6	175.0
	163.3	180.3	173.9	175.6	165.5	172.0

Source: Dr. J. Anderson, Beckman Instruments, Inc.

Statistical Model for Nested and Crossed Factors

The statistical model for this particular experiment is

$$y_{ijkl} = \mu + \alpha_i + b_j + c_{k(j)} + (ab)_{ij} + (ac)_{ik(j)} + e_{ijkl} \tag{7.14}$$

$$i = 1, 2, \ldots, a \quad j = 1, 2, \ldots, b \quad k = 1, 2, \ldots, c \quad l = 1, 2, \ldots, r$$

where α_i is the fixed effect for Concentration, b_j is the random effect for Day, $c_{k(j)}$ is the random effect for Runs nested within Day, $(ab)_{ij}$ is the random effect for Concentration by Day interaction, $(ac)_{ik(j)}$ is the random effect for Concentration by Run interaction nested within Day, and e_{ijkl} is the random experimental error. The assumptions for the effects are consistent with those indicated in previous sections on random, mixed, and nested models.

The model effects for the two crossed factors, Concentration and Day, follow the usual convention with main effects and interaction. The model effect for Runs within Day follows the usual convention for one factor nested within another. The interaction effect for Concentration by Run nested within Day is the one new feature in the model. Since each concentration is evaluated on each run the two factors constitute a complete factorial arrangement on each day. Therefore, the Concentration by Run interaction can be evaluated on each day and can be nested or pooled over days since the Runs nested within Day are unique runs on each day.

Expected Mean Squares

The model shown in Equation (7.14) is a mixed model with one of the crossed factors random and the other fixed. The expected mean squares for the analysis of variance are affected by the model assumed for the crossed factors and the manner by which the other factors are nested in the experiment. Consequently, a variety of expected mean square patterns are possible. The expected mean squares for models with A and B fixed or with A fixed and B random are shown in Table 7.13.

Table 7.13 Expected mean squares for a nested factorial with A and B crossed and C nested within B

Source of Variation	Degrees of Freedom	Expected Mean Square	
		A Fixed, B and C Random	A and B Fixed, C Random
Total	$abcr - 1$		
A	$a - 1$	$\sigma^2 + r\sigma^2_{ac(b)} + cr\sigma^2_{ab} + bcr\theta^2_a$	$\sigma^2 + r\sigma^2_{ac(b)} + bcr\theta^2_a$
B	$b - 1$	$\sigma^2 + r\sigma^2_{ac(b)} + ar\sigma^2_{c(b)} + cr\sigma^2_{ab} + acr\sigma^2_b$	$\sigma^2 + r\sigma^2_{ac(b)} + ar\sigma^2_{c(b)} + acr\theta^2_b$
AB	$(a-1)(b-1)$	$\sigma^2 + r\sigma^2_{ac(b)} + cr\sigma^2_{ab}$	$\sigma^2 + r\sigma^2_{ac(b)} + cr\theta^2_{ab}$
C/B	$b(c-1)$	$\sigma^2 + r\sigma^2_{ac(b)} + ar\sigma^2_{c(b)}$	$\sigma^2 + r\sigma^2_{ac(b)} + ar\sigma^2_{c(b)}$
AC/B	$b(a-1)(c-1)$	$\sigma^2 + r\sigma^2_{ac(b)}$	$\sigma^2 + r\sigma^2_{ac(b)}$
Error	$abc(r-1)$	σ^2	σ^2

Degrees of Freedom

The degrees of freedom for the sources of variation in the analysis of variance follow the usual conventions for the crossed factors, A and B (Section 6.4). The degrees of freedom for C nested within B follow the convention for one factor nested within another (Section 7.3). The degrees of freedom for the AC interaction nested within B follows the nesting degrees of freedom convention. Only in this case the interaction measured at each level of B has $(a-1)(c-1)$ degrees of freedom; therefore, when nested over b levels of B there are $b(a-1)(c-1)$ degrees of freedom. The degrees of freedom for error are the degrees of freedom for each cell, $r-1$, pooled over the abc cells in the experiment.

Analysis for Nested and Crossed Factors

The analysis of variance for the observations from the spectrophotometer evaluation of Example 7.5 is shown in Table 7.14. Four different mean squares are required as denominators for the F_0 statistics to test hypotheses about the effects or variance components in the model with one fixed factor and two random factors (see Tables 7.13 and 7.14). The mean square for Error is the denominator of F_0 to test concentration $\times$ runs within day, CR/D. The mean square for CR/D is the denominator of F_0 required to test runs within day, R/D, and concentration $\times$ day interaction, CD. The mean square CD is the denominator of F_0 required to test differences among the concentration means, C. Synthesized mean squares for F_0 required to test day variation, D, are

$$MSN = MS(D) + MS(CR/D) = 42.4$$

for the numerator and

$$MS = MS(CD) + MS(R/D) = 131.8$$

for the denominator. Obviously, $F_0 = MSN/MS = 42.4/131.8 = 0.32$ will not be significant. Many programs automatically use the mean square listed for experimental error for all of the F_0 statistics, which is not always the correct denominator mean square. Those programs must be given special instructions to compute the correct F_0 statistics if they have the capability.

Table 7.14   Analysis of variance for spectrophotometer glucose measurements from a factorial with nested and crossed factors

Source of[†] Variation	Degrees of Freedom	Sum of Squares	Mean Squares	F	Pr > F
Total	35	108,934.1			
C	2	108,263.6	54,131.8	1,227.48	.000
D	2	24.9	12.4	*	*
CD	4	176.4	44.1	1.47	.321
R/D	3	263.1	87.7	2.92	.122
CR/D	6	180.2	30.0	21.43	.000
Error	18	25.8	1.4		

*Test with synthesized Mean Square
[†]C = Concentration, D = Day, R/D = Run nested in Days

As expected the concentration differences were significant, $F_0 = 1227.48$ with $Pr > F = .000$. The concentration $\times$ day interaction was not significant, $Pr > F = .321$, which indicates relatively consistent performance of the instrument from day to day with respect to concentration measurements. However, concentration $\times$ run within day was significant, $F_0 = 21.43$ and $Pr > F = .000$. Runs within days with $F_0 = 2.92$ and $Pr > F = .122$ was not significant. The run-to-run consistency of the instrument across concentrations requires some inspection. The inconsistency could be due to the operation of the instrument or the lack of consistency in preparation of the samples for each of the concentrations from run to run.

7.5 How Many Replications?

Random Models

Replication numbers to detect desired significant contributions from a component of variance require a value for the λ constant where the F_0 statistic has the central F_{ν_1,ν_2} distribution multiplied by λ^2 (see Section 5.8). The value for λ^2 may be evaluated in general as follows. Let F_0 be the ratio of mean squares $F_0 = MSN/MSD$, where MSN and MSD, respectively, designate the mean squares for the numerator and denominator of F_0. The constant λ^2 is the ratio of the expected mean squares, or $\lambda^2 = E(MSN)/E(MSD)$. Technical details may be found in Graybill (1961). The charts in Appendix Table X are used as described in Section 5.8.

Consider the two-factor random effects model in Section 7.1. A test of H_0: $\sigma_a^2 = 0$ requires $F_0 = MSA/MS(AB)$, so that

$$\lambda^2 = \frac{E(MSA)}{E[MS(AB)]} = \frac{\sigma^2 + r\sigma_{ab}^2 + rb\sigma_a^2}{\sigma^2 + r\sigma_{ab}^2} = 1 + \frac{rb\sigma_a^2}{\sigma^2 + r\sigma_{ab}^2} \qquad (7.15)$$

Mixed Models

The detection of prescribed fixed-factor effects, say factor A, for the two-factor mixed-model experiment requires

$$\Phi^2 = \frac{br \sum_{i=1}^{a} \alpha_i^2}{a(\sigma^2 + r\sigma_{ab}^2)} \qquad (7.16)$$

for the charts in Appendix Table IX for fixed effects. The value of the constant λ for tests about σ_b^2 and σ_{ab}^2 can be determined as shown for the random model.

7.6 Expected Mean Square Rules

The rules for expected mean squares given in this section apply to most balanced designs with equal replication numbers. The number of levels for any factor do not vary within the balanced designs. The designs include crossed factorials, nested factorials, and mixtures of crossed and nested factors. The rules are adapted from those given in various publications, including Bennett and Franklin (1954) and Mason, Gunst, and Hess (1989). Many computer programs have commands to produce expected mean squares for an analysis of variance.

Rules Illustrated with the Unrestricted Mixed Model

The rules are exemplified with a two-factor mixed-model experiment, A fixed and B random, with r replications of each treatment combination.

1. Write out the linear model for the design:

$$y_{ijk} = \mu + \alpha_i + b_j + (ab)_{ij} + e_{k(ij)}$$

$$i = 1, 2, \ldots, a \quad j = 1, 2, \ldots, b \quad k = 1, 2, \ldots, r$$

   Note the replication subscript k is nested within the ijth treatment combination.

2. Construct a two-way table with
   (a) a row for each term in the model, excluding μ, labeled with the model term and
   (b) a column for each subscript used in the model.

3. Over each column subscript write the number of factor levels for the subscript and write "R" if the factor is random and "F" if the factor is fixed.

4. Add another column with entries as the appropriate fixed or random variance component for the effect represented by that row in the table.

Source		F a i	R b j	R r k	Component
A	α_i				θ_a^2
B	b_j				σ_b^2
AB	$(ab)_{ij}$				σ_{ab}^2
Error	$e_{k(ij)}$				σ^2

5. For each row, if the column subscript does not appear in the row effect, enter the number of levels corresponding to the subscript.

Source		F a i	R b j	R r k	Component
A	α_i		b	r	θ_a^2
B	b_j	a		r	σ_b^2
AB	$(ab)_{ij}$			r	σ_{ab}^2
Error	$e_{k(ij)}$				σ^2

6. If a subscript is bracketed in a row effect, place a 1 in cells under those subscripts that are inside the brackets.

Source		F a i	R b j	R r k	Component
A	α_i		b	r	θ_a^2
B	b_j	a		r	σ_b^2
AB	$(ab)_{ij}$			r	σ_{ab}^2
Error	$e_{k(ij)}$	1	1		σ^2

7. (a) For each row, if any row subscript matches the column subscript, enter a 0 if the column represents a fixed factor F and there is a fixed component of variance for the effect represented by the row.

   (b) Enter a 1 in the remaining cells.

Source		F a i	R b j	R r k	Component
A	α_i	0	b	r	θ_a^2
B	b_j	a	1	r	σ_b^2
AB	$(ab)_{ij}$	1	1	r	σ_{ab}^2
Error	$e_{k(ij)}$	1	1	1	σ^2

8. To determine the expected mean square for a specific source of variation:

   (a) Include σ^2 with a coefficient of 1 in all expected mean squares.

   (b) Of the remaining variance components include only those whose corresponding model terms include the subscripts of the effect under consideration. For $E(MS\boldsymbol{B})$ the b_j effect, include σ_{ab}^2 and σ_b^2 in addition to σ^2.

   (c) Cover the columns containing non-bracketed subscripts for the effect under consideration. For α_i cover i and for $e_{k(ij)}$ cover k.

   (d) The coefficient for each component in the $E(MS)$ is the product of the remaining columns of the row for that effect. For $E(MS\boldsymbol{B})$ the column with j is covered so that only the values in columns i and k are visible. For the $(ab)_{ij}$ row the visible values are 1 and r so that the coefficient for σ_{ab}^2 is $1 \cdot r = r$. For the b_j row the visible values are a and r so that the coefficient for σ_b^2 is $a \cdot r$.

Source		F a i	R b j	R r k	Component	$E(MS)$
A	α_i	0	b	r	θ_a^2	$\sigma^2 + r\sigma_{ab}^2 + br\theta_a^2$
B	b_j	a	1	r	σ_b^2	$\sigma^2 + r\sigma_{ab}^2 + ar\sigma_b^2$
AB	$(ab)_{ij}$	1	1	r	σ_{ab}^2	$\sigma^2 + r\sigma_{ab}^2$
Error	$e_{k(ij)}$	1	1	1	σ^2	σ^2

Illustration: The complete table for expected mean square determination of a mixed-model factorial with crossed and nested factors follows. Factors A and B are fixed and crossed, and factor C is random and nested within B across A. The model is

$$y_{ijkl} = \mu + \alpha_i + \beta_j + (\alpha\beta)_{ij} + c_{k(j)} + (ac)_{ik(j)} + e_{l(ijk)}$$

| | | F | F | R | R | | |
| | | a | b | c | r | | |
Source		i	j	k	l	Component	E(MS)
A	α_i	0	b	c	r	θ_a^2	$\sigma^2 + r\sigma_{ac(b)}^2 + bcr\theta_a^2$
B	β_j	a	0	c	r	θ_b^2	$\sigma^2 + r\sigma_{ac(b)}^2 + ar\sigma_{c(b)}^2 + acr\theta_b^2$
AB	$(\alpha\beta)_{ij}$	0	0	c	r	θ_{ab}^2	$\sigma^2 + r\sigma_{ac(b)}^2 + cr\theta_{ab}^2$
C/B	$c_{k(j)}$	a	1	1	r	$\sigma_{c(b)}^2$	$\sigma^2 + r\sigma_{ac(b)}^2 + ar\sigma_{c(b)}^2$
(AC)/B	$(ac)_{ik(j)}$	1	1	1	r	$\sigma_{ac(b)}^2$	$\sigma^2 + r\sigma_{ac(b)}^2$
Error	$e_{l(ijk)}$	1	1	1	1	σ^2	σ^2

Alteration of the Rules for Restricted Mixed Models

One alteration of the method in Step 7 is required if the restricted mixed model is used—that is, the model in which the interaction effects are correlated and summation of interaction effects over the fixed effects subscripts is restricted to zero. Step 7 (a) will be "*For each row, if any row subscript matches the column subscript, enter a 0 if the column subscript represents a fixed factor F.*" Step 7 (b) remains unchanged.

Other Estimation Methods for Variance Components

Only analysis of variance estimators have been considered for estimation of variance components in Chapters 5 and 7. Estimation of variance components by the analysis of variance method is relatively straightforward with balanced data (that is, all data cells contain the same number of observations). Estimation of variance components is much more difficult when data are unbalanced. The first major breakthrough for variance component estimation was made by Henderson (1953), who presented three different adaptations of the analysis of variance method for estimating variance components with unbalanced data and random or mixed models.

Since that time, other methods of estimation have been developed, including maximum likelihood (ML) estimation by Hartley and Rao (1967) and a modification of maximum likelihood known as restricted maximum likelihood (REML) attributed to work by Thompson (1962) and Patterson and Thompson (1971). The MINQUE method to find minimum variance quadratic unbiased estimators can be traced to a variety of authors. These methods of estimation are usually available in many of the more comprehensive statistical packages. A comprehensive treatment of variance component models and estimation can be found in Searle et al. (1992).

EXERCISES FOR CHAPTER 7

---

1. Cholesterol was measured in the serum samples of five randomly selected patients from a pool of patients. Two independent replicate tubes were prepared for each patient for each of four runs on a spectrophotometer. The objective of the study was to determine whether the relative cholesterol measurements for patients were consistent from run to run in the clinic. The data are mg/dl of cholesterol in the replicate samples from each patient on each run.

			Patient		
Run	1	2	3	4	5
1	167.3	186.7	100.0	214.5	148.5
	166.7	184.2	107.9	215.3	148.5
2	179.6	193.8	111.6	228.9	158.6
	175.3	198.9	114.4	220.4	154.7
3	169.4	179.4	105.9	208.2	144.7
	165.9	177.6	104.1	207.1	145.9
4	177.7	190.4	113.4	221.0	156.1
	177.1	192.4	114.6	219.7	151.0

Source: Dr. J. Anderson, Beckman Instruments, Inc.

a. Write a linear model for the experiment assuming patients and runs are random effects, explain the terms, and conduct an analysis of variance for the data.
b. Show the expected mean squares for the analysis of variance.
c. Estimate the components of variance for runs, patients, and interaction.
d. State the null and alternate hypotheses for main effects and interaction, test each of the null hypotheses, and interpret your results.

2. An animal scientist conducted an experiment to study the effect of water quality on feedlot performance of steer calves. Four water quality treatments were used for the experiment. The water sources were designated as normal (N) and saline (S). The saline water was formulated to approximate the mineral concentrations in some underground water sources utilized in practice for watering livestock. Four combinations of water used in two consecutive 56-day periods of the experiment were N-N, N-S, S-N, and S-S. The feeding trial consisted of the four water treatments with two replicate pens of animals for each treatment in a completely randomized design. The trial was conducted on two separate occasions (two consecutive summers). The resulting design is a factorial arrangement of four water treatments and two summers. The water treatments are considered fixed effects and summers are considered random effects, so that a mixed model is appropriate for the study. The data are the average daily gains for the 16 pens of steers.

		Water		
Summer	*N-N*	*N-S*	*S-N*	*S-S*
1	2.65	2.46	2.56	2.43
	2.53	2.36	2.38	2.50
2	2.25	1.95	2.01	2.14
	2.20	2.25	1.98	2.37

Source: Dr. D. Ray, Department of Animal Sciences, University of Arizona.

a. Write a linear model for the experiment, explain the terms, and conduct an analysis of variance for the data.
b. Prepare a table of cell and marginal means and their respective standard errors.
c. Show the expected mean squares for the analysis of variance.
d. Test the null hypotheses for main effects and interaction, and interpret your results.
e. The water treatments have a 2×2 factorial arrangement. The first factor (A) is normal or saline water in the first 56-day period, and the second factor (B) is normal or saline water in the second 56-day period. Write the linear model for the experiment with this arrangement, considering summers as random effects and factors A and B as fixed effects. Repeat parts (a) through (d) of this exercise with the new model.

3. Three formulations of an alloy were prepared with four separate castings for each formulation. Two bars from each casting were tested for tensile strength. The data are tensile strengths of the individual bars. There are four castings nested within each alloy.

		Castings		
Alloys	*1*	*2*	*3*	*4*
A	13.2	15.2	14.8	14.6
	15.5	15.0	14.2	15.1
B	17.1	16.5	16.1	17.4
	16.7	17.3	15.4	16.8
C	14.1	13.2	14.5	13.8
	14.8	13.9	14.7	13.5

a. Write a linear model for the experiment, assuming alloys as fixed effects and castings within alloys and bars within castings as random effects. Explain the terms, and compute the analysis of variance.
b. Show the expected mean squares for the analysis.
c. Test the null hypothesis for alloy effects, and interpret your results.
d. Compute the estimated means and the 95% confidence interval estimates for the means of each alloy.
e. Estimate the components of variance for castings and bars.

4. A traffic engineering study was conducted to evaluate the effects of three traffic signal types on traffic delay at intersections. The study was also designed to evaluate two methods for measuring

traffic delay. The three signal types were pre-timed, semi-actuated, and fully actuated signals. The two methods, point-sample and path-trace, estimated stopped time per vehicle at an intersection.

Two intersections were used for each signal type. Measurements were made during rush hour and nonrush hour periods. The three crossed factors in the study were signal type, method, and time of day (rush hour and nonrush hour). The intersections were nested within signal types but crossed with method and time of day since both methods were used on the same intersection during both times of day. The data were traffic delay measured as seconds per vehicle.

| Signal | Intersection | Point-Sample | | Path-Trace | |
		Rush	Nonrush	Rush	Nonrush
Pretimed	1	61.7	57.4	53.1	36.5
	2	35.8	18.5	35.5	15.9
Semi-actuated	3	20.0	24.6	17.0	21.0
	4	2.7	3.1	1.5	1.1
Fully actuated	5	35.7	26.8	35.4	20.7
	6	24.3	25.9	27.5	23.3

Source: W. Reilly, C. Gardner, J. Kell (1976), A technique for measurement of delay at intersections. *Technical Report*, FHWA–RD–76-135, Federal Highway Administration, Office of R&D, Washington, D.C.

a. Consider only one method of measuring stopped delay (point-sample or path-trace). Write a linear model for the study, assuming signal type and time of day are fixed effects. Sketch the analysis of variance table, including source of variation, degrees of freedom, and expected mean squares.

b. Suppose you suspect an intersection × time-of-day interaction. Will you be able to test a null hypothesis about the interaction?

c. Suppose there is an interaction between intersection and time of day. Which hypotheses can you test from the analysis of variance?

d. Compute the analysis of variance for the point-sample data. State your assumptions about the model, and test the hypotheses that can be tested with your stated model.

e. Now suppose you want to include the factor for method of measurement into the analysis— that is, point-sample versus path-trace as a fixed effect factor. Write a linear model for the analysis, and sketch the analysis of variance table, including sources of variation, degrees of freedom, and expected mean squares.

f. What assumptions are necessary about the interaction between intersection and the other factors for you to test some hypotheses from the analysis of variance?

g. Compute the analysis of variance for the entire data set. State your assumptions about the model, and test the hypotheses that can be tested about model effects.

5. An experiment was conducted to compare the accuracy of two mass spectrometers in measuring the ratio of ^{14}N to ^{15}N. Two soil samples were taken from each of three plots of land treated with ^{15}N. Two subsamples of each sample were analyzed on each of the two machines. The resulting design has machines crossed with plots and samples. However, the samples are nested within plots. The data are ratios of ^{14}N to ^{15}N (after multiplying by 1000).

Plot	1		2		3	
Sample	1	2	3	4	5	6
Machine A	3.833	3.819	3.756	3.882	3.720	3.729
	3.866	3.853	3.757	3.871	3.720	3.768
Machine B	3.932	3.884	3.832	3.917	3.776	3.833
	3.943	3.888	3.829	3.915	3.777	3.827

Source: D. Robinson (1987), Estimation and use of variance components. *The Statistician* 36, 3–14.

a. Write a linear model for the experiment, assuming machines with fixed effects and plots and samples with random effects; explain the terms; and compute the analysis of variance for the data.

b. Show the expected mean squares.

c. Test the null hypothesis of no difference between the means for the two machines.

6. Use the rules given in Section 7.6 to derive the expected mean squares for the following studies or models:

a. the cholesterol study in Exercise 7.1

b. the cattle-feeding trial in Exercise 7.2

c. the alloy casting experiment in Exercise 7.3

d. the traffic study in Exercise 7.4

e. the soils study in Exercise 7.5

f. a four-stage nested design with the model

$$y_{ijkl} = \mu + a_i + b_{j(i)} + c_{k(ij)} + d_{l(ijk)}$$

$$i = 1, 2, 3, 4 \quad j = 1, 2, 3 \quad k = 1, 2 \quad l = 1, 2, 3$$

g. a model with nested and crossed factors written as

$$y_{ijklm} = \mu + \alpha_i + \beta_j + (\alpha\beta)_{ij} + c_{k(ij)} + \delta_l + (\alpha\delta)_{il} + (\beta\delta)_{jl}$$

$$+ (\alpha\beta\delta)_{ijl} + (cd)_{kl(ij)} + e_{m(ijkl)}$$

$$i = 1, 2, \dots, a \quad j = 1, 22, \dots, b \quad k = 1, 2, \dots, c \quad l = 1, 2, \dots, d \quad m = 1, 2, \dots, r$$

where α_i, β_j, δ_k, and their interactions are fixed effects and $c_{k(ij)}$, $(cd)_{kl(ij)}$, and $e_{m(ijkl)}$ are random effects

7. How would your statistical inference change if the model with restrictions on the interactions had been used for Example 7.5?

8   Complete Block Designs

Experiment designs to improve the precision of results from research studies are the topics of discussion in this chapter and others to follow. Blocking was introduced in Chapter 1 as a method to reduce experimental error variation. Blocking groups the experimental units into homogeneous blocks to compare treatments within a more uniform environment. The designs in this chapter use either one grouping criterion in a randomized complete block design or two grouping criteria in Latin square arrangements. The features, randomization, analysis, and evaluation of these designs are discussed. Extensions of the designs include factorial treatment designs, multiple experimental units per treatment in each block, and subsampling. Conducting analysis when some observations are missing is discussed. The topic for the final section in this chapter is combining the results from several repetitions of the same experiment at several places or several times.

8.1   Blocking to Increase Precision

Our objective is to have precise comparisons among treatments in our research studies. Blocking is a means to reduce and control experimental error variance to achieve more precision.

Previous chapters concentrated on treatment designs and the associated statistical methods for efficient analysis of research hypotheses. All of the illustrations utilized completely randomized designs. However, outside of appropriate experimental unit selection and good research techniques, the completely randomized designs provide no control over experimental error variance. The experimental units are assumed to be relatively homogeneous with respect to the measured response variable in completely randomized designs. However, sometimes sufficient numbers of homogeneous units do not exist for a complete experiment with these designs.

Any factor that affects the response variable and that varies among the experimental units will increase the experimental error variance and decrease the precision of the experimental results. Factors such as age or weight of animals, different batches of reagents or manufactured material, gender of human subjects, and physical separation of field plot locations are examples of variables external to the treatments that can increase the variation among observations on the response variable.

Blocking stratifies experimental units into homogeneous groups, or like units. A successful choice of blocking criteria results in less variation among the units within the blocks than that among units from different blocks. General categories of successful blocking criteria are (1) proximity (neighboring field plots), (2) physical characteristics (age or weight), (3) time, and (4) management of tasks in the experiment.

A group of neighboring field plots forms a block in agronomic field experiments. Animals grouped by weight, stage of lactation, or litter form blocks of homogeneous experimental units. The engineer uses a single batch of manufactured material to form a block or homogeneous group of experimental units for the treatments. Laboratory experiments use technicians as a blocking factor to eliminate variation among technicians. Each technician performs one replication of the treatments as a block.

8.2 Randomized Complete Block Designs Use One Blocking Criterion

The randomized complete block design is the simplest of the blocking designs used to control and reduce experimental error. The experimental units are stratified into blocks of homogeneous units. Each treatment is assigned randomly to an equal number (usually one) of experimental units in each block. More precise comparisons are possible among treatments within the homogeneous set of experimental units in a block. Blocking turned out to be very beneficial in the following study.

Example 8.1   Timing of Nitrogen Fertilization for Wheat

Current nitrogen fertilization recommendations for wheat included applications of specified amounts at specified stages of plant growth. The recommendations were developed through the use of periodic stem tissue analysis for nitrate content of the plant. Stem tissue analysis was thought to be an effective means to monitor the nitrogen status of the crop and provide a basis for predicting required nitrogen for optimum production.

Research Objective: In certain situations, however, the stem nitrate tests were found to overpredict the required nitrogen amounts. Consequently, the researcher wanted to evaluate the effect of several different fertilization timing schedules on the stem tissue nitrate amounts and wheat production to refine the recommendation procedure.

Treatment Design: The treatment design included six different nitrogen application timing and rate schedules that were thought to provide the range of conditions necessary to evaluate the process. For comparison, a control treatment of no nitrogen was included as was the current standard recommendation.

Experiment Design: The experiment was conducted in an irrigated field with a water gradient along one direction of the experimental plot area as a result of irrigation. Since plant responses are affected by variability in the amount of available moisture, the field plots were grouped into blocks of six plots such that each block occurred in the same part of the water gradient. Thus, any differences in plant responses caused by the water gradient could be associated with the blocks. The resulting experiment design was a randomized complete block design with four blocks of six field plots to which the nitrogen treatments were randomly allocated.

The layout of the experimental plots in the field is shown in Display 8.1. The observed nitrate nitrogen content (ppm $\times$ 10^{-2}) from a sample of wheat stems is shown for each plot along with the treatment numbers, which appear in the small box of each plot.

Display 8.1   Arrangement of Experimental Plots for the Wheat Experiment in a Randomized Complete Block Design

Irrigation Gradient
$\downarrow$

Block 1	2	5	4	1	6	3
	40.89	37.99	37.18	34.98	34.89	42.07

Block 2	1	3	4	6	5	2
	41.22	49.42	45.85	50.15	41.99	46.69

Block 3	6	3	5	1	2	4
	44.57	52.68	37.61	36.94	46.65	40.23

Block 4	2	4	6	5	3	1
	41.90	39.20	43.29	40.45	42.91	39.97

Source: Dr. T. Doerge, Department of Soil and Water Science, University of Arizona.

How to Randomize the Design

The random allocation of treatments to experimental units is restricted in the randomized complete block design such that each treatment must occur an equal number of times (one or more) within each block. Randomization is illustrated with the wheat experiment of Example 8.1.

A random permutation of the order in which the treatments are placed with the units in each block provides a random allocation of treatments to units. There are $6! = 720$ possible permutations of the six treatments. One permutation is randomly selected for each block since a separate randomization is required for each block.

Assign the treatment label, such as A, B, C, D, E, F, to the respective integer values 1, 2, 3, 4, 5, 6. Obtain a random permutation of the integer values from a computer program or a table of permutations found in Appendix Table XIII. One such permutation is shown in Display 8.2. Given the permutation 2, 5, 4, 1, 6, 3, assign treatment B to unit 1, treatment E to unit 2, treatment D to unit 3, and so forth in block 1.

| Display 8.2 | Assignment of Treatments to Experimental Units in a Complete Block |

Permutation	2	5	4	1	6	3
Treatment	B	E	D	A	F	C
Experimental unit	1	2	3	4	5	6

Separate random permutations are required for each of the remaining three blocks. Given additional random permutations, say $(1, 3, 4, 6, 5, 2)$, $(6, 3, 5, 1, 2, 4)$, and $(2, 4, 6, 5, 3, 1)$, the final assignment of treatments to units within each block is shown in Figure 8.1.

Statistical Model and Analysis for Randomized Complete Block Design

The linear model for an experiment in a randomized complete block design requires a term to represent the variation identifiable in the observations as a consequence of blocking. The response of the unit with the ith treatment in the jth block is written as

$$y_{ij} = \mu + \tau_i + \rho_j + e_{ij} \tag{8.1}$$

$$i = 1, 2, \ldots, t \qquad j = 1, 2, \ldots, r$$

where μ is the general mean, τ_i is the treatment effect, and e_{ij} is the experimental error. The block effect ρ_j represents the average deviation of the units in block j from the general mean. The treatment and block effects are assumed to be additive. *Additivity* means there is no interaction between treatments and blocks. The experimental errors are assumed independent with zero means and common variance σ^2. The independence assumption is justified through random assignment of treatments to the experimental units.

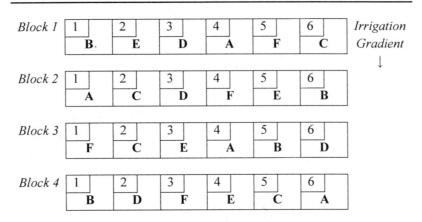

Figure 8.1 Randomized assignment of treatments in a randomized complete block design

Blocks Add a Sum of Squares Partition to the Analysis of Variance

The basic data table for a randomized complete block design is shown in Table 8.1. The deviation of any observation from the grand mean in Table 8.1, $y_{ij} - \overline{y}_{..}$, may be written as the algebraic identity

$$y_{ij} - \overline{y}_{..} = (\overline{y}_{i.} - \overline{y}_{..}) + (\overline{y}_{.j} - \overline{y}_{..}) + (y_{ij} - \overline{y}_{i.} - \overline{y}_{.j} + \overline{y}_{..}) \qquad (8.2)$$

Table 8.1 Data table for a randomized complete block design

	Block				Treatment
Treatment	1	2	...	r	Means
1	y_{11}	y_{12}	$\cdots$	y_{1r}	$\overline{y}_{1.}$
2	y_{21}	y_{22}	$\cdots$	y_{2r}	$\overline{y}_{2.}$
$\vdots$	$\vdots$	$\vdots$		$\vdots$	$\vdots$
t	y_{t1}	y_{t2}	$\cdots$	y_{tr}	$\overline{y}_{t.}$
Block means	$\overline{y}_{.1}$	$\overline{y}_{.2}$	$\cdots$	$\overline{y}_{.r}$	$\overline{y}_{..}$

The terms on the right-hand side of Equation (8.2) are

- a treatment deviation $(\overline{y}_{i.} - \overline{y}_{..})$
- a block deviation $(\overline{y}_{.j} - \overline{y}_{..})$
- experimental error $(y_{ij} - \overline{y}_{i.} - \overline{y}_{.j} + \overline{y}_{..})$

For example, the means for the wheat experiment in Example 8.1 (shown in Table 8.3) are used to illustrate the deviations for treatment 1 in block 2, y_{12}, as

- the treatment 1 deviation: $\bar{y}_{1.} - \bar{y}_{..} = 38.28 - 42.07 = -3.79$
- the block 2 deviation: $\bar{y}_{.2} - \bar{y}_{..} = 45.89 - 42.07 = 3.82$
- the experimental error for y_{12}:

$$y_{12} - \bar{y}_{1.} - \bar{y}_{.2} + \bar{y}_{..} = 41.22 - 38.28 - 45.89 + 42.07 = -0.88$$

and the sum of the three deviations, $-3.79 + 3.82 - 0.88 = -0.85$, is the same as the total deviation $y_{12} - \bar{y}_{..} = 41.22 - 42.07 = -0.85$, as it should be.

On closer inspection the last two terms of Equation (8.2) compose an algebraic identity for the deviation of the observation from the treatment mean

$$y_{ij} - \bar{y}_{i.} = (\bar{y}_{.j} - \bar{y}_{..}) + (y_{ij} - \bar{y}_{i.} - \bar{y}_{.j} + \bar{y}_{..}) \tag{8.3}$$

The experimental error deviation from the completely randomized design, $y_{ij} - \bar{y}_{i.}$, is partitioned into two components. The first term is identified with the blocking criteria as $\bar{y}_{.j} - \bar{y}_{..}$. The second term is identified only as a residual or experimental error, $y_{ij} - \bar{y}_{i.} - \bar{y}_{.j} + \bar{y}_{..}$.

Squaring and summing both sides of Equation (8.2) results in

$$\sum_{i=1}^{t} \sum_{j=1}^{r} (y_{ij} - \bar{y}_{..})^2 = r \sum_{i=1}^{t} (\bar{y}_{i.} - \bar{y}_{..})^2 + t \sum_{j=1}^{r} (\bar{y}_{.j} - \bar{y}_{..})^2$$

$$+ \sum_{i=1}^{t} \sum_{j=1}^{r} (y_{ij} - \bar{y}_{i.} - \bar{y}_{.j} + \bar{y}_{..})^2 \tag{8.4}$$

or

$$SS\ \text{Total} = SS\ \text{Treatment} + SS\ \text{Blocks} + SS\ \text{Error}$$

The sum of any crossproducts from the right-hand side is zero. Table 8.2 summarizes the sum of squares partition.

Table 8.2 Analysis of variance for an experiment in a randomized complete block design

Source of Variation	Degrees of Freedom	Sum of Squares	Mean Square	Expected Mean Square
Total	$rt - 1$	$\sum_i \sum_j (y_{ij} - \bar{y}_{..})^2$		
Blocks	$r - 1$	$t \sum_j (\bar{y}_{.j} - \bar{y}_{..})^2$	MSB	
Treatments	$t - 1$	$r \sum_i (\bar{y}_{i.} - \bar{y}_{..})^2$	MST	$\sigma^2 + r\theta_t^2$
Error	$(r - 1)(t - 1)$	$SS\ \text{Error}$	MSE	σ^2

The observations on stem nitrate nitrogen along with treatment and block means for the wheat fertilization study are shown in Table 8.3. The analysis of variance for the stem nitrate data is shown in Table 8.4.

Table 8.3 Nitrate nitrogen in wheat plant stems (ppm $\times$ 10^{-2}) for six nitrogen-timing treatments in each of four blocks of a randomized complete block design

Nitrogen Timing Schedule	*Block*				*Treatment Means* ($\bar{y}_{i.}$)
	1	*2*	*3*	*4*	
Control	34.98	41.22	36.94	39.97	38.28
2	40.89	46.69	46.65	41.90	44.03
3	42.07	49.42	52.68	42.91	46.77
4	37.18	45.85	40.23	39.20	40.62
5	37.99	41.99	37.61	40.45	39.51
6	34.89	50.15	44.57	43.29	43.23
Block Means ($\bar{y}_{.j}$)	38.00	45.89	43.11	41.29	$\bar{y}_{..} = 42.07$

Table 8.4 Analysis of variance for wheat stem nitrate nitrogen

Source of Variation	*Degrees of Freedom*	*Sum of Squares*	*Mean Square*	*F*	*Pr > F*
Total	23	506.33			
Blocks	3	197.00	65.67	9.12	
Nitrogen	5	201.32	40.26	5.59	.004
Error	15	108.01	7.20		

As a consequence of blocking, a sum of squares for blocks is partitioned out of what would have been the sum of squares for experimental error with the completely randomized design. The blocked design will markedly improve the precision on the estimates of the treatment means if the reduction in SS Error with blocking is substantial. The reduction in SS Error can be offset by a reduction in degrees of freedom, since $r - 1$ of the degrees of freedom must be allocated to SS Blocks. A measure of relative efficiency, shown later in this section, is necessary to evaluate the full benefit of blocking.

Standard Errors for Treatment Means

The standard error estimate for a treatment mean is

$$s_{\bar{y}_{i.}} = \sqrt{\frac{MSE}{r}} = \sqrt{\frac{7.20}{4}} = 1.34 \tag{8.5}$$

and a 95% confidence interval estimate of any treatment mean in Table 8.3 is $\bar{y}_{i.} \pm t_{.025,15}(s_{\bar{y}_{i.}})$, where $t_{.025,15} = 2.131$. The standard error of a difference between any two treatment means is estimated by

$$s_{(\bar{y}_{i.} - \bar{y}_{j.})} = \sqrt{\frac{2MSE}{r}} = \sqrt{\frac{2(7.20)}{4}} = 1.90 \qquad (8.6)$$

Tests of Hypothesis About Treatment Means

The F_0 statistic to test the null hypothesis of no differences among the treatment means is

$$F_0 = \frac{MST}{MSE} = \frac{40.26}{7.20} = 5.59 \qquad (8.7)$$

which exceeds the critical value of $F_{.05,5,15} = 2.90$. The observed significance level is $Pr > F = .004$ (Table 8.4). There are significant differences among the nitrogen treatments with respect to stem nitrate nitrogen at this stage of plant development.

There is little interest in formal inferences about block effects, and the F_0 statistic generally is not computed for this purpose even though it may appear in the output of a computer program. The extent to which blocking increased the efficiency of the design to utilize existing resources is discussed later in this section.

Interpretations with Multiple Comparisons

Schedule 4 was the standard fertilizer recommendation for wheat. The nitrate nitrogen in the stem of the wheat plant measured throughout the growing season is used to assess nitrogen requirements for optimum wheat yields. The investigator would be interested in differences between any of the individual nitrogen-timing treatments and the current recommendation at each stage of growth. The Dunnett method (Chapter 3) can be used to compare the standard recommendation with each of the other timing treatments, including the no-nitrogen control treatment. The no-nitrogen control provides a means of evaluating the nitrogen available without fertilization in these particular plots. (See Chapter 1 on the use of control treatments.)

The Dunnett 95% simultaneous confidence intervals require the standard error of the difference, $s(\bar{y}_{i.} - \bar{y}_{j.}) = 1.90$, and the Dunnett statistic from Appendix Table VI, $d_{.05,5,15} = 2.82$, for a two-sided comparison.

The 95% SCI for the difference between the mean of any other schedule for nitrogen application and schedule 4 requires $D(5, .05) = d_{.05,5,15}[s(\bar{y}_{i.} - \bar{y}_{4.})] = 2.82(1.90) = 5.36$ and are computed as $\bar{y}_{i.} - \bar{y}_{4.} \pm 5.36$. The intervals are given in Table 8.5, along with the results of the confident inequalities test in which the absolute difference is declared significantly different from 0 if the difference exceeds $D(5, .05) = 5.36$. Schedule 3 is the only treatment that has a mean nitrate nitrogen level significantly different from the current recommended schedule 4. It has a nitrate content greater than that of schedule 4 since the lower bound of the SCI is greater than 0. The SCI for all other treatment comparisons include 0 and have upper and lower bounds considerably removed from 0.

Table 8.5 Results of the Dunnett method for treatments vs. control (Example 8.1)

Schedule	Mean	$\bar{y}_i - \bar{y}_c$	95% SCI (L, U)
4	$\bar{y}_c = 40.62$	—	—
1	38.28	− 2.34	(− 7.70, 3.02)
2	44.03	3.41	(− 1.95, 8.77)
3	46.77	6.15*	(0.79, 11.51)
5	39.51	− 1.11	(− 6.47, 4.25)
6	43.23	2.61	(− 2.75, 7.97)

* $| \bar{y}_i - \bar{y}_c |$ exceeds $D(5, .05) = 5.36$ and is significantly different from schedule 4.

If Treatments Interact with Blocks

The assumption of no treatment $\times$ block interaction implies that the treatment and block effects are additive. The differences among the treatments are assumed to be relatively constant from block to block as a consequence of additive block and treatment effects even though the use of blocking may result in large differences in responses between units from different blocks. The nonadditivity test, introduced in Chapter 6 (Tukey, 1949b), can be conducted to detect nonadditivity of the multiplicative form $\lambda\tau_i\rho_j$. If the environmental conditions are sufficiently different in one or more of the blocks the relative performance of the treatments may be affected. For example, if the residual nutrient base in the soil is quite different from block to block in a field crops fertility trial there can be little or no response to fertilizers in some blocks, whereas the responses can be quite substantial in other blocks.

Multiple Experimental Units per Treatment in Each Block

A more general nonadditivity is represented by the general interaction term $(\tau\rho)_{ij}$. To test the existence of interaction the experiment must have more than one experimental unit for each treatment within each of the blocks. The linear model for an experiment with u experimental units for each treatment in each of r blocks is

$$y_{ijk} = \mu + \tau_i + \rho_j + (\tau\rho)_{ij} + e_{ijk} \tag{8.8}$$

$$i = 1, 2, \ldots, t \quad j = 1, 2, \ldots, r \quad k = 1, 2, \ldots, u$$

where e_{ijk} are the random, independent experimental errors with means 0 and variance σ^2. The variance σ^2 is the variability among experimental units within a block that have received the same treatment. The computations for sum of squares partitions and the test for interaction are the same as for a two-factor factorial arrangement shown in Chapter 6.

Subsamples of Experimental Units

There are occasions wherein two or more samples from the experimental units are required for data collection. Situations requiring subsampling of the units were discussed in Chapter 5. The linear model for an experiment in a randomized complete block design with n subsamples of the experimental units is

$$y_{ijk} = \mu + \tau_i + \rho_j + e_{ij} + d_{ijk} \qquad (8.9)$$

$$i = 1, 2, \ldots, t \qquad j = 1, 2, \ldots, r \qquad k = 1, 2, \ldots, n$$

where the d_{ijk} are the random effects for subsamples with mean 0 and variance σ_d^2. The other terms are as described in Equation (8.1) for the randomized complete block design. The adjustments to the analysis are those shown in Chapter 5.

Gates (1995) provided a detailed discussion on experimental error estimation in block designs with different configurations of experimental units and sampling units.

Analysis of Residuals to Evaluate Assumptions

The assumptions regarding the experimental errors of the linear model for the randomized complete block design can be evaluated with an analysis of the residuals (discussed in Chapter 4). The residuals are computed from the experimental error deviation component shown in Equation (8.2) as $\widehat{e}_{ij} = y_{ij} - \overline{y}_{i.} - \overline{y}_{.j} + \overline{y}_{..}$. For example, the residual for the control treatment in block 1 of Example 8.1 is

$$\widehat{e}_{11} = y_{11} - \overline{y}_{1.} - \overline{y}_{.1} + \overline{y}_{..}$$

$$= 34.98 - 38.28 - 38.00 + 42.07 = 0.77$$

The plot of residuals versus the estimated values and the normal probability plot of the residuals are shown in Figures 8.2a and 8.2b. The assumptions of homogeneous variance (Figure 8.2a) and normal distribution (Figure 8.2b) appear to hold for these data.

Did Blocking Increase Precision?

The expectation of increased precision in the estimates of treatment means motivates us to use the randomized complete block design. Planning and conducting an experiment with the randomized complete block design requires extra effort relative to the completely randomized design. The relative efficiency measure (discussed in Chapter 1) evaluates the benefits of blocking for a particular experiment.

The efficiency of a randomized complete block design is compared with that of a completely randomized design. The estimate of σ^2, say s_{rcb}^2, is the mean square

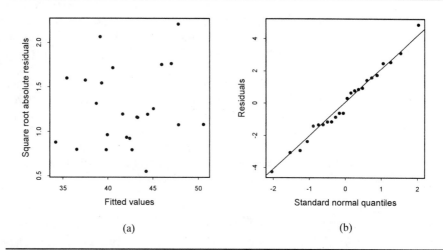

(a)   (b)

Figure 8.2   Residual plots from the analysis of variance for data on nitrate nitrogen in Example 8.1: (a) square root of the absolute residuals vs estimated values and (b) normal probability plot of residuals

for experimental error from the analysis of variance for the current experiment in a randomized complete block design. An estimate of σ^2 from the completely randomized design not used, say s_{cr}^2, is required for the relative efficiency measure. The computation of s_{cr}^2 from the information in the randomized complete block analysis of variance is (Cochran & Cox, 1957; Kempthorne, 1952)

$$s_{cr}^2 = \frac{SS\ \text{Blocks} + r(t-1)MSE}{rt - 1} \tag{8.10}$$

The estimate of σ_{cr}^2 for the wheat fertilization study is

$$s_{cr}^2 = \frac{197.0 + 20(7.2)}{23} = 14.8$$

and the estimate of σ_{rcb}^2 is $s_{rcb}^2 = MSE = 7.2$.

The relative efficiency estimate, without degrees of freedom correction for estimates of σ^2, is

$$RE = \frac{s_{cr}^2}{s_{rcb}^2} = \frac{14.8}{7.2} = 2.06 \tag{8.11}$$

The correction for estimation of σ^2 by s^2 is

$$\frac{(f_{rcb} + 1)(f_{cr} + 3)}{(f_{rcb} + 3)(f_{cr} + 1)} = \frac{16 \cdot 21}{18 \cdot 19} = 0.98$$

where $f_{rcb} = 15$ and $f_{cr} = 18$ are the error degrees of freedom for the randomized complete block and completely randomized design, respectively. The correction reduces the RE to $0.98(2.06) = 2.02$. The correction has little effect with moderately sized degrees of freedom for experimental error variance estimates.

The randomized complete block design for the wheat experiment is estimated to be slightly more than twice as efficient as a completely randomized design. The completely randomized design would require twice as many replications of the treatments as the randomized complete block design. Eight replications are required by the completely randomized design to have equivalent variances for the treatment means under the same experimental conditions in that field. Blocking on the irrigation gradient was effective as a measure to control and reduce the experimental error variance estimate in this instance. Future experiments of this same nature would likely benefit from the blocking practice.

A Quick Check for Effective Blocking

Lentner, Arnold, and Hinkelmann (1989) discuss a relationship between relative efficiency and the ratio $H = MSB/MSE$. They point out that, although the ratio H is equivalent to an F_0 statistic, a valid test for block effects does not exist. From Equations (8.10) and (8.11), the relative efficiency measure can be expressed as

$$RE = \frac{s_{cr}^2}{s_{rcb}^2} = \frac{(r-1)MSB + r(t-1)MSE}{(rt-1)MSE} \tag{8.12}$$

With some rewriting, the expression becomes

$$RE = k + (1-k)H \tag{8.13}$$

where $H = MSB/MSE$ and $k = r(t-1)/(rt-1)$. The following relationships for RE and H can be determined from Equation (8.13):

$$RE < 1 \quad \text{if and only if} \quad H < 1$$
$$RE = 1 \quad \text{if and only if} \quad H = 1$$
$$RE > 1 \quad \text{if and only if} \quad H > 1$$

H can be used to evaluate the effectiveness of blocking even though it is not a valid F statistic for testing block effects. For example, if $RE > 1$, then $H > 1$; blocking has been effective in reducing experimental error. Fewer replications are required for the randomized complete block design relative to the completely randomized design. The value of H does not provide the complete information about relative efficiency but only the implication of greater, lesser, or equal efficiency. H is a quick check for whether blocking was effective.

Random Blocks

The blocks of units often constitute a random sample of blocks available to the investigator. Sites used as blocks in ecological, forestry, or wildlife studies can be random samples of many sites available for the study. Plots may be established in each of the sites for the treatments. Manufactured batches of material used as blocks for experimental treatments are random batches. The batch of material (such as fabric, asphalt, or chemical product) is divided into smaller experimental unit batches to which the experimental treatments are administered. Schools used as blocking criteria in education studies can be random representatives of available schools in the school district. Classrooms within the schools serve as the experimental units for treatments.

The base of inference for treatments in a study with random blocks extends to a population of blocks from which the blocks in the study are a random sample. As a consequence of random blocks, the standard errors for treatment means will be different from those for an experiment with fixed blocks. The linear model with random block effects is

$$y_{ij} = \mu + \tau_i + b_j + e_{ij} \tag{8.14}$$

$$i = 1, 2, \ldots, t \qquad j = 1, 2, \ldots, r$$

where μ is the general mean, τ_i is the fixed treatment effect, b_j is the random block effect with mean 0 and variance σ_b^2, and e_{ij} is the random experimental error with mean 0 and variance σ^2. Under the mixed model an observation has expected value $E(y_{ij}) = \mu + \tau_i$ and variance $\sigma^2 + \sigma_b^2$. There is also a covariance of σ_b^2 between any two observations in the same block with random blocks.

The variance of a treatment mean with random blocks in the randomized complete block design is

$$\sigma_{\bar{y}_{i.}}^2 = \frac{1}{r}(\sigma^2 + \sigma_b^2) \tag{8.15}$$

The variance of a treatment mean with random blocks includes the component of variance for blocks, σ_b^2, and will be larger than the variance with fixed block effects. The variance of the difference between two treatment means is not affected by the random block effects and will be the same as that shown for fixed effects in Equation (8.6).

8.3 Latin Square Designs Use Two Blocking Criteria

Two Blocking Factors May Be Necessary

Recognition of a factor, other than planned treatments, that influences the response variable was important in the experiment on wheat fertilization in Example 8.1. Blocking the experimental plots on the basis of the irrigation gradient doubled the efficiency of the experiment.

In some experimental settings two factors, other than treatments, may influence the response variable. Even more precision can be achieved if we can block the units on the basis of the two factors. If a second factor is a candidate for a blocking criterion, the Latin square arrangement can be used for the experiment design.

The Latin square arrangement derives from an arrangement of the Latin letters A, B, C, ... into a square array such that each letter appears once in each column and once in each row of the square. In applications to experiments, the rows and columns of the array are identified with the two blocking criteria and the Latin letters are identified with the treatments. One such application occurred in the following example.

Example 8.2   Relationship Between Wheat Yield and Seeding Rate

Wheat cultivation practices such as seeding rate, row spacing, and date of planting have direct effects on the crop yield. Cultivation practices to optimize production are established with experiments on newly introduced wheat cultivars.

Research Objective: In one such instance, a researcher wanted to determine the optimum seeding rate for a newly introduced durum wheat with a high semolina extract important to the making of pasta.

Treatment Design: Five seeding rates (30, 80, 130, 180, and 230 lb/acre) were used for the treatment design. Based on other cultivars common to the area, these seeding rates should have bracketed the rate for optimum production.

Experiment Design: The experiment was conducted in an irrigated field with the water gradient along one direction of the experimental area. In addition, the experimental fields on this farm were known to have soil gradients created by the grading required to make the land suitable for irrigation. These soil differences were generally perpendicular to the irrigation runs.

The researcher blocked the field plots in a row and column arrangement to control for soil and water gradients in two directions on the experimental field. The seeding rate treatments were randomized to the field plots in a 5 × 5 Latin square arrangement.

The layout of the experimental plots in a Latin square design after randomization is shown in Display 8.3. The grain yield for each plot in hundredweight (100 lb) per acre is shown in each plot along with a treatment letter.

The field row blocks coincided with the irrigation gradient, and the column row blocks coincided with the soil gradient perpendicular to the irrigation gradient. The treatments are arrayed in a Latin square arrangement with each of the treatments appearing once in each row block and once in each column block.

Display 8.3   Arrangement of Experimental Plots for the Wheat Experiment in a Latin Square Design

Field Row	Column 1	Column 2	Column 3	Column 4	Column 5	Irrigation
1	59.45(E)	47.28(A)	54.44(C)	50.14(B)	59.45(D)	Gradient ↓
2	55.16(C)	60.89(D)	56.59(B)	60.17(E)	48.71(A)	
3	44.41(B)	53.72(C)	55.87(D)	47.99(A)	59.45(E)	
4	42.26(A)	50.14(B)	55.87(E)	58.74(D)	55.87(C)	
5	60.89(D)	59.45(E)	49.43(A)	59.45(C)	57.31(B)	

Soil Gradient ⟶

Source: Dr. M. Ottman, Department of Plant Sciences, University of Arizona.

Other Applications of Latin Squares

An experiment to test automobile tire treatments is a classic example used to illustrate the Latin square design. The experiment tests four automobile tire treatments (A, B, C, D) on four automobiles. Each tire treatment appears on one of the four tire positions of each automobile. The row and column blocking criteria are the automobiles and tire positions, respectively, for the design in Display 8.4.

Display 8.4   Latin Square Arrangement for Automobile Tire Treatments

	Tire Position			
Auto	1	2	3	4
1	A	B	C	D
2	B	C	D	A
3	C	D	A	B
4	D	A	B	C

Each treatment (A, B, C, or D) appears once in each row (auto) and once in each column (tire position). The rationale for the blocking criteria is that the wear on the tires may differ among the automobiles and the positions in which the tires are mounted on the automobile.

The blocking arrangement does not have to be rectangular for a Latin square arrangement to be useful for error variance reduction. A linear arrangement of treatments in a greenhouse experiment or a linear arrangement of treatments processed over time may be ordered according to the Latin square assignment. Display

8.5 illustrates a linear arrangement of four treatments with a Latin square assignment.

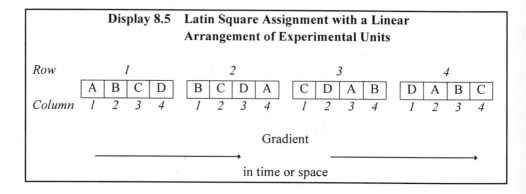

Display 8.5   **Latin Square Assignment with a Linear Arrangement of Experimental Units**

Use Standard Latin Squares to Generate the Designs

All Latin squares of a specified size can be generated from the *standard squares*. A standard square has the treatment symbols (A, B, C, ...) written in alphabetical order in the first row and in the first column of the array. Each treatment symbol occurs once in each column and once in each row of the array. Only one standard square exists for $t = 2$ or 3 treatments. There are 4 standard squares with $t = 4$ treatments and 56 standard squares with $t = 5$. The number of standard squares increases dramatically with the number of treatments since there are 9408 standard squares with 6 treatments.

The standard squares for $t = 2$, 3, and 4 treatments and samples of squares up to $t = 10$ treatments are shown in Appendix 8A. Fisher and Yates (1963) published the complete set of Latin squares for $t = 4$ through $t = 6$ treatments along with samples of squares for up to $t = 12$ treatments.

A standard square of any size can be generated by writing the first row of letters in alphabetical order. The second row is obtained from the first by shifting the first row one letter to the left, moving the letter A to the extreme right-hand position of row 2. The third row is obtained by shifting the second row one letter to the left, placing the letter B in the extreme right-hand position of row 3. This process is continued for the remaining rows. A standard 6×6 Latin square constructed in this manner is

$$
\begin{array}{cccccc}
A & B & C & D & E & F \\
B & C & D & E & F & A \\
C & D & E & F & A & B \\
D & E & F & A & B & C \\
E & F & A & B & C & D \\
F & A & B & C & D & E
\end{array}
$$

The Latin square is a restrictive design because it requires the number of treatments, rows, and columns to be equal values. The requirement can be difficult to satisfy in some experimental settings that require two blocking criteria. Latin squares with $t = 4$ or fewer treatments have few degrees of freedom for estimating experimental error variance; thus, their value can be limited with small experiments unless two or more repetitions of the design are possible. With treatment numbers in excess of $t = 8$ to 10 the required number of experimental units can become prohibitive, depending on the circumstances of the contemplated experiments. The suitable size for many experiments in the Latin square arrangement includes those with $t = 5$ to 7 treatments.

Preece (1983) provides a historical background of the Latin square and a general discussion of variations on the Latin square design suggested for experimental work with row–column blocking design.

How to Randomize the Design

If all standard Latin squares of size $t \times t$ are available, randomization is accomplished with the following steps:

Step 1.   Randomly select one of the standard squares.

Step 2.   Randomly order all but the first row.

Step 3.   Randomly order all columns.

Step 4.   Randomly assign treatments to the letters.

All possible randomizations can be generated without including the first row in Step 2 if a standard square is randomly selected. If all standard squares are not available for selection, then it is recommended in Step 2 that all rows be included in the randomization. Not all possible Latin squares are generated in this way but the number of possibilities is increased considerably. Suppose the standard square selected at Step 1 for the 4×4 Latin square experiment with automobile tires is

$$
\begin{array}{cccc}
A & B & C & D \\
B & C & D & A \\
C & D & A & B \\
D & A & B & C
\end{array}
$$

Step 2. Obtain a random permutation of numbers to order the last three rows:

Permutation	Original Row
3	2
1	3
2	4

The placement of the rows for the standard square with row 1 in its original position is

Original Row				
1	A	B	C	D
3	C	D	A	B
4	D	A	B	C
2	B	C	D	A

Step 3. Obtain a random permutation of numbers to order the four columns from Step 2.

Permutation	Original Column
1	1
4	2
3	3
2	4

The placement of the columns for the standard square is

Original Column			
1	4	3	2
A	D	C	B
C	B	A	D
D	C	B	A
B	A	D	C

Step 4. Obtain a random permutation to assign treatments to the letters. This assignment is not necessary if the standard square has been selected at random from all possible standard squares. The method of assignment is shown here for illustration. Suppose the treatment labels are W, X, Y, and Z:

Permutation	Treatment
4 = D	W
2 = B	X
3 = C	Y
1 = A	Z

The treatment labels W, X, Y, Z replace the Latin square letters in the order D, B, C, A in the randomized arrangement. The final placement of the tire treatments on the automobiles and tire positions is

	Tire Position			
Auto	*1*	*2*	*3*	*4*
1	Z	W	Y	X
2	Y	X	Z	W
3	W	Y	X	Z
4	X	Z	W	Y

Statistical Model and Analysis for Latin Square Designs

The linear statistical model for an experiment with t treatments in a $t \times t$ Latin square design is

$$y_{ij} = \mu + \rho_i + \gamma_j + \tau_k + e_{ij} \tag{8.16}$$

$$i, j, k = 1, 2, \ldots, t$$

where y_{ij} is the observation on the experimental unit in the ith row and jth column of the design. The row and column effects are ρ_i and γ_j, respectively; τ_k is the effect of the kth treatment; and the e_{ij} are random, independent experimental errors with mean 0 and variance σ^2. It is assumed there is no interaction between the treatments and the rows and columns.

The notation for totals and means of observations for rows and columns follows the usual convention with $y_{i.} = \sum_j^t y_{ij}$ for a row total and $y_{.j} = \sum_i^t y_{ij}$ for a column total. The treatment total will be represented simply as y_k, implying a sum of the observations over the t experimental units receiving treatment k. Likewise, $\bar{y}_k$ will represent the mean of the observations on the kth treatment.

Two Sum of Squares Partitions for Blocking

The sum of squares partitions can be derived from the algebraic identity

$$y_{ij} - \bar{y}_{..} = (\bar{y}_{i.} - \bar{y}_{..}) + (\bar{y}_{.j} - \bar{y}_{..}) + (\bar{y}_k - \bar{y}_{..})$$

$$+ (y_{ij} - \bar{y}_{i.} - \bar{y}_{.j} - \bar{y}_k + 2\bar{y}_{..}) \tag{8.17}$$

The deviation of an observation from the grand mean $y_{ij} - \bar{y}_{..}$ is expressed as an additive sum of:

- a row deviation $(\bar{y}_{i.} - \bar{y}_{..})$
- a column deviation $(\bar{y}_{.j} - \bar{y}_{..})$
- a treatment deviation $(\bar{y}_k - \bar{y}_{..})$
- experimental error $(y_{ij} - \bar{y}_{i.} - \bar{y}_{.j} - \bar{y}_k + 2\bar{y}_{..})$

For example, the means for the wheat experiment in Example 8.2 (shown in Table 8.7) are used to illustrate deviations for the observation in row 1 and column 1 with treatment 5 = **E**, as

- the row 1 deviation: $\bar{y}_{1.} - \bar{y}_{..} = 54.15 - 54.53 = -0.38$

- the column 1 deviation: $\bar{y}_{.1} - \bar{y}_{..} = 52.43 - 54.53 = -2.10$

- the treatment **E** deviation: $\bar{y}_5 - \bar{y}_{..} = 58.88 - 54.53 = 4.35$

- the experimental error for y_{11}:

$$y_{11} - \bar{y}_{1.} - \bar{y}_{.1} - \bar{y}_5 + 2\bar{y}_{..} = 59.45 - 54.15 - 52.43 - 58.88 + 2(54.53) = 3.05$$

The sum of the four deviations is $-0.38 - 2.10 + 4.35 + 3.05 = 4.92$, which is the same as the total deviation $y_{11} - \bar{y}_{..} = 59.45 - 54.53 = 4.92$.

Squaring both sides of Equation (8.17) and summing the terms leads to an additive partition of

$$SS \text{ Total} = SS \text{ Rows} + SS \text{ Columns} + SS \text{ Treatment} + SS \text{ Error}$$

Table 8.6 summarizes the sums of squares in an analysis of variance with expected mean squares for fixed treatment effects.

Table 8.6 Analysis of variance for experiments in a Latin square design

Source of Variation	Degrees of Freedom	Sum of Squares	Mean Square	Expected Mean Square
Total	$t^2 - 1$	$\sum_i \sum_j (y_{ij} - \bar{y}_{..})^2$		
Rows	$t - 1$	$t \sum_i (\bar{y}_{i.} - \bar{y}_{..})^2$	MSR	
Columns	$t - 1$	$t \sum_j (\bar{y}_{.j} - \bar{y}_{..})^2$	MSC	
Treatments	$t - 1$	$t \sum_k (\bar{y}_k - \bar{y}_{..})^2$	MST	$\sigma^2 + t\theta_t^2$
Error	$(t - 1)(t - 2)$	SS Error	MSE	σ^2

The sum of squares for experimental error has been reduced from that of the randomized complete block design by an amount equal to SS Rows or SS Columns at a cost of $t - 1$ degrees of freedom. The mean square for experimental error as an estimate of σ^2 has very few degrees of freedom with a small number of treatments. Considerable power is lost in the tests of hypotheses for treatment comparisons unless the reduction in error sum of squares due to blocking by both row and column criteria is substantial.

The effectiveness of blocking by either criterion evaluated with the relative efficiency measure is demonstrated with the analysis of grain yield from the wheat seeding rate experiment in Example 8.2.

The observations along with row, column, and treatment means are shown in Table 8.7 in the Latin square arrangement. The data are the grain yield for each plot in hundredweight (100 lb) per acre. The analysis of variance is shown in Table 8.8.

Table 8.7 Grain yield of a wheat variety for five different seeding rates in a Latin square design [Treatment label (**A, B, C, D,** or **E**) in parentheses following yield value]

Field Row	Field Column 1	2	3	4	5	Row Means ($\bar{y}_{i.}$)
1	59.45(**E**)	47.28(**A**)	54.44(**C**)	50.14(**B**)	59.45(**D**)	54.15
2	55.16(**C**)	60.89(**D**)	56.59(**B**)	60.17(**E**)	48.71(**A**)	56.30
3	44.41(**B**)	53.72(**C**)	55.87(**D**)	47.99(**A**)	59.45(**E**)	52.29
4	42.26(**A**)	50.14(**B**)	55.87(**E**)	58.74(**D**)	55.87(**C**)	52.58
5	60.89(**D**)	59.45(**E**)	49.43(**A**)	59.45(**C**)	57.31(**B**)	57.31
Column Means ($\bar{y}_{.j}$)	52.43	54.30	54.44	55.30	56.16	$\bar{y}_{..} =$ 54.53
Treatment		**A**	**B**	**C**	**D**	**E**
Seed rate		30	80	130	180	230
Mean ($\bar{y}_k$)		47.13	51.72	55.73	59.17	58.88

Table 8.8 Analysis of variance for grain yield of a wheat variety at five seeding rates in a 5 × 5 Latin square design

Source of Variation	Degrees of Freedom	Sum of Squares	Mean Square	F	Pr > F
Total	24	716.61			
Row	4	99.20	24.80		
Column	4	38.48	9.62		
Seed Rate	4	522.30	130.57	27.67	.000
Error	12	56.63	4.72		

Standard Errors for Treatment Means

The standard error estimate for a treatment mean is

$$s_{\bar{y}_k} = \sqrt{\frac{MSE}{t}} = \sqrt{\frac{4.72}{5}} = 0.97 \qquad \textbf{(8.18)}$$

and the standard error estimate for a difference between two treatment means is

$$s_{(\bar{y}_k - \bar{y}_m)} = \sqrt{\frac{2MSE}{t}} = \sqrt{\frac{2(4.72)}{5}} = 1.37 \qquad \textbf{(8.19)}$$

Tests of Hypotheses About Treatment Means

The F_0 statistic to test the null hypothesis of no differences among treatment means,

$$F_0 = \frac{MST}{MSE} = \frac{130.57}{4.72} = 27.67 \qquad \textbf{(8.20)}$$

exceeds the critical value $F_{.05,4,12} = 3.26$ with observed significance level $Pr > F = .000$ (Table 8.8).

Interpretations of the Quantitative Treatment Factor with Regression Contrasts

The treatment factor for this experiment is a quantitative factor with five levels. An analysis for the regression of grain yield on seeding rate with orthogonal polynomial contrasts for seeding rate will provide a good description of the effect of seeding rate on grain yield. The coefficients for the orthogonal polynomial contrasts are found in Appendix Table XI. Orthogonal coefficients and the sums of squares for the linear and quadratic polynomial regression contrasts are shown in Table 8.9.

Table 8.9   Linear and quadratic polynomial regression sums of squares partitions for seeding rate (Example 8.2)

Seeding Rate	30	80	130	180	230	SS*
Mean ($\bar{y}_k$)	47.13	51.72	55.73	59.17	58.88	
Linear (P_{1k})	-2	-1	0	1	2	478.95
Quadratic (P_{2k})	2	-1	-2	-1	2	38.11

*$SS = r[\Sigma P_{ck}\bar{y}_k]^2/\Sigma P_{ck}^2$

The respective F_0 statistics to test the null hypotheses for the linear and quadratic contrasts are $F_0 = 478.95/4.72 = 101.47$ and $F_0 = 38.11/4.72 = 8.07$. Both ratios exceed the critical value $F_{.05,1,12} = 4.75$. The sum of squares for deviations from linear and quadratic regression is

$$SS \text{ Seed(deviations)} = SS \text{ Seed} - SS \text{ Seed(linear)} - SS \text{ Seed(quadratic)}$$

$$= 522.30 - 478.95 - 38.11 = 5.24$$

with 2 degrees of freedom; a test of hypothesis would indicate no significant deviations from the quadratic equation.

The computed quadratic polynomial regression equation to estimate grain yield ($\widehat{y}$) from seeding rate (R), using the techniques described in Chapter 3, is

$$\widehat{y} = 43 + 0.14R - 0.0003R^2$$

A plot of the equation is shown in Figure 8.3. The maximum estimated grain yield occurs for a seeding rate of $R = 233$ pounds per acre, which is an extrapolation beyond the highest rate used in the experiment. Another experiment with a seeding rate above 230 pounds per acre would be required to safely estimate the maximum seeding rate.

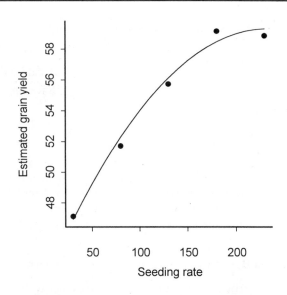

Figure 8.3 Estimated response between grain yield and seeding rate

Analysis of Residuals to Evaluate Assumptions

Residuals can be used to evaluate the assumptions of the model (as discussed in Chapter 4). From Equation (8.17), the residual for the observations on the kth treatment in the ith row and jth column is

$$\widehat{e}_{ij} = y_{ij} - \overline{y}_{i.} - \overline{y}_{.j} - \overline{y}_{k} + 2\overline{y}_{..}$$

The residual plots are left as an exercise for the reader.

If Treatments Interact with Blocks

Tukey (1955) devised a test for the assumption of additive treatment, row, and column effects. From Equation (8.17), the additivity of treatment, row, and column effects results in the estimated value of an observation as

$$\widehat{y}_{ij} = \overline{y}_{..} + (\overline{y}_{i.} - \overline{y}_{..}) + (\overline{y}_{.j} - \overline{y}_{..}) + (\overline{y}_k - \overline{y}_{..}) = \overline{y}_{i.} + \overline{y}_{.j} + \overline{y}_k - 2\overline{y}_{..}$$

The computation of a 1 degree of freedom sum of squares for nonadditivity requires the values of $\widehat{y}_{ij}$ for each experimental unit and the residual for each experimental unit, $\widehat{e}_{ij} = y_{ij} - \widehat{y}_{ij}$. The sum of squares for nonadditivity is

$$SS(\text{nonadditivity}) = \frac{\left[\sum\limits_{i=1}^{t} \sum\limits_{j=1}^{t} \widehat{y}_{ij}^2 \widehat{e}_{ij} \right]^2}{SS} \tag{8.21}$$

where SS is the sum of squares for error from the Latin square analysis of variance on $\widehat{y}_{ij}^2$. The F_0 statistic to test the hypothesis for nonadditivity is

$$F_0 = \frac{SS(\text{nonadditivity})}{MS\,\text{Residual}} \tag{8.22}$$

where

$$MS\,\text{Residual} = \frac{[SSE - SS(\text{nonadditivity})]}{\nu}$$

has $\nu = (t-1)(t-2) - 1$ degrees of freedom and SSE is the sum of squares for error from the Latin square analysis of variance for the original observations y_{ij}. It is recommended that the values for $\widehat{y}_{ij}$ be coded as $k(\widehat{y}_{ij} - \overline{y}_{..})$, where k is a constant to scale the size of the values for computational convenience.

Did Both Blocking Factors Increase Precision?

The efficiency of the Latin square design with two blocking criteria is determined relative to the randomized complete block design with only one blocking criterion. Relative efficiency measures can be computed separately for the row and column blocking criteria of the Latin square.

Relative Efficiency of Column Blocking

If only the row blocking criterion is used for blocking in a randomized complete block design, the estimated mean square for error is

$$s_{rcb}^2 = \frac{MS\,\text{Columns} + (t-1)MSE}{t} \tag{8.23}$$

where MSE is the mean square for error from the current Latin square analysis of variance.

The estimated value for Example 8.2 is $s_{rcb}^2 = [9.62 + 4(4.72)]/5 = 5.70$ and $s_{ls}^2 = MSE = 4.72$. The relative efficiency of column blocking for the experiment is

$$RE_{col} = \frac{s_{rcb}^2}{s_{ls}^2} = \frac{5.70}{4.72} = 1.21$$

There is a 21% gain in efficiency over the randomized complete block design in which only the row criterion of the Latin square design is used for blocking. Thus, the column blocks for soil gradients across the field effectively reduced the error variance by 21%. The randomized block design without the column blocks for soil gradients would require $1.21(5) = 6$ replications to have an estimated variance of the treatment mean equal to that from the Latin square design.

Relative Efficiency of Row Blocking

If only the column criterion is used for blocking in a randomized complete block design, the estimated mean square for error is

$$s_{rcb}^2 = \frac{MS \text{ Rows} + (t-1)MSE}{t} \tag{8.24}$$

For Example 8.2, $s_{rcb}^2 = [24.80 + 4(4.72)]/5 = 8.74$ and $RE_{row} = 8.74/4.72 = 1.85$. There is an 85% gain in efficiency with row blocking for the irrigation gradient in the experiment. Without row blocking for irrigation gradient the experiment would require $1.85(5) = 9.25$ or 10 replications of each treatment in the randomized complete block design to have an estimated variance of the treatment equal to that from the current Latin square design.

Correction for Estimating σ^2

The correction for estimating σ^2 by s^2 is

$$\frac{(f_{ls} + 1)(f_{rcb} + 3)}{(f_{ls} + 3)(f_{rcb} + 1)} = \frac{13 \cdot 19}{15 \cdot 17} = 0.97$$

where $f_{ls} = 12$ and $f_{rcb} = 16$ are the error degrees of freedom for the Latin square and randomized complete block design, respectively. The correction reduces the RE from 1.21 to $0.97(1.21) = 1.17$ for column blocking and from 1.85 to $0.97(1.85) = 1.79$ for row blocking. The correction has a small effect on the efficiency estimates.

Multiple Latin Squares and Latin Rectangles

The Latin square design with two blocking criteria and four or less treatments is very restrictive and provides too few degrees of freedom for an effective estimate of experimental error variance. It is common to repeat the experiment with more than one square under these circumstances.

There are two distinct forms of the design with multiple squares. The first form has distinct row and column identification for each square in the experiment. The second form has either the row or the column identification common to all squares. An example of the first form occurs in agricultural field trials in which two Latin squares are utilized in separated areas of the research farm. The second form is exemplified with the automobile tire experiment in which two groups of four automobiles are used for two Latin square arrangements, with the eight automobiles representing column blocking and the four tire positions common to both squares representing row blocking. This latter form of the design is known as the *Latin rectangle*. The two forms are illustrated in Figure 8.4.

Figure 8.4a is representative of the agricultural field experiment with multiple unique Latin squares and Figure 8.4b is representative of the automobile tire experiment with positions as rows and automobiles as columns in a Latin rectangle.

					Column			
	Row	*1*	*2*	*3*	*4*	*5*	*6*	*7*	*8*
	1	A	B	C	D				
	2	B	C	D	A				
	3	C	D	A	B				
(a)	*4*	D	A	B	C				
	5					A	B	C	D
	6					B	A	D	C
	7					C	D	B	A
	8					D	C	A	B

					Column			
	Row	*1*	*2*	*3*	*4*	*5*	*6.*	*7*	*8*
(b)	*1*	A	B	C	D	A	B	C	D
	2	B	C	D	A	B	A	D	C
	3	C	D	A	B	C	D	B	A
	4	D	A	B	C	D	C	A	B

Figure 8.4 Multiple Latin squares with (a) rows and columns unique for each square and (b) the Latin rectangle with rows common to both squares and columns unique to each square

Randomization is performed separately for each square when row and column blocking is unique for each of the s squares. With the Latin rectangle it is also possible to consider each square unique and randomize accordingly. When the row criterion is consistent across columns it is only necessary to require that each treatment occur s times in each row and one time in each column. The randomization then consists of one randomization of the t rows and a separate randomization of the entire set of st columns.

The linear model for the Latin rectangle with unique row and column blocking criteria in each Latin square is

$$y_{ijl} = \mu + \kappa_l + \rho_{i(l)} + \gamma_{j(l)} + \tau_k + e_{ijl} \qquad (8.25)$$

$$l = 1, 2, \ldots, s \quad i, j, k = 1, 2, \ldots, t$$

where κ_l is the square effect, and $\rho_{i(l)}$ and $\gamma_{j(l)}$ are the row and column effects nested in the squares, respectively. It may be necessary to consider a square × treatment interaction component in the model if it is suspected that the treatment comparisons may differ from one square to another. A discussion of an experiment by treatment interaction is found in Section 8.6. An outline of the analysis of variance with unique row and column blocking is shown in Table 8.10.

Table 8.10 Analysis of variance for an experiment repeated with s unique Latin square arrangements

Source of Variation	Degrees of Freedom	Sum of Squares
Total	$st^2 - 1$	$\sum_l \sum_i \sum_j (y_{ijl} - \bar{y}_{...})^2$
Squares	$s - 1$	$t^2 \sum_l (y_{..l} - \bar{y}_{...})^2$
Rows within squares	$s(t - 1)$	$t \sum_l \sum_i (y_{i.l} - \bar{y}_{..l})^2$
Columns within squares	$s(t - 1)$	$t \sum_l \sum_j (\bar{y}_{.jl} - \bar{y}_{..l})^2$
Treatments	$t - 1$	$st \sum_k (\bar{y}_k - \bar{y}_{...})^2$
Error	$(st - s - 1)(t - 1)$	By subtraction

The linear model for the Latin rectangle that has common row blocking criteria for s complete Latin squares and unique column blocking is

$$y_{ij} = \mu + \rho_i + \gamma_j + \tau_k + e_{ij} \qquad (8.26)$$

$$j = 1, 2, \ldots, st \quad i, k = 1, 2, \ldots, t.$$

An outline of the analysis of variance for common row and unique column blocking is shown in Table 8.11. The standard error estimates for treatment means and differences between two treatment means, respectively, are $\sqrt{MSE/st}$ and $\sqrt{2MSE/st}$.

8.4 Factorial Experiments in Complete Block Designs

The treatment design used to address the research hypothesis can be placed in any compatible experiment design. The appropriate sum of squares partitions can be computed for the analysis of variance as long as the randomization restrictions are followed for the experiment design in question.

Table 8.11   Analysis of variance for an experiment with t treatments in a Latin rectangle with t rows and st columns

Source of Variation	Degrees of Freedom	Sum of Squares
Total	$st^2 - 1$	$\sum_i \sum_j (y_{ijl} - \overline{y}_{..})^2$
Rows	$t - 1$	$st \sum_i (\overline{y}_{i.} - \overline{y}_{..})^2$
Columns	$st - 1$	$t \sum_j (\overline{y}_{.j} - \overline{y}_{..})^2$
Treatments	$t - 1$	$st \sum_k (\overline{y}_k - \overline{y}_{..})^2$
Error	$(st - 2)(t - 1)$	By subtraction

The two-factor factorial in a randomized complete block design has all ab treatment combinations an equal number of times in each block. With the Latin square design each of the ab treatment combinations appears one time in each row and one time in each column. A Latin rectangle design with a common row criterion for s squares requires each treatment combination to appear one time in each column and s times in each row.

The linear model for a two-factor factorial, factor A with a levels and factor B with b levels, in a randomized complete block design with r blocks is

$$y_{ijk} = \mu + \rho_k + \alpha_i + \beta_j + (\alpha\beta)_{ij} + e_{ijk} \tag{8.27}$$

$$i = 1, 2, \ldots, a \quad j = 1, 2, \ldots, b \quad k = 1, 2, \ldots, r$$

The sum of squares partitions for a two-factor factorial in a randomized complete block design are illustrated in Table 8.12. The treatment sum of squares with $(t - 1) = (ab - 1)$ degrees of freedom is partitioned into sums of squares for the main effects of factors A and B and interaction effects as described in Table 6.5. The sum of squares for blocks and experimental error are analogous to those shown for the randomized complete block design in Table 8.2. Standard errors are computed according to the conventions outlined in Chapter 6 for the factorial experiment in a completely randomized design.

A similar pattern of analysis is followed for the Latin square and Latin rectangle in which the square, row, and column sums of squares are partitioned according to the analyses in Tables 8.6, 8.10, or 8.11 with $t = ab$. The sum of squares partitions for factorial main effects and interactions are computed as described in Chapter 6.

Table 8.12 Analysis of variance for a two-factor treatment design in a randomized complete block experiment design

Source of Variation	Degrees of Freedom	Sum of Squares	Mean Square	Expected Mean Square
Total	$rab - 1$	SS Total		
Blocks	$r - 1$	SS Blocks		
A	$a - 1$	SSA	MSA	$\sigma^2 + rb\theta_a^2$
B	$b - 1$	SSB	MSB	$\sigma^2 + ra\theta_b^2$
AB	$(a - 1)(b - 1)$	SS(AB)	$MS(AB)$	$\sigma^2 + r\theta_{ab}^2$
Error	$(ab - 1)(r - 1)$	SSE	MSE	σ^2

*See Chapter 6 for computational formulae of SSA, SSB, and $SS(AB)$. SSE is obtained by subtraction. SS Blocks $= ab \sum (\bar{y}_{.k} - \bar{y}_{..})^2$.

8.5 Missing Data in Blocked Designs

Missing data in research studies were discussed with factorial treatment designs in Chapter 6. Missing observations affect the relationships between blocks and treatments just as they affected the relationships between factors in the factorials.

The effects of treatments and blocks are nonorthogonal when data are missing. A contrast for one set of effects conveys some information about the other set of effects. Thus, the effects for blocks must be considered when computing the sum of squares partition for treatments.

Full and reduced alternative models are used for the complete block design to (1) compute unbiased sum of squares partitions for treatments and experimental error and (2) compute unbiased least squares estimates of treatment means and their standard errors from the available data. The procedures for least squares solutions to the normal equations and the sums of squares for the full and reduced models were outlined in Chapter 6. Many of the present-day computer programs for the analysis of variance are capable of performing this analysis.

Analysis with Missing Data in Complete Block Designs

The analysis with missing data for the randomized complete block design will differ slightly from that for the factorial design described in Chapter 6 because the randomized complete block model assumes no interaction effects. Thus, the analysis omits any test for interaction prior to estimating the treatment effects.

Solutions to the normal equations for the full model, $y_{ij} = \mu + \tau_i + \rho_j + e_{ij}$, are used to compute the estimates, $\hat{y}_{ij} = \hat{\mu} + \hat{\tau}_i + \hat{\rho}_j$, and the experimental error sum of squares for the full model,

$$SSE_f = \sum (y_{ij} - \hat{\mu} - \hat{\tau}_i - \hat{\rho}_j)^2$$

Solutions to the normal equations for the reduced model, $y_{ij} = \mu + \rho_j + e_{ij}$, are used to compute the estimates, $\widehat{y}_{ij} = \widehat{\mu} + \widehat{\rho}_j$, and the experimental error sum of squares for the reduced model,

$$SSE_r = \Sigma(y_{ij} - \widehat{\mu} - \widehat{\rho}_j)^2$$

The treatment sum of squares is computed as

$$SS \text{ Treatment (adjusted)} = SSE_r - SSE_f \qquad (8.28)$$

and it represents the reduction in sum of squares as a result of including τ_i in the full model. It is referred to as the adjusted treatment sum of squares, implying that block effects are also considered when estimating the treatment effects in the full model.

The orthogonal sum of squares partitions with m missing observations are shown in Table 8.13. The reduced model ignores the treatment classification, and SS Blocks (unadjusted) from the reduced model is the sum of squares due to differences among block means ignoring treatments.

The application of an analysis with missing observations is reserved for an exercise at the end of the chapter.

Table 8.13 Analysis of variance for a randomized complete block design with m missing observations

Source of Variation	Degrees of Freedom	Sum of Squares
Total	$tr - m - 1$	$\sum_i \sum_j (y_{ij} - \overline{y}_{..})^2$
Blocks (unadjusted)*	$r - 1$	$\sum_j n_j(\overline{y}_{.j} - \overline{y}_{..})^2$
Treatment (adjusted)	$t - 1$	$SSE_r - SSE_f$
Error	$(r-1)(t-1) - m$	SSE_f

*SS Blocks (unadjusted) $= SS$ Total $- SSE_r$; $n_j =$ number of observations in jth block

8.6 Experiments Performed Several Times

Experiments are repeated at several places or on several different occasions for various reasons. Repetition over time or space provides a form of replication to increase the precision of treatment mean estimates or increase the degrees of freedom for estimates of experimental error. Repeated experiments can provide an expanded inference base to evaluate treatments over a broader range of conditions. In other cases, the magnitude of treatment comparisons is expected to differ among places or times. The series of experiments are used to examine the variation in treatment differences relative to the environmental changes.

Whatever the reason for conducting the same experiment over a series of times or places, a certain amount of discretion must be exercised before we automatically combine the data from the series to conduct a single overall analysis. The following example will help clarify some caveats with a series of experiments.

Example 8.3   Efficiency of Water Use by Bermuda Grass

Bermuda grass is used extensively for lawns, parks, and golf courses in warm and dry climates. In dry climate areas maintenance of the Bermuda grass requires regular irrigation. Turf managers want species of plants that efficiently utilize water to reduce maintenance costs and conserve water. Considerable variation in water use efficiency exists among species and within species.

Research Objective: A plant breeder wanted to determine the amount of variability in water use efficiency of Bermuda grass that he could attribute to genetic differences. Given sufficient genetic variability he could initiate a breeding program to develop a water-use efficient hybrid Bermuda grass.

Treatment Design: The plant breeder made all possible (30) hybrid crosses between six Bermuda grass cultivars in what is known as a *diallel mating design*. This particular mating design allows the plant breeder to evaluate the genetic potential of specific cultivars or the population represented by the cultivars in the designs.

Experiment Design: He grew the progeny and the six parental cultivars in a randomized complete block design with two blocks at each of four separate locations in a field plot experiment. Two of the crosses did not produce progeny, so there were 34 plots in each block of each experiment.

  The plant breeder measured the total dry matter produced by the plants in each plot and the amount of water utilized on the plot to produce the plant material. The measurement he utilized for analysis was the ratio of water used to total dry matter production on each plot. The analysis of variance with source of variation, degrees of freedom, and mean square for each of the four experiments is shown in Table 8.14.

Table 8.14   Analysis of variance for water use in each of four experiments from a diallel mating design with six Bermuda grass cultivars

Source of Variation	Degrees of Freedom	Mean Squares by Location			
		1	2	3	4
Block	1	2.80	20.17	2.53	1.81
Genotypes	33	1.08	17.85	1.92	1.39
Error	33	1.61	10.56	1.07	0.74

Source: Dr. W. Kneebone, Department of Plant Sciences, University of Arizona.

Location 2 Is Different

The most noticeable feature of the analyses in Table 8.14 is that the mean squares for all sources of variation at location 2 are considerably larger than those for the other three locations. In particular, the error variance at location 2, $MSE = 10.56$, is considerably larger than the error variance at the other locations. The greater variability at location 2 is an indication that the experimental conditions there may have been quite different from other locations.

Homogeneous Variances Required to Combine Results of Several Experiments

There were four experiments, each with two replications, for a total of eight replications. If the four experiments were combined into one analysis the estimates of the genotype means would be much more precise.

Homogeneity of error variances is required for a combined analysis of variance for the four experiments. A homogeneity of variance test for the error variances with the F Max test (Chapter 4) rejects the hypothesis of equal error variances for the four locations. One possible solution is a transformation to a logarithmic scale to achieve homogeneous variances prior to a combined analysis of variance. The analyses of variance for each location after the logarithmic transformation are shown in Table 8.15.

Table 8.15 Analysis of variance for $10[\log_{10}(\text{water use})]$ in each of four experiments from a diallel mating design with six Bermuda grass cultivars

Source of Variation	Degrees of Freedom	Mean Squares by Location			
		1	*2*	*3*	*4*
Block	1	1.68	3.28	1.42	1.01
Genotypes	33	0.56	2.51	1.09	1.11
Error	33	0.82	1.54	0.56	0.51

The logarithmic transformation considerably reduced the disparity in the error variances. However, the observed mean squares for all sources of variation are still somewhat larger at location 2. The combined analysis still should be approached with caution, and an interpretation of the results initially should be made from each of the separate experiments. The F_0 statistics at each location to test the hypothesis of no differences among the genotypes, $F_0 = MS \text{ Genotypes}/MS \text{ Error}$, result in the rejection of the null hypothesis at the .05 level of significance for locations 3 and 4. The null hypothesis is not rejected at locations 1 and 2. The same conclusions are reached from the F_0 statistics with the untransformed data in Table 8.14.

Statistical Model and Analysis of Variance for Combined Analysis of Experiments

The results of the F tests indicate a differential performance among the genotypes at the four locations. The combined analysis should include the possibility of a genotype × location interaction. The linear statistical model for the combined analysis with random genotype and location effects is

$$y_{ijk} = \mu + p_i + b_{j(i)} + g_k + (gp)_{ik} + e_{ijk} \qquad (8.29)$$

where μ is the general mean, p_i is the random location effect, $b_{j(i)}$ is the random block within location effect, g_k is the random genotype effect, $(gp)_{ik}$ is the random effect for genotype $\times$ location interaction, and e_{ijk} is the random error effect. The combined analysis of variance for the transformed data from the four experiments is shown in Table 8.16 with expected mean squares for random genotype effects and either fixed or random location effects.

Table 8.16 Analysis of variance for $10[\log_{10}(\text{water use})]$ for combined experiments from a diallel mating design of six Bermuda grass cultivars

Source of Variation	Degrees of Freedom	Sum of Squares	Mean Square	Expected Mean Square
Locations (L)	3	433.63	144.54	
Blocks/L	4	7.39	1.85	
Genotypes (G)	33	81.11	2.46	$\sigma^2 + r\sigma_{gl}^2 + rl\sigma_g^2$
$G \times L$	99	93.01	0.94	$\sigma^2 + r\sigma_{gl}^2$
Pooled error	132	113.19	0.86	σ^2

The mean square for error in the combined analysis is the pooled error from the four experiments (average of the mean squares for error from the four experiments). The blocks are unique to each experiment and they are a nested-factor effect for the analysis of variance. The sum of squares for blocks within locations is the sum of the block sum of squares from the individual analyses of variance.

The expected mean squares for the random or mixed model were derived using guidelines given in Chapter 7. If genotypes (treatments) are fixed effects, then σ_g^2 is replaced by the equivalent component for fixed effects, θ_g^2. When both locations and genotypes (or treatments) are fixed, then σ_{gl}^2 is also replaced by the equivalent component for fixed effects, θ_{gl}^2 in Table 8.16.

Whether the repetitions of experiments over time and places are random or fixed effects depends upon the objective of the repetition. If the repetitions are chosen to investigate the treatment responses to deliberate changes in environment, then fixed effects models for places or time, seem appropriate. If the repetitions can be justified as legitimate random representatives of places or time, then the random effects model can be used for places or time. It is perhaps most difficult to consider the repetitions random if only a limited number of places are available for experiments or a sequence of successive weeks, months, or years represent the time repetition.

Tests of Hypotheses in Combined Analysis

The hypothesis of no genotype $\times$ location interaction is tested with the F_0 statistic in Table 8.16 as $F_0 = MS(\boldsymbol{G} \times \boldsymbol{L})/MSE = 0.94/0.86 = 1.09$, and critical value

$F_{.05,99,132} = 1.36$. The null hypothesis of no genotype $\times$ location is not rejected. The hypothesis of no genotype effects is tested with the F_0 statistic $F_0 = MSG/MSE = 2.46/0.94 = 2.62$ with critical value $F_{.05,33,99} = 1.55$. The null hypothesis of no genotype effects is rejected.

Dissecting Treatment $\times$ Experiment Interaction

The nonsignificance of genotype $\times$ location interaction should not be taken lightly since an analysis of the separate experiments revealed significant genotype differences at the .05 level of significance at only two of the locations. It is entirely possible that certain treatment comparisons interact with the environment while other comparisons are relatively constant across environments. Three sets of comparisons important in the Bermuda grass diallel mating design are

- comparisons among the six parent cultivars

- comparisons among the 28 crosses

- a contrast between the mean of the six parent cultivars and that of the 28 crosses

A separate analysis of variance for the parent cultivars provides the sums of squares for the first set of comparisons. A separate analysis of variance for the crosses provides the sums of squares for the second set of comparisons. The linear model in Equation (8.29) is used for the analysis. It is the same model used when all genotypes were included in the analysis. Only those sums of squares required for the comparisons of interest are shown in Table 8.17. The sums of squares for locations and blocks within location are not shown.

Table 8.17 Separate analyses of variance for parents and crosses from a diallel mating design of six Bermuda grass cultivars

(1) *Parents Analysis*			
Source of Variation	*Degrees of Freedom*	*Sum of Squares*	*Mean Square*
Parents	5	3.45	0.69
Parents $\times$ locations	15	4.87	0.32
Error (P)	20	10.04	0.50
(2) *Crosses Analysis*			
Source of Variation	*Degrees of Freedom*	*Sum of Squares*	*Mean Square*
Crosses	27	46.06	1.71
Crosses $\times$ locations	81	79.56	0.98
Error (C)	108	101.22	0.94

The sum of squares for the contrast between the means of the parents and crosses and the sum of squares for the interaction between the contrast and locations can be computed by subtraction. Utilizing the sums of squares from Tables 8.16 and 8.17, they are

(1) Parents versus Crosses

$$SS(\boldsymbol{P} \text{ versus } \boldsymbol{C}) = SS\boldsymbol{G} - SS\boldsymbol{P} - SS\boldsymbol{C} = 81.11 - 3.45 - 46.06$$
$$= 31.60$$

(2) (Parents versus Crosses) $\times$ Locations

$$SS[(\boldsymbol{P} \text{ versus } \boldsymbol{C}) \times \boldsymbol{L}] = SS(\boldsymbol{G} \times \boldsymbol{L}) - SS(\boldsymbol{P} \times \boldsymbol{L}) - SS(\boldsymbol{C} \times \boldsymbol{L})$$
$$= 93.01 - 4.87 - 79.56$$
$$= 8.58$$

The separate analyses of variance for parent cultivars and the crosses also provide separate sums of squares for experimental error that can be identified with each set of comparisons. A partition of the error sum of squares can be useful because the experimental errors can differ considerably among the several comparisons. The experimental error mean squares for parents and crosses, Error ($\boldsymbol{P}$) and Error ($\boldsymbol{C}$), from the separate analyses are found in Table 8.17.

The sum of squares for experimental error associated with the contrast between parents and crosses is found from the error sums of squares in Tables 8.16 and 8.17 as

$$SS \text{ Error}(\boldsymbol{P} \text{ versus } \boldsymbol{C}) = SS \text{ Error} - SS \text{ Error}(\boldsymbol{P}) - SS \text{ Error}(\boldsymbol{C})$$
$$= 113.19 - 10.04 - 101.22$$
$$= 1.93$$

A summary of the analysis of variance with all of the sum of squares partitions is shown in Table 8.18. The three mean squares for the partition of experimental error shown at the bottom of Table 8.18 have similar values. The pooled error with 132 degrees of freedom probably can be used with some confidence that the error variances of the three groups of comparisons are relatively homogeneous.

To Pool or Not to Pool Variances

The decision to use the partitioned error term or the pooled error can have an effect on the tests of hypotheses. Tests with the pooled error have larger degrees of freedom and, therefore, greater ability (power) to detect differences than the partitioned mean square error. This effect can be seen in Table 8.18 with a test of the ($\boldsymbol{P}$ vs. $\boldsymbol{C}$) $\times \boldsymbol{L}$ interaction. The test statistic with the pooled error is $F_0 = 2.86/0.86 = 3.33$ with critical value $F_{.05,3,132} = 2.67$. The test statistic with the partitioned mean square error is $F_0 = 2.86/0.48 = 5.96$ with critical value $F_{.05,3,4} = 6.59$. In the latter test with fewer degrees of freedom for mean square error the null hypothesis is not rejected. The hypothesis is rejected with the pooled

Table 8.18   Analysis of variance for $10[\log_{10}(\text{water use})]$ for combined experiments from a diallel mating design of six Bermuda grass cultivars

Source of Variation	Degrees of Freedom		Sum of Squares		Mean Square	
Locations (L)	3		433.63		144.54	
Blocks/L	4		7.39		1.85	
Genotypes (G)	33		81.11		2.46	
Parents (P)		5		3.45		0.69
Crosses (C)		27		46.06		1.71
P vs C		1		31.60		31.60
G × L	99		93.01		0.94	
P × L		15		4.87		0.32
C × L		81		79.56		0.98
(P vs. C) × L		3		8.58		2.86
Pooled Error	132		113.19		0.86	
Error (P)		20		10.04		0.50
Error (C)		108		101.22		0.94
Error (P vs. C)		4		1.93		0.48

mean square error. It should be recognized that the null hypothesis would not be rejected for either of the tests at the 0.01 significance level. The difference between the two tests is not dramatic in this particular case, but it is necessary to be aware of the two possibilities for the test.

In this particular example the results of other F tests are not affected by which error term is used, partitioned or pooled. Neither of the other $G \times L$ interaction partitions is significant. Although the overall genotype × location interaction was not significant, there appears to be some evidence that a component of the interaction was significant. This result may explain the differences in significance levels of the tests for genotypes from the separate location analyses in Table 8.15.

More Details to be Found

McIntosh (1983) provided analysis of variance tables with sources of variation, degrees of freedom, and appropriate F_0 statistics to test hypotheses for an extensive array of experiments with fixed, random, and mixed models combined over time, places, or a combination of time and places. Carmer, Nyquist, and Walker (1989) provided formulae for the estimation of variances for pairwise mean differences from combined experiments with two- and three-factor treatment designs. The experiments had randomized complete block designs with fixed treatment effects and random time and place effects.

EXERCISES FOR CHAPTER 8

1. An irrigation experiment was conducted in a randomized complete block design in a Valencia orange grove. Six irrigation treatments were used in eight blocks of trees. The data that follow are the weight in pounds of harvested fruit from each plot.

	Block							
Method	*1*	*2*	*3*	*4*	*5*	*6*	*7*	*8*
Trickle	450	469	249	125	280	352	221	251
Basin	358	512	281	58	352	293	283	186
Spray	331	402	183	70	258	281	219	46
Sprinkler	317	423	379	63	289	239	269	357
Sprinkler + Spray	479	341	404	115	182	349	276	182
Flood	245	380	263	62	336	282	171	98

Source: Dr. R. Roth and Dr. B. Gardner, Department of Soil and Water Science, University of Arizona.

a. Write a linear model for the experiment, explain the terms, and compute the analysis of variance.
b. What are the assumptions necessary for the analysis of variance to be valid? How do they relate to the experiment?
c. Compute the standard error estimate for an irrigation treatment mean and the difference between two irrigation treatment means.
d. Consider the flood irrigation method to be the standard practice. Use the Dunnett method to test the difference between the flood irrigation and each of the other methods.
e. Compute the relative efficiency of this design relative to a completely randomized design. What are your conclusions?
f. Obtain the residual plots from the analysis, and interpret them.

2. A fertilizer trial on a range grass, blue grama, was conducted in a randomized complete block design by a management scientist. Five fertilizer treatments were randomly assigned to plots in each of five blocks. The data are $100 \times$ (percent phosphorus) in a plant tissue sample from each plot.

	Block				
Treatment	*1*	*2*	*3*	*4*	*5*
No fertilizer	7.6	8.1	7.3	7.9	9.4
50 lb. nitrogen	7.3	7.7	7.7	7.7	8.2
100 lb. nitrogen	6.9	6.0	5.6	7.4	7.0
50 lb nitrogen + 75 lb P_2O_5	10.8	11.2	9.0	12.9	11.6
100 lb nitrogen + 75 lb P_2O_5	9.6	9.3	12.0	10.6	10.4

Source: Dr. P. Ogden, Range Management, University of Arizona.

a.   Write a linear model for this experiment, explain the terms, and compute the analysis of variance.

b.   Compute the 1 degree of freedom sum of squares for each contrast that follows, and test the null hypothesis for each. The four orthogonal contrasts among the five treatments are (1) no fertilizer versus the four fertilizer treatments, (2) the main effect of nitrogen, (3) the main effect of P_2O_5, and (4) the interaction between nitrogen and P_2O_5.

c.   Compute the standard error for each of the contrasts in part (b).

d.   Compute the relative efficiency of the randomized complete block design.

e.   Obtain the residual plots from the analysis, and interpret them.

3.   The self-inductance of coils with iron-oxide cores was measured under different temperature conditions of the measuring bridge. The coil temperature was held constant. Five coils were used for the experiment. The self-inductance of each coil was measured for each of four temperatures (22°, 23°, 24°, and 25°) for the measuring bridge. The temperatures were utilized in a random order for each coil. The data are percentage deviations from a standard.

	Coil				
Temperature	1	2	3	4	5
22	1.400	0.264	0.478	1.010	0.629
23	1.400	0.235	0.467	0.990	0.620
24	1.375	0.212	0.444	0.968	0.495
25	1.370	0.208	0.440	0.967	0.495

Source: H. Hamaker (1955), Experimental design in industry. *Biometrics* 11, 257–286.

a.   Write a linear model for this experiment, explain the terms, and compute the analysis of variance.

b.   What are the assumptions necessary for the analysis of variance to be valid? How do they relate to the experiment?

c.   Compute the orthogonal polynomial regression contrasts for temperature and their sums of squares. Determine the best-fitting equation for the data.

d.   Compute the relative efficiency of using coils as blocks.

e.   Obtain the residual plots from the analysis, and interpret them.

4.   A traffic engineer conducted a study to compare the total unused red light time for five different traffic light signal sequences. The experiment was conducted with a Latin square design in which the two blocking factors were (1) five randomly selected intersections and (2) five time periods. In the data table the five signal sequence treatments are shown in parentheses as A, B, C, D, E, and the numerical values are the unused red light times in minutes.

| | | Time Period | | | |
Intersection	*1*	*2*	*3*	*4*	*5*
1	15.2(**A**)	33.8(**B**)	13.5(**C**)	27.4(**D**)	29.1(**E**)
2	16.5(**B**)	26.5(**C**)	19.2(**D**)	25.8(**E**)	22.7(**A**)
3	12.1(**C**)	31.4(**D**)	17.0(**E**)	31.5(**A**)	30.2(**B**)
4	10.7(**D**)	34.2(**E**)	19.5(**A**)	27.2(**B**)	21.6(**C**)
5	14.6(**E**)	31.7(**A**)	16.7(**B**)	26.3(**C**)	23.8(**D**)

Source: Mason, Gunst, and Hess (1989), 393.

a. Write a linear model for this experiment, explain the terms, and compute the analysis of variance.

b. Compute the standard error for a signal sequence treatment mean and for the difference between two signal sequence treatment means.

c. Use the Multiple Comparisons with the Best procedure to select the set of signal sequences with the shortest unused red light time.

d. What is the relative efficiency of blocking by time periods?

e. Obtain the residual plots from the analysis, and interpret them.

5. A research engineer studied the time efficiency of four construction methods (A, B, C, D) for an electronic component. Four technicians were selected for the study. The construction process produces fatigue such that the required construction time by the technicians increases as they change from one method to another regardless of the order of construction methods. The engineer used a Latin square design with columns as "technician" and rows as "time period." The construction methods were randomized to the technicians and time periods according to the Latin square arrangement. The values are construction times in minutes required for the component with the construction method indicated in parentheses.

| | | Technician | | |
Time Period	*1*	*2*	*3*	*4*
1	90(**C**)	96(**D**)	84(**A**)	88(**B**)
2	90(**B**)	91(**C**)	96(**D**)	88(**A**)
3	89(**A**)	97(**B**)	98(**C**)	98(**D**)
4	104(**D**)	100(**A**)	104(**B**)	106(**C**)

a. Write a linear model for this experiment, explain the terms, and compute the analysis of variance.

b. Compute the standard error for a construction method mean and the difference between two construction method means.

c. Use the Tukey method to make all pairwise comparisons between the means of the construction method times.

d. Use the relative efficiency measure to determine whether the time period was a critical blocking factor to reduce experimental error variance.

6. The experiment in Example 8.2 on the relationship between wheat yield and seeding rate was conducted in a 5 × 5 Latin square design. A second replication of the experiment was conducted in an

adjacent field of the experimental farm. Consequently, there are two repetitions of the experiment with unique row and column blockings for each of the two experiments. The data for the second experiment follow. The seeding rates were 30, 80, 130, 180, and 230 for A through E, respectively. The grain yield for each plot is expressed in hundredweight (100 pounds) per acre.

Field	Field Column				
Row	1	2	3	4	5
1	26.88(A)	38.40(D)	35.33(E)	34.56(C)	24.57(B)
2	37.63(E)	24.57(A)	36.09(C)	23.81(B)	32.25(D)
3	29.95(C)	29.18(B)	33.02(D)	22.27(A)	33.02(E)
4	32.25(D)	31.49(C)	21.50(B)	33.02(E)	18.43(A)
5	26.11(B)	36.09(E)	23.81(A)	29.95(D)	29.95(C)

Source: Dr. M. Ottman, Department of Plant Sciences, University of Arizona.

a. Write the linear model for the second experiment, explain the terms, and compute the analysis of variance.
b. Compute the standard error estimates for a seeding rate mean and the difference between two seeding rate means.
c. Compute the relative efficiency of row blocking for this experiment. Would more replications be required for a randomized complete block design using only the columns as blocks? If so, how many more would you recommend?
d. Compute the linear and quadratic orthogonal polynomial regression sum of squares partitions for seeding rate, and test their null hypotheses. Are the deviations from a linear or a quadratic relationship significant?
e. Do you think it is reasonable to perform an analysis of variance for the two experiments combined? Explain.
f. Compute the analysis of variance for the two experiments combined.
g. Compute the standard error estimates for a seeding rate mean and the difference between two seeding rates from the combined experiments.
h. Is there an interaction between seeding rate linear or quadratic contrasts and experiments?

7. A horticulturalist conducted a nitrogen fertility experiment for lettuce in a randomized complete block design. Five rates of ammonium nitrate treatments (0, 50, 100, 150, and 250 lb/acre) were randomly assigned to each of two plots in each of two blocks for a total of four plots for each level of nitrogen. Each block consisted of ten plots, two plots for each treatment in each block. The data are the number of lettuce heads from each plot.

Nitrogen	Block 1		Block 2	
0	104	114	109	124
50	134	130	154	164
100	146	142	152	156
150	147	160	160	163
200	133	146	156	161

Source: Dr. W.D. Pew, Department of Plant Sciences, University of Arizona.

a. Write the linear model for the experiment, explain the terms, and compute the analysis of variance. Note that there are multiple plots for each treatment in each block. How does this affect your estimates of experimental error from the analysis of variance?
b. Test the assumption of no block × treatment interaction.
c. Compute the linear and quadratic polynomial regression contrasts sum of squares partitions for nitrogen, and test the null hypotheses. Interpret the results.
d. Are cubic deviations significant?

8. A field plot experiment was conducted to evaluate the interaction between timing of nitrogen application to the soil (early, optimum, late) and two levels of a nitrification inhibitor (none, .5 lb/acre). The inhibitor delays conversion of ammonium forms of nitrogen into a more mobile nitrate form to reduce leaching losses of fertilizer-derived nitrates. The nitrogen was supplied as pulse-labeled ^{15}N through a drip irrigation system at an early, an optimum, and a late date of application. The data are percent of ^{15}N taken up by sweet corn plants grown on the plots.

	Nitrogen Inhibitor					
	None			.5 lb/acre		
Block	Early	Optimum	Late	Early	Optimum	Late
1	21.4	50.8	53.2	54.8	56.9	57.7
2	11.3	42.7	44.8	47.9	46.8	54.0
3	34.9	61.8	57.8	40.1	57.9	62.0

Source: Dr. T. Doerge, Department of Soil and Water Science, University of Arizona.

a. Write the linear model for the experiment, explain the terms, and compute the analysis of variance.
b. Compute the standard error estimates for the marginal means of nitrogen inhibitor and timing of nitrogen application and the cell means.
c. Test the null hypotheses of no interaction effects and no main effects for the two factors.
d. Compute the relative efficiency of the randomized complete block design.
e. Obtain the residual plots from the analysis, and interpret them.

9. Use the data from Exercise 8.1 for the Valencia orange grove irrigation experiment. Assume the plots for trickle irrigation in block 1 (450) and flood irrigation in block 5 (336) were lost from the experiment.
a. Use an appropriate computer program to compute the orthogonal analysis of variance according to the following partitions:

Source of Variation: SS Blocks (unadjusted) = 432,384

SS Irrigation Treatments (adjusted for blocks) = 51,923

SS Error = 130,402

b. Test the hypothesis of no treatment differences.
c. Show the least squares estimates of the treatment means and their standard error estimates if the computer program is capable of producing the estimates. For treatment means with no

missing observations, the standard errors should be the same as those computed in the usual manner.

d. How have the standard errors for the trickle and flood irrigation treatment means been affected by the lost plots?
e. If the computer program is capable, compute the standard error of the difference between
    (i)   the trickle and flood irrigation means
    (ii) the trickle and basin irrigation means
    (iii) the flood and sprinkler irrigation means
    (iv) the basin and sprinkler irrigation means
f. How were the standard errors in part (e) affected by the lost plots? What effect do the missing plots have on associated tests of differences between pairs of treatment means?

10. An animal scientist conducted a beef animal-feeding trial with four treatments composed of different qualities of drinking water for the animals in a completely randomized design with two replications. The experiment was conducted in the spring months and the winter months on two successive years. Each of the four feeding trials lasted 112 days. The data that follow are the average daily gains for each pen of animals in each of the trials.

	Year 1		Year 2	
Treatment	Spring	Winter	Spring	Winter
1	1.81	2.14	2.06	2.17
	1.88	2.32	1.91	2.55
2	1.77	2.27	1.57	2.06
	1.60	2.02	1.32	2.20
3	1.85	2.13	1.51	2.25
	1.59	1.93	1.49	1.94
4	1.51	1.85	1.31	1.83
	1.56	1.95	1.20	2.15

Source: Dr. D. Ray, Department of Animal Sciences, University of Arizona.

a. Compute the analysis of variance for each of the four trials as a completely randomized design.
b. Determine whether the experimental error variances are homogeneous among the experiments.
c. Compute the combined analysis of variance for the four trials with year, season, treatment, and all interaction effects in the model. The experimental error is the pooled experimental error from the four analyses of variance from the separate trials.
d. Assume years are random while seasons and treatments are fixed effects. Test the null hypotheses for seasons, treatments, and interaction effects. What are your conclusions?

11. Use the data from Exercise 8.4, the study on traffic signals. Assume the observations at intersection 1 during time period 2 (33.8) and at intersection 4 during time period 5 (21.6) are missing.
a. Use an appropriate computer program to compute the orthogonal analysis of variance according to the following partitions:

Source of Variation
Intersections (unadjusted)
Time periods (adjusted for intersections)
Signal sequence (adjusted for intersections and periods)
Error

b. Test the hypothesis of no signal sequence differences.

c. Show the least squares estimates of the signal sequence means and their standard error estimates if the computer program is capable of producing the estimates.

d. How do the lost data affect the standard errors for the sequence **B** and **C** treatment means?

e. If the computer program is capable, compute the standard error of the difference between
   (i) the sequence **B** and **C** means
   (ii) the sequence **B** and **A** means
   (iii) the sequence **C** and **D** means
   (iv) the sequence **A** and **D** means

f. How were the standard errors in part (e) affected by the lost data? What effect do the missing data have on associated tests of differences between pairs of treatment means?

12. An experiment is to be conducted in a randomized complete block design with $t = 6$ treatments in $r = 4$ blocks.

a. Randomize the six treatments to the experimental units in a randomized complete block design. Show details of your randomization procedure.

b. How many different arrangements of the treatments are possible in each of the blocks?

c. How many different arrangements are possible for the entire experiment?

13. An experiment is to be conducted in a Latin square arrangement with $t = 5$ treatments. Choose one of the standard Latin squares from Appendix 8A, and randomize the five treatments to the experimental units in a Latin square arrangement. Show details of your randomization procedure.

14. Construct a standard square for a 7×7 Latin square arrangement.

a. Randomize the seven treatments to the experimental units in a Latin square arrangement. Show details of your design construction and randomization.

b. How many ways can the Latin square columns be arranged on the column blocks?

c. How many ways can the Latin square rows be arranged on the row blocks?

d. How many ways can the Latin square letters be arranged on the treatment labels?

e. How many arrangements of columns, rows, and treatments are possible for the entire experiment?

15. An experiment is to be conducted in a 4×8 Latin rectangle arrangement with four treatments. Choose two standard Latin squares from Appendix 8A, and randomize the four treatments to the experimental units in the Latin rectangle arrangement. Show details of your randomization procedure.

16. An experiment is to be conducted on accelerated failure tests with small electric motors at five different temperatures. A maximum of five tests can be conducted in one day. There are 20 motors available for testing. Design an experiment with a randomized assignment of temperatures to the motors. Show a sketch of the final set of tests that are to be conducted.

17. You are designing a study to investigate soil–plant relationships in oak–pine mixed forests. The factor under study is the percent of oak in the forest mix. You have identified three suitable replicate sites for each of the following oak percentages in the forest mix: (1) 0%, (2) 20%–30%, (3) 45%–55%, (4) 70%–80%, and (5) 100%. You must collect soil and forest floor litter samples from each site and conduct laboratory chemical analyses of the samples. The labor is to be divided equally among three people because of the amount of work involved in the field and the laboratory.
    a. Design the study to control the experimental error variation with the 15 chosen sites using yourself and two other people as workers.
    b. Sketch the analysis of variance for data from the study. Include the source of variation and degrees of freedom for each sum of squares partition.
    c. Suppose you take two samples from each site. Repeat part (b).

18. You are conducting an in-vitro feedstuff digestion trial in flasks that must be inoculated with CO_2 and rumen microorganisms obtained from a steer just prior to inoculating the flasks in the lab. Oxygen and temperatures less than 37°C adversely affect the microorganisms. Given the time required to add the CO_2 and microorganisms to the flasks even under the best of controlled conditions the first flasks receive warm healthy microorganisms, but the later flasks receive microorganisms with reduced activity.
    a. Suppose you have five treatments and 25 flasks to inoculate in sequence by yourself. Set up a complete block design for the study to control variation caused by reduced microorganism activity due to exposure.
    b. Sketch the analysis of variance for data from the study. Include the source of variation and degrees of freedom for each sum of squares partition.
    c. Suppose you have 50 flasks available for the five treatments. How could you design the study? Repeat part (b).

19. You are going to conduct a study to determine the contamination of stream water by human activity in a national forest. You have located four streams, each of which has a small permanent community located near the stream. The communities each have a waste disposal plant in the watershed of the stream. Also, each stream has a large recreational camp site located five to ten miles downstream from the community.

    You want to take a water sample at each of four locations on each stream: a sample upstream from the community, a sample one mile below the community, and one sample each from immediately above and below the recreational campsite.

    You are also required to take a sample on each of four days of the week: Friday, Sunday, Monday, and Wednesday. Your resources are limited such that you can only take 4 water samples from each stream for a total of 16 water samples for the entire study.
    a. Set up a complete block design to acquire the water samples with "stream location" as the treatment factor.
    b. Sketch the analysis of variance for the data showing source of variation and degrees of freedom for each sum of squares partition.
    c. Suppose you take two water samples each time you sample a stream location. Repeat part (b).

8A   Appendix: Selected Latin Squares

4 × 4

```
A B C D        A B C D        A B C D        A B C D
B A D C        B C D A        B D A C        B A D C
C D B A        C D A B        C A D B        C D A B
D C A B        D A B C        D C B A        D C B A
```

5 × 5

```
A B C D E      A B C D E      A B C D E
B A E C D      B A D E C      B A D E C
C D A E B      C E B A D      C D E A B
D E B A C      D C E B A      D E B C A
E C D B A      E D A C B      E C A B D
```

```
A B C D E      A B C D E      A B C D E
B C D E A      B C E A D      B C A E D
C E A B D      C A D E B      C E D A B
D A E C B      D E B C A      D A E B C
E D B A C      E D A B C      E D B C A
```

6 × 6

```
A B C D E F      A B C D E F
B F D C A E      B A F E C D
C D E F B A      C F B A D E
D A F E C B      D C E B F A
E C A B F D      E D A F B C
F E B A D C      F E D C A B
```

```
A B C D E F      A B C D E F
B C F A D E      B A E C F D
C F B E A D      C F B A D E
D E A B F C      D E F B C A
E A D F C B      E D A F B C
F D E C B A      F C D E A B
```

7×7

A	B	C	D	E	F	G		A	B	C	D	E	F	G
B	E	A	G	F	D	C		B	E	A	G	F	D	C
C	F	G	B	D	A	E		C	F	G	B	D	A	E
D	G	E	F	C	B	A		D	G	E	F	B	C	A
E	D	B	C	A	G	F		E	D	B	C	A	G	F
F	C	D	A	G	E	B		F	C	D	A	G	E	B
G	A	F	E	B	C	D		G	A	F	E	C	B	D

8×8

A	B	C	D	E	F	G	H		A	B	C	D	E	F	G	H
B	C	D	E	F	G	H	A		B	C	A	E	F	D	H	G
C	D	E	F	G	H	A	B		C	A	D	G	H	E	F	B
D	E	F	G	H	A	B	C		D	F	G	C	A	H	B	E
E	F	G	H	A	B	C	D		E	H	B	F	G	C	A	D
F	G	H	A	B	C	D	E		F	D	H	A	B	G	E	C
G	H	A	B	C	D	E	F		G	E	F	H	C	B	D	A
H	A	B	C	D	E	F	G		H	G	E	B	D	A	C	F

9×9

A	B	C	D	E	F	G	H	I
B	C	D	E	F	G	H	I	A
C	D	E	F	G	H	I	A	B
D	E	F	G	H	I	A	B	C
E	F	G	H	I	A	B	C	D
F	G	H	I	A	B	C	D	E
G	H	I	A	B	C	D	E	F
H	I	A	B	C	D	E	F	G
I	A	B	C	D	E	F	G	H

10 × 10

```
A  B  C  D  E  F  G  H  I  J
B  C  D  E  F  G  H  I  J  A
C  D  E  F  G  H  I  J  A  B
D  E  F  G  H  I  J  A  B  C
E  F  G  H  I  J  A  B  C  D
F  G  H  I  J  A  B  C  D  E
G  H  I  J  A  B  C  D  E  F
H  I  J  A  B  C  D  E  F  G
I  J  A  B  C  D  E  F  G  H
J  A  B  C  D  E  F  G  H  I
```

9 Incomplete Block Designs: An Introduction

It is sometimes necessary to block experimental units into groups smaller than a complete replication of all treatments with a randomized complete block or Latin square design as illustrated in Chapter 8. The incomplete block design is utilized to decrease experimental error variance and provide more precise comparisons among treatments than is possible with a complete block design. A general description of some major groups of incomplete block designs is presented in this chapter. The method of randomization and basic analysis methods are demonstrated for balanced and partially balanced incomplete block designs. The efficiency of the designs is also considered.

9.1 Incomplete Blocks of Treatments to Reduce Block Size

Experiments can require a reduction in block size for one of several reasons. Complete block designs can reduce the estimate of experimental error variances, but sometimes the reduction is insufficient. Alternatively, the number of treatments may be so large as to render a complete block design impractical for reducing experimental error variance. Also, the natural grouping of experimental units into blocks can result in fewer units per block than required by the number of treatments for a complete block design. In the following example, limited numbers of environmental control chambers prevented the performance of a complete replication of all treatments in one run of the available chambers.

Example 9.1 Tomato Seed Germination at Constant High Temperature

Tomatoes often are produced during the winter months in arid and tropical regions. Winter production requires seeding during late summer when soil temperatures can exceed 40°C, which surpasses the suggested maximum germination temperature of 35°C for tomato seed.

Research Objective: A plant scientist wanted to determine in what temperature range she could expect inhibition of tomato seed germination for a group of tomato cultivars.

Treatment Design: Four temperatures were chosen to represent a temperature range common for the cultivation area under consideration. They were 25°C, 30°C, 35°C, and 40°C. The tomato seed was to be subjected to a constant temperature in a controlled environment chamber.

Experiment Design: A single chamber would be an experimental unit since true replication of any temperature treatment required a separate run of the temperature treatment in a chamber. Any number of factors could contribute to variation in response between runs since the entire experimental setup had to be repeated for a replicate run. Thus, blocking on runs was considered essential.

One complete block and replication of the experiment required four chambers; however, only three chambers were at the disposal of the plant scientist. Since the natural block of one run had fewer chambers (experimental units) than treatments, she constructed an incomplete block design.

A diagram of the design is shown in Display 9.1. Three different temperatures were tested in each of the four runs. The runs represent incomplete blocks of three temperature treatments. The treatments were randomly assigned to the chambers in each run. Some special features of this design are discussed in the next section.

Display 9.1 An Incomplete Block Design with Four Treatments in Blocks of Three Units

	Chamber					Chamber		
	1	*2*	*3*			*1*	*2*	*3*
Run 1	25°	30°	40°	*Run 2*		35°	30°	25°

	Chamber					Chamber		
	1	*2*	*3*			*1*	*2*	*3*
Run 3	40°	25°	35°	*Run 4*		40°	30°	35°

Source: Dr. J. Coons, Department of Botany, Eastern Illinois University.

Other Examples: Batches of material for industrial research serve as blocks, but insufficient material may exist in a single batch for all experimental treatments. The criteria for matching subjects may result in insufficient numbers of like subjects or cohorts in each group to accommodate the treatments planned for the study. Agronomic variety trials often contain large numbers of varieties for test and complete blocks are not practical for reduction of error variance. Incomplete block designs are suitable choices for the experiments in each of these examples.

The guiding principle for block size is to have a homogeneous set of experimental units for more precise treatment comparisons. Incomplete block designs were introduced by Yates (1936a, 1936b) for experiments in which the number of experimental units per block are less than the number of treatments. The designs were developed from a need for experiments that included the relevant set of treatments to address the research hypotheses yet were constrained to sensible block sizes.

The incomplete block designs may be classified into two major groups of designs: those arranged in randomized incomplete blocks with one blocking criterion and those with arrangements based on Latin squares with two blocking criteria. The designs may be balanced wherein each treatment is paired an equal number of times with every other treatment in the same blocks somewhere in the experiment. A partially balanced design occurs when different treatment pairs occur in the same blocks an unequal number of times or some treatment pairs never occur together in the same block. An overview of incomplete block designs will be presented in this chapter along with an introduction to the analysis of data from these designs.

9.2 Balanced Incomplete Block (BIB) Designs

The BIB Design Compares All Treatments with Equal Precision

The **balanced incomplete block design** is arranged such that all treatments are equally replicated and each treatment pair occurs in the same block an equal number of times somewhere in the design. The balance obtained from equal occurrence of all treatment pairs in the same block results in equal precision for all comparisons between pairs of treatment means.

The incomplete block design has r replications of t treatments in b blocks of k experimental units with $k < t$. The total number of experimental units is $N = rt = bk$. The design for the tomato experiment in Display 9.1 has $b = 4$ blocks of $k = 3$ experimental units. Each of the $t = 4$ treatments is replicated $r = 3$ times. There are a total of $N = bk = 4 \cdot 3 = 12$ or $N = rt = 3 \cdot 4 = 12$ experimental units. Upon inspection it can be seen that each treatment pair occurs together in the same block twice. For example, the treatment pair $(25°, 30°)$ occurs in blocks 1 and 2 and the treatment pair $(30°, 35°)$ occurs in blocks 2 and 4.

The number of blocks in which each pair of treatments occurs together is $\lambda = r(k - 1)/(t - 1)$, where $\lambda < r < b$. The integer value λ derives from the fact that each treatment is paired with the other $t - 1$ treatments somewhere in the design λ times. There are $\lambda(t - 1)$ pairs for a particular treatment in the experiment. The same treatment appears in r blocks with $k - 1$ other treatments, and each treatment appears in $r(k - 1)$ pairs. Therefore,

$$\lambda(t - 1) = r(k - 1) \quad \text{or} \quad \lambda = r(k - 1)/(t - 1)$$

For the tomato experiment design in Display 9.1, $\lambda = 3(3 - 1)/(4 - 1) = 2$.

A balanced incomplete block design can be constructed by assigning the appropriate combinations of treatments to each of $b = \binom{t}{k}$ blocks to achieve a balanced design. Frequently, balance is possible with less than $\binom{t}{k}$ blocks.

There is no single method for constructing all classes of balanced incomplete block designs. Methods do exist for constructing certain classes of incomplete block designs. The topic of design construction has been the subject of much mathematical research, resulting in a vast array of balanced and partially balanced incomplete block designs.

Plans for small numbers of treatments are given in Appendix 9A.1. Additional tables of designs useful for many practical situations can be found in Cochran and Cox (1957) and Fisher and Yates (1963). Several important categories of traditional balanced incomplete block designs will be illustrated in more detail in Chapter 10. An introduction to the basic structure of incomplete block designs with illustrations of their application and analysis is the focus of this chapter.

9.3   How to Randomize Incomplete Block Designs

After the basic design has been constructed with the treatment code numbers, the steps in randomization follow:

Step 1.   Randomize the arrangement of the blocks of treatment code number groups.

Step 2.   Randomize the arrangement of the treatment code numbers within each block.

Step 3.   Randomize the assignment of treatments to the treatment code numbers in the plan.

Randomization is illustrated with the basic design plan for $t = 4$ treatments in $b = 4$ blocks of $k = 3$ experimental units each. Prior to randomization the plan is

Block			
1	1	2	3
2	1	2	4
3	1	3	4
4	2	3	4

Step 1. Suppose the blocks for the experiment are the runs of three growth chambers with three temperature treatments used in Example 9.1. The treatment groups $(1,2,3), (1,2,4), (1,3,4)$, and $(2,3,4)$ must be randomly assigned to the runs. Choose a random permutation of the numbers 1 to 4, and assign the four blocks to the four runs. With the random permutation $2, 4, 1, 3$ the assignment is

Run				Original Block
1	1	2	4	2
2	2	3	4	4
3	1	2	3	1
4	1	3	4	3

Step 2. Assign random treatment code numbers to the three growth chambers in each run. Choose a random permutation of the numbers 1 to 4 for each chamber, and omit the treatment number absent in the run. Four random permutations along with the assignment to each chamber (A, B, C) in each run follow:

	Chamber			Permutation			
Run	A	B	C				
1	2	4	1	2	4	3̸	1
2	3	4	2	3	4	1̸	2
3	1	2	3	4̸	1	2	3
4	1	4	3	1	4	3	2̸

Step 3. Suppose the treatments are temperatures of 25°C, 30°C, 35°C, and 40°C. A random permutation $2, 4, 3, 1$ gives a random assignment of temperatures to treatment code numbers with the replacements $2 \to 25°C, 4 \to 30°C, 3 \to 35°C$, and $1 \to 40°C$ in the previous display.

	Chamber		
Run	A	B	C
1	25°C	30°C	40°C
2	35°C	30°C	25°C
3	40°C	25°C	35°C
4	40°C	30°C	35°C

9.4 Analysis of BIB Designs

Statistical Model for BIB Designs

The linear statistical model for a balanced incomplete block design is

$$y_{ij} = \mu + \tau_i + \rho_j + e_{ij} \tag{9.1}$$

$$i = 1, 2, \ldots, t \quad j = 1, 2, \ldots, b$$

where μ is the general mean, τ_i is the fixed effect of the ith treatment, ρ_j is the fixed effect of the jth block, and the e_{ij} are independent, random experimental errors with mean 0 and variance σ^2.

Recall that there are r replications of the t treatments in b incomplete blocks of k experimental units. The total number of observations is $N = rt = bk$, with each treatment pair appearing together in $\lambda = r(k-1)/(t-1)$ blocks in the experiment.

The effects of treatments and blocks are not orthogonal in the incomplete block design because all treatments do not appear in each of the blocks. Therefore, the sum of squares partition for treatments computed in the manner of complete block designs will not be correct for the incomplete block designs, nor will the observed treatment means provide unbiased estimates of $\mu_i = \mu + \tau_i$. The parameter estimates and treatment sum of squares for the balanced incomplete block designs can be computed with relatively straightforward formulae. A derivation of the least squares estimates for the balanced incomplete block design is given in Appendix 9A.3.

Sum of Squares Partitions for BIB Designs

The sum of squares partitions can be derived by considering alternative full and reduced models for the design. Solutions to the normal equations are obtained for the full model, $y_{ij} = \mu + \tau_i + \rho_j + e_{ij}$, with estimates, $\widehat{y}_{ij} = \widehat{\mu} + \widehat{\tau}_i + \widehat{\rho}_j$, to compute the experimental error sum of squares for the full model,

$$SSE_f = \sum_i \sum_j (y_{ij} - \widehat{y}_{ij})^2 = \sum_i \sum_j (y_{ij} - \widehat{\mu} - \widehat{\tau}_i - \widehat{\rho}_j)^2 \tag{9.2}$$

Solutions to the normal equations for the reduced model, $y_{ij} = \mu + \rho_j + e_{ij}$, with estimates, $\widehat{y}_{ij} = \widehat{\mu} + \widehat{\rho}_j$, are used to compute the experimental error sum of squares for the reduced model,

$$SSE_r = \sum_i \sum_j (y_{ij} - \widehat{y}_{ij})^2 = \sum_i \sum_j (y_{ij} - \widehat{\mu} - \widehat{\rho}_j)^2 \tag{9.3}$$

The difference, $SSE_r - SSE_f$, is the reduction in sum of squares as a result of including τ_i in the full model. It is the sum of squares due to treatments after block effects have been considered in the model. It is referred to as SS Treatment(adjusted), implying that block effects are also considered when estimating the treatment effects in the full model. For balanced incomplete block designs the sum of squares for treatments adjusted, $SSE_r - SSE_f$, can be computed directly as

$$SST\text{(adjusted)} = \frac{k \sum_{i=1}^{t} Q_i^2}{\lambda t} \tag{9.4}$$

with $t - 1$ degrees of freedom. The quantity Q_i is an adjusted treatment total computed as

$$Q_i = y_{i.} - \frac{1}{k} B_i \tag{9.5}$$

where $B_i = \sum_j^b n_{ij} y_{.j}$ is the sum of all block totals that include the ith treatment, and $n_{ij} = 1$ if treatment i appears in block j and $n_{ij} = 0$ otherwise. This correction to the treatment total has the net effect of removing the block effects from the treatment total.

The sum of squares for blocks is derived from the reduced model with treatments ignored in the model as $SSB\text{(unadjusted)} = SS \text{ Total} - SSE_r$. Treatment effects are not considered when estimating the block effects, and the sum of squares is called an *unadjusted* sum of squares.

The additive partition of SS Total is

$$SS \text{ Total} = SSB\text{(unadjusted)} + SST\text{(adjusted)} + SSE_f \tag{9.6}$$

The analysis of variance outline for the sum of squares partitions is shown in Table 9.1. Many computer programs are capable of computing the correct sum of squares partitions, least squares estimates of treatment means, contrasts among least squares means, and their standard errors for incomplete block designs.

Table 9.1   Analysis of variance for a balanced incomplete block design

Source of Variation	Degrees of Freedom	Sum of Squares	Mean Square
Total	$N - 1$	$\sum (y_{ij} - \bar{y}_{..})^2$	
Blocks	$b - 1$	$k \sum (\bar{y}_{.j} - \bar{y}_{..})^2$	$MSB\text{(unadj.)}$
Treatments	$t - 1$	$\dfrac{k \sum Q_i^2}{\lambda t}$	$MST\text{(adj.)}$
Error	$N - t - b + 1$	By subtraction	MSE

Example 9.2   Vinylation of Methyl Glucoside

The addition of acetylene to methyl glucoside in the presence of a base under high pressure had been found to result in the production of several monovinyl ethers, a process known as *vinylation*. The monovinyl ethers are suitable for polymerization in many industrial applications.

Research Objective: The chemists wanted to obtain more specific information about the effect of pressure on percent conversion of methyl glucoside to monovinyl isomers.

Treatment Design: Based on previous work, pressures were selected within the range thought to produce maximum conversion. Five pressures were selected to estimate a response equation: 250, 325, 400, 475, and 550 psi.

Experiment Design: Only three high-pressure chambers were available for one run of the experimental conditions. It was necessary to block on runs because there could be substantial run-to-run variation produced by new setups of the experiment in the high-pressure chambers. The chemists set up a balanced incomplete block design with ten blocks (runs) each with three experimental units (pressurized chambers). Three different pressures were used in each run. The resulting design had six replications of each pressure treatment.

The pressures used in each run and the percent conversions to monovinyl isomers are shown in Table 9.2. The additive sum of squares partition of methyl glucoside is shown in Table 9.3.

Table 9.2 Percent conversion of methyl glucoside by acetylene under high pressure in a balanced incomplete block design

Run	Pressure (psi)					$y_{.j}$
	250	325	400	475	550	
1	16	18	–	32	–	66
2	19	–	–	46	45	110
3	–	26	39	–	61	126
4	–	–	21	35	55	111
5	–	19	–	47	48	114
6	20	–	33	31	–	84
7	13	13	34	–	–	60
8	21	–	30	–	52	103
9	24	10	–	–	50	84
10	–	24	31	37	–	92
$y_{i.}$	113	110	188	228	311	950
*$B_{i.}$	507	542	576	577	648	
†Q_i	– 56.0	– 70.7	– 4.0	35.7	95.0	

*Example: $B_1 = y_{.1} + y_{.2} + y_{.6} + y_{.7} + y_{.8} + y_{.9}$

†$Q_1 = y_{1.} - \frac{1}{3}B_1 = 113 - \frac{1}{3}(507) = -56.0$

Source: Drs. J. Berry and A. Deutschman, University of Arizona.

Inferences for Treatment Means

The least squares estimates of the treatment means and their estimated standard errors are shown in Table 9.4. The least squares estimate for a treatment mean μ_i is $\widehat{\mu}_i = \widehat{\mu} + \widehat{\tau}_i$, where

Table 9.3 Analysis of variance for percent conversion of methyl glucoside in a balanced incomplete block design

Source of Variation	Degrees of Freedom	Sum of Squares	Mean Squares	F	Pr > F
Total	29	5576.67			
Blocks (unadj)	9	1394.67	154.96		
Pressure(adj)	4	3688.58	922.14	29.90	.000
Error	16	493.42	30.84		

$$\widehat{\mu} = \overline{y}_{..} \quad \text{and} \quad \widehat{\tau}_i = \frac{kQ_i}{\lambda t} \tag{9.7}$$

For example, from Table 9.2, $Q_1 = -56.00$ and $\widehat{\mu} = \overline{y}_{..} = 950/30 = 31.67$, so that

$$\widehat{\tau}_1 = \frac{kQ_1}{\lambda t} = \frac{3}{(3)(5)}(-56.00) = -11.20 \quad \text{and} \quad \widehat{\mu}_1 = 31.67 - 11.20 = 20.47$$

Standard Errors for Treatment Means

The standard error for a treatment mean estimate is

$$s_{\widehat{\mu}_i} = \sqrt{\frac{MSE}{rt}\left(1 + \frac{kr(t-1)}{\lambda t}\right)} = \sqrt{\frac{30.84}{(6)(5)}\left(1 + \frac{(3)(6)(4)}{(3)(5)}\right)} = 2.44 \tag{9.8}$$

A 95% confidence interval estimate of a treatment mean in Table 9.4 is $\widehat{\mu}_i \pm t_{.025,16}(s_{\widehat{\mu}_i})$, where $t_{.025,16} = 2.120$. The standard error of the estimated difference between two treatment means, $\widehat{\mu}_i - \widehat{\mu}_j$, is

$$s_{(\widehat{\mu}_i - \widehat{\mu}_j)} = \sqrt{\frac{2kMSE}{\lambda t}} = \sqrt{\frac{(2)(3)30.84}{(3)(5)}} = 3.51 \tag{9.9}$$

Table 9.4 Least squares estimates of pressure means for percent conversion of methyl glucoside in a balanced incomplete block design

Pressure (psi)	$\widehat{\mu}_i$	$s_{\widehat{\mu}_i}$
250	20.47	2.44
325	17.53	2.44
400	30.87	2.44
475	38.80	2.44
550	50.67	2.44

Tests of Hypotheses About Treatment Means

The F_0 statistic to test the null hypothesis of no differences among the treatment means is

$$F_0 = \frac{MST(\text{adj.})}{MSE} = \frac{922.14}{30.84} = 29.90 \tag{9.10}$$

with $Pr > F = .000$ (Table 9.3). The critical value at the .05 level of significance is $F_{.05,4,16} = 3.01$. There are significant differences among the pressures with respect to conversion of methyl glucoside to vinylation products.

Contrasts Among Treatment Means

Treatment contrasts are estimated with least squares estimates of the treatment means $\widehat{\mu}_i$ as

$$c = \sum_{i=1}^{t} d_i \widehat{\mu}_i \tag{9.11}$$

with standard error estimate

$$s_c = \sqrt{\frac{kMSE}{\lambda t} \left(\sum_{i=1}^{t} d_i^2 \right)} \tag{9.12}$$

The statistic $t_0 = c/s_c$ can be used to test the null hypothesis $H_0: C = 0$ with a critical value based on the Student t statistic with $N - t - b + 1$ degrees of freedom.

The 1 degree of freedom sum of squares for the contrast can be computed with the least squares means $\widehat{\mu}_i$ as

$$SSC = \frac{\lambda t (\sum d_i \widehat{\mu}_i)^2}{k \sum d_i^2} \tag{9.13}$$

The statistic $F_0 = SSC/MSE$ is used to test the null hypothesis $H_0: C = 0$ with critical value $F_{\alpha,1,(N-t-b+1)}$.

The pressure treatment for this experiment is a quantitative factor with five levels. The regression of percent conversion of methyl glucoside on pressure using orthogonal polynomial contrasts for pressure will describe the effect of pressure on conversion rate. The regression analysis is left as an exercise at the end of the chapter.

Recovering Treatment Information from Block Comparisons

The analyses for incomplete block designs illustrated thus far estimate treatment effects based on treatment information contained within the blocks; this is referred to as *intrablock* analysis. The incomplete block designs are nonorthogonal because not all treatments appear in all blocks, and comparisons among the blocks contain some information about treatment comparisons. Yates (1940a) showed this *inter-block* information can be recovered with an interblock analysis and combined with the information from the intrablock analysis.

Block Contrasts Contain Treatment Contrasts

The information on treatment comparisons contained in a comparison between blocks can be illustrated simply with the first two runs of the chemistry experiment in Example 9.2. The data for percent conversion of methyl glucoside in the first two runs of the experiment are

Run	250	325	400	475	550	Mean
1	16	18	—	32	—	22
2	19	—	—	46	45	37

(Column group header: Presssure (psi) spanning 250–550)

A contrast between the two runs means, 37 and 22, is also a contrast between two of the pressure treatments, 550 psi and 325 psi. Similar treatment comparisons are contained in all of the block comparisons. The objective is to recover this between-block, or interblock, information about treatments and combine it with the within-block, or intrablock, information about treatments.

The subject of interblock information recovery is mentioned here only to indicate the availability of the method. A thorough treatment on the recovery of interblock information can be found in Kempthorne (1952), Cochran and Cox (1957), John (1971), John (1987), and John and Williams (1995).

Information from the interblock analysis is incorporated into the intrablock analysis with an estimator of the treatment effects that combines the intrablock and interblock estimators of the treatment effects.

If blocking has been very effective in reducing experimental error, the interblock estimate contributes only a small amount of information to the combined estimate. On the other hand, if the block effects are small the information recovered from the interblock estimate can be substantial. If blocking is not effective, then the analysis with recovery of interblock information virtually reduces to the ordinary analysis with no block adjustments.

9.5 Row–Column Designs for Two Blocking Criteria

When the need exists to control variation with more than one blocking criterion the Latin square design is a complete block design used to control variation among the experimental units with two blocking factors. Latin square designs may be impractical in some situations since the number of experimental units required by the design, $N = t^2$, can exceed the constraints of the experimental material or treatment numbers can exceed available block sizes.

Row–column designs can be used with either rows or columns or both rows and columns as incomplete blocks when two blocking criteria are required for the experiment. The designs are arranged in p rows and q columns of experimental units. Consider the classical experiment to test four automobile tire treads on the four positions of four automobiles in a Latin square design. Suppose the research team wanted to evaluate $t = 7$ tire treatments. The requirement to control variation due to tire position and automobile is still necessary. However, the automobiles only have four positions on which to test the seven treatments. The row–column design with an incomplete set of treatments in each column shown in Display 9.2 can be used for the experiment.

Display 9.2   A 4 × 7 Row–Column Balanced Incomplete Block Design

Position	Automobile						
	(1)	*(2)*	*(3)*	*(4)*	*(5)*	*(6)*	*(7)*
(1)	3	4	5	6	7	1	2
(2)	5	6	7	1	2	3	4
(3)	6	7	1	2	3	4	5
(4)	7	1	2	3	4	5	6

The automobiles are used as incomplete blocks with four treatments evaluated on the four tire positions. The positions are complete blocks since each treatment is evaluated at each of the positions. Upon close inspection it can be seen that each treatment pair occurs on the same automobile two times somewhere in the experiment. Therefore, the incomplete block design on automobiles is balanced. The design is naturally balanced with respect to the positions because they constitute complete blocks. The example is clearly a case where an incomplete block design was necessary because there was an insufficient number of positions available to test all treatments at one time.

Row Orthogonal Designs Have a Complete Replication in Each Row

Given the row–column design in Display 9.2 is a complete block design for rows and a balanced incomplete block design for columns the design is referred to as a row orthogonal design (John, 1987). Since each treatment occurs in each row, the treatments are orthogonal to the rows.

Youden (1937, 1940) developed incomplete Latin square arrangements, now known as *Youden squares,* by omitting two or more rows from the Latin square design. The design parameters are $t = b$, $r = k$, and $\lambda = k(k-1)/(t-1)$. Plans of some Youden squares for small experiments are shown in Appendix 9A.2. Additional plans can be found in Cochran and Cox (1957) and Petersen (1985).

The row orthogonal design in Display 9.2 is a Youden square that has $r = k = 4$ rows as replications, $b = 7$ columns, and $t = 7$ treatments, with $\lambda = 2$

for the incomplete column blocks. The columns are incomplete blocks of $k = r = 4$ units, and the rows are complete blocks containing each of the $t = 7$ treatments.

Randomization in row–column designs is accomplished in the same manner as that described for Latin square designs in Chapter 8. There is a separate permutation of each of the row and column groups of treatments to the actual blocks, and the treatments are randomly assigned to the treatment labels of the design.

The Analysis Outline for Row–Column Designs

The linear model for the row–column design is

$$y_{ijm} = \mu + \tau_i + \rho_j + \gamma_m + e_{ijm} \tag{9.14}$$

$$i = 1, 2, \ldots, t \quad j = 1, 2, \ldots, k \quad m = 1, 2, \ldots, b$$

where μ is the general mean, τ_i is the treatment effect, ρ_j is the row effect, γ_m is the column effect, and e_{ijm} is the random experimental error.

Each row contains a complete replication of all of the treatments, and the treatments are orthogonal to the rows. The columns are also orthogonal to the rows. The treatment totals are adjusted only for the incomplete column blocks to provide unbiased estimates of treatment means and a valid F test for treatment effects. The intrablock analysis of variance outline is shown in Table 9.5. The analysis of variance for the row orthogonal designs only differs from the balanced incomplete block designs by the addition of a sum of squares partition for the rows. All other aspects of the analysis remain the same.

Table 9.5   Intrablock analysis for a row–column balanced incomplete block design with treatments orthogonal to rows

Source of Variation	Degrees of Freedom	Sum of Squares
Total	$N - 1$	$\sum (y_{ijm} - \bar{y}_{...})^2$
Rows (replications)	$k - 1$	$t \sum (\bar{y}_{.j.} - \bar{y}_{...})^2$
Columns (unadjusted)	$b - 1$	$k \sum (\bar{y}_{..m} - \bar{y}_{...})^2$
Treatments (adjusted)*	$t - 1$	$k \sum Q_i^2 / \lambda t$
Error	$(t - 1)(k - 1) - (b - 1)$	By subtraction

*$Q_i = y_{i..} - (B_i / k)$ where B_i is the sum of column block totals that include treatment i.

9.6   Reduce Experiment Size with Partially Balanced (PBIB) Designs

Balanced designs cannot be constructed for every experimental situation requiring incomplete blocks. In some cases the required number of replications may become prohibitive. Frequently, partially balanced designs requiring much less replication can be constructed. The minimum number of replications required for the balanced design is $r = \lambda(t - 1)/(k - 1)$. Suppose an experiment with $t = 6$ treatments requires blocks of size $k = 4$. The balanced design shown in Appendix 9A.1 requires $r = 10$ replications or $rt = 60$ experimental units. It is completely possible that 60 experimental units are not available or that the cost of the experiment with 60 units is prohibitive.

Unequal Occurrences of Treatment Pairs

The partially balanced incomplete block design was introduced by Bose and Nair (1939). The partially balanced design has some treatment pairs occurring in more blocks than other treatment pairs, and consequently some treatment comparisons will be made with greater precision than others. It is more straightforward to use a balanced design that provides equal precision for all comparisons between treatments. However, if resources are limited and sufficient replication is possible, the partially balanced design is an attractive alternative to the balanced design that requires excessive numbers of experimental units.

Consider the partially balanced incomplete block design for six treatments in blocks of four units shown in Display 9.3. The balanced design in Appendix 9A.1 requires ten replications for balance with $\lambda = 6$. The partially balanced design shown in Display 9.3 has two replications and requires twelve experimental units in three blocks of four experimental units.

Display 9.3	A Partially Balanced Incomplete Block Design with Six Treatments in Blocks of Four Units			
Block 1	1	4	2	5
Block 2	2	5	3	6
Block 3	3	6	1	4

Some pairs of treatments occur together in two blocks, whereas other pairs occur together in only one block. Treatment pairs $(1, 4), (2, 5)$, and $(3, 6)$ occur together in two blocks. All other treatment pairs occur together in only one block. Those treatment pairs occurring together in two blocks will be compared with somewhat greater precision than those pairs that occur together in only one block. The differing precisions for treatment comparisons is the sacrifice made for the smaller experiment. However, the difference in precision is not so great as to inhibit the use of partially balanced designs.

Increasing the replication number can increase the precision on the treatment comparisons. Another repetition of the same experiment will provide four replications if the two replications provided by the initial experiment are insufficient. If four replications are sufficient, there is still a gain with the partially balanced design in terms of reduced costs over the fully balanced design.

Associate Classes for Treatment Pair Occurrences

Each treatment is a member of two or more associate classes in a partially balanced design. An *associate class* is a group of treatments wherein each treatment pair occurs together in λ_i blocks. Treatment pairs that occur together in λ_i blocks are known as *i*th associates.

The design in Display 9.3 has two associate classes. The treatment pairs $(1, 4), (2, 5)$, and $(3, 6)$ are first associates with $\lambda_1 = 2$. Each pair occurs together in two blocks. Each treatment has $n_1 = 1$ first associates. The remaining treatment pairs are second associates with $\lambda_2 = 1$. There are $n_2 = 4$ second associates for each treatment. For example, the second associates of treatment 1 are treatments 2, 3, 5, and 6. They occur with treatment 1 somewhere in one block of the design. The complete sets of associates are shown in Display 9.4.

**Display 9.4   First and Second Associates for Six Treatments
in the Partially Balanced Design of Display 9.3**

	Associates	
Treatment	*First*	*Second*
1	4	2, 3, 5, 6
2	5	1, 3, 4, 6
3	6	1, 2, 4, 5
4	1	2, 3, 5, 6
5	2	1, 3, 4, 6
6	3	1, 2, 4, 5

A catalog for some of the major groups of partially balanced designs with two associate classes can be found in Bose, Clatworthy, and Shrikhande (1954) and Clatworthy (1973).

Notes on the Analysis of PBIB Designs

One advantage of the balanced designs over the partially balanced designs is that the manual calculations for the analysis of variance are somewhat simpler. Before the advent of modern-day statistical programs it was imperative that realistic manual computational formulae be available for an analysis of the data. Now it is possible to compute the appropriate sum of squares partitions for the analysis of variance as well as unbiased estimates of treatment means and their standard errors for most designs with available statistical programs. Therefore, the partially balanced design is a realistic alternative provided there are reasonably precise comparisons between all treatment pairs.

Sum of Squares Partitions for PBIB Designs

The linear model for the partially balanced incomplete block design is the same as that for the balanced incomplete block design shown in Equation (9.1). Since the treatments are not orthogonal to blocks, the orthogonal sum of squares partitions are again derived by fitting the full model, $y_{ij} = \mu + \tau_i + \rho_j + e_{ij}$, to obtain SSE_f and the alternative reduced model without treatment effects, $y_{ij} = \mu + \rho_j + e_{ij}$, to obtain SSE_r.

The sum of squares for treatments adjusted for blocks is derived as SS Treatment (adjusted) $= SSE_r - SSE_f$. The block sum of squares unadjusted for treatments is the same as that shown in Table 9.1 for the balanced design.

The formulae to compute least squares estimates of treatment effects, treatment means, and the adjusted treatment sum of squares are not so straightforward as those for the balanced designs. Because there is more than one associate class for each of the treatments, more complex adjustments must be made to the treatment totals. Fortunately, many of the statistical programs available can perform the computations. Thus, it is not necessary to utilize the cumbersome manual calculation formulae for a thorough analysis of the data. Details of manual calculations can be found in Cochran and Cox (1957). Detailed derivations can be found in John (1971) and John (1987).

9.7   Efficiency of Incomplete Block Designs

The efficiency of one design relative to another is measured by comparing the variance for the estimates of treatment mean differences in the two designs. For example, the variance of the difference between two treatment means for the *randomized complete block* (RCB) design is $2\sigma_{rcb}^2/r$. Is that variance less than its counterpart from the completely randomized design $2\sigma_{cr}^2/r$? The relative efficiency of the RCB design to the completely randomized design can be determined because an estimate of σ^2 for the latter can be obtained from the data in the RCB design. It is then possible to evaluate the effectiveness of blocking to reduce experimental error variance.

No such luxury exists if we want to determine the efficiency of a balanced incomplete block (BIB) design relative to the RCB design. It is not possible to compute an estimate of σ^2 for the RCB design from the data in the BIB design. We would like to determine whether the smaller block size of the BIB design resulted in a smaller experimental error variance. The ratio of variances for a difference between two treatment means is still used to compare the BIB design with the RCB design. The difference is that it measures only the *potential* efficiency of the BIB design since we are unable to estimate σ^2 for the RCB design.

The Efficiency Factor for Incomplete Block Designs

The variance of the difference between two treatment means in the BIB design is $2k\sigma_{bib}^2/\lambda t$. If the BIB design and the RCB design have the *same number of treatments and replications* the efficiency of the BIB design relative to the RCB design is the ratio of the variances

$$\text{Efficiency} = \frac{(2\sigma_{rcb}^2/r)}{(2k\sigma_{bib}^2/\lambda t)} = \frac{\sigma_{rcb}^2}{\sigma_{bib}^2} \cdot \frac{\lambda t}{rk} \qquad (9.15)$$

The quantity $E = \lambda t/rk$ is called the *efficiency factor* for the balanced incomplete block design. It provides an indication of the loss caused by using an incomplete block design without reducing σ^2. The BIB design is more precise for the comparison of two treatment means than the RCB design if

$$\frac{\sigma_{bib}^2}{\sigma_{rcb}^2} < \frac{\lambda t}{rk} \qquad (9.16)$$

The efficiency factor is a lower limit to the efficiency of the BIB design relative to the RCB design for an experiment with the same number of replications and the same error variance σ^2. Given a BIB design with $t = 4, r = 3, k = 3$, and $\lambda = 2$ and an RCB design with $t = 4$ and $r = 3$, both designs require the same number of experimental units but they have different blocking arrangements. The BIB design is more precise than the RCB design if

$$\frac{\sigma_{bib}^2}{\sigma_{rcb}^2} < \frac{2(4)}{3(3)} = 0.89$$

In other words, σ_{bib}^2 would have to be about 11% smaller than σ_{rcb}^2 for the BIB design to be as precise as the RCB design with the same number of replications.

The intent of an incomplete block design is to reduce the error variance and increase the precision of the comparisons between treatment means. The goal would be to reduce σ^2 for the BIB design so that the inequality shown in Equation (9.16) would be achieved. A successful incomplete block design will reduce the error variance, and σ_{bib}^2 will be smaller than σ_{rcb}^2. Some elements that enter into a successful blocking strategy are considered in the next chapter.

EXERCISES FOR CHAPTER 9

---

1. A horticulturalist studied the germination of tomato seed with four different temperatures (25°C, 30°C, 35°C, and 40°C) in a balanced incomplete block design because there were only two growth chambers available for the study. The experiment was conducted in a balanced incomplete block design. Each run of the experiment was an incomplete block consisting of the two growth chambers as the experimental units ($k = 2$). Two experimental temperatures were randomly assigned to the chambers for each run. The data that follow are germination rates of the tomato seed.

Run	25 °C	30 °C	35 °C	40 °C
1	24.65	——	——	1.34
2	——	24.38	——	2.24
3	29.17	21.25	——	——
4	——	——	5.90	1.83
5	28.90	——	18.27	——
6	——	25.53	8.42	——

Source: Dr. J. Coons, Department of Botany, Eastern Illinois University.

a. How many times did each treatment pair occur together in the same block?
b. What is the efficiency factor for this design?
c. Write a linear model for the experiment, explain the terms, compute the intrablock analysis of variance, and test the null hypothesis for temperature effects.
d. Compute the least squares estimates of the temperature means and their standard error.
e. Compute the standard error of the difference between two least squares estimates of temperature means.
f. Compute the 1 degree of freedom sum of squares partitions for linear and quadratic orthogonal polynomial regression, and test their null hypotheses.
g. Are the deviations from quadratic regression significant?

2. A company is developing an insulating compound. An accelerated life testing procedure is conducted to determine the breakdown time in minutes after subjection to elevated voltages. Four voltages could be tested on a given run of the test. A balanced incomplete block design was used to test the compound with $t = 7$ voltages in $b = 7$ runs of $k = 4$ tests. The minutes to compound breakdown were recorded for each test.

Block	Voltage (kv)						
	24	28	32	36	40	44	48
1	—	—	38.19	—	5.44	1.96	0.55
2	220.22	—	—	7.66	—	2.54	0.67
3	270.85	200.67	—	—	6.24	—	0.76
4	360.14	170.52	45.43	—	—	3.22	—
5	—	220.12	56.74	9.32	—	—	0.61
6	300.66	—	55.34	10.41	7.19	—	—
7	—	190.78	—	8.74	6.92	2.21	—

a. How many times did each treatment pair occur together in the same block?
b. What is the efficiency factor for this design?
c. Write a linear model for the experiment, explain the terms, and compute the intrablock analysis of variance.
d. Compute the least squares residuals for each observation, and analyze the residuals according to the procedures discussed in Chapter 4. A typical transformation utilized for data of this nature is the natural logarithm. Does that seem appropriate here?

e.   Transform the data, compute the additive sum of squares partitions, and summarize in an analysis of variance table. Describe how the transformation has changed the nature of the residuals.

f.   Compute the least squares estimates of the voltage means and their standard error.

g.   Compute the standard error of the difference between two least squares estimates of voltage means.

h.   Compute the 1 degree of freedom sum of squares partition for linear regression of time to breakdown (or a suitable transformation) on voltage, and test the null hypotheses.

i.   Are the deviations from linear regression significant?

3.   A study was conducted to evaluate a method to measure traffic delay at urban street intersections in seven cities. Seven types of intersections were chosen for the study based on their geometry and traffic light configurations. Four of the intersection types were measured in each city in an incomplete block design. Four observers were used to measure traffic delay, one at each of the intersections for a specified period of time during peak traffic loads. The design was laid out as a Youden square with $t = 7$ intersection type treatments. The row–column blocking consisted of $k = 4$ observers as rows and $b = 7$ cities as columns. The order in which the cities were visited and the sequence of intersection type treatments each of the observers measured were randomized according to the Latin square procedure. The intersection types were randomly assigned to the treatment labels of the basic Youden square and are shown in parentheses in the data table for time in queue (seconds per vehicle).

Observer	City						
	1	2	3	4	5	6	7
1	(4)	(1)	(3)	(6)	(2)	(7)	(5)
	45.8	66.4	27.0	92.6	32.7	34.6	44.1
2	(2)	(6)	(1)	(4)	(7)	(5)	(3)
	28.6	94.6	64.6	39.9	34.5	45.7	23.7
3	(7)	(4)	(6)	(2)	(5)	(3)	(1)
	32.3	40.7	82.6	29.3	47.6	31.2	74.7
4	(3)	(7)	(2)	(5)	(1)	(6)	(4)
	28.0	31.1	25.5	41.9	68.7	68.1	38.7

a.   Write a linear model for this study, describe the terms, and conduct the intrablock analysis of variance for this experiment.

b.   Compute the standard error of the difference between least squares means of two intersection means.

c.   What is the efficiency factor for this design?

d.   Use the Multiple Comparisons with the Best procedure to select the set of intersection types with the minimum delay.

4.   (*Note*: This exercise presumes you have available a program to compute the analysis of variance, least squares estimates of treatment means with their standard errors, and contrasts with their standard errors.) A study was conducted in a partially balanced incomplete block design to evaluate $t = 9$ feed rations on nitrogen balance in ruminants. The study required expensive digestion stalls

and equipment, and only three such stalls were available for the study. The design consisted of $r = 3$ replications of the 9 treatments in $b = 9$ blocks of $k = 3$ animals. There were $n_1 = 6$ first associates for each treatment with $\lambda_1 = 1$ and $n_2 = 2$ second associates for each treatment with $\lambda_2 = 0$. The observed responses are shown in the table with treatment numbers in parentheses.

Block			
1	33.72 (1)	37.80 (2)	42.25 (3)
2	38.58 (1)	45.39 (4)	47.75 (6)
3	34.55 (1)	34.82 (5)	38.29 (7)
4	42.95 (2)	45.35 (4)	48.84 (9)
5	45.12 (2)	36.36 (5)	40.58 (8)
6	43.04 (3)	45.22 (6)	37.26 (8)
7	40.64 (3)	30.49 (7)	36.34 (9)
8	38.53 (4)	34.58 (7)	40.81 (8)
9	36.40 (5)	33.28 (6)	38.46 (9)

Source: J. L. Gill (1978), *Design and analysis of experiments in the animal and medical sciences*, Vol 2. Ames, Iowa: Iowa State University Press.

a. Make a table of first and second associates for each treatment.
b. Write a linear model for the experiment, and explain the terms.
c. Use your available computer program to compute the analysis of variance to obtain the adjusted treatment mean square and the experimental error variance. Test the null hypothesis for rations.
d. Compute the least squares estimates of the ration means and their standard errors.
e. Compute the standard error of the difference between two least squares estimates of ration means for first associates and second associates.
f. Frequently, a weighted average standard error is utilized for all comparisons among two treatment means in a partially balanced design. With $n_1 = 6$ first associates and $n_2 = 2$ second associates, the weighted average variance of the difference between two treatment means for this study is

$$\text{average variance} = \frac{n_1 s_1^2 + n_2 s_2^2}{n_1 + n_2}$$

where s_1^2 and s_2^2 are the estimated variances of the differences between first and second associates, respectively. Compute a weighted average standard error of the difference for this experiment.

5. An incomplete block design consists of the following arrangement of blocks (1, 2, 3, 4, 5) and treatments (A, B, C, D, E).

Block	
1	(B, C, D, E)
2	(A, B, D, E)
3	(A, C, D, E)
4	(A, B, C, D)
5	(A, B, C, E)

    a.   What are the design parameters t, r, k, and b?
    b.   Verify that the design is balanced.

6.   An incomplete block design consists of the following arrangement of blocks 1 through 6 and treatments (A, B, C, D, E, F).

Block	1	(A, B, C)	Block	4	(A, B, D)
	2	(C, D, E)		5	(C, D, F)
	3	(B, E, F)		6	(A, E, F)

    a.   What are the design parameters t, r, k, and b?
    b.   Is the design balanced? Explain.

7.   Suppose you are to test $t = 6$ automobile fuels, and you have $k = 3$ engines available for the tests. It is therefore necessary to use an incomplete block design for the tests with runs of the three engines as blocks. Find the appropriate balanced incomplete block design in Appendix 9A.1 to conduct the experiment. Randomize the treatment groups to blocks of runs and automobile fuel treatments to the engines.

8.   A paired comparisons study is to be conducted to evaluate campsites in a national park. There are $t = 10$ campsite designs, which range from primitive to full-facility sites. Visitors to the national park will be selected at random during the month of June and shown photographs of $k = 2$ of the ten campsites. They are to indicate their preference for one of the two campsites based on the photographs. Construct a balanced incomplete block design for the study. How many visitors will be required for the study? How many replications of each campsite will there be in the study? How many times will each pair of campsites be viewed by a visitor? How would the study design change if each visitor was asked to view $k = 3$ campsites?

9A.1   Appendix: Selected Balanced Incomplete Block Designs

Plan 9A.1 $t = 5, k = 3, r = 6, b = 10, \lambda = 3, E = 0.83$
(1,2,3), (1,2,5), (1,4,5), (2,3,4), (3,4,5) reps 1,2,3
(1,2,4), (1,3,4), (1,3,5), (2,3,5), (2,4,5) reps 4,5,6

Plan 9A.2 $t = 6, k = 3, r = 5, b = 10, \lambda = 2, E = 0.80$
(1,2,5), (1,2,6), (1,3,4), (1,3,6), (1,4,5),
(2,3,4), (2,3,5), (2,4,6), (3,5,6), (4,5,6)

Plan 9A.3 $t = 6, k = 4, r = 10, b = 15, \lambda = 6, E = 0.90$
(1,2,3,4), (1,4,5,6), (2,3,5,6) reps 1,2
(1,2,3,5), (1,2,4,6), (3,4,5,6) reps 3,4
(1,2,3,6), (1,3,4,5), (2,4,5,6) reps 5,6
(1,2,4,5), (1,3,5,6), (2,3,4,6) reps 7,8
(1,2,5,6), (1,3,4,6), (2,3,4,5) reps 9,10

Plan 9A.4 $t = 7, k = 3, r = 3, b = 7, \lambda = 1, E = 0.78$
(1,2,4), (2,3,5), (3,4,6), (4,5,7), (1,5,6),
(2,6,7), (1,3,7)

Plan 9A.5 $t = 7, k = 4, r = 4, b = 7, \lambda = 2, E = 0.88$
Replace each block in Plan 9A.4 by a block containing the remaining treatments.

Plan 9A.6 $t = 8, k = 4, r = 7, b = 14, \lambda = 3, E = 0.86$
(1,2,3,4), (5,6,7,8) rep 1   (1,2,7,8), (3,4,5,6) rep 2
(1,3,6,8), (2,4,5,7) rep 3   (1,4,6,7), (2,3,5,8) rep 4
(1,2,5,6), (3,4,7,8) rep 5   (1,3,5,7), (2,4,6,8) rep 6
(1,4,5,8), (2,3,6,7) rep 7

Plan 9A.7 $t = 9, k = 3, r = 4, b = 12, \lambda = 1, E = 0.75$
(1,2,3), (1,4,7), (1,5,9), (1,6,8), (2,4,9), (2,5,8)
(2,6,7), (3,4,8), (3,5,7), (3,6,9), (4,5,6), (7,8,9)

Plan 9A.8 $t = 9, k = 4, r = 8, b = 18, \lambda = 3, E = 0.84$
(1,2,3,4), (1,3,8,9), (1,4,6,7), (1,5,7,8), (2,3,6,7),
(2,4,5,8), (2,6,8,9), (3,5,7,9), (4,5,6,9), reps 1,2,3,4
(1,2,4,9), (1,2,5,7), (1,3,6,8), (1,5,6,9), (2,3,5,6),
(2,7,8,9), (3,4,5,8), (3,4,7,9), (4,6,7,8), reps 5,6,7,8

Plan 9A.9 $t = 9, k = 5, r = 10, b = 18, \lambda = 5, E = 0.90$
Replace each block of Plan 9A.8 by a block containing the remaining treatments.

Plan 9A.10 $t = 9, k = 6, r = 8, b = 12, \lambda = 5, E = 0.94$
Replace each block of Plan 9A.7 by a block containing the remaining treatments.

Plan 9A.11 $t = 10, k = 3, r = 9, b = 30, \lambda = 2, E = 0.74$
(1,2,3), (1,4,6), (1,7,9), (2,5,8), (2,8,10)
(3,4,7), (3,9,10), (4,6,9), (5,6,10), (5,7,8) reps 1,2,3
(1,2,4), (1,5,7), (1,8,10), (2,3,6), (2,5,9),
(3,4,8), (3,7,10), (4,5,9), (6,7,10), (6,8,9) reps 4,5,6
(1,3,5), (1,6,8), (1,9,10), (2,4,10), (2,6,7),
(2,7,9), (3,5,6), (3,8,9), (4,5,10), (4,7,8) reps 7,8,9

Plan 9A.12 $t = 10, k = 4, r = 6, b = 15, \lambda = 2, E = 0.83$
(1,2,3,4), (1,2,5,6), (1,3,7,8), (1,4,9,10), (1,5,7,9),
(1,6,8,10), (2,3,6,9), (2,4,7,10), (2,5,8,10), (2,7,8,9),
(3,5,9,10), (3,6,7,10), (3,4,5,8), (4,5,6,7), (4,6,8,9)

Plan 9A.13 $t = 10, k = 5, r = 9, b = 18, \lambda = 4, E = 0.89$
(1,2,3,4,5), (1,2,3,6,7), (1,2,4,6,9), (1,2,5,7,8), (1,3,6,8,9), (1,3,7,8,10),

(1,4,5,6,10), (1,4,8,9,10), (1,5,7,9,10), (2,3,4,8,10), (2,3,5,9,10), (2,4,7,8,9), (2,5,6,8,10), (2,6,7,9,10), (3,4,6,7,10), (3,4,5,7,9), (3,5,6,8,9), (4,5,6,7,8)

Plan 9A.14 $t = 10, k = 6, r = 9, b = 15, \lambda = 5, E = 0.93$
Replace each block in Plan 9A.12 by a block containing the remaining treatments.

Plan 9A.15 $t = 11, k = 5, r = 5, b = 11, \lambda = 2, E = 0.88$
(1,2,3,5,8), (1,2,4,7,11), (1,3,6,10,11), (1,4,8,9,10),
(1,5,6,7,9), (2,3,4,6,9), (2,5,9,10,11), (2,6,7,8,10),
(3,4,5,7,10), (3,7,8,9,11), (4,5,6,8,11)

Plan 9A.16 $t = 11, k = 6, r = 6, b = 11, \lambda = 3, E = 0.92$
Replace each block in Plan 9A.15 by a block containing the remaining treatments.

9A.2   Appendix: Selected Incomplete Latin Square Designs

Table 9A   Plans for incomplete Latin square designs derived from complete Latin squares

t	k	r	b	λ	E	Plan
4	3	3	4	2	.89	*
	3	6	8	4	.89	**
	3	9	12	6	.89	**
5	4	4	5	3	.94	*
	4	8	10	6	.94	**
6	5	5	6	4	.96	*
	5	10	12	8	.96	**
7	6	6	7	5	.97	*
8	7	7	8	6	.98	*
9	8	8	9	7	.98	*
10	9	9	10	8	.99	*
11	10	10	11	9	.99	*

* Constructed from a $t \times t$ Latin square by omission of the last column
** By repetition of the plan for $r = t - 1$, which is constructed by taking a $t \times t$ Latin square and omitting the last column

Plans for other incomplete Latin square designs:

Plan 9B.1 $t = 5, k = 2, r = 4, b = 10, \lambda = 1, E = .63$

Reps I and II					*Reps III and IV*				
1	2	3	4	5	1	2	3	4	5
2	5	4	1	3	3	4	2	5	1

Plan 9B.2 $t = 5, k = 3, r = 6, b = 10, \lambda = 3, E = .83$

	Reps I, II, and III					*Reps IV, V, and VI*			
1	2	3	4	5	1	2	3	4	5
2	1	4	5	3	2	3	4	5	1
3	5	2	1	4	4	5	1	2	3

Plan 9B.3 $t = 7, k = 2, r = 6, b = 21, \lambda = 1, E = .58$

Reps I and II							*Reps III and IV*							*Reps V and VI*						
1	2	3	4	5	6	7	1	2	3	4	5	6	7	1	2	3	4	5	6	7
2	6	4	7	1	5	3	3	4	5	6	7	1	2	4	3	6	5	2	7	1

Plan 9B.4 $t = 7, k = 3, r = 3, b = 7, \lambda = 1, E = .78$

7	1	2	3	4	5	6
1	2	3	4	5	6	7
3	4	5	6	7	1	2

Plan 9B.5 $t = 7, k = 4, r = 4, b = 7, \lambda = 2, E = .88$

3	4	5	6	7	1	2
5	6	7	1	2	3	4
6	7	1	2	3	4	5
7	1	2	3	4	5	6

Plan 9B.6 $t = 9, k = 2, r = 8, b = 36, \lambda = 1, E = .56$

Reps I and II									*Reps V and VI*								
1	2	3	4	5	6	7	8	9	1	2	3	4	5	6	7	8	9
2	8	4	7	6	1	3	9	5	4	6	2	5	7	8	9	1	3

Reps III and IV									*Reps VII and VIII*								
1	2	3	4	5	6	7	8	9	1	2	3	4	5	6	7	8	9
3	5	6	9	8	7	1	4	2	5	4	8	6	3	9	2	7	1

Plan 9B.7 $t = 9, k = 4, r = 8, b = 18, \lambda = 3, E = .84$

Reps I, II, III, and IV									*Reps V, VI, VII, and VIII*								
1	2	3	4	5	6	7	8	9	1	2	3	4	5	6	7	8	9
4	6	8	1	7	9	3	2	5	2	3	4	9	1	8	6	5	7
6	8	9	3	1	4	2	5	7	5	6	7	2	9	1	4	3	8
7	9	1	2	8	5	6	4	3	7	5	9	1	6	3	8	4	2

Plan 9B.8 $t = 9,\ k = 5,\ r = 10,\ b = 18,\ \lambda = 5,\ E = .90$

Reps I, II, III, IV, and V										*Reps VI, VII, VIII, IX, and X*								
1	2	3	4	5	6	7	8	9		1	2	3	4	5	6	7	8	9
2	6	8	3	1	4	9	5	7		2	6	5	3	7	8	4	9	1
3	8	5	9	7	2	1	4	6		3	5	1	2	9	7	8	4	6
7	4	9	2	3	5	6	1	8		5	1	4	8	2	3	9	6	7
8	1	2	6	4	7	3	9	5		9	8	6	7	4	5	1	3	2

Plan 9B.9 $t = 10,\ k = 3,\ r = 9,\ b = 30,\ \lambda = 2,\ E = .74$

Reps I, II, and III

1	2	3	4	5	6	7	8	9	10
2	5	7	1	8	4	9	10	3	6
3	8	4	6	7	9	1	2	10	5

Reps IV, V, and VI

1	2	3	4	5	6	7	8	9	10
2	3	4	9	7	8	10	1	5	6
4	6	8	5	1	9	3	10	2	7

Reps VII, VIII, and IX

1	2	3	4	5	6	7	8	9	10
3	7	8	2	6	1	9	4	10	5
5	6	9	10	3	8	2	7	1	4

Plan 9B.10 $t = 11,\ k = 2,\ r = 10,\ b = 55,\ \lambda = 1,\ E = .55$

Reps I and II

1	2	3	4	5	6	7	8	9	10	11
2	11	10	5	6	7	1	3	4	9	8

Reps III and IV

1	2	3	4	5	6	7	8	9	10	11
3	6	5	10	9	8	2	1	7	11	4

Reps V and VI

1	2	3	4	5	6	7	8	9	10	11
4	3	7	6	10	9	11	2	1	8	5

Reps VII and VIII

1	2	3	4	5	6	7	8	9	10	11
5	9	6	2	7	10	8	4	11	1	3

Reps IX and X

1	2	3	4	5	6	7	8	9	10	11
6	5	4	7	8	11	10	9	3	2	1

Plan 9B.11 $t = 11, k = 5, r = 5, b = 11, \lambda = 2, E = .88$

1	7	9	11	10	8	2	6	3	5	4
2	1	8	9	11	7	6	3	4	10	5
3	6	1	7	5	2	4	11	10	9	8
4	10	6	1	8	3	11	5	9	2	7
5	3	2	4	1	11	10	9	8	7	6

Plan 9B.12 $t = 11, k = 6, r = 6, b = 11, \lambda = 3, E = .92$

6	5	4	3	2	1	9	8	7	11	10
7	8	5	10	3	6	1	2	11	4	9
8	4	7	2	9	10	3	1	5	6	11
9	11	3	6	7	4	5	10	1	8	2
10	2	11	5	4	9	8	7	6	1	3
11	9	10	8	6	5	7	4	2	3	1

Plan 9B.13 $t = 13, k = 3, r = 6, b = 26, \lambda = 1, E = .72$

Reps I, II, and III

1	2	3	4	5	6	7	8	9	10	11	12	13
3	4	5	6	7	8	9	10	11	12	13	1	2
9	10	11	12	13	1	2	3	4	5	6	7	8

Reps IV, V, and VI

2	3	4	5	6	7	8	9	10	11	12	13	1
6	7	8	9	10	11	12	13	1	2	3	4	5
5	6	7	8	9	10	11	12	13	1	2	3	4

Plan 9B.14 $t = 13, k = 4, r = 4, b = 13, \lambda = 1, E = .81$

13	1	2	3	4	5	6	7	8	9	10	11	12
1	2	3	4	5	6	7	8	9	10	11	12	13
3	4	5	6	7	8	9	10	11	12	13	1	2
9	10	11	12	13	1	2	3	4	5	6	7	8

Plan 9B.15 $\quad t = 13, \ k = 9, \ r = 9, \ b = 13, \ \lambda = 6, \ E = .96$

2	3	4	5	6	7	8	9	10	11	12	13	1
5	6	7	8	9	10	11	12	13	1	2	3	4
6	7	8	9	10	11	12	13	1	2	3	4	5
7	8	9	10	11	12	13	1	2	3	4	5	6
9	10	11	12	13	1	2	3	4	5	6	7	8
10	11	12	13	1	2	3	4	5	6	7	8	9
11	12	13	1	2	3	4	5	6	7	8	9	10
12	13	1	2	3	4	5	6	7	8	9	10	11
13	1	2	3	4	5	6	7	8	9	10	11	12

Plan 9B.16 $\quad t = 15, \ k = 7, r = 7, \ b = 15, \ \lambda = 3, \ E = .92$

13	5	15	12	4	11	1	2	8	10	9	14	7	3	6
8	14	12	11	5	9	2	3	6	4	13	7	15	1	10
12	10	11	6	8	7	3	1	4	5	14	13	9	2	15
6	7	5	9	1	2	4	13	15	11	10	3	12	8	14
7	12	8	2	14	13	5	15	10	1	6	4	3	9	11
1	2	3	4	9	15	6	14	13	12	5	8	10	11	7
9	8	6	14	15	5	7	12	2	13	3	11	4	10	1

Plan 9B.17 $\quad t = 15, \ k = 8, \ r = 8, \ b = 15, \ \lambda = 4, \ E = .94$

11	4	9	15	7	6	12	10	5	8	1	2	13	14	3
4	1	2	3	13	10	9	11	14	15	7	6	8	5	12
2	3	14	1	10	12	13	9	7	6	11	15	5	4	8
5	15	4	10	12	1	14	7	3	2	8	9	11	6	13
10	13	7	8	11	4	15	6	1	3	2	5	14	12	9
3	11	1	13	2	14	10	8	9	7	4	12	6	15	5
14	6	10	7	3	8	11	5	12	9	15	1	2	13	4
15	9	13	5	6	3	8	4	11	14	12	10	1	7	2

9A.3 Appendix: Least Squares Estimates for BIB Designs

The linear model for the balanced incomplete block design is

$$y_{ij} = \mu + \tau_i + \rho_j + e_{ij} \tag{9A.1}$$

$$i = 1, 2, \ldots, t \quad j = 1, 2, \ldots, b$$

where μ is the general mean, τ_i is the treatment effect, ρ_j is the block effect, and the e_{ij} are random, independent experimental errors with mean 0 and variance σ^2.

Let n_{ij} be an indicator variable with $n_{ij} = 1$ if the ith treatment is in the jth block and $n_{ij} = 0$ otherwise.

The least squares estimates for the parameters in the full model minimize the sum of squares for experimental error:

$$Q = \sum_{i=1}^{t} \sum_{j=1}^{b} n_{ij} e_{ij}^2 = \sum_{i=1}^{t} \sum_{j=1}^{b} n_{ij}(y_{ij} - \mu - \tau_i - \rho_j)^2 \qquad (9A.2)$$

The normal equations for the balanced incomplete block design are

$$\mu: \quad N\widehat{\mu} + r \sum_{i=1}^{t} \widehat{\tau}_i + k \sum_{j=1}^{b} \widehat{\rho}_j = y_{..} \qquad (9A.3)$$

$$\tau_i: \quad r\widehat{\mu} + r\widehat{\tau}_i + \sum_{j=1}^{b} n_{ij}\widehat{\rho}_j = y_{i.} \qquad i = 1, 2, \dots, t \qquad (9A.4)$$

$$\beta_j: \quad k\widehat{\mu} + \sum_{i=1}^{t} n_{ij}\widehat{\tau}_i + k\widehat{\rho}_j = y_{.j} \qquad j = 1, 2, \dots, b \qquad (9A.5)$$

The set of t equations for the τ_i and the set of b equations for the ρ_j each sum to the equation for μ, resulting in two linear dependencies among the equations. The constraints $\Sigma_i \widehat{\tau}_i = 0$ and $\Sigma_j \widehat{\rho}_j = 0$ can be used to provide a unique solution.

After the constraints are imposed, the estimate of μ is $\widehat{\mu} = y_{..}/N$ from Equation (9A.3). The equations for ρ_j, Equations (9A.5), are used to eliminate the $\widehat{\rho}_j$ from the equations for the τ_i in Equations (9A.4). After elimination the equations with $\widehat{\tau}_i$ are

$$rk\widehat{\tau}_i - r\widehat{\tau}_i - \sum_{j=1}^{b} \sum_{\substack{p=1 \\ p \neq i}}^{t} n_{ij} n_{pj} \widehat{\tau}_p = ky_{i.} - \sum_{j=1}^{b} n_{ij} y_{.j} \qquad (9A.6)$$

The right-hand side of Equation (9A.6) is kQ_i, where Q_i is the adjusted treatment total used to compute the adjusted treatment sum of squares. Since

$$\sum_{j=1}^{b} n_{ij} n_{pj} = \lambda, \text{ for } p \neq i \qquad (9A.7)$$

and $n_{pj}^2 = n_{pj}$ ($n_{pj} = 0$ or 1), Equation (9A.6) can be written as

$$r(k-1)\widehat{\tau}_i - \lambda \sum_{\substack{p=1 \\ p \neq i}}^{t} \widehat{\tau}_p = kQ_i \qquad (9A.8)$$

Since condition $\Sigma_i \widehat{\tau}_i = 0$ has been imposed on the solution, the substitution

$$-\widehat{\tau}_i = \sum_{\substack{p=1 \\ p \neq i}}^{t} \widehat{\tau}_p$$

can be made in Equation (9A.8). With the equality $\lambda(t-1) = r(k-1)$, the equation for $\widehat{\tau}_i$ is

$$\lambda t \widehat{\tau}_i = kQ_i \tag{9A.9}$$

$$\widehat{\tau}_i = \frac{kQ_i}{\lambda t} \quad i = 1, 2, \ldots, t$$

10 Incomplete Block Designs: Resolvable and Cyclic Designs

A general description of and analyses for incomplete block designs were introduced in Chapter 9. Several major classes of useful incomplete block designs illustrated in this chapter include resolvable designs with blocks grouped in complete replications of treatments. Cyclic designs that can be constructed without the use of extensive tabled plans and the α designs that extend the number of available resolvable designs for experiments are also discussed.

10.1 Resolvable Designs to Help Manage the Experiment

Resolvable designs have blocks grouped such that each group of blocks constitutes one complete replication of the treatments. The grouping into complete replications is useful in the management of an experiment.

One of the early applications for resolvable designs occurred with plant-breeding trials placed in field plots on experimental farmland. Plant breeders wanted to test a large number of genetic lines and to make all comparisons among pairs of lines with equal precision. Arrangement of field plots into blocks smaller than a complete replication was necessary to reduce experimental error variance more so than was possible with the complete block designs. The resolvable incomplete block design was attractive because it not only reduced the block sizes for greater precision, but it also allowed the researcher to manage these large studies in the field on a replication-by-replication basis.

The resolvable designs can be useful in practice when the entire experiment cannot be completed at one time. The experiment in a resolvable design can be conducted in stages, with one or more replications completed at each stage. In addition, should the experiment for any reason be terminated prematurely there will be equal replication of all treatments.

The resolvable designs are arranged in r replicate groups of s blocks with k units per block. The number of treatments is a multiple of the number of units per block, $t = sk$, and the total number of blocks satisfies the relationship $b = rs \geq t + r - 1$ for balanced resolvable incomplete block designs.

Example 10.1   A Resolvable Design for Food Product Tests

Consider an experiment to test nine variations of a food product on the same day with human subjects as judges. Several factors associated with this experiment make an incomplete block design attractive. First, a judge can be expected to adequately discriminate at most four or five food product samples at any one time. Due to scheduling problems, it is not possible to have a sufficient number of judges available on any one day for adequate replication of the treatments. New food preparations must be made each day of the test, and it is necessary to ensure day-to-day variation in food preparation does not interfere with treatment comparisons in the experiment.

A test designed for three judges to each evaluate three of the food product variations on a given day is shown in Display 10.1. Three judges are scheduled on each of four test days. The three judges each constitute one incomplete block of three treatments. Each judge is randomly assigned to one of the blocks of three treatments and randomly presented with three food products for evaluation.

The design is resolvable with one complete replication of the experiment conducted on any given day. The design is balanced since each treatment occurs with every other treatment one time in the same block somewhere in the experiment. The judges only evaluate three food products at one time, reducing the within-block variation considerably from what would have resulted if a judge been required to evaluate nine products at one time.

Display 10.1   A Resolvable Balanced Incomplete Block Design to Evaluate Food Products

	Day I	Judge	Day II	Judge	Day III	Judge	Day IV
Judge							
1	(1, 2, 3)	4	(1, 4, 7)	7	(1, 5, 9)	10	(1, 6, 8)
2	(4, 5, 6)	5	(2, 5, 8)	8	(2, 6, 7)	11	(2, 4, 9)
3	(7, 8, 9)	6	(3, 6, 9)	9	(3, 4, 8)	12	(3, 5, 7)

Balanced Lattice Designs

The **balanced lattice designs** are a well-known group of resolvable designs introduced by Yates (1936b). The number of these square lattice designs available is limited because the number of treatments must be an exact square $t = k^2$. The designs require $r = (k + 1)$ replications and $b = k(k + 1)$ blocks for complete balance with $\lambda = 1$. Each replicate group contains $s = k$ blocks of k experimental units each. The number of units per block k must be a prime number or a power of a prime number. Plans for balanced lattice designs with $t = 9, 16, 25, 49, 64$, and 81 can be found in Cochran and Cox (1957) or Petersen (1985).

The design in Display 10.1 is a balanced lattice with nine treatments in blocks of three units. The design has four replication groups for a total of 12 blocks. Note that each treatment pair occurs together in the same block $\lambda = 1$ time somewhere in the balanced design.

Complete balance requires $(k + 1)$ replications with the lattice designs. Since they are resolvable designs, one or more of the replicate groups can be eliminated to produce a partially balanced lattice design. Designs with $r = 2$, 3, or 4 replicate groups are known as *simple, triple,* or *quadruple* lattices, respectively.

The average efficiency factor for a balanced lattice design is

$$E = \frac{(k + 1)(r - 1)}{r(k + 2) - (k + 1)}$$

and for simple and triple lattices the average efficiency factors are, respectively, $(k + 1)/(k + 3)$ and $(2k + 2)/(2k + 5)$.

Rectangular Lattice Designs

The restriction on numbers of treatments or varieties with square lattices is quite severe. The **rectangular lattice designs** developed by Harshbarger (1949) provided resolvable designs with treatment numbers intermediate to those provided by the square lattice designs.

The rectangular lattice designs accommodate $t = s(s - 1)$ treatments in blocks of size $k = (s - 1)$ units. The treatment numbers fall about midway between those provided by the square lattices. Cochran and Cox (1957) have tabled plans for rectangular lattices with $t = 12, 20, 30, 42, 56, 72$, and 90 treatments. Designs with $r = 2$ and $r = 3$ replicates are known as *simple* and *triple* rectangular lattices, respectively.

Other resolvable balanced incomplete block designs exist for numbers of treatments without the restrictions of the square or rectangular lattices; however, they exist for a limited set of block sizes and treatment numbers. Some balanced resolvable designs for $t \leq 11$ treatments are cataloged in Appendix 9A.1. Other balanced resolvable designs for $t > 11$ are given by Cochran and Cox (1957).

Analysis Outline for Resolvable Designs

The linear model for the resolvable design is

$$y_{ijm} = \mu + \tau_i + \gamma_j + \rho_{m(j)} + e_{ijm} \tag{10.1}$$

$$i = 1, 2, \ldots, t \quad j = 1, 2, \ldots, r \quad m = 1, 2, \ldots, s$$

where μ is the general mean, τ_i is the treatment effect, γ_j is the replicate group effect, $\rho_{m(j)}$ is the block nested within replication effect, and e_{ijm} is the random experimental error.

The sequential fit of alternative models outlined in Chapter 9 provides an orthogonal sum of squares partition. An additional sum of squares partition is required for replicate groups. The blocks are nested within replicate groups and the sums of squares for blocks from the replicate groups are pooled as SS(Blocks/Reps). The intrablock analysis of variance outline is shown in Table 10.1.

Table 10.1 Intrablock analysis for a resolvable balanced incomplete block design

Source of Variation	Degrees of Freedom	Sum of Squares
Total	$N - 1$	$\sum (y_{ijm} - \bar{y}_{...})^2$
Replications	$r - 1$	$sk \sum (\bar{y}_{.j.} - \bar{y}_{...})^2$
Blocks (unadjusted) within replications	$r(s - 1)$	$k \sum (\bar{y}_{.jm} - \bar{y}_{...})^2$
Treatments (adjusted)	$t - 1$	$\dfrac{k \sum Q_i^2}{\lambda t}$
Error	$N - t - rs + 1$	By subtraction

$Q_i = y_{i..} - (B_i/k)$, and B_i is the sum of block totals that include the ith treatment.

The treatments are orthogonal to the replicate groups since each treatment appears one time in each of the replicate groups. The treatment totals adjusted only for block effects provide unbiased estimates of the treatment means and a valid F test for treatment effects.

10.2 Resolvable Row–Column Designs for Two Blocking Criteria

Some research settings require two blocking criteria and do not permit complete blocks of treatments for either the row or column blocks required by the row orthogonal Youden squares. Row–column designs have the t treatments placed into blocks of $k = pq$ units. The treatments in each block are then arranged in a $p \times q$ row–column design.

A limited number of balanced row–column incomplete block designs exists for experiments that require block sizes of both rows and columns to be less than the number of treatments. The designs require a large number of blocks to meet the requirement of balance. See John (1971) for an extended bibliography of designs.

Example 10.2 A Nested Row–Column Design to Sample Insect Populations

Consider an epidemiological study on the transmission of a microbial parasite to humans by an insect in an agricultural area of the tropics. The microbial parasite is passed to the human when the insect bites the human. In turn, if the insect is not infected by the parasite it may become infected when it bites an infected human. A public health project is planning to test the effectiveness of a drug that is known to disrupt the parasite's life cycle in humans. An entomologist working with the project is going to sample the insect populations to monitor the effect the program has on reducing their infection rate.

Nine types of habitats in and around each of four agricultural plantations are scheduled for monitoring in the project. The habitats consist of locations such as flowing streams, villages on the plantation, the agricultural fields, stagnant water ponds, and so forth. The entomologist can only collect insect samples at three such habitat sites in a single day. A row–column sampling design is set up to control the potential variation caused by sampling on different days and different times of day at each of four plantations. The design is shown in Display 10.2. The design has nine habitat treatments sampled in a 3×3 row–column design at four plantations. The three rows in the 3×3 array are the time of day the samples are collected, and the three columns are the three days for collection at each plantation.

Upon inspection of each habitat treatment pair it can be verified that each is sampled once at the same time and once on the same day in the design. Each treatment pair must occur together one time in the same row *and* one time in the same column somewhere in the design to have a balanced row–column design. The design is a resolvable design with plantations as replications and one complete replication of the nine habitat types at each plantation. It is a nested design wherein the rows and columns are nested within the replicate groups.

Balanced Lattice Square Designs

These designs, introduced by Yates (1940b), are resolvable nested row–column designs developed early in the history of incomplete block designs. The example exhibited in Display 10.2 is such a design. The **balanced lattice square** has $t = k^2$ treatments laid out in a $k \times k$ row–column design. One $k \times k$ array consists of one complete replication of the treatments with $p = q = k$. The balanced design requires $r = (k + 1)$ replications for complete balance such that every treatment pair occurs together one time in the same row *and* one time in the same column

---

Display 10.2 **A Nested 3 × 3 Row–Column Incomplete Block Design to Sample Insect Populations**

	Day				Day		
Time	(1)	(2)	(3)	Time	(4)	(5)	(6)
(1)	1	2	3	(4)	1	4	7
(2)	4	5	6	(5)	2	5	8
(3)	7	8	9	(6)	3	6	9
	Plantation I				Plantation II		

	Day				Day		
Time	(7)	(8)	(9)	Time	(10)	(11)	(12)
(7)	1	6	8	(10)	1	9	5
(8)	9	2	4	(11)	6	2	7
(9)	5	7	3	(12)	8	4	3
	Plantation III				Plantation IV		

---

somewhere in the experiment. When k is an odd value, a replication value of $r = \frac{1}{2}(k+1)$ will provide semi-balance such that any pair of treatments occurs together one time in the same row *or* one time in the same column.

Plans for lattice square designs with $t = 9, 16, 25, 49, 64, 81, 121,$ and 169 may be found in Cochran and Cox (1957) and Petersen (1985). Extensive discussions of the designs are found also in Kempthorne (1952) and Federer (1955).

Analysis Outline for Nested Row–Column Designs

The linear model for any resolvable nested row–column design is

$$y_{ijlm} = \mu + \beta_m + \rho_{j(m)} + \gamma_{l(m)} + \tau_i + e_{ijlm} \tag{10.2}$$

$$i = 1, 2, \ldots, t \quad j = 1, 2, \ldots, p \quad l = 1, 2, \ldots, q \quad m = 1, 2, \ldots, r$$

where μ is the general mean, τ_i is the treatment effect, β_m is the replicate group effect, $\rho_{j(m)}$ and $\gamma_{l(m)}$ are, respectively, the row and column nested in replicate group effect, and e_{ijlm} is the random experimental error.

Each of the treatments occurs in each replicate group, and the treatments are orthogonal to replicate groups. The rows and columns also are orthogonal to replicate groups. The sums of squares partitions for replications, rows within replications, and columns within replications can be computed in the usual manner, as shown in the analysis of variance outline in Table 10.2. The treatments are not orthogonal to rows or columns of the replicate groups. The treatments must be adjusted for both row and column effects.

The adjusted treatment totals required for least squares estimates of treatment effects and the adjusted treatment sum of squares are

Table 10.2   Intrablock analysis for a nested row–column incomplete block design

Source of Variation	Degrees of Freedom	Sum of Squares
Total	$N-1$	$\sum(y_{ijlm}-\bar{y}_{....})^2$
Groups (replications)	$r-1$	$t\sum(\bar{y}_{...m}-\bar{y}_{....})^2$
Columns (unadjusted) within groups	$r(q-1)$	$p\sum(\bar{y}_{.lm}-\bar{y}_{...m})^2$
Rows (unadjusted) within groups	$r(p-1)$	$q\sum(\bar{y}_{.j.m}-\bar{y}_{...m})^2$
Treatments (adjusted)	$t-1$	$\dfrac{pq(t-1)}{rt(p-1)(q-1)}\sum Q_i^2$
Error	$r(p-1)(q-1)-(t-1)$	By subtraction

$$Q_i = y_{i...} + r\bar{y}_{....} - \frac{1}{p}R_i - \frac{1}{q}C_i \tag{10.3}$$

where $y_{i...}$ is the total for the ith treatment, $\bar{y}_{....}$ is the grand mean for the experiment, R_i is the sum of row totals that includes the ith treatment, and C_i is the sum of column totals that include the ith treatment.

The least squares estimate of the treatment effect is

$$\hat{\tau}_i = \frac{pq(t-1)Q_i}{rt(p-1)(q-1)} \tag{10.4}$$

and the least squares estimate of the treatment mean is $\hat{\mu}_i = \hat{\mu} + \hat{\tau}_i$, where $\hat{\mu} = \bar{y}_{....}$ is the estimate of the general mean.

The efficiency factor for the nested row–column design is

$$E = \frac{t(p-1)(q-1)}{pq(t-1)} \tag{10.5}$$

where for the balanced lattice square the efficiency factor is $E = (k-1)/(k+1)$, with $p = q = k$ and $t = k^2$. For example, with $k = 3$ in Display 10.2, the efficiency factor is $E = 2/4 = .50$. The balanced lattice square would have to reduce experimental error variance by 50% to estimate comparisons between treatment means as precisely as a complete block design of the same size.

10.3   Cyclic Designs Simplify Design Construction

Extensive tables of design plans must be available to use the incomplete block designs discussed to this point. Setting up any design in the physical facility required for the experiment demands constant vigilance. Care must be taken to avoid mistakes when treatments are assigned to experimental units and when data are recorded. Complex incomplete block designs can add further complication to the process if there is a need for constant reference to the design layout.

Cyclic designs offer a degree of simplicity in the construction of the design and implementation in the experimental process. Cyclic designs are generated from an initial block of treatments so it is not necessary to table complete plans of the experimental layout. Once the initial block is known the experimental plan can be generated, and the treatments can be randomly assigned to the numerical labels in the design.

How to Construct a Cyclic Design

The *cyclic* designation refers to the method used to construct the designs as well as to the association scheme among the treatments. The treatments are assigned to the blocks through a cyclic substitution of treatment labels from an initial generating block.

The method requires that the treatments be labeled as $(0, 1, 2, \ldots, t-1)$. The cycle begins with an initial block of size k required by the experiment. Succeeding blocks are obtained by adding 1 to each treatment label in the previous block. If a treatment label exceeds $(t-1)$ it is reduced by modulo t. The cycle continues until it returns to the configuration of the initial block.

Consider the design for six treatments in blocks of three units generated by cyclic substitution in Display 10.3. The six treatments are labeled $(0, 1, 2, 3, 4, 5)$ for the design. A cyclic design is constructed with the initial block $(0, 1, 3)$. The second block is constructed by adding 1 to each treatment label in the first block which gives $(1, 2, 4)$ as the configuration for the second block. Add 1 to each treatment label in the second block to provide $(2, 3, 5)$ as the treatments for the third block.

	Design 10.3 A Cyclic Design for Six Treatments in Blocks of Three Units
Block	
1	$(0, 1, 3)$ initial block
2	$(1, 2, 4)$
3	$(2, 3, 5)$
4	$(3, 4, 0)$
5	$(4, 5, 1)$
6	$(5, 0, 2)$

The transition from block 3 to block 4 in the cycle requires one label in block 4 to be reduced by modulo $t = 6$. If 1 is added to each label in block 3 the configuration of block 4 will be $(3, 4, 6)$ where $5 + 1 = 6$ exceeds $(t-1) = 5$. Therefore, 6 must be reduced modulo 6, which is $6 = 0(\text{mod } 6)$, and 0 replaces 6 in the block 4 configuration to give $(3, 4, 0)$. A similar replacement occurs in the transition from block 5 to block 6.

A full set of $b = 6$ blocks is constructed in the complete cycle shown in the display. The numbers make one complete cycle through the treatment labels down each of the columns in the display. The treatment labels in the first position of the blocks cycle from 0 to 5 while the second position cycle begins with the label 1 and the third position cycle begins with the label 3. The design is complete with blocks 1 through 6 since the seventh member of the cycle would be the initial block configuration, $(0, 1, 3)$.

Construct Row–Column Designs with the Cyclic Method

Cyclic designs can be used for row–column designs if the experiment requires two blocking criteria. The only requirement is that the number of replications be some multiple of the block size, or $r = ik$, where i is an integer value. The design in Display 10.3 with $t = 6$ and $r = k = 3$ generated from the initial block $(0, 1, 3)$ can be used as a row–column design with six rows and three columns.

The design is a column orthogonal design since each treatment appears in each of the columns. If the row and column integrity is maintained upon randomization, then a sum of squares for columns and rows is partitioned out of the total sum of squares. The treatments are adjusted for row blocks in the analysis.

Tables to Construct Cyclic Designs

A compact table of initial blocks required for relatively efficient cyclic designs with $4 \leq t \leq 15$ treatments, from John (1987), can be found in Appendix Table 10A.1. An additional table for $16 \leq t \leq 30$ treatments also was provided by John (1987). Other tables can be found in John (1966).

For a given block size k, the first $k - 1$ treatments in the initial block are all taken to be the same for any number of treatments t. The final treatment in the block is taken as the corresponding value in the column for t in Appendix Table 10A.1. For example, with $k = 3$, $r = 3$, the initial block for $t = 5$ is $(0, 1, 2)$ and for $t = 6$ the initial block is $(0, 1, 3)$.

If a design with more replications is required ($r > k$), a second- or third-generating block is given in the table. A set of blocks is constructed from each by cyclic substitution. The design with $t = 6$ treatments in blocks of size $k = 3$ generated with the initial block $(0, 1, 3)$ had $r = 3$ replications. The design with $r = 6$ replications is constructed by adding the initial block $(0, 2, 1)$ found in Table 10A.1 on the line with $k = 3$ and $r = 6$. The resulting design has $b = 12$ blocks, half of them generated from each of the initial blocks as shown in Display 10.4. The design is partially balanced with $\lambda_1 = 3$ and $\lambda_2 = 2$ for the two associate classes.

> ### Display 10.4   A Cyclic Design Generated from
> ### Two Initial Blocks
>
> | Block | | Block | |
> |---|---|---|---|
> | 1 | (0, 1, 3) initial block | 7 | (0, 2, 1) initial block |
> | 2 | (1, 2, 4) | 8 | (1, 3, 2) |
> | 3 | (2, 3, 5) | 9 | (2, 4, 3) |
> | 4 | (3, 4, 0) | 10 | (3, 5, 4) |
> | 5 | (4, 5, 1) | 11 | (4, 0, 5) |
> | 6 | (5, 0, 2) | 12 | (5, 1, 0) |

α Designs for Versatile Resolvable Designs from Cyclic Construction

The α designs are resolvable designs developed by Patterson and Williams (1976) in response to the requirements for statutory field trials of agricultural field crop varieties in the United Kingdom. The number of varieties and replications were fixed by statutory requirements and were not under the control of the experimenters. The large number of varieties necessitated incomplete block designs, and resolvable designs were required for proper management in the field. Existing resolvable designs did not always accommodate the number of varieties or block sizes required by the trials.

α Design Features

The designs have no limitation on block size except for the constraint that the number of treatments, t, must be a multiple of block size k, so that $t = sk$, to have a resolvable design with equal block sizes. Tabled plans of the complete designs are unnecessary since they can be generated by cyclic substitution from an initial array of numbers much like the cyclic designs.

The development of the α designs greatly increased the number and flexibility of resolvable designs available for experimental trials. The designs originally developed by Patterson and Williams (1976) for the statutory trials included designs with $r = 2$, 3, or 4 replications with $t \leq 100$ treatments and block sizes $4 \leq k \leq 16$.

The designs are partially balanced with two or three associate classes. The designs with two associate classes having $\lambda_1 = 0$ and $\lambda_2 = 1$ are denoted $\alpha(0, 1)$ designs, and designs with three associate classes having $\lambda_1 = 0$, $\lambda_2 = 1$, and $\lambda_3 = 2$ are denoted $\alpha(0, 1, 2)$ designs. Most of the designs considered for the variety trials were $\alpha(0, 1)$ designs. Because of the manner in which the designs are constructed neither balanced designs nor $\alpha(1, 2)$ designs exist.

The upper bound for the efficiency of the α designs provided by Patterson and Williams (1976) is

$$E = \frac{(t-1)(r-1)}{(t-1)(r-1) + r(s-1)} \qquad (10.6)$$

How to Construct α Designs

The construction of a design for $t = sk$ treatments in r replications begins with a $k \times r$ array designated as the *generating* α array. Each column of the generating α array is utilized to construct $s - 1$ additional columns by cyclic substitution. The new $k \times rs$ array is denoted as the *intermediate* α^* array.

Consider a 4×3 generating α array for $t = 12$ treatments, $k = 4$ units per block, $r = 3$ replications, and $s = 3$ blocks per replication.

Generating α Arrays

Column		
1	*2*	*3*
0	0	0
0	0	2
0	2	1
0	1	1

Each column is used to generate $s - 1 = 2$ additional columns by cyclic substitution with mod$(s) =$ mod(3) to derive the intermediate α^* array. The result will be $r = 3$ arrays of size $k \times s$. In this example, there will be three 4×3 arrays. The intermediate α^* arrays generated from each of the three columns of the α array are

Intermediate α* Arrays

Array Generated From								
Column 1			*Column 2*			*Column 3*		
0	1	2	0	1	2	0	1	2
0	1	2	0	1	2	2	0	1
0	1	2	2	0	1	1	2	0
0	1	2	1	2	0	1	2	0

Finally, treatment labels are obtained by adding $s = 3$ to each element of the second row of the intermediate α^* array, $2s = 6$ to each element of the third row, and $3s = 9$ to the fourth row. The columns of the arrays are the blocks of the design, and each set of s columns generated from a generating α array column constitutes a complete replication. The resulting $\alpha(0, 1, 2)$ design is shown in Display 10.5.

The treatment labels for the design are $(0, 1, 2, \ldots, t - 1)$, and the actual treatments are randomly allocated to the labels. In addition, randomization proceeds with a random allocation of the design blocks to the actual blocks within the

**Design 10.5   An α Design for 12 Treatments
in Three Replication Groups**

| Replicate | I | | | II | | | III | | |
Block	*1*	*2*	*3*	*1*	*2*	*3*	*1*	*2*	*3*
	0	1	2	0	1	2	0	1	2
	3	4	5	3	4	5	5	3	4
	6	7	8	8	6	7	7	8	6
	9	10	11	10	11	9	10	11	9

replicate groups and a random allocation of the treatments to the actual experimental units within the physical blocks.

A table of 11 basic α arrays reproduced from Patterson, Williams, and Hunter (1978) is shown in Appendix Table 10A.2. A total of 147 α designs can be generated from the 11 arrays. The arrays can be used to generate designs for treatment numbers $20 \leq t \leq 100$ with $r = 2$, 3, or 4 replications and the usual constraint that $t = sk$. The arrays will produce designs with block sizes $k \geq 4$ with the condition $k \leq s$.

Consider the first array listed in Appendix Table 10A.2 for $s = k = 5$. The array has $k = 5$ rows and $r = 4$ columns.

$$
\begin{array}{cccc}
0 & 0 & 0 & 0 \\
0 & 1 & 4 & 2 \\
0 & 2 & 3 & 4 \\
0 & 3 & 2 & 1 \\
0 & 4 & 1 & 3
\end{array}
$$

This particular array can be used to generate an α design with $r = 2$ replications by utilizing only the first two columns of the array. Likewise, designs for three or four replications can be generated by using the first three or all four columns, respectively. A design with block size $k = 5$ is obtained by utilizing all five rows of the array. A design with block size $k = 4$ is constructed by utilizing the first four rows of the array. With $s = 5$ a design may then be constructed for $t = 25$ treatments in blocks of $k = 5$ units with $r = 2$, 3, or 4 replications. Also with $s = 5$ a design may be constructed for $t = 20$ treatments in blocks of $k = 4$ units with $r = 2$, 3, or 4 replications.

Latinized Resolvable Incomplete Block Designs

The resolvable designs discussed to this point have a nested blocking structure. The incomplete blocks are randomized independently within each replicate whether the designs have a single blocking criterion or two blocking criteria with rows and columns.

In some instances, the blocks of one replicate may have a relationship to blocks in another replicate. These relationships could exist in experiments wherein the blocks in different replicates are contiguous in space, as illustrated in Display 10.6.

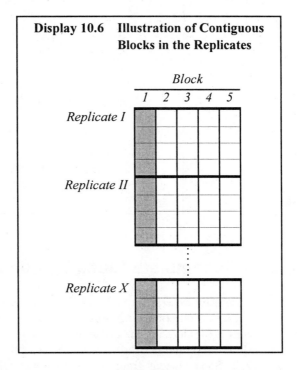

Display 10.6   Illustration of Contiguous Blocks in the Replicates

If the spatial layout is that shown in Display 10.6, then any single block—e.g., Block 1—in each replicate occurs in the same spatial line as Block 1 in each of the other replicates. The result is the occurrence of a "long" block in the vertical direction of the layout in Display 10.6.

If blocks are randomized separately in each replicate, it is entirely possible for all replications of one or more of the treatments to occur in the same "long" block. This could create an undesirable circumstance in the case of unforeseen disturbances which negated the use of the "long" block. To avoid this possibility Harshbarger and Davis (1952) introduced *Latinized* rectangular lattice designs. The rectangular lattice was structured such that each treatment occurred only once in each long block.

Williams (1986) presented results on more general designs in which treatments do not occur more than one time in the long blocks and extended the Latinized designs to include row–column blocking. The α design can be used to generate the more general Latinized designs as illustrated by John and Williams (1995) with a Latinized row–column design shown in Display 10.7. Notice that treatments appear no more than one time in each of the long blocks (columns). John and Williams (1995) provide a thorough discussion of cyclic and resolvable designs.

Display 10.7	A Latinized Row–Column Design			

	Block			
	1	*2*	*3*	*4*
Replicate I	0	1	2	3
	8	5	10	7
	4	9	6	11
Replicate II	6	0	3	5
	1	7	4	2
	11	10	9	8
Replicate III	9	8	0	1
	7	6	5	4
	2	3	11	10

10.4 Choosing Incomplete Block Designs

A successful experiment will include the correct treatment design to address specific research questions and provide precise estimates of the treatment means and contrasts of interest to the investigator. The incomplete block design often is necessary to obtain the best possible conditions for comparisons among the treatments with as homogeneous groups of experimental units as possible.

The investigator's knowledge of the experimental conditions and material is indispensable in constructing suitable block sizes and composition. Each experiment presents its own challenges and conditions, and no set rules provide the correct design for every experiment. A knowledgeable investigator can craft a successful experiment with the intelligent application of sound design principles to the experimental material.

The guiding principle in selecting or constructing a design for a specific research problem is to make every attempt to choose a design that accommodates the research problem. In most cases, appropriate designs are available to accommodate the requirements of the experiments because such a large number of published plans for balanced incomplete block designs are available in the literature as a result of the research on design construction.

The systematic methods to construct balanced incomplete block designs require mathematics beyond the scope of this book. Discussions of research on design existence and construction can be found in John (1971), John (1987), and John and Williams (1995).

Where to Find Design Plans

Computer programs provide the most convenient method to develop an incomplete block design plan; CycDesigN (Whitaker, Williams, & John, 1998) and ALPHA+ (Williams & Talbot, 1993) are two examples of programs that generate designs discussed in Chapters 9 and 10.

Tabled plans for small experiments with $t \leq 11$ treatments are found in Appendix 9A.1. Plans for other numbers of treatments are given in Cochran and Cox (1957), Cox (1958), Fisher and Yates (1963), Box, Hunter, and Hunter (1978), and Mason, Gunst, and Hess (1989). Plans for balanced and partially balanced lattice, rectangular lattice, and lattice square designs may be found in Cochran and Cox (1957) and Petersen (1985). Plans for designs with row–column blocking as row orthogonal incomplete Latin squares may be found in Youden (1940), Cochran and Cox (1957), Cox (1958), and Petersen (1985). Some of the plans for row orthogonal designs are given in Appendix 9A.2. A large number of partially balanced incomplete block designs can be found in Clatworthy (1973) and Bose, Clatworthy, and Shrikande (1954).

An Informal Approach to Design Construction

In the cases where a standard design does not accommodate the research problem, it may be necessary to find an innovative design solution to avoid alterations in the treatment design. The need for occasional innovation argues for a more informal approach to experiment design (Mead, 1988). The primary reason for advocating the informal design is to avoid the trap of changing the treatment design or the research problem to fit some particular tabled design. Changing a research problem to fit the experiment design would be unwise under any circumstances. However, it is advantageous to understand some of the more formal properties of experiment design to fully appreciate innovative approaches and avoid possible pitfalls of the informal approach. The investigator must be quite familiar with the experimental conditions and material to construct a design appropriate to the study.

The construction can be quite informal if great care is taken in the placement of treatments with experimental units to avoid mishaps. Particular attention must be paid to the amount of information available for various treatment comparisons as the design is constructed. The relative differences in variances of the pertinent treatment comparisons must be evaluated prior to actual experimentation. Such an approach is advocated by Mead (1988).

The Efficiency Factor Increases with Block Size

We expect the experimental error variance to decrease as the block size decreases. The decrease in σ^2 with smaller block sizes depends on the success we have in placing the experimental units in the correct blocking arrangement. However, a close inspection of the efficiency factor suggests taking as large a block size as possible. The efficiency measure for a balanced incomplete block design can be expressed as

$$E = \frac{\lambda t}{rk} = \frac{t}{t-1} \left(1 - \frac{1}{k}\right) \tag{10.7}$$

The value of E increases with increasing block size k. A cursory scan of the list of balanced incomplete block designs in Appendix 9A.1 will confirm this relationship. A similar relationship is true for the partially balanced incomplete block designs.

Let Resources Dictate Block Size

The available resources dictate the choice of block size. In some experiments the block size is constrained to the available instruments for the experiment as was the case for the methyl glucoside vinylation experiment in Example 9.2. The investigator only had three chambers available to test five pressure treatments. With $t = 5, r = 6, k = 3$, and $\lambda = 3$, the efficiency factor for the balanced incomplete block design was $E = 3(5)/6(3) = 0.83$. Equivalent precision to a complete block design with five pressure chambers would require a 17% reduction in σ^2 for the incomplete block design. However, the only choice available to the investigator was the incomplete block design with three treatments per block.

Some experimental conditions allow greater flexibility in the choice of block sizes. Variety yield trials are placed in field plots on experimental farmland. In many of these situations the block sizes can be of any convenient size. The guiding principle is to have a sensible block size that provides as much homogeneity among the plots within blocks as is necessary for precise comparisons among the varieties. In this case the block size should be as large as possible while maintaining the required precision for variety comparisons.

Partially Balanced Designs Increase Efficient Use of Resources

Partially balanced designs can be used in place of balanced designs to utilize experimental resources more efficiently. Consider the partially balanced design in Display 9.3. The block sizes in the experiment were constrained to four experimental units. A balanced design would have required ten replications of the six treatments and 15 blocks. The 3 blocks of the partially balanced design provide two replications. If more replications are required the partially balanced design can be repeated. Unless ten replications are required the partially balanced design requires fewer resources than the balanced design.

The investigator had to make some sacrifice to have the smaller experiment. The partially balanced design provides more precision for some comparisons than others because not all the treatment pairs occur in the same number of blocks. However, with a prudent choice of design it is still possible to have a partially balanced design that provides near balance in all comparisons.

The partially balanced design in Display 9.3 had some treatment pairs occurring in $\lambda_1 = 2$ blocks and other treatment pairs occurring in $\lambda_2 = 1$ block. The efficiency for the comparisons among first and second associates can be shown to be $E_1 = 1.00$ and $E_2 = 0.86$, respectively, with an average efficiency of $E = 0.88$

(Clatworthy, 1973). There is near equal precision for both sets of associates, and only a small amount of precision is sacrificed with the partially balanced design.

α Designs Relieved Constraints for Large Experiments

The incomplete block designs originated with experimental trials that included relatively large numbers of treatments such as variety trials. The primary objectives of the incomplete block designs were to reduce the block sizes to reduce experimental error variance and to have experiments that could be managed on a replicate group basis. In agricultural field trials the block sizes were somewhat flexible as long as a reduction in block size managed to reduce the error variance. The original square and rectangular lattice designs provided a degree of relief for these situations. The variety numbers for these trials were still quite restrictive even though it often was possible for the investigators to adjust the number of included varieties without dire consequences.

The α designs relieved the constraint on the number of treatments for resolvable incomplete block designs to a considerable extent. They do still restrict the number of treatments to be a product of the number of units per block and the number of blocks in a replication group. However, it is now possible to have a resolvable design for a greater variety of treatment, block, and replication sizes. In addition, the α designs can be generated from a relatively simple set of tables.

Factorial Treatment Designs Require Special Attention

Structured treatment designs such as factorials require special attention when constructing incomplete block designs. The construction of incomplete block designs with factorial effects confounded with blocks will be considered in Chapter 11. These designs are particularly useful for the 2^n series of factorials that have many factors because no great sacrifice results from completely confounding large order interactions with blocks.

Cyclic Designs Are Helpful with Unstructured Treatment Designs

Balanced designs or resolvable designs can in some circumstances strain available resources because of the required number of replications. The more flexible partially balanced designs can be used in place of the balanced design with little loss in precision. However, the tables of partially balanced designs may not be readily available because many of them appear in older publications. Many of the tabled designs found in these publications can be constructed by cyclic substitution. Cyclic designs may be one of the best alternatives for producing an efficient design. They provide designs of many different sizes and are constructed easily from one or two initial blocks of treatments. The tables of initial blocks for cyclic designs appear in somewhat more recent publications than the other classes of designs.

EXERCISES FOR CHAPTER 10

1. A horticulturalist conducted a field test of $t = 8$ broccoli varieties with a resolvable balanced incomplete block design in a field experiment. The design had $k = 4$ plots per block and $s = 2$ blocks per replication. There were $r = 7$ replicate groups in the design. The data are pounds of broccoli harvested per plot. (R = replication, B = block within replication, T = variety)

R	B	T	Yield	R	B	T	Yield	R	B	T	Yield
1	1	1	46.5	3	2	2	52.7	6	1	1	45.6
1	1	2	55.7	3	2	4	52.5	6	1	3	37.0
1	1	3	37.7	3	2	5	46.4	6	1	5	49.5
1	1	4	50.3	3	2	7	46.3	6	1	7	45.4
1	2	5	43.1	4	1	1	40.7	6	2	2	48.2
1	2	6	47.6	4	1	4	53.0	6	2	8	54.0
1	2	7	35.5	4	1	6	45.0	6	2	4	47.4
1	2	8	45.9	4	1	7	38.0	6	2	6	53.8
2	1	1	40.1	4	2	2	56.1	7	1	1	44.6
2	1	2	55.8	4	2	3	39.0	7	1	4	52.4
2	1	7	39.7	4	2	5	54.7	7	1	5	50.2
2	1	8	51.7	4	2	8	48.5	7	1	8	52.0
2	2	3	41.2	5	1	1	44.1	7	2	2	56.8
2	2	4	61.7	5	1	2	56.6	7	2	3	37.8
2	2	5	49.8	5	1	5	44.7	7	2	6	45.7
2	2	6	53.6	5	1	6	51.7	7	2	7	42.6
3	1	1	42.3	5	2	3	39.0				
3	1	3	43.8	5	2	4	47.8				
3	1	6	45.6	5	2	7	41.6				
3	1	8	51.0	5	2	8	49.4				

a. Write a linear model for this experiment, describe the terms, and conduct the intrablock analysis.
b. Compute the standard error estimate of the difference between least squares estimates of two treatment means.
c. What is the efficiency factor for this design?
d. Write a summary of your analysis results and evaluation of the design.

2. An agronomist conducted an alfalfa variety trial in a quadruple lattice design. There were 25 varieties grown in four replicate groups. Each replicate group consisted of five blocks of five varieties in a square lattice. The yield data that follow are pounds of alfalfa hay harvested per plot. The variety numbers appear in parentheses.

Block	Replicate I									
1	19.1	(2)	20.4	(1)	23.2	(4)	19.3	(3)	21.4	(5)
2	18.6	(8)	18.3	(6)	21.3	(10)	12.0	(9)	19.3	(7)
3	20.8	(15)	19.5	(11)	20.8	(13)	20.3	(12)	19.0	(14)
4	21.0	(17)	19.4	(16)	19.7	(19)	17.5	(18)	20.2	(20)
5	19.6	(24)	19.0	(23)	19.4	(21)	20.6	(25)	20.3	(22)

Block	Replicate II									
1	19.4	(11)	19.8	(21)	18.4	(6)	21.5	(1)	19.7	(16)
2	18.6	(17)	20.8	(2)	20.9	(12)	19.7	(7)	20.0	(22)
3	16.6	(18)	19.8	(3)	18.2	(23)	19.1	(8)	20.7	(13)
4	21.8	(4)	20.6	(14)	20.0	(19)	16.8	(9)	19.5	(24)
5	21.3	(15)	19.6	(20)	20.1	(25)	20.4	(10)	20.7	(5)

Block	Replicate III									
1	21.1	(1)	20.1	(13)	20.3	(7)	20.8	(25)	22.2	(19)
2	19.5	(21)	19.1	(14)	20.5	(20)	20.5	(2)	20.3	(8)
3	13.5	(9)	23.1	(22)	20.5	(3)	21.2	(15)	18.6	(16)
4	22.6	(10)	17.2	(23)	22.2	(4)	17.0	(17)	17.8	(11)
5	20.0	(24)	17.7	(18)	18.7	(5)	19.8	(12)	16.4	(6)

Block	Replicate IV									
1	20.2	(12)	22.6	(1)	19.5	(23)	22.4	(20)	13.6	(9)
2	22.5	(24)	21.9	(16)	22.5	(10)	21.1	(13)	19.3	(2)
3	19.2	(25)	20.0	(3)	17.1	(6)	21.0	(17)	18.9	(14)
4	21.7	(15)	20.0	(21)	18.2	(7)	17.3	(18)	20.9	(4)
5	17.3	(8)	19.3	(22)	20.7	(5)	19.2	(11)	16.9	(19)

Source: Dr. M. Ottman, Department of Plant Sciences, University of Arizona.

a. Write a linear model for the experiment, explain the terms, and conduct the intrablock analysis for a simple, triple, or quadruple lattice design (instructor discretion).

b. Compute the standard error estimates of the difference between intrablock estimates of two variety means that are first associates and between those that are second associates. Also compute an average standard error.

c. What is the efficiency factor for the simple or triple lattice design? Interpret the efficiency factor.

d. Write a summary of your analysis results and evaluation of the design.

3. An agronomist conducted a wheat variety test in a balanced lattice square design. There were $t = 9$ varieties in a 3×3 balanced lattice square in $r = 4$ replicate groups. The wheat yields are shown in the table that follows in the row–column arrangements for each replicate group. Variety numbers appear in parentheses.

	Replicate I	
53.5 (6)	53.2 (4)	57.7 (5)
53.1 (3)	58.6 (1)	53.9 (2)
57.2 (9)	55.0 (7)	51.5 (8)

	Replicate II	
53.7 (4)	53.6 (2)	57.8 (9)
54.5 (3)	52.8 (7)	53.3 (5)
48.9 (8)	53.5 (6)	56.7 (1)

	Replicate III	
49.4 (8)	54.7 (4)	55.6 (3)
54.5 (6)	54.2 (2)	54.4 (7)
59.7 (1)	55.7 (9)	54.1 (5)

	Replicate IV	
54.0 (7)	57.2 (1)	53.2 (4)
56.9 (9)	54.8 (3)	55.4 (6)
48.9 (8)	53.4 (2)	55.9 (5)

a. Write a linear model for this experiment, describe the terms, and conduct the intrablock analysis for this experiment.

b. Compute the standard error estimate of the difference between least squares estimates of two treatment means.

c. What is the efficiency factor for this design?

d. Write a summary of your analysis results and evaluation of the design.

4. A variety trial was conducted in an $\alpha(0, 1, 2)$ resolvable design. There were $t = 18$ varieties in $r = 4$ replicate groups. There were $s = 3$ blocks of $k = 6$ varieties in each replicate. The yield data follow with the variety numbers in parentheses. Varieties 1 and 5 were control varieties.

Block	*Replicate I*											
1	88.2	(5)	82.5	(10)	84.3	(15)	87.0	(6)	84.5	(12)	88.9	(8)
2	82.4	(1)	82.9	(14)	83.1	(3)	84.7	(13)	83.3	(16)	89.0	(4)
3	93.1	(2)	82.7	(11)	88.9	(17)	88.6	(18)	84.1	(9)	87.5	(7)

Block	*Replicate II*											
1	85.4	(4)	73.0	(11)	84.2	(7)	80.3	(14)	79.6	(10)	86.0	(6)
2	87.9	(8)	85.1	(9)	79.4	(18)	80.7	(13)	89.3	(5)	81.5	(3)
3	82.4	(1)	88.5	(2)	87.0	(12)	85.4	(17)	85.9	(15)	79.1	(16)

Block	*Replicate III*											
1	83.6	(6)	79.4	(17)	81.3	(4)	80.5	(9)	80.9	(8)	79.3	(1)
2	80.4	(7)	88.2	(5)	82.3	(14)	88.0	(12)	90.0	(2)	83.6	(3)
3	81.4	(18)	84.8	(15)	81.0	(10)	81.2	(13)	79.1	(11)	83.8	(16)

Block	*Replicate IV*											
1	80.5	(16)	77.1	(11)	84.4	(17)	90.4	(6)	82.9	(14)	83.0	(12)
2	87.9	(8)	78.9	(18)	81.4	(1)	83.5	(2)	82.2	(15)	79.0	(3)
3	84.2	(7)	83.0	(10)	87.6	(9)	81.7	(13)	91.3	(5)	87.4	(4)

Source: P. Seeger, Department of Statistics, The Swedish University of Agricultural Sciences.

a. Identify the first, second, and third associates of variety 1, where $\lambda_1 = 0$, $\lambda_2 = 1$, and $\lambda_3 = 2$. How many of each are there, and which varieties are in each group of associates?

b. Write a linear model for the experiment, explain the terms, and conduct the intrablock analysis to obtain intrablock estimates of the variety means and their standard error estimates.

c. The α designs have a multiplicity of standard errors for the differences between pairs of estimated treatment means. If your program is capable, compute the standard error estimates of the difference between intrablock estimates of the control variety means, varieties 1 and 5, and the other varieties; the average standard error for the first, second, and third associates; and an overall average standard error.

d. Compute the standard error estimates of the difference between the variety means; an average standard error of the difference for first, second, and third associates; and an overall average standard error.

e. Write a summary of your analysis results and evaluation of the design.

5. A human factors study is to be conducted on speed of perception relative to object shape in statistical graphs. Eight shapes will be used as treatments in the study. It is thought that a subject should evaluate no more than four shapes in any one sitting. Researchers have decided to use an incomplete block design with subjects as blocks.

a. Construct a cyclic design with $r = 4$ replications of $t = 8$ object shape treatments. Use human subjects as blocks with $k = 4$ treatments per subject.

b. Randomize the blocks of treatment labels to subjects, and randomize the actual object shape treatments to the treatment labels.

c. Suppose the order of presentation to the subjects could be an important source of variation. Construct the design as a row–column design, and randomize accordingly.

d. Suppose eight replications were required for the study; construct the cyclic design with eight replications.

6. Generate an α design for three replications of 35 treatments in blocks of five plots.

a. Show the generating α array, the intermediate α arrays, and the final alpha design with treatment labels.

b. Randomize the incomplete blocks of treatment labels to actual blocks, and randomize the treatment labels to the units within a block.

c. Randomly assign the actual treatments to the treatment labels.

10A.1 Appendix: Plans for Cyclic Designs

Table 10A.1 Initial blocks to generate efficient cyclic partially balanced incomplete block designs for $4 \leq t \leq 15, r \leq 10$

											kth treatment, t =											
k	r	*First k − 1 treatments*									4	5	6	7	8	9	10	11	12	13	14	15
	2	0									1	1	1	1	1	1	1	1	1	1	1	1
	4	0									2	2	2	3	3	3	3	3	3	5	4	4
2	6	0									1	3	3	2	2	2	2	5	5	2	6	2
	8	0									1	4	5	4	4	4	4	2	2	4	3	7
	10	0									2	1	4	5	5	5	5	4	4	3	5	5
	3	0	1								2	2	3	3	3	3	4	4	4	4	4	4
3	6	0	2								1	3	1	3	7	6	7	7	5	7	7	8
	9	0	1								3	2	3	3	4	5	3	3	6	4	6	5
4	4	0	1	3							–	2	2	6	7	7	6	7	7	9	7	7
	8	0	1	4							–	2	2	6	7	8	2	6	6	6	6	6
	5	0	1	2	4						–	–	5	5	7	7	7	7	7	7	9	10
5	10	0	2	3	4						–	–	5	5	7	8	9	8				
		0	2	3	6														7	11	12	10
6	6	0	1	2	3	6					–	–	–	5	5	5	5	10	10	10	10	10
7	7	0	1	2	3	4	7				–	–	–	–	5	5	9	9	9	9	9	10
8	8	0	1	2	3	4	6	8			–	–	–	–	–	5	9	9	9	9	11	11
9	9	0	1	2	3	4	5	7	9		–	–	–	–	–	–	8	8	8	10	10	10
10	10	0	1	2	3	4	5	6	9	10	–	–	–	–	–	–	–	7	7	7	12	12

Reproduced with permission from J. A. John (1981), "Efficient Cyclic Designs." *Journal of the Royal Statistical Society B*, 43, 76–80.

10A.2 Appendix: Generating Arrays for α Designs

Table 10A.2 Generating Arrays ($k \times r$) for α Designs

$s = k = 5$				$s = k = 6$				$s = k = 7$			
0	0	0	0	0	0	0	0	0	0	0	0
0	1	4	2	0	1	5	4	0	1	3	2
0	2	3	4	0	3	2	5	0	2	6	4
0	3	2	1	0	2	3	1	0	4	5	1
0	4	1	3	0	4	1	2	0	3	2	6
				0	5	1	3	0	5	1	3
								0	6	4	5

$s = k = 8$				$s = k = 9$				$s = k = 10$			
0	0	0	0	0	0	0	0	0	0	0	0
0	1	2	6	0	1	8	7	0	1	9	5
0	3	7	1	0	3	6	4	0	3	6	9
0	5	3	4	0	7	2	3	0	5	7	2
0	2	5	3	0	2	3	5	0	4	5	6
0	4	1	6	0	4	1	6	0	6	3	1
0	6	0	2	0	5	7	2	0	7	2	4
0	7	6	5	0	6	5	1	0	8	4	7
				0	8	4	7	0	9	8	2
								0	2	6	3

$s = 11, k = 9$				$s = 12, k = 8$				$s = 13, k = 7$			
0	0	0	0	0	0	0	0	0	0	0	0
0	1	6	7	0	1	2	3	0	1	4	10
0	4	8	1	0	7	5	1	0	3	8	11
0	9	7	5	0	9	6	4	0	9	2	1
0	2	3	6	0	4	11	8	0	12	10	6
0	5	1	3	0	11	3	10	0	8	5	12
0	6	5	10	0	10	4	7	0	6	7	8
0	3	9	4	0	5	1	6				
0	7	4	1								

$s = 14, k = 7$				$s = 15, k = 6$			
0	0	0	0	0	0	0	0
0	1	8	10	0	1	8	7
0	9	10	7	0	3	12	14
0	11	13	2	0	7	2	5
0	2	6	1	0	10	13	11
0	5	11	12	0	14	3	8
0	3	1	11				

Reproduced with permission of Cambridge University Press from H. D. Patterson, E. R. Williams, and E. A. Hunter (1978), Block designs for variety trials. *Journal of Agricultural Science* 90, Table 2, p. 399.

11 Incomplete Block Designs: Factorial Treatment Designs

The versatility of factorial treatment designs is explored in this chapter and the next with emphasis on 2^n and 3^n factorials. The effects and notations specific to factorial treatments are used to develop methods to construct incomplete block designs for factorial treatment designs in this chapter.

11.1 Taking Greater Advantage of Factorial Treatment Designs

The factorial treatment design was discussed in Chapters 6 and 7 as a means for investigating the effects of several treatment factors in the same experiment. The primary advantage of the factorial arrangement resides in the ability to determine whether the factors act independently or interact with one another as they affect the experimental units.

The 2^n and 3^n factorial treatment designs are of great practical importance and widely used in research studies. The 2^n factorials have n factors at two levels, and the 3^n factorials have n factors at three levels. As the number of factors increases, the number of treatment combinations rapidly escalates so that incomplete block designs are required to control experimental error. The factorial arrangement in each of these designs may be exploited to provide effective incomplete block designs for the investigation of factor effects and to facilitate the analysis of factor effects.

11.2 2^n Factorials to Evaluate Many Factors

The requirements of a research program may demand the investigation of many factors and their interrelationships as they affect the outcome of a process. Consider a large continuous production process in a chemical manufacturing plant that may not have as high a percent yield of the final product as projected by the plant construction specifications. The chemical yield of the final product is affected by factors at each of several separate reaction steps in the process. The problem is to identify factors that affect the final chemical yield and to determine the level of those factors that optimize the yield. The factors can include such items as the concentration of catalysts, the concentrations of reactants in solvents, and the ratio of one reactant to another, as well as temperature, pressure, agitation rates, and residence time in the reaction chambers.

Clearly, the engineers who are attempting to improve the plant performance have a formidable task before them. They have many factors to investigate with the potential that only a few are of major importance. They must screen the factors to determine the ones that warrant more detailed study, and at the same time they must control the cost of the experiments.

The 2^n factorial with many factors, each at a "low" level and a "high" level, can be used to detect the important factors in the process with a minimum of experimental units. Major trends can be detected with factors at two levels to identify potentially important factors. Consequently, 2^n factorials are often used at the early stage of experimentation to detect potential candidate factors for more detailed investigation. The following example with three factors illustrates the features of 2^n factorial treatment designs.

Example 11.1   Truck Leaf Spring Manufacture

An experiment described by Pignatiello and Ramberg (1985) was designed to investigate the effects of factors on a manufacturing process for leaf springs used on trucks. The assembled leaf spring was passed through a high-temperature furnace. Afterward, it was put into a forming machine to induce curvature in the spring by holding the spring in a high-pressure press for a short length of time.

The factorial treatment design for the experiment consisted of three factors, each at two levels. They were furnace temperature (A), heating time in the furnace (B), and transfer time between the furnace and the forming machine (C). The eight treatment combinations for the three factors in a 2^3 factorial are shown in Table 11.1, along with the observed measure of product quality (y) for one replication of each treatment.

New Treatment Labels

In general, the low level of a quantitative factor is denoted by a "0" and the high level by a "1." Equivalently, the two categories of a qualitative factor can be coded

Table 11.1 Truck leaf spring quality observations from a 2^3 factorial experiment

A Furnace Temperature (°F)	B Heating Time (sec)	C Transfer Time (sec)	y
1840	23	10	32
1880	23	10	35
1840	25	10	28
1880	25	10	31
1840	23	12	48
1880	23	12	39
1840	25	12	28
1880	25	12	29

as "0" and "1." Another useful notation for treatment combinations in 2^n factorials is illustrated with a 2^3 factorial in Display 11.1.

The uppercase letters A, B, and C represent factors. The "Treatment" label uses corresponding lowercase letters $a, b,$ and c. The lowercase letter is present if the factor is at level 1. The lowercase letter is absent if the factor is at level 0. The Treatment label is "(1)" if all factors are at level 0. The correspondence between the Treatment labels and the $(0, 1)$ designations for factor levels are shown in Display 11.1, along with the actual levels of each factor in the truck spring experiment.

Evaluating the 2^n Factorial Effects

The *effect* of a factor with 2^n factorials corresponds to a change in the response from the low level to the high level of the factor. The simple effects, main effects, and interaction effects for factorials were discussed in Chapter 6. For illustration,

Display 11.1 Treatment Labels for the Truck Spring Experiment

Treatment	A	B	C	Furnace Temperature (°F)	Heating Time (sec)	Transfer Time (sec)	y
(1)	0	0	0	1840	23	10	32
a	1	0	0	1880	23	10	35
b	0	1	0	1840	25	10	28
ab	1	1	0	1880	25	10	31
c	0	0	1	1840	23	12	48
ac	1	0	1	1880	23	12	39
bc	0	1	1	1840	25	12	28
abc	1	1	1	1880	25	12	29

consider the observations from the truck spring experiment. The simple effects of Furnace Temperature as it changes from 1840° to 1880° F with factors Heating Time and Transfer Time held constant are shown in Table 11.2.

Table 11.2 Simple effects of Furnace Temperature at constant levels of Heating Time and Transfer Time for the truck spring experiment

Heating Time (B)	Transfer Time (C)	Furnace Temperature (A)			B	C
		1840	1880	Simple Effect		
23	10	32	35	$35 - 32 = 3$	0	0
25	10	28	31	$31 - 28 = 3$	1	0
23	12	48	39	$39 - 48 = -9$	0	1
25	12	28	29	$29 - 28 = 1$	1	1

The main effect of a factor is the average effect of moving from its 0 level to its 1 level. Thus, the main effect of Furnace Temperature is the average of its simple effects

$$A = \frac{1}{4}[3 + 3 + (-9) + 1] = -0.5 \tag{11.1}$$

Equivalent calculations produce the main effects $B = -9.5$ and $C = 4.5$.

Two factors, say A and B, interact if the effect of A is different at the two levels of B. When Heating Time is 23 seconds, $B = 0$ and the effect of Furnace Temperature is

$$(A \mid B = 0) = \frac{1}{2}[3 + (-9)] = -3$$

However, when Heating Time is 25 seconds, $B = 1$ and the effect of Furnace Temperature is

$$(A \mid B = 1) = \frac{1}{2}(3 + 1) = 2$$

If the furnace temperature is increased from 1840° to 1880° F the product quality is *reduced* three units if the heating time is 23 seconds. However, product quality is *increased* by two units if the heating time is 25 seconds. The response to the furnace temperature is different for the two heating times, implying the potential for an interaction between Furnace Temperature and Heating Time.

The interaction between two factors, say A and B, is defined as one-half the difference between the effect of A at $B = 1$ and $B = 0$. The estimate of interaction between Furnace Temperature and Heating Time is then

$$AB = \frac{1}{2}\left\{(A \mid B = 1) - (A \mid B = 0)\right\} = \frac{1}{2}[2 - (-3)] = 2.5$$

If the roles of factors A and B are interchanged the resulting interaction value is unchanged. Calculations for the other two-factor interactions give $AC = -3.5$ and $BC = -5.5$.

The three-factor interaction ABC arises if a two-factor interaction, say AB, is different for $C = 0$ and $C = 1$. The AB interaction is one-half the difference between the effect of A at $B = 1$ and $B = 0$. When transfer time is 10 seconds, $C = 0$ and the estimate of interaction between Furnace Temperature and Heating Time is

$$(AB \mid C = 0) = \frac{1}{2}(3 - 3) = 0$$

The response to A is three units regardless of the level for B. The zero value for AB interaction when $C = 0$ is represented graphically as two parallel response lines in Figure 11.1a.

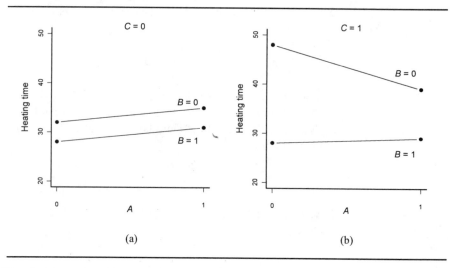

Figure 11.1   Graphic representation of ABC interaction depicted by different AB interaction at $C = 0$ and $C = 1$

When transfer time is 12 seconds, $C = 1$ and the interaction between Furnace Temperature and Heating Time is

$$(AB \mid C = 1) = \frac{1}{2}[1 - (-9)] = 5$$

The response to A is different for $B = 0$ and $B = 1$. The positive measure of AB interaction when $C = 1$ is represented as two different response lines in Figure 11.1b.

The interaction between Furnace Temperature and Heating Time differs with the transfer time. The estimate of AB interaction is 0 when transfer time is 10 seconds, and the estimate is 5 when transfer time is 12 seconds.

One-half the difference between these two evaluations of the AB interaction is the three-factor interaction. The three-factor interaction estimate is

$$ABC = \frac{1}{2}\left\{(AB \mid C = 1) - (AB \mid C = 0)\right\} = \frac{1}{2}(5 - 0) = 2.5$$

The existence of a three-factor interaction translates into the two different graphic representations of the AB interaction in Figure 11.1. The same value for ABC interaction is attained if the AC interaction is evaluated at $B = 0$ and $B = 1$ or the BC interaction is evaluated at $A = 0$ and $A = 1$.

A Table of Contrasts to Summarize the Effects

The effects for 2^n factorials can be defined as contrasts with coefficients of $+1$ or -1 for each of the treatment combinations. A table of $+$ and $-$ signs determines the correct coefficient for any treatment combination in the contrast, and the table provides a systematic method for calculating factorial effects.

The full complement of $+$ and $-$ signs required for the contrasts in the 2^3 factorial for the truck spring experiment is shown in Table 11.3. The table of $+$ and $-$ signs is generated in the following manner:

Table 11.3 Coefficients for contrasts in a 2^3 factorial treatment design

Treatment	*Factorial Effects*								
	I	*A*	*B*	*C*	*AB*	*AC*	*BC*	*ABC*	*y*
(1)	+	−	−	−	+	+	+	−	32
a	+	+	−	−	−	−	+	+	35
b	+	−	+	−	−	+	−	+	28
ab	+	+	+	−	+	−	−	−	31
c	+	−	−	+	+	−	−	+	48
ac	+	+	−	+	−	+	−	−	39
bc	+	−	+	+	−	−	+	−	28
abc	+	+	+	+	+	+	+	+	29
Divisor	8	4	4	4	4	4	4	4	
Effect	33.8	− 0.5	− 9.5	4.5	2.5	− 3.5	− 5.5	2.5	
SS		0.5	180.5	40.5	12.5	24.5	60.5	12.5	

- The column denoted I, containing all $+$ signs, is used to estimate the grand mean with a divisor of 2^n.

- The next three columns, labeled with factors A, B, and C, have the $+$ and $-$ signs in *standard order*. The standard order has factor levels arranged such that column A has successive pairs of $-$ and $+$ signs. Column B has pairs of $-$ signs followed by pairs of $+$ signs. Column C has four $-$ signs followed by four $+$ signs. In general the kth column has 2^{k-1} of the $-$ signs followed by an equal number of $+$ signs.

- The coefficients for any two-factor interaction are obtained as a product of the columns of coefficients for the corresponding main effects. For example, the coefficients for the AB column are the products of the corresponding elements in the A column and the B column.

- The coefficients for the three-way interaction ABC are obtained from the product of the coefficients for any set of columns whose symbol product is equal to ABC. The coefficients for ABC can be obtained from any of the symbolic products $A \times B \times C$, $AB \times C$, $AC \times B$, or $BC \times A$.

Column I often is referred to as the *Identity* column. Each column, except I, has an equal number of $+$ and $-$ signs.

Estimates of effects in Table 11.3 are computed by multiplying the corresponding column sign by the response y in each row of the table, summing the products, and dividing the sum by the appropriate divisor. The calculation of the main effect for Furnace Temperature in Equation (11.1) is the result of using the $+$ and $-$ coefficients in column A in Table 11.3 for the corresponding value of the response y and dividing by 4, or

$$A = \frac{1}{4}(-32 + 35 - 28 + 31 - 48 + 39 - 28 + 29) = -0.5$$

Similarly, utilizing the $+$ and $-$ codes under column B the main effect of Heating Time is

$$B = \frac{1}{4}(-32 - 35 + 28 + 31 - 48 - 39 + 28 + 29) = -9.5$$

The AB interaction effect is calculated using the $+$ and $-$ signs under the AB column to give

$$AB = \frac{1}{4}(32 - 35 - 28 + 31 + 48 - 39 - 28 + 29) = 2.5$$

In general, the contrast among treatment means for any main effect or interaction is

$$l_{AB...} = \sum_i k_i \bar{y}_i \tag{11.2}$$

where the coefficients for the contrasts are $k_i = \pm 1$. Following the convention for calculating contrasts, standard errors, and sums of squares in Chapter 3, the estimate of the effect for any contrast among treatment means in a complete 2^n factorial can be expressed as

$$AB\ldots = \frac{1}{2^{n-1}}(l_{AB\ldots}) \tag{11.3}$$

The standard error estimate for an effect estimate is

$$s_{AB\ldots} = \sqrt{\frac{4\sigma^2}{r2^n}} \tag{11.4}$$

where r is the number of replications for each treatment. The 1 degree of freedom sum of squares for the effects shown at the bottom of Table 11.3 is obtained with

$$SS(\boldsymbol{AB}\ldots) = \frac{r}{2^n}(l_{AB\ldots})^2 \tag{11.5}$$

For the example in Table 11.3, $r = 1$ because each treatment combination only occurs one time. The 1 degree of freedom sum of squares for A is

$$SSA = \frac{1}{8}(-2)^2 = 0.5$$

while those for B and AB are

$$SSB = \frac{1}{8}(-38)^2 = 180.5 \quad \text{and} \quad SS(\boldsymbol{AB}) = \frac{1}{8}(10)^2 = 12.5$$

11.3 Incomplete Block Designs for 2^n Factorials

The use of incomplete block designs to reduce experimental error variance was introduced in Chapter 9. The performance of a complete replication for the 2^n factorials with many factors may not be possible in a single complete block. If there is insufficient raw material in a manufactured batch to accommodate all of the treatments, each batch of raw materials can be used as an incomplete block. If experimental error is too large with complete block designs in agricultural field experiments, the variation among field plots can be controlled in the experiment with reduced block sizes for more homogeneous groups of experimental plots. Blocks of reduced size for 2^n factorials can be devised by exploiting the construction of effect contrasts in the 2^n factorials.

Sacrifice Treatment Information to Increase Precision

Incomplete block designs for 2^n factorials are constructed such that one or more treatment contrasts are identical to block contrasts. The treatment effect is said to be completely **confounded** with blocks. The confounded treatment effect is indistinguishable from the effect of the blocks with which it is confounded.

Ordinarily the highest order interaction effect in a 2^n factorial is chosen to be confounded with blocks. Main effects, two-factor interactions, and other lower order interactions, in the case of experiments with many factors, are those effects of most interest. By confounding the highest order interaction the other effects are estimated without penalty.

The construction of an incomplete block design is illustrated with a 2^3 factorial. A complete block design requires eight experimental units per block. Half of the treatments have a $+$ coefficient and half have a $-$ coefficient for every effect. A resolvable incomplete block design with two blocks of four units each per replication can be constructed using the contrast l_{ABC}.

From Table 11.3, the ABC interaction is estimated with the comparison

$$l_{ABC} = abc + a + b + c - ab - ac - bc - (1)$$

Put the treatment combinations with a $+$ coefficient—abc, a, b, and c—in one block and the treatment combinations with a $-$ coefficient—ab, ac, bc, and (1)—in the other block. The two incomplete blocks of treatments are shown in Display 11.2.

Display 11.2 ABC Interaction Confounded in Two Blocks of Four Experimental Units Each

	Block	
	1	*2*
	abc	ab
	a	ac
	b	bc
	c	(1)
ABC	$+1$	-1

The comparison (block 1 − block 2) is the contrast required to estimate the ABC interaction. As a consequence, the estimate of the ABC interaction effect is completely confounded with a comparison between the blocks. It will not be possible to estimate the three-factor interaction independent of block effects.

On the other hand, the other six factorial effects are not confounded with blocks and can be estimated in the usual way. For example, the main effect of factor A estimated from

$$l_A = (abc + a - b - c) + (ab + ac - bc - (1))$$

contains two $+$ coefficients and two $-$ coefficients from the four units in each block. Any differences among the blocks will not affect the estimate.

Block Construction with the Evens–Odds Rule

Half of the treatments have a + coefficient and half have a − coefficient for every effect in the 2^n factorials. The treatments can be divided into the two groups with the *Evens–Odds* rule.

Any treatment combination that has an even number of letters from the factorial effect receives one of the coefficients (+ or −). Treatments with an odd number of letters from the factorial effect receive the other coefficient.

The ABC interaction has the associated letters a, b, and c. Treatments (1), ab, ac, and bc have an even number of letters from the ABC factorial effect. Treatment (1) with zero letters has an even number of letters, since zero is considered an even number. Treatments a, b, c, and abc have an odd number of letters. The treatment combinations with the + coefficient for the ABC interaction effect are

$$a, \ b, \ c, \text{ and } abc$$

and the treatment combinations with the − coefficient for the ABC interaction effect are

$$(1), \ ab, \ ac, \text{ and } bc$$

The treatment combination that contains all of the factor letters, abc in this case, will always have the + coefficient for any factorial effect. All of the treatment combinations in that group will also have the + coefficient. The treatment combinations in the other group will have the − coefficient for the factorial effect. This assignment of + and − coefficients for the contrast coincides with that shown in Table 11.3.

Suppose the $ABCD$ interaction from a 2^4 factorial is to be confounded with blocks for an experiment with two blocks of eight experimental units per block in each replication. The letters associated with the four-factor interaction effect $ABCD$ are a, b, c, and d. The treatment combinations with an even number of letters are

$$(1), \ ab, \ ac, \ ad, \ bc, \ bd, \ cd, \text{ and } abcd$$

and the treatment combinations with an odd number of letters are

$$a, \ b, \ c, \ d, \ abc, \ abd, \ acd, \text{ and } bcd$$

The treatments in the group containing the treatment combination $abcd$ receive a + coefficient since the treatment combination $abcd$ contains the letters of all four factors. The other group of treatments all have a − coefficient for the $ABCD$ interaction effect and are placed in a separate block from the first group.

Analysis of Variance Outline for a Completely Confounded Design

The sums of squares are computed as usual for the analysis of variance except for the exclusion of a sum of squares partition for the interaction effect confounded with blocks. The block sum of squares will include the confounded factorial effect.

The sources of variation and degrees of freedom for the analysis of variance are outlined for the 2^3 factorial with $b = 2$ incomplete blocks in each of $r = 2$ replicates in Table 11.4. Since ABC is confounded with blocks the sum of squares for blocks includes the ABC effect.

Table 11.4   Analysis of variance for a 2^3 factorial with $b = 2$ incomplete blocks in each of $r = 2$ replicate groups

Source of Variation	Degrees of Freedom
Replicates	$r - 1 = 1$
Blocks within replicates	$r(b - 1) = 2$
Treatments	6
A	1
B	1
C	1
AB	1
AC	1
BC	1
Error	6
Total	15

Retain Some Treatment Information with Partial Confounding

Block size was reduced in the previous section by confounding the highest order interaction with blocks. However, any gain that may occur with a reduced experimental error has its price. In the previous designs, the loss of total information on the confounded factorial effect was the cost for a possible reduction in the experimental error variance.

Any factorial effect can be confounded with blocks. To avoid the loss of all information on any one factorial effect a different effect can be confounded in each replication group of the resolvable design. In this way a factorial effect is only confounded in one of the replications and thus is said to be *partially confounded* with blocks.

A 2^3 factorial with a different effect confounded in each of three replications is used to illustrate the principle in Example 11.2.

Example 11.2   Partial Confounding in a 2^3 Factorial

The purity of a chemical product was thought to be influenced by three factors—agitation rate (A), base component concentration (B), and concentration of reagent (C). The chemist set up an experiment using a factorial treatment design with each of the factors at two levels for a 2^3 factorial arrangement.

Experiment Design: The chemist wanted three replications of the experiment, but only four runs of the chemical process could be conducted in a single day. Therefore, each replication had to be run in two incomplete blocks (days).

An effect contrast for a 2^3 factorial consists of four treatment combinations with a + coefficient and four treatment combinations with a − coefficient. Thus, it was possible to construct the incomplete block design by confounding a 2^3 factorial effect with blocks. To avoid complete confounding of one effect, the chemist confounded a different two-factor interaction in each replication. The treatment combinations required for each block are shown in the diagram in Table 11.5.

Table 11.5 Observed purity of a chemical product in a partially confounded 2^3 factorial

BC Confounded				AC Confounded				AB Confounded			
+ 1		− 1		+ 1		− 1		+ 1		− 1	
(1)	25	ab	43	abc	39	bc	38	(1)	26	a	43
bc	34	c	30	b	29	a	37	c	32	b	34
abc	42	ac	40	(1)	27	ab	46	ab	52	ac	40
a	25	b	33	ac	40	c	34	abc	51	bc	36
Block 1		Block 2		Block 3		Block 4		Block 5		Block 6	
Replicate I				Replicate II				Replicate III			

The BC, AC, and AB interactions were each confounded in one replication of the experiment. The two-factor interaction confounded with the two blocks in each replication is shown above the blocks. The observed purity rates are to the right of the treatment combinations in each block.

The treatment combinations required in each of the incomplete blocks can be determined with the Evens–Odds rule. In replicate I the BC interaction is confounded by placing treatment combinations with an even number of the letters b and c together in block 1. They are (1), a, bc, and abc. The treatments with an odd number of the letters b and c are placed in block 2. They are c, b, ab, and ac. The treatments in block 1 have the + 1 coefficient for the BC interaction effect because the treatment combination abc contains all of the factor letters. The treatments in block 2 have the − 1 coefficient for the BC interaction effect. Consequently, the difference between the observations in block 1 and block 2 will have the estimate of the BC interaction effect confounded with the difference between the effects of blocks 1 and 2. The treatment combinations in each of the blocks of replicates II and III can be determined in a similar manner with the AC effect confounded in replicate II and the AB effect confounded in replicate III.

Computing the Sums of Squares for Partially Confounded Designs

The sums of squares for the effects not confounded with blocks, A, B, C, and ABC, may be calculated in the usual manner. The sums of squares can be calculated from Equation (11.5) using the coefficients for the contrasts in Table 11.3. The effects can be estimated in all replications, thus the sum of squares is $SS = r(\Sigma k_i \bar{y}_i)^2/2^n$, where $r = 3$, $2^n = 8$, and $k_i = \pm 1$. The calculations are shown in Table 11.6.

Table 11.6 Computations for factorial effects not confounded with blocks

	(1)	a	b	ab	c	ac	bc	abc	$\Sigma k_i \bar{y}_i$	2^n	SS
Means:	26	35	32	47	32	40	36	44			
A	−	+	−	+	−	+	−	+	40	8	600.00
B	−	−	+	+	−	−	+	+	26	8	253.50
C	−	−	−	−	+	+	+	+	12	8	54.00
ABC	−	+	+	−	+	−	−	+	− 6	8	13.50

The sum of squares for each of the partially confounded two-factor interactions must be calculated from the replications in which they are not confounded. The sum of squares for AB must be calculated from replicates I and II, for AC from replicates I and III, and for BC from replicates II and III. Totals are used instead of the means for these calculations.

The sum of squares can be computed conveniently by determining the effect contrast from all of the observations and subtracting the value of the contrast represented by the difference of block totals for the replications in which the effect is confounded. The three confounded contrasts, AB, AC and BC, computed from the totals of all treatment observations are

Treatments	(1)	a	b	ab	c	ac	bc	abc	$\Sigma k_i y_i$
Totals	78	105	96	141	96	120	108	132	
AB	+	−	−	+	+	−	−	+	18
AC	+	−	+	−	−	+	−	+	− 24
BC	+	+	−	−	−	−	+	+	− 30

The totals and difference computed for each pair of blocks are

	Replicate I BC Confounded		Replicate II AC Confounded		Replicate III AB Confounded	
	Block 1	Block 2	Block 3	Block 4	Block 5	Block 6
Totals	126	146	135	155	161	153
Difference	− 20		− 20		8	

To estimate the AB effect contrast only from replicates I and II, the difference between blocks 5 and 6 (8) is subtracted from the AB contrast obtained from the totals of all observations, 18. The corrected estimate of the contrast is

$$l_{AB} = \frac{18 - 8}{2} = 5$$

where the divisor is $r = 2$ to put the result on a mean basis. The sum of squares for AB interaction is

$$SS(\mathbf{AB}) = \frac{2(5)^2}{8} = 6.25$$

Similarly, the estimate of the AC contrast is

$$l_{AC} = \frac{-24 - (-20)}{2} = -2$$

with sum of squares

$$SS(\mathbf{AC}) = \frac{2(-2)^2}{8} = 1$$

Finally,

$$l_{BC} = \frac{-30 - (-20)}{2} = -5$$

and

$$SS(\mathbf{BC}) = \frac{2(-5)^2}{8} = 6.25$$

The analysis of variance is shown in Table 11.7. The design is a resolvable incomplete block design with blocks of treatments in complete replication groups. The total sum of squares and the sums of squares for replicates and for blocks within replicates are computed in the usual manner. The sum of squares for experimental error is found by subtracting the sums of squares for all treatment effects, replicates, and blocks within replicates from the total sum of squares.

Tests of Hypotheses About Factor Effects

The critical value for a test of hypothesis for any factorial effect is $F_{.05,1,11} = 4.84$. The agitation rate (A) and base component concentration (B) main effects are significant. None of the other effects had a significant effect on the purity of the chemical product.

From Equations (11.3) and (11.4), the estimated effect of agitation rate is

$$A = \frac{(40)}{4} = 10$$

Table 11.7 Analysis of variance for purity of chemical product in a partially confounded 2^3 factorial

Source of Variation	Degrees of Freedom	Sum of Squares	Mean Square	F_0
Replicates	2	111.00	55.50	
Blocks within reps	3	108.00	36.00	
A	1	600.00	600.00	40.6
B	1	253.50	253.50	17.2
C	1	54.00	54.00	3.7
ABC	1	13.50	13.50	< 1
AB (Reps I, II)	1	6.25	6.25	< 1
AC (Reps I, III)	1	1.00	1.00	< 1
BC (Reps II, III)	1	6.25	6.25	< 1
Error	11	162.50	14.77	
Total	23	1,316.00		

with standard error

$$s_A = \sqrt{\frac{4(14.77)}{3(8)}} = 1.57$$

The estimated effect of the base component concentration is

$$B = \frac{26}{4} = 6.5$$

with standard error $s_B = 1.57$. The purity of the chemical product is increased by 10 units if the agitation rate is increased from the low level to the high level. The purity is increased by 6.5 units if the concentration of the base component is increased from the low level to the high level. The reagent concentration C had no significant effect on the purity of the product, and there was no significant interaction among any of the factors.

Confound Another Interaction to Further Reduce Block Size

The number of experimental units per block can be reduced further by choosing a second factorial effect to confound with blocks. The design will have four blocks per replication if two factorial effects are used for confounding.

The technique is illustrated with the 2^3 factorial to produce four blocks of two experimental units. If the interaction effects AB and AC are chosen as confounding effects, the assignment of + and − coefficients to the treatments by the Evens–Odds rule is

	(1)	abc	c	ab	b	ac	a	bc
AB	$+$	$+$	$+$	$+$	$-$	$-$	$-$	$-$
AC	$+$	$+$	$-$	$-$	$+$	$+$	$-$	$-$

Four blocks of two experimental units are formed by taking treatment pairs that have the same configurations of $+$ and $-$ coefficients for the AB and AC interactions.

There are four configurations of $+$ and $-$ coefficients for the joint AB and AC interaction contrasts. Each configuration contains two treatments. The two treatments are each assigned to a separate block. The configurations and treatment assignments to blocks are

AB	AC	Treatment		Block
$+$	$+$	(1)	abc	1
$+$	$-$	c	ab	2
$-$	$+$	b	ac	3
$-$	$-$	a	bc	4

A Third Effect Is Automatically Confounded

If two effects are confounded with blocks in a 2^n factorial, then a third effect is confounded. There are 3 degrees of freedom for block comparisons, and three treatment effects are confounded with blocks. The AB interaction is confounded with the block contrast

$$AB = l_1 = B_1 + B_2 - B_3 - B_4$$

where B_1, B_2, B_3, and B_4 represent block means. The AC interaction is confounded with the block contrast

$$AC = l_2 = B_1 - B_2 + B_3 - B_4$$

The third contrast among the blocks that is orthogonal to $AB = l_1$ and $AC = l_2$ is

$$l_3 = B_1 - B_2 - B_3 + B_4$$

$$= (1) + abc - c - ab - b - ac + a + bc$$

If the contrasts in Table 11.3 are checked it can be seen that the third contrast is equal in the BC interaction.

The Third Confounded Effect Is a Generalized Interaction

The third effect confounded is known as the **generalized interaction** of the first two confounded effects. The generalized interaction of two effects is obtained by forming the symbolic product of the two effects and striking out any letters that appear twice in the product.

For example, AB and AC were chosen as the confounding effects. Their symbolic product is $ABAC$. Striking out the letter A, which appears twice, the result is $ABAC = \cancel{A}B\cancel{A}C = BC$. The generalized interaction is BC. Suppose ABC and AB are chosen as confounding effects. The generalized interaction is $ABCAB = \cancel{A}BC\cancel{A}B = C$. The main effect C is confounded with blocks along with ABC and AB. Care must be taken to avoid confounding effects that are of particular interest in the study, especially main effects.

A table is given in Appendix 11A to aid in the construction of useful incomplete block designs for 2^n factorials. A general method of confounding that can be used with factorial systems other than the 2^n series is presented in the next section.

11.4   A General Method to Create Incomplete Blocks

The allocation of treatment combinations to incomplete blocks for the 2^n factorials has been accomplished thus far on the basis of the $\pm$ signs of the effects confounded with blocks. However, the method becomes rather cumbersome with many factors due to the large number of treatment combinations involved in defining the contrasts. In addition, the use of $\pm$ signs for the confounding system does not carry over to other systems such as the 3^n factorials.

A general method to construct incomplete block designs with chosen factorial effects confounded with blocks utilizes the mathematics of *residues modulo m* or *residues mod m*. For an integer k the residue mod m is the remainder when k is divided by m. The residue r for the integer k mod m is written as $k = r(\text{mod } m)$.

With 2^n factorials we work with residues of (mod 2). Any integer divided by $m = 2$ leaves a remainder of 0 or 1. Thus, the values for the residues (mod 2) are 0 and 1. The even integers (mod 2) have residue 0 and the odd integers (mod 2) have residue 1. For example, $7 = 1(\text{mod } 2)$ since 7 divided by 2 is 3 with a remainder of 1. Also, $4 = 0(\text{mod } 2)$ since 4 divided by 2 is 2 with a remainder of 0.

The levels of a factor (0 or 1) are used as the values of a variable x_i representing the ith factor. A treatment combination for the 2^n factorial is represented as the sequence $x_1 x_2 x_3 \cdots x_n$. For a 2^3 factorial the eight treatments written in standard order are

x_1	x_2	x_3	*Treatment*
0	0	0	(1)
1	0	0	a
0	1	0	b
1	1	0	ab
0	0	1	c
1	0	1	ac
0	1	1	bc
1	1	1	abc

A Defining Contrast for Two Blocks

A general method to determine the allocation of treatment contrasts to incomplete blocks is accomplished with a linear function

$$L = \alpha_1 x_1 + \alpha_2 x_2 + \cdots + \alpha_n x_n \tag{11.6}$$

where L is the *defining contrast,* or the contrast confounded with blocks. The value of α_i is 1 if the *i*th factor is present in the defining contrast and 0 if the factor is absent in the defining contrast. The defining contrast function L is evaluated for each treatment combination. The value of x_i is the level of the *i*th factor (0 or 1) in any treatment combination under consideration for allocation to an incomplete block.

Suppose the defining contrast for a 2^3 factorial in two blocks of four units each is the two-factor interaction AB. Factors A and B are present in the defining contrast so that $\alpha_1 = 1$ and $\alpha_2 = 1$. Since factor C is not in the defining contrast, $\alpha_3 = 0$. The defining contrast function for allocation of treatments is

$$L = x_1 + x_2 \tag{11.7}$$

The values of x_1 and x_2 for each treatment combination are substituted into L and the residue for L modulo 2, $r(\text{mod }2)$, is determined. For example, the value of $L = x_1 + x_2$ for the treatment combination bc, $x_1 x_2 x_3 = (011)$, is $L = 0 + 1$. The residue for $L = 1$ is $1(\text{mod }2)$. The values of L and residues, $L = r(\text{mod }2)$, are

Treatment	x_1	x_2	$L = x_1 + x_2$	$r(mod\ 2)$
(1)	0	0	0	0
a	1	0	1	1
b	0	1	1	1
ab	1	1	2	0
c	0	0	0	0
ac	1	0	1	1
bc	0	1	1	1
abc	1	1	2	0

The treatment combinations with $L = 0(\text{mod }2)$ are assigned to one block, and the treatment combinations with $L = 1(\text{mod }2)$ are assigned to the other block. Thus, the treatment assignments are

Block 1 | (1) ab c abc | with $L = 0 \ (\text{mod }2)$

Block 2 | a b ac bc | with $L = 1 \ (\text{mod }2)$

Let B_1 and B_2 represent the block totals. The block with residue 0 for L, block 1, contains the treatment combinations with a $+$ sign for the AB contrast, while block 2 with residue 1 for L contains the treatment combinations with a $-$ sign for the AB contrast. Thus, the contrast l among block totals equivalent to the AB contrast is

$$l_{AB} = B_1 - B_2$$

Use Two Defining Contrasts for Four Blocks

The designs to this point have been constructed for 2^n factorials with two incomplete blocks of 2^{n-1} experimental units, each with one defining contrast confounded with blocks. The use of two blocks with one factorial effect confounded may not reduce block size sufficiently for a 2^n factorial when there are many factors, say $n \geq 4$. For example, a 2^5 factorial with 32 treatments may require block sizes no larger than eight units per block with four blocks per replication. Further reductions in block size can be accomplished by confounding an additional defining contrast with blocks.

Suppose a 2^4 factorial is to have the 16 treatment combinations placed in four blocks of $2^{4-2} = 2^2 = 4$ experimental units each. Two defining contrasts are required to construct the four incomplete blocks. If AB and CD are chosen to be confounded with blocks, then the defining contrasts are

$$L_1 = x_1 + x_2 \qquad\qquad \textbf{(11.8)}$$

$$L_2 = x_3 + x_4$$

where L_1 represents AB and L_2 represents CD. Each treatment combination will provide a pair of residues modulo 2 for the pair (L_1, L_2). There are four pairs of residues possible—(0,0), (0,1), (1,0), and (1,1). Treatment combinations with the same values for a pair of residues modulo 2 are placed in the same incomplete block. The values for the defining contrasts, L_1 and L_2 in Equation (11.8), and the residue pairs for each of the 16 treatment combinations are shown in Display 11.3 along with the block assignments.

The 3 degrees of freedom among the blocks represent three orthogonal contrasts among the blocks. Two of the contrasts are known to include the two-factor interactions, AB and CD, chosen to be confounded with blocks. The contrast for the AB interaction is the difference between the blocks with residues 0 and 1 for L_1. Let B_1, B_2, B_3, and B_4 represent the block totals. The block contrast for AB is

$$l_{AB} = B_1 + B_2 - B_3 - B_4$$

because blocks 1 and 2 have residue 0 and blocks 3 and 4 have residue 1 for L_1. Similarly, the blocks with residues 0 and 1 for L_2 define the contrast for the CD interaction. The block contrast for CD is

$$l_{CD} = B_1 - B_2 + B_3 - B_4$$

because blocks 1 and 3 have residue 0 and blocks 2 and 4 have residue 1 for L_2.

Display 11.3   Incomplete Block Design for a 2^4 Factorial in Four Blocks of Four Units with Defining Contrasts AB and CD

Block	Treatment	x_1	x_2	x_3	x_4	L_1	L_2	Residue
1	(1)	0	0	0	0	0	0	(0,0)
	ab	1	1	0	0	2	0	(0,0)
	cd	0	0	1	1	0	2	(0,0)
	$abcd$	1	1	1	1	2	2	(0,0)
2	c	0	0	1	0	0	1	(0,1)
	d	0	0	0	1	0	1	(0,1)
	abc	1	1	1	0	2	1	(0,1)
	abd	1	1	0	1	2	1	(0,1)
3	a	1	0	0	0	1	0	(1,0)
	b	0	1	0	0	1	0	(1,0)
	acd	1	0	1	1	1	2	(1,0)
	bcd	0	1	1	1	1	2	(1,0)
4	ac	1	0	1	0	1	1	(1,1)
	bc	0	1	1	0	1	1	(1,1)
	ad	1	0	0	1	1	1	(1,1)
	bd	0	1	0	1	1	1	(1,1)

$$L_1 = x_1 + x_2 \qquad L_2 = x_3 + x_4$$

Determining the Generalized Interaction

The $ABCD$ interaction is the *generalized interaction* confounded as a consequence of purposely confounding AB and CD with blocks. The generalized interactions can be determined by more formal algebraic rules than those described in the previous section. Form the product of the symbols for the defining contrasts with the exponent of any symbol reduced modulo 2. The product of AB by CD is

$$AB \times CD = ABCD$$

The generalized interaction of AB and CD as determined by the symbol product is $ABCD$ since all exponents of the symbol product are 1(mod 2).

Suppose the contrasts ABC and BCD had been chosen for the original defining contrasts. The product is $ABC \times BCD = ABBCCD = AB^2C^2D = AD$, where B^2 and C^2 do not appear in the generalized interaction because their exponents are 0(mod 2). Therefore, the generalized interaction is AD when ABC and BCD are the defining contrasts.

Because main effects and two-factor interactions are of greatest interest in preliminary studies of factor effects, we try to avoid confounding them with blocks. The number of experimental units in an incomplete block design for 2^n factorials is equal to the powers of 2—2, 4, 8, and so forth—up to 2^{n-1}. In general, if blocks of size $k = 2^q$ are required there will be $2^n/2^q = 2^{n-q}$ blocks in a complete replication. Consequently, $n - q$ defining contrasts must be chosen.

Consider the 2^7 factorial with 128 treatment combinations. A design with $2^4 = 16$ experimental units per block requires $2^{7-4} = 2^3 = 8$ blocks and $(7 - 4) = 3$ defining contrasts. Choose ABG, CDE, and EFG as the defining contrasts to reduce block sizes. The first contrast, ABG, is used to reduce block sizes to 64 units. The second, CDE, is used to reduce block sizes to 32 units, and the third contrast, EFG, is used to reduce block sizes to 16 units. There will be four generalized interactions also confounded with the eight blocks. They are

$$ABG \times CDE = ABCDEG$$

$$ABG \times EFG = ABEFG^2 = ABEF$$

$$CDE \times EFG = CDE^2FG = CDFG$$

and

$$ABG \times CDE \times EFG = ABCDE^2FG^2 = ABCDF$$

As before, the treatment combinations are assigned to the blocks according to the residues of the defining contrasts

$$
\begin{array}{ccl}
ABG & \longrightarrow & L_1 = x_1 + x_2 + x_7 \\
CDE & \longrightarrow & L_2 = x_3 + x_4 + x_5 \\
EFG & \longrightarrow & L_3 = x_5 + x_6 + x_7
\end{array}
$$

Each treatment contrast will have a triplet of values (L_1, L_2, L_3). There will be eight triplets of residues—(0, 0, 0), (0, 0, 1), (0, 1, 0), (1, 0, 0), (0, 1, 1), (1, 0, 1), (1, 1, 0), and (1, 1, 1)—each representing the residue triplet for a block assignment of the treatment combinations. A treatment combination with the residue triplet (0, 0, 0) is assigned to one block, a treatment combination with the residue triplet (0, 0, 1) is assigned to a second block, and so forth.

Table 11A.1 in the Appendix lists defining contrasts and their generalized interactions to construct incomplete block designs for 2^n factorials with block sizes $k \geq 4$ for $n = 4, 5, 6$, and 7 factors. Notice that some two-factor interactions will be confounded with blocks when blocks have four or less units. Designs with only three-factor and larger interactions confounded are always possible with block sizes of eight or more. If the defining contrasts are properly chosen it will be possible to avoid confounding any two-factor interactions or main effects.

11.5 Incomplete Block Designs for 3^n Factorials

The 2^n factorials are useful designs to detect factors with major effects on the measured responses in an experiment. The 3^n factorials have three levels for each factor, making it possible to estimate linear and quadratic trends for quantitative factors and to provide more detailed descriptions of qualitative factor effects. However, the number of experimental units required by 3^n factorials increases by powers of 3 as more factors are added. Thus, incomplete block designs can be very useful with these treatment designs. The construction of incomplete blocks designs for 3^n factorials is discussed briefly in this section.

Some 3^n Basics

Notation for 3^n Factorials

The levels of a factor are represented by $x_i = 0$, 1, 2. For example, the nine treatment combinations for a 3^2 factorial with factors A and B are

		A		
		0	1	2
	0	00	10	20
B	1	01	11	21
	2	02	12	22

The three columns of the array represent treatment combinations for the three levels of factor A, and the three rows represent treatment combinations for the three levels of factor B.

Three Incomplete Blocks Required for Confounded 3^n Factorials

The construction of incomplete block designs for 3^n factorials requires three blocks to have blocks of equal size. There will be 2 degrees of freedom between blocks, and a treatment effect with 2 degrees of freedom must be confounded with blocks.

The 3^n factorials have 2 degrees of freedom for main effects, 2^2 degrees of freedom for two-factor interactions, and so forth. We would not want to confound the main effects with blocks. Instead, the interactions are partitioned into two orthogonal components, each of which has 2 degrees of freedom. The orthogonal component can then be used as the defining contrast in the construction of the incomplete block design.

Construct Incomplete Blocks with Defining Contrasts

The allocation of treatment combinations to blocks utilizes the defining contrast function $L = \alpha_1 x_1 + \alpha_2 x_2 + \cdots + \alpha_n x_n$ introduced in Section 11.4. The value of x_i for the 3^n factorial is $x_i = 0$, 1, 2. As a mathematical convenience, the values of

$\alpha_i = 0, 1, 2$ are used for factors included in the 2 degrees of freedom interaction used as the defining contrast. For example, with two factors, A and B, the AB interaction with 4 degrees of freedom is split into defining contrasts labeled AB and AB^2, each with 2 degrees of freedom. By convention, the first letter of the interaction expression always has a power of 1 so that A^2B or A^2B^2 are not used as defining contrast expressions. The powers of A and B are the coefficients α_1 and α_2 in the defining contrast function $L = \alpha_1 x_1 + \alpha_2 x_2$.

The modulo 3 residues of $L = \alpha_1 x_1 + \alpha_2 x_2 + \cdots + \alpha_n x_n$ are determined for 3^n factorials since there are three levels for each of the factors. The residues $r(\text{mod } 3)$ are 0, 1 and 2. The defining contrast function for AB with $\alpha_1 = \alpha_2 = 1$ is

$$L_1 = x_1 + x_2$$

and for AB^2 with $\alpha_1 = 1$ and $\alpha_2 = 2$ the defining contrast function is

$$L_2 = x_1 + 2x_2$$

The treatments with $L_i = 0(\text{mod } 3)$ are assigned to one block, those with $L_i = 1(\text{mod } 3)$ are assigned to a second block, and those with $L_i = 2(\text{mod } 3)$ are assigned to the third block. The treatments are assigned to blocks according to the value of the residues for L_1 or L_2. The common practice is to have one-half of the replications assigned according to each of the defining contrast functions. The treatment assignments for L_1 and L_2 are shown in Display 11.4.

Display 11.4   Incomplete Block Design for a 3^2 Factorial with the AB or AB^2 Component Confounded with Blocks

AB confounded
$$L_1 = x_1 + x_2$$

Block 1			Block 2			Block 3		
00	12	21	01	10	22	02	11	20
$L_1 = 0(\text{mod } 3)$			$L_1 = 1(\text{mod } 3)$			$L_1 = 2(\text{mod } 3)$		

AB^2 confounded
$$L_2 = x_1 + 2x_2$$

Block 1			Block 2			Block 3		
00	11	22	02	10	21	01	12	20
$L_2 = 0(\text{mod } 3)$			$L_2 = 1(\text{mod } 3)$			$L_2 = 2(\text{mod } 3)$		

The residues of L_1 and L_2 for the nine treatment combinations are

x_1	x_2	L_1	$r(\bmod 3)$	L_2	$r(\bmod 3)$
0	0	0	0	0	0
0	1	1	1	2	2
0	2	2	2	4	1
1	0	1	1	1	1
1	1	2	2	3	0
1	2	3	0	5	2
2	0	2	2	2	2
2	1	3	0	4	1
2	2	4	1	6	0

Confounding with Three or More Factors

The 3^3 factorial requires 27 experimental units for a single replication. An incomplete block design with three blocks of nine experimental units each can be constructed by confounding a three-factor interaction component with blocks. The three-factor interaction with 8 degrees of freedom has four components, each with 2 degrees of freedom. For the purpose of obtaining a defining contrast the four components are designated ABC, ABC^2, AB^2C, and AB^2C^2. The defining contrast for each of these components is

$$ABC \quad \longrightarrow \quad L = x_1 + x_2 + x_3$$

$$ABC^2 \quad \longrightarrow \quad L = x_1 + x_2 + 2x_3$$

$$AB^2C \quad \longrightarrow \quad L = x_1 + 2x_2 + x_3$$

$$AB^2C^2 \quad \longrightarrow \quad L = x_1 + 2x_2 + 2x_3$$

Any component of the three-factor interaction can be used to generate three blocks of nine experimental units each. If the component AB^2C is used, the defining contrast is

$$L = x_1 + 2x_2 + x_3$$

The three blocks are constructed with treatment combinations $x_1 x_2 x_3$ that provide the residues for the defining contrast of $L = 0(\bmod 3)$, $L = 1(\bmod 3)$, and $L = 2(\bmod 3)$, respectively.

Residues are determined for each of the 27 treatment combinations—000, 001, 002, 010, ... , 222. For example, the treatment combination 000 has residue $L = 0 + 2(0) + 0 = 0(\bmod 3)$ and the treatment combination 021 has residue $L = 0 + 2(2) + 1 = 2(\bmod 3)$. Each is placed in its respective block with other treatments that have the same residues for the defining contrast. A different component of the three-factor interaction can be confounded in each replication of the experiment.

Four-factor interaction components are confounded with blocks in 3^4 factorials. The eight components each with 2 degrees of freedom are $ABCD$, AB^2CD, ABC^2D, $ABCD^2$, AB^2C^2D, AB^2CD^2, ABC^2D^2, and $AB^2C^2D^2$. Any of these components can be used to block the 81 experimental units into three blocks of 27 experimental units each.

Two Generalized Interactions Occur with Further Reductions in Block Size

The number of incomplete blocks remains a multiple of 3 with the 3^n factorials upon reduction of block sizes. A second defining contrast is confounded with blocks to reduce block sizes from nine experimental units to three experimental units for a 3^3 factorial. In the case of 3^n factorials, there are two generalized interactions for each pair of defining contrasts.

If two defining contrasts are chosen, say X and Y, then the generalized interactions are given as the symbolic products XY and XY^2. The exponents of the symbolic products are reduced modulo 3 to derive the generalized interactions. Suppose the two-factor interaction components $X = AB$ and $Y = AC^2$ are used as defining contrasts.

The XY symbolic product is

$$XY = AB \times AC^2 = A^2BC^2$$

Since the exponent of the first term of the symbolic product A is not unity, upon reduction modulo 3, the product is squared and the exponents are reduced modulo 3 as follows:

$$(A^2BC^2)^2 = A^4B^2C^4 = AB^2C$$

where the last term results from reducing the exponents of the previous term modulo 3.

The XY^2 symbolic product is

$$XY^2 = AB \times (AC^2)^2 = AB \times A^2C^4 = A^3BC^4$$

and upon reducing the exponents of the last term modulo 3, the result is

$$XY^2 = BC$$

Upon choosing the two defining contrasts, AB and AC^2, two generalized interactions, AB^2C and BC, are also confounded with the nine blocks of three experimental units each. Care must be exercised in the choice of defining contrasts to avoid confounding a main effect with blocks. For example, the choice of two three-factor interaction components, ABC and AB^2C^2, results in a generalized interaction of

$$XY = ABC \times AB^2C^2 = A^2B^3C^3 \longrightarrow (A^2B^3C^3)^2 = A^4B^6C^6$$

When the exponents of the squared symbolic product are reduced modulo 3 the generalized interaction is the main effect A.

Suppose the two interaction components AB and AC^2 are chosen to construct nine blocks of three units each for the 3^3 factorial. The defining contrasts are

$$L_1 = x_1 + x_2$$

and

$$L_2 = x_1 + 2x_3$$

There will be nine pairs of residues modulo 3 from the two defining contrasts, and three treatment combinations will be associated with each of the pairs of residues. For example, the residues $(L_1, L_2) = (2, 2)$ occur with treatment combinations (021), (112), and (200). Those three treatment combinations are placed in the same block.

11.6 Concluding Remarks

As we have seen in this chapter, the factorials are versatile treatment designs that can be adapted to a variety of experimental blocking conditions. The use of incomplete block designs results in some factorial effects confounded with blocks, either completely or partially, and increased complexity in the analysis of the results with the partially confounded designs. The goal of reducing block size is to reduce the experimental error variance. If the reduction in the estimate of experimental error variance is sufficient to overcome the loss in some information on the confounded factorial effects, then the use of an incomplete block design has been worthwhile. Cochran and Cox (1957) have tabled a number of incomplete block designs for symmetrical and asymmetrical factorials and provide some examples of the analysis with several of these designs. Detailed information on the underlying principles of design construction can be found in Kempthorne (1952), John (1987), and John and Williams (1995).

EXERCISES FOR CHAPTER 11

---

1. Construct one replication of an incomplete block design for a 2^4 factorial in two blocks of eight experimental units each with $ABCD$ confounded with blocks. Randomly assign the treatments to the eight experimental units in each block.

2. Suppose an incomplete block design is required for a 2^5 factorial, and the blocks cannot exceed eight experimental units in size.
    a. How many blocks will there be?
    b. How many defining contrasts are required for one replication of the experiment?
    c. How many other effects, generalized interactions, will be confounded with blocks?

d. Choose some defining contrast(s) to construct one replication of the design, and determine what other interaction(s), if any, also will be confounded with blocks.

3. Four replications of a 2^4 factorial experiment are required. The design must be conducted with incomplete blocks of eight experimental units. Construct a design such that no effect is completely confounded with blocks in the experiment.

4. A 2^7 factorial experiment is to be conducted in eight blocks of 16 experimental units, each using $ABCD$, $ABEF$, and EFG as defining contrasts. What other effects are confounded with blocks?

5. Suppose a 2^6 factorial experiment was conducted in eight blocks of eight experimental units each, and $BCDE$, $ABDE$, and ADE were used as defining contrasts.
   a. What other effects were confounded with blocks?
   b. Could there have been a better choice of defining contrasts for this design? Explain.
   c. If you have decided that a better choice is possible, then choose a set that will improve the design and defend your choice.

6. An animal scientist conducted a study on the effects of heat stress and dietary intake of protein and saline water on laboratory mice. The three factors were each used at two levels in a 2^3 factorial arrangement. The levels for the factors were (A) protein (low, high); (B) water (normal, saline); and (C) heat stress (room temperature, heat stress). An incomplete block design was used with four mice from an individual litter used in each block. Each mouse was put in an individual cage and assigned one of the treatments. One replication of the experiment consisted of two litters of mice. The weight gains (grams) for the mice are shown for each mouse next to the treatment identification.

Litter 1		Litter 2		Litter 3		Litter 4		Litter 5		Litter 6	
(1)	27.5	ab	24.3	bc	19.5	abc	19.7	(1)	24.5	a	33.1
bc	20.6	c	24.3	a	24.1	b	19.5	c	23.0	b	20.5
abc	22.0	ac	22.8	ab	22.4	(1)	22.5	ab	23.4	ac	19.8
a	28.6	b	24.6	c	22.0	ac	18.8	abc	21.7	bc	18.5

| Replicate I | Replicate II | Replicate III |

a. Which treatment effects are confounded with blocks (litters)?
b. Estimate the factor effects and interactions and their standard errors, and compute the analysis of variance for these data.
c. Interpret the results.

7. An experiment was conducted to investigate the effects of four factors on the operation of a metal lathe. The four factors each at two levels were (A) speed of lathe rotation $(60, 75)$; (B) angle of cut $(30, 45)$; (C) frequency of lubrication (10 sec, 30 sec); and (D) alloy for cutting tip (1 and 2). The factor levels were used in a 2^4 factorial arrangement for the experiment. Only eight cutting trials could be run in a single day. An incomplete block design with two blocks (days) in each of two replications was set up with $ABCD$ as the defining contrast. The cutting tip wear for each of

the treatments in each block follows. Days 1 and 2 constitute the first replication and Days 3 and 4 constitute the second replication.

Cutting Tip Wear

Day 1		Day 2		Day 3		Day 4	
(1)	40	a	24	(1)	43	a	28
ab	33	b	31	ab	30	b	35
ac	31	c	27	ac	30	c	28
bc	38	abc	23	bc	32	abc	20
ad	22	d	48	ad	26	d	44
bd	37	abd	35	bd	33	abd	36
cd	49	acd	29	cd	40	acd	25
abcd	30	bcd	37	abcd	31	bcd	34

a. Compute the analysis of variance for the data.
b. Compute the factor effects and their standard errors.
c. Interpret the results.

8. Design an incomplete block design for a 3^3 factorial that has nine experimental units per block in four replications of three blocks each. Confound a different component of the three-factor interaction in each replication.

9. An investigator needs an incomplete block design for a 3^3 factorial that has three experimental units in each block. Consider the two pairs of defining contrasts to construct the design:
a. ABC, AB^2C
b. AB, AC^2
Which is the better pair of defining contrasts to construct the design? Explain.

11A   Appendix: Incomplete Block Design Plans for 2^n Factorials

Table 11A.1   Number of factors and blocks, block sizes, defining contrasts, and generalized interactions to construct incomplete block designs with 2^n factorials

Factors n	Blocks 2^{n-q}	Block Size $k = 2^q$	Defining Contrasts	Generalized Interactions
4	2	8	ABCD	
	4	4	ABC, ABD	CD
5	2	16	ABCDE	
	4	8	ABC, CDE	ABDE
	8	4	ABC, ACD, ADE	BD, CE, ABE, BCDE
6	2	32	ABCDEF	
	4	16	ABCD, CDEF	ABEF
	8	8	ACE, ABEF, ABCD	ADF, BCF, BED, CDEF
	16	4	ABF, ACF, CDF, DEF	AD, CE, BC, BE, AEF, BDF, ABCD, ABDE, ACDE, ABCEF, BCDEF
7	2	64	ABCDEFG	
	4	32	ABCDE, ABEFG	CDFG
	8	16	ABG, CDE, EFG	ABEF, CDFG, ABCDF, ABCDEG
	16	8	ABC, ADG, CDE, DEFG	AEF, BDF, BEG, CFG, ABDE, ABFG, ACDF, ACEG, BCDG, BCEF, ABCDEFG
	32	4	ABG, BCG, CDG, DEG, EFG	AC, BD, BF, DG, CE, ADG, AFG, BEG, CFG, ABCD, ABEF, ABFG, ACDF, ABDE, ACEG, ADEF, BCDE, BCEF, CDEF, ABCEG, ABDFG, ACDEG, ACEFG, BCDFG, BDEFG, ABCDEFG

12 Fractional Factorial Designs

Discussions on the versatility of 2^n factorial treatment designs are continued in this chapter. Experiments using only a fractional replication of the factorial arrangement are proposed as a means of effectively obtaining information on factors in the early stages of experimentation. Methods for constructing the designs are described and the analysis to estimate and test significance of factorial effects is illustrated for large 2^n factorial experiments without complete replication.

12.1 Reduce Experiment Size with Fractional Treatment Designs

The 2^n factorial treatment designs are useful for conducting preliminary studies with many factors to identify the more important factors and factor interactions. However, the number of experimental units increases geometrically with the number of factors in the study.

Fractional factorial designs use only one-half, one-fourth, or even smaller fractions of the 2^n treatment combinations. They are used for one or more of several reasons, including

- the number of treatments required exceeds resources

- information is required only on main effects and low-order interaction

- screening studies are needed to check on many factors

- an assumption is made that only a few effects are important

Some industrial research and development studies can exceed the capacity of the research facility if all treatment combinations are included in the experiment for a 2^n factorial. For example, one complete replication of a 2^7 study requires 128 runs of a process. The observations from the 128 runs provide estimates of 7 main effects, 21 two-factor interactions, 35 three-factor interactions, and 64 interactions including four or more factors. The 128 treatment combinations provide a large amount of information, perhaps even more than is necessary for high-order interactions.

Reasons for Fractional Factorial Use

Design Redundancy in the Absence of High-Order Interactions

The experiences with studies involving many factors have led to the observation that at some point higher order interactions tend to become negligible and may be ignored in the overall scheme of preliminary investigations. In practice main effects generally tend to be larger than two-factor interactions, which in turn tend to be larger than three-factor interactions, and so forth.

The complete factorial has a degree of redundancy if the investigator can be reasonably assured that the high-order interactions are negligible. Under these circumstances estimates of main effects and low-order interactions can be obtained from a fraction of the full factorial treatment design.

The Factor Sparsity Hypothesis

The use of fractional factorial designs in industrial research (Diamond, 1989) or biotechnology (Haaland, 1989) rests to a large degree on a *factor sparsity hypothesis* (Box & Meyer, 1986). Factor sparsity assumes a small fraction of factor effects are large and of significance to a process, while the remaining effects are inert for all practical purposes. Thus, a large fraction of the variation is associated with only a few factors.

Screening Studies with Fractional Factorials

The fractional factorial also is useful in screening studies that include "major" factors and "minor" factors. The effects and interactions associated with major factors are the primary objective of the study. However, the study can include a number of minor factors that must be checked for their effects even though most or all are expected to be negligible. The features, construction, and analysis of fractional factorials are explored in this chapter beginning with the half fraction experiment in the next section.

12.2 The Half Fraction of the 2^n Factorial

The half fraction design is referred to as a 2^{n-1} fractional factorial design because $\frac{1}{2}2^n = 2^{n-1}$. The notation indicates the design includes n factors each at two levels that use only 2^{n-1} experimental units.

When one replication of 2^n factorial was placed into two incomplete blocks in Chapter 11 a defining contrast was used to separate the treatment combinations into two sets. Each of the two sets was placed into one of the incomplete blocks according to the $+$ and $-$ coefficients of the treatment combinations in the defining contrast. Each of the blocks was half of a complete replication of the treatments. Although the defining contrast was confounded with blocks, it was possible to estimate the remaining effects.

The same principle is used to construct fractional factorial designs. A half replicate of the design consists of all treatment combinations with the $+$ coefficient of the defining contrast.

Example 12.1 Truck Leaf Spring Manufacture Revisited

Recall from Example 11.1 that an assembled truck leaf spring was passed through a high-temperature furnace. Afterward, it was put into a forming machine to induce curvature in the spring by holding the spring in a high-pressure press for a short length of time. The factors and levels for the 2^3 factorial were (A) furnace temperature at 1840° and 1880° F, (B) heating time at 23 and 25 seconds, and (C) transfer time at 10 and 12 seconds.

The eight treatment combinations, spring quality observations, and coefficients to estimate the factorial effects are shown in Table 12.1. Notice the eight treatments are divided into two groups of four treatments using the defining contrast based on the ABC, three-factor interaction. This particular division of treatments could have been used to construct an incomplete block design with the ABC interaction confounded with blocks.

Table 12.1 Treatment combinations required for a 2^3 factorial treatment design

Treatment	I	A	B	C	AB	AC	BC	ABC	y
				Factorial Effect					
a	$+$	$+$	$-$	$-$	$-$	$-$	$+$	$+$	35
b	$+$	$-$	$+$	$-$	$-$	$+$	$-$	$+$	28
c	$+$	$-$	$-$	$+$	$+$	$-$	$-$	$+$	48
abc	$+$	$+$	$+$	$+$	$+$	$+$	$+$	$+$	29
(1)	$+$	$-$	$-$	$-$	$+$	$+$	$+$	$-$	32
ab	$+$	$+$	$+$	$-$	$+$	$-$	$-$	$-$	31
ac	$+$	$+$	$-$	$+$	$-$	$+$	$-$	$-$	39
bc	$+$	$-$	$+$	$+$	$-$	$-$	$+$	$-$	28

If the engineers wanted to construct a half replicate fractional factorial design they would use the four treatments in the top half of Table 12.1 with a + coefficient for the ABC factorial effect.

Gains and Losses with the Fractional Factorial

Each Treatment Effect Has an Alias

The gain in reduced size of the experiment comes with a price—the loss of some information on the treatment effects. If a half replicate of the 2^3 factorial is used, we lose the ability to estimate the three-factor interaction and each main effect is confounded or aliased with a two-factor interaction. The contrasts for the main effects for A, B, and C are

$$l_A = a - b - c + abc = 35 - 28 - 48 + 29 = -12$$

$$l_B = -a + b - c + abc = -35 + 28 - 48 + 29 = -26$$

$$l_C = -a - b + c + abc = -35 - 28 + 48 + 29 = 14$$

Contrasts for the BC, AC, and AB interaction effects are

$$l_{BC} = a - b - c + abc = 35 - 28 - 48 + 29 = -12$$

$$l_{AC} = -a + b - c + abc = -35 + 28 - 48 + 29 = -26$$

$$l_{AB} = -a - b + c + abc = -35 - 28 + 48 + 29 = 14$$

Notice that $l_A = l_{BC}$ and the contrast $l = (a - b - c + abc)$ estimates the combined effect $A + BC$, making it impossible to differentiate between the main effect of temperature (A) and the interaction between heating time and transfer time (BC). Two effects estimated by the same contrast are known as *aliases*.

Similar relationships exist with the linear contrasts for the main effects of heating time (B) and transfer time (C), $l_B = l_{AC}$ and $l_C = l_{AB}$. The linear contrast l_B estimates $B + AC$ and l_C estimates $C + AB$. Therefore, B and AC are aliases and C and AB are aliases. The alias relationships are

$$A = BC$$
$$B = AC$$
$$C = AB$$

The ABC interaction used as the defining contrast is known as the *design generator*. The three-factor interaction ABC used as the defining contrast cannot be estimated from that half replication because it has the same coefficients ($+$) in

Table 12.1 as the estimate of the mean, or the identity column I. The identity relationship

$$I = ABC$$

is known as the *defining relation* for the design.

Generate the Design with Either Half of the Design Generator

Either half of the ABC contrast can be used for the design generator. If we use the treatment combinations with a $-$ coefficient for the ABC interaction from the bottom half of Table 12.1 as the half replication, the defining relation for the design is $I = -ABC$. The contrast for the main effect of temperature (A) is

$$l_A = [-(1) + ab + ac - bc]$$

The contrast for the interaction between heating time and transfer time (BC) is

$$l_{BC} = [(1) - ab - ac + bc]$$

Thus, $l_A = -l_{BC}$; the temperature main effect is the negative of the heating time and transfer time interaction effect.

Similarly, the main effects of heating time (B) and transfer time (C) are the negative of the AC and AB interactions, respectively. The relationships between the main effects and two-factor interactions are $A = -BC$, $B = -AC$, and $C = -AB$. Regardless of which design half is used, the main effects are aliased with the two-factor interactions.

The alias for any treatment effect can be determined readily from its generalized interaction with the defining contrast ABC. (The determination of generalized interactions was explained in Chapter 11.) The aliases for the main effects are determined from their respective products with the defining contrast as

$$A \times ABC = A^2BC = BC$$
$$B \times ABC = AB^2C = AC$$
$$C \times ABC = ABC^2 = AB$$

Had the defining relation $I = -ABC$ been used, the aliases would be

$$A \times (-ABC) = -A^2BC = -BC$$
$$B \times (-ABC) = -AB^2C = -AC$$
$$C \times (-ABC) = -ABC^2 = -AB$$

Redundancy in 2^n Factorials

When there are no two-factor or three-factor interactions in the 2^3 factorial, the main effects for A, B, and C can be estimated from a half replicate of the design defined by either $I = ABC$ or $I = -ABC$. Thus, in the absence of interaction there is a redundancy in the complete design such that main effects can be estimated with two different contrasts, each from a different half of the design. The half fraction derived from $I = ABC$ is commonly known as the *principal fraction*, while the half fraction derived from $I = -ABC$ is the *complementary fraction*.

The two half fractions form a complete 2^3 design. If each is run on a separate occasion the resulting design is an incomplete block design with two blocks of four treatments each. The ABC interaction is confounded with blocks, but all main effects and two-factor interactions can be estimated.

How to Construct Half Replicate 2^{n-1} Designs

The half fraction design is constructed with the highest order interaction as the design generator. The treatment combinations are identified as follows:

- Write the $+$ and $-$ coefficients in standard order for the factors in a 2^{n-1} factorial.

- Identify the $\pm$ coefficients for the nth factor by equating them to the coefficients for the highest order interaction in the 2^{n-1} factorial.

The construction of a half replicate 2^{4-1} design is illustrated in Table 12.2. The $-$ and $+$ coefficients for the main effects of A, B, and C for the $2^{4-1} = 2^3$ factorial are written in standard order. The coefficients for the ABC interaction are taken as the level of the fourth factor, D, to be combined with the levels of the three other factors. The coefficients for the ABC interaction contrast are obtained as the product of the coefficients for the three main effects, as shown in the fourth column of Table 12.2.

Table 12.2 Construction of a 2^{4-1} fractional factorial design

A	B	C	$D = ABC$	$Treatment$
$-$	$-$	$-$	$-$	(1)
$+$	$-$	$-$	$+$	ad
$-$	$+$	$-$	$+$	bd
$+$	$+$	$-$	$-$	ab
$-$	$-$	$+$	$+$	cd
$+$	$-$	$+$	$-$	ac
$-$	$+$	$+$	$-$	bc
$+$	$+$	$+$	$+$	$abcd$

For example, the product of the coefficients for A, B, and C in the first row of coefficients is a $-$ sign. Thus, the first row of coefficients is $(- - - -)$ for the treatment combination (1). The second row of coefficients is $(+ - - +)$ for the treatment combination ad.

Build the Half Replicate with the Evens–Odds Rule

The highest order interaction is the design generator on the basis of the Evens–Odds rule. The resulting treatment combinations required for the half replicate are shown in the right-hand column of Table 12.2. They are the treatment combinations that receive a $+$ coefficient in the $ABCD$ interaction by the Evens–Odds rule since each has an even number of effect letters.

A table of design generators for fractional factorials is given in the Appendix. The generator for the 2^{4-1} design in Table 12.2 is listed in Appendix Table 12A.1 as $D = ABC$ to indicate how the levels for the fourth factor, D, are determined from the levels of the other factors.

The design in Table 12.2 is a complete factorial for A, B, and C if factor D is omitted from the design. Regardless of which factor is omitted from the half fraction design, the resulting design is a complete factorial in the remaining effects.

Choose the Highest Order Interaction for the Design Generator

The highest order interaction of least interest ordinarily is used to generate the half replicate because the defining contrast chosen for the design generator cannot be estimated. Suppose a half replicate of a 2^5 factorial is generated with the $ABCDE$ contrast. The design requires $2^{5-1} = 16$ experimental units. The complete set of aliases is shown in Table 12.3.

Table 12.3 Aliases for main effects and two factor interactions in a half fraction of the 2^5 factorial design

Main Effects	Alias	Two-Factor Interactions	Alias
A	$BCDE$	AB	CDE
B	$ACDE$	AC	BDE
C	$ABDE$	AD	BCE
D	$ABCE$	AE	BCD
E	$ABCD$	BC	ADE
		BD	ACE
		BE	ACD
		CD	ABE
		CE	ABD
		DE	ABC

Each main effect has a four-factor interaction as an alias, and each two-factor interaction has a three-factor interaction as an alias. For example, the alias of B is found as the generalized interaction $B \times ABCDE = AB^2CDE = ACDE$, and the alias of DE is $DE \times ABCDE = ABCD^2E^2 = ABC$.

Use the Fractional Factorial to Guide the Experimental Process

A sound experimental strategy in industrial research will utilize fractional factorial designs. For example, a half replicate of the 2^5 factorial requires 16 runs of an experimental process. Suppose the first 16 required runs were completed. The second fraction of 16 runs can always be run later in a second block if necessary to complete the factorial. There may be sufficient information on main effects and two-factor interactions from the first fraction that signals the need to change levels of some factor(s) or allows elimination of some inert factor(s). Thus, the investigator can move on to the next stage of experiments. The resources previously needed for the second set of 16 runs can be more wisely expended on the new experiments.

Randomization with fractional factorials is achieved by performing the actual runs of the treatment combinations in a random order if the treatments must be tested in sequence. If all treatments are tested on a set of physical experimental units they are randomly assigned to the units.

12.3 Design Resolution Related to Aliases

Fractional factorial designs are grouped into classes according to existing alias relationships in the design. The groups are identified by their **resolution**. The most common designs are those of Resolution III, IV, and V.

Resolution III: A design in which no main effect is confounded with any other main effect, but main effects are confounded with two-factor interactions and two-factor interactions are confounded with other two-factor interactions.

Resolution IV: A design in which no main effect is confounded with any other main effect or two-factor interaction, but two-factor interactions are confounded with one another.

Resolution V: A design in which no main effect or two-factor interaction is confounded with any other main effect or two-factor interaction, but two-factor interactions are confounded with three-factor interactions.

The resolution of a design is determined by the smallest number of characters appearing in the design generator. The 2^{3-1} design generated from the ABC contrast with three characters was a design of Resolution III, because main effects were not confounded with each other but were confounded with two-factor interactions. The 2^{5-1} design generated from the $ABCDE$ contrast with five characters was a design of Resolution V because there was no confounding among the main effects or two-factor interactions, but the two-factor interactions were confounded with three-factor interactions. The 2^{4-1} design generated from the $ABCD$ interaction with four characters is a design of Resolution IV.

In general, any fractional factorial design of Resolution R has no q-factor interaction confounded with any effect consisting of less than $R - q$ factors. The notation used to identify a design along with its resolution contains a Roman subscript indicating the design resolution. Thus, the notation 2_V^{5-1} identifies a 2^{5-1} fractional factorial with Resolution V.

12.4 Analysis of Half Replicate 2^{n-1} Designs

The analysis of replicated factorial experiments was discussed and illustrated in Chapters 6, 7, and 11. The replicated designs provided an estimate of experimental error variance from an analysis of variance. However, with fractional replication of the experiment no direct estimate of experimental error is available to evaluate the significance of factor effects and interactions. A strategy commonly used to estimate experimental error from fractional replications of 2^n experiments is illustrated with a 2_V^{5-1} fractional factorial experiment in Example 12.2.

Example 12.2 Polymer Coatings for Aluminum Cases

Research Objective: A company had received a contract to manufacture aluminum cases with a plastic polymer coating on the case for transceivers. A research team was given the responsibility to develop a process to adhere the polymer coating to the aluminum. The team identified five factors in the process that they thought had any potential to affect adhesion of the polymer to the aluminum. They were the type of aluminum alloy (*A*), the type of solvent used to clean the aluminum (*S*), the molecular structure of the coating polymer (*M*), the percent catalyst used in the adhesion process (*C*), and the curing temperature for the process (*T*).

Treatment Design: A 2^5 factorial design was chosen so they could take advantage of the factorial treatment structure to evaluate important interactions along with the main effects. They used two aluminum alloys, two types of solvents, two molecular structures for the polymer, 10% or 15% of the catalyst, and 150° or 175° as a curing temperature.

Experiment Design: With this preliminary experiment, the team wanted to identify any main effects and two-factor interaction effects that were important to the process. They also wanted to reduce the number of experimental units and time required for this first experiment.

They chose to use a 2_V^{5-1} fractional factorial or a half replicate design that required only 16 experimental units. The five-factor interaction $ASMCT$ was the defining contrast used to generate the design. Thus, they could run the second half of the design later if it became necessary to complete the factorial. Likewise, the half fraction might provide sufficient information to signal the need to alter the levels of some factors or eliminate some with no effect.

The treatment combinations, the coefficients for main effect and two-factor interaction contrasts, and the responses from the experimental runs are shown in Table 12.4. The force required to remove the plastic coating from the aluminum case was used as the response variable.

Table 12.4 Force required to remove plastic coating from an aluminum surface for treatments in a 2_V^{5-1} fractional factorial experiment

A	S	M	C	T	AS	AM	AC	AT	SM	SC	ST	MC	MT	CT	Treatment	Force[*]
−	−	−	−	+	+	+	+	+	−	+	+	−	+	−	−	t	41.5
+	−	−	−	−	−	−	−	−	+	+	+	+	+	+	a	39.6
−	+	−	−	−	−	+	+	+	−	−	−	+	+	+	s	43.9
+	+	−	−	+	+	−	−	+	−	−	+	+	−	−	ast	38.8
−	−	+	−	−	+	−	+	+	−	+	+	−	−	+	m	48.7
+	−	+	−	+	−	+	−	+	−	+	−	−	+	−	amt	52.0
−	+	+	−	+	−	−	+	−	+	−	+	−	+	−	smt	55.8
+	+	+	−	+	+	−	−	+	−	−	−	−	+	asm	43.2
−	−	−	+	−	+	+	−	+	+	−	+	−	+	−	c	39.5
+	−	−	+	+	−	−	+	+	+	−	−	−	−	+	act	42.6
−	+	−	+	+	−	+	−	−	−	+	+	−	−	+	sct	44.0
+	+	−	+	−	+	−	+	−	−	+	−	−	+	−	asc	33.8
−	−	+	+	+	+	−	−	−	−	−	−	+	+	+	mct	53.6
+	−	+	+	−	−	+	+	−	−	−	+	+	−	−	amc	48.1
−	+	+	+	−	−	−	−	+	+	+	−	+	−	−	smc	51.3
+	+	+	+	+	+	+	+	+	+	+	+	+	+	+	asmct	48.7

[*]Example computation for effect estimates [see Equation (12.1)]:

$A = 2(-41.5 + 39.6 - 43.9 + 38.8 - 48.7 + 52.0 - 55.8 + 43.2 - 39.5 + 42.6 - 44.0 + 33.8 - 53.6 + 48.1 - 51.3 + 48.7)/16 = 2(-31.5)/16 = -3.94$

The + and − coefficients for the first four factors (A, S, M, and C) of a $2^{5-1} = 2^4$ factorial are written in standard order in Table 12.4. Coefficients for the $ASMC$ interaction are used to identify the levels of the fifth factor (T) to be used in the treatment combinations. The 16 treatment combinations shown in Table 12.4 all have a + coefficient in the five-factor interaction, $ASMCT$, because they have an odd number of effect letters.

The 2_V^{5-1} fractional factorial design has no main effect or two-factor interaction aliased with any other main effect or two-factor interaction (see Table 12.3). The main effects are aliased with four-factor interactions (for example, $l_A = l_{SMCT}$), and two-factor interactions are aliased with three-factor interactions (for example, $l_{AS} = l_{MCT}$). Assuming the three-factor and four-factor interactions are negligible, estimates of the main effects and two-factor interactions can be used to identify the candidate factors for more in-depth investigation in follow-up studies if necessary.

The effect of a factor in a fractional factorial is estimated as

$$AB... = \frac{2(l_{AB...})}{N}$$ (12.1)

where $l_{AB...}$ is the contrast of the treatment combinations and N is the total number of observations in the experiment. The 1 degree of freedom sum of squares for an effect in a 2^{n-1} fractional factorial is

$$SS(AB...) = \frac{1}{2^{n-1}}(l_{AB...})^2 \qquad (12.2)$$

The estimates of the main effects and two-factor interaction effects are listed in order of increasing value from -3.94 to $+9.71$ in Table 12.5. An example calculation for the effect estimate of factor A is given at the bottom of Table 12.4. The estimates of the other effects can be verified by using Equation (12.1).

Table 12.5 Estimates of main effects and two-factor interactions with their normal scores sum of squares from the 2^{5-1} fractional factorial data in Table 12.4

Effect	Estimate	Normal Quantile	Sum of Squares
A	− 3.94	− 1.74	62.02
AS	− 3.69	− 1.24	54.39
S	− 0.76	− 0.94	2.33
SC	− 0.74	− 0.71	2.18
AM	− 0.41	− 0.51	0.68
C	− 0.24	− 0.33	0.23
SM	− 0.09	− 0.16	0.03
AC	0.14	0.00	0.08
ST	0.16	0.16	0.11
CT	0.44	0.33	0.77
AT	0.74	0.61	2.18
MC	0.74	0.61	2.18
MT	1.09	0.94	4.73
T	3.61	1.24	52.20
M	9.71	1.74	377.33

Examine Effects with a Normal Probability Plot

The observations from the experiment are assumed to be normally distributed with a constant variance σ^2. The observations represent random normal variation about a fixed mean if the changes in the levels of the factors have no real effect on the response. Thus, the estimated factorial effects from these observations are normally distributed about a mean of 0 with variance σ^2 when the factors have no effect. The estimated effects will plot on a straight line in the normal probability plot (see Chapter 4). Estimated effects that do not fit on the straight line cannot be explained as random chance variation.

The corresponding normal quantile computed for each effect is shown in Table 12.5. Recall that the *normal quantile* is the expected value for the normally distributed variable with a mean of 0 and variance of 1 found in that position of the ordered effects (see Chapter 4).

The normal probability plot of main effects and two-factor interaction effects is shown in Figure 12.1. The main effects that do not plot on the straight line of the normal probability plot are the effects of the aluminum alloy (A), molecular structure of the polymer (M), and the curing temperature (T). The effect estimate for the interaction between aluminum alloy and type of solvent (AS) also deviates considerably from the straight line of the normal probability plot. All other effects appear to be representative of random experimental error.

Interpretations of the Largest Effects

Graphs of the effects are shown in Figure 12.2. The positive estimate for T, 3.61, in Figure 12.2a indicates a high level of curing temperature produced stronger adhesion. Likewise, the positive estimate for M, 9.71, in Figure 12.2b indicates the molecular structure represented by the "high" level produced stronger adhesion.

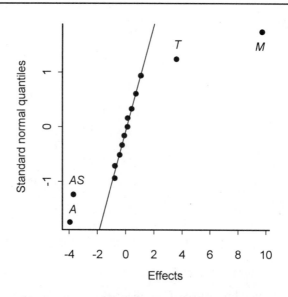

Figure 12.1 Normal probability plot of estimated factor effects and interactions from the 2^{5-1} fractional factorial in Example 12.2

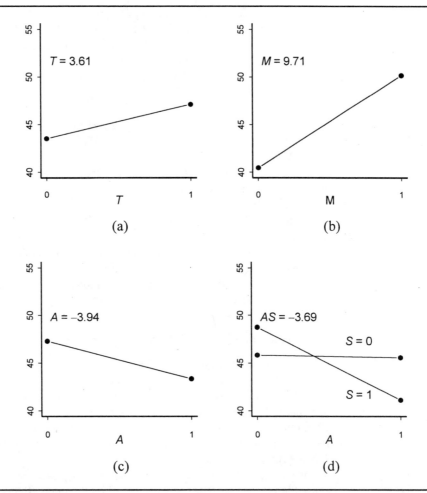

Figure 12.2 Graphs for the main effects of T, M, and A and the AS interaction

The negative estimate for A, -3.94, in Figure 12.2c suggests the "low" level of the aluminum alloy factor produced a stronger adhesion. However, the large estimate for AS interaction, -3.69, may alter the inference about the main effect of A, the aluminum alloy.

An investigation of the interaction between the aluminum alloy (A) and the solvent used to clean the aluminum (S) requires the averages of the four treatment combinations involving A and S. From Table 12.4 the values are

Treatments	A	S	Total	Mean
$t + m + c + mct$	0	0	183.3	45.8
$a + amt + act + amc$	1	0	182.3	45.6
$s + smt + sct + smc$	0	1	195.0	48.8
$ast + asm + asc + asmct$	1	1	164.5	41.1

The effect of A when $S = 0$ is $45.6 - 45.8 = -0.2$, whereas the effect of A when $S = 1$ is $41.1 - 48.8 = -7.7$. Therefore, the difference between the two alloys with the second solvent ($S = 1$) is much greater than that with the first solvent ($S = 0$), as shown in Figure 12.2d.

Some Cautions About the Assumptions

It is important to remember that because of aliasing the estimated effects judged important for the process are really $A + SMCT$, $M + ASCT$, $T + ASMC$, and $AS + MCT$. Thus, the assumption of negligible three-factor and higher interactions is crucial to the decisions regarding the importance of the three main effects and one two-factor interaction to the bonding process.

Daniel (1959) pointed out some possible misinterpretations with a normal plot of estimated effects in unreplicated factorials. A perceived smooth line of plotted effect estimates in one portion of the normal plot thought to indicate no evidence of factor effects could lead to an overestimate of error variance. An extremely irregular line could lead to an opposite error of judging effects real when in fact they only represent random experimental error. The plot is most effective when only a small proportion of the factorial effects are important.

One Method to Estimate Experimental Error Variance

A test for the significance of these effects requires an estimate of experimental error variance. One general strategy to obtain an estimate involves several steps. The first step selects those effects that appear to be negligible on the normal probability plot and pools their sums of squares to estimate experimental error. Based on Figure 12.1, only the effects for A, M, T, and AS appear to be significant and all other effects appear to be negligible and representative of random experimental variation. The analysis of variance for a model containing effects only for A, M, T, S, and AS is shown in Table 12.6. The main effect for S was included in the analysis because the interaction effect AS is defined on the main effects in the model (Chapter 6) and should properly be included in the model to obtain correct sums of squares

Table 12.6 Analysis of variance for the reduced model with A, M, T, and AS effects for the 2^{5-1} fractional factorial (Example 12.2)

Source of Variation	Degrees of Freedom	Sum of Squares	Mean Square	F	Pr > F
Total	15	561.41			
A	1	62.02	62.02	44.11	0.000
M	1	377.33	377.33	268.36	0.000
T	1	52.20	52.20	37.12	0.000
S	1	2.33	2.33	1.77	0.213
AS	1	54.39	54.39	38.68	0.000
Error	10	13.14	1.31		

values. Notice that S is not significant and could be pooled with the sum of squares for error following the general strategy. The pooled sum of squares for the effects that appear negligible is $SSE = SS(\text{Error}) + SSS = 13.14 + 2.33 = 15.47$ from the analysis of variance in Table 12.6 with 11 degrees of freedom. Thus, an estimate of experimental error variance is $s^2 = 15.47/11 = 1.41$.

No Substitute for Replication

A replicated experiment would have provided a legitimate estimate of σ^2. The validity of any significance test is questionable when the variance estimate is based on subjective judgments. The method just described to estimate experimental error variance from an unreplicated experiment is very subjective.

A number of more formal procedures have been developed to address the problem of variance estimation in unreplicated fractional factorials. However, each of them requires a degree of subjective judgment on the part of the investigator. None of them are discussed here but references are given for further reading.

Berk and Picard (1991) reviewed and evaluated some of the methods including one of their own. More computational intensive procedures were proposed by Box and Meyer (1986) and Zahn (1975a, 1975b). The Zahn methods are extensions of the original half normal plot methods of Daniel (1959). Some of the procedures (Lenth, 1989; Voss, 1988) proposed simpler methods based on standard analysis of variance procedures.

On the basis of simulation studies, Berk and Picard (1991) concluded most of the methods produced nearly identical error rates and most were prone to declare null effects to be real.

The dangers inherent in unreplicated fractional factorials cannot be ignored. Only replication can protect against selecting spurious effects as legitimate effects. Fractional factorials are legitimate studies if they are recognized as preliminary screening studies to prepare the way for more rigorous replicated experiments.

Standard Errors and Tests of Hypotheses About Effects

The variance of an effect estimate in a fractional factorial is

$$\text{variance} = \frac{4\sigma^2}{N} \tag{12.3}$$

where σ^2 is the experimental error variance and N is the number of observations in the experiment. The estimated experimental error variance is $s^2 = 1.41$ with 11 degrees of freedom from which a standard error estimate can be calculated for each of the effects in Table 12.5. The variance estimate for any effect is

$$\frac{4s^2}{N} = \frac{4(1.41)}{16} = 0.35$$

and the estimated standard error of the effect estimates is $\sqrt{0.35} = 0.59$. The significance of each effect can be determined with the Student t test as a ratio of the effect estimate to the standard error or with the F test as shown in Table 12.6. All of the suspected effects are significant according to the F tests in Table 12.6.

A Residual Plot to Evaluate the Model

The estimate for experimental error variance was found by assuming all effects other than those in the model were negligible. If effects other than those in the model are negligible the residual plots for the fitted model should provide evidence for correctness of model choice. The residuals were computed for the model including only the A, M, T, S, and AS effects. The plot of residuals versus predicted values and the normal probability plot of the residuals are shown in Figure 12.3. The residuals all appear to lie on a relatively straight line in the normal probability plot, and there does not appear to be any reason to suspect heterogeneous variances in the residuals versus the fitted values plot. Also, no major effects appear to be absent from the model since there are no evident outliers in the residual plots.

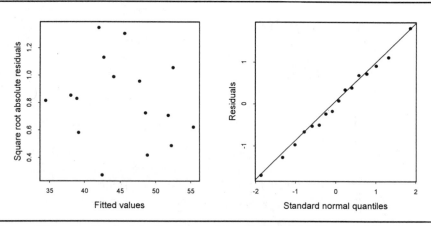

Figure 12.3 Residual plots for the model with A, M, T, S, and AS effects for Example 12.2

The catalyst, factor C, had no detectable effects and appears to be an inert factor. Thus, the design was effectively a single replicate of a complete 2^4 factorial in the other factors A, S, M, and T. The research team now can consider an in-depth study of the other factors and their interactions.

12.5 The Quarter Fractions of 2^n Factorials

Smaller fractions for initial runs of the 2^n factorial may provide sufficient information for critical decisions about the effectiveness of factors, especially when a moderately large number of factors are under consideration or individual runs are quite expensive. The quarter replicate of the 2^n design is designated as a $\frac{1}{4}2^n = 2^{n-2}$ fractional factorial.

A quarter replicate of a 2^6 design is a 2^{6-2} fractional factorial that requires only 16 runs of the 64 runs required for the complete 2^6 design. The quarter replicate corresponds to one block of the incomplete block design for a 2^6 design in four blocks of 16 units per block (see Chapter 11). Any of the four blocks could be used for the 2^{6-2} design.

Use Two Defining Contrasts for Quarter Fraction Designs

Two defining contrasts are required to construct the four incomplete blocks, and they along with their generalized interaction are confounded with blocks. These same defining contrasts can be used to construct the 2^{6-2} design. Suppose $ABDE$ and $ABCF$ are chosen as the design generators. The generalized interaction of the design generator is

$$ABDE \times ABCF = A^2B^2CDEF = CDEF$$

The design generators along with their generalized interaction complete the defining relationship for the design as

$$I = ABDE = ABCF = CDEF$$

The treatment combinations required for the design can be identified in a manner analogous to that demonstrated for the half replicate design in Table 12.2. The $+$ and $-$ coefficients for the main effects of A, B, C, and D are written in standard order for a $2^{6-2} = 2^4$ complete factorial as shown in Table 12.7.

The aliases of E and F that contain any of the factor letters A through D are used to determine coefficients for E and F in the treatment combinations. Given the design generators $ABDE$, $ABCF$, and their generalized interaction $CDEF$, the aliases of E and F, respectively, are

$$E = ABD = CDF = ABCEF$$

and

$$F = ABC = CDE = ABDEF$$

The equivalencies $E = ABD$ and $F = ABC$ can be used to determine the coefficients for E and F. That is, the coefficients for ABD are taken as the coefficients for E, and the coefficients for ABC are taken as the coefficients for F. The treatments required for the design are shown in Table 12.7.

The defining contrasts for the quarter replicate design have four letters, so the resulting design is of Resolution IV, or a 2^{6-2}_{IV} fractional factorial. No main effect is aliased with any other main effect or two-factor interaction, but two-factor interactions are aliased with one or more other two-factor interactions. The complete set of aliases are shown in Table 12.8.

Table 12.7 Construction of a 2^{6-2}_{IV} fractional factorial design

			Design			
A	B	C	D	E = ABD	F = ABC	Treatment
−	−	−	−	−	−	(1)
+	−	−	−	+	+	aef
−	+	−	−	+	+	bef
+	+	−	−	−	−	ab
−	−	+	−	−	+	cf
+	−	+	−	+	−	ace
−	+	+	−	+	−	bce
+	+	+	−	−	+	abcf
−	−	−	+	+	−	de
+	−	−	+	−	+	adf
−	+	−	+	−	+	bdf
+	+	−	+	+	−	abde
−	−	+	+	+	+	cdef
+	−	+	+	−	−	acd
−	+	+	+	−	−	bcd
+	+	+	+	+	+	abcdef

Table 12.8 Alias relationships for the 2^{6-2}_{IV} fractional factorial with defining relationship $I = ABDE = ABCF = CDEF$

	Alias from		
Effect	ABDE	ABCF	CDEF
A	BDE	BCF	ACDEF
B	ADE	ACF	BCDEF
C	ABCDE	ABF	DEF
D	ABE	ABCDF	CEF
E	ABD	ABCEF	CDF
F	ABDEF	ABC	CDE
AB	DE	CF	ABCDEF
AC	BCDE	BF	ADEF
AD	BE	BCDF	ACEF
AE	BD	BCEF	ACDF
BC	ACDE	AF	BDEF
CD	ABCE	ABDF	EF
CE	ABCD	ABEF	DF
ACD	BCE	BDF	AEF
ACE	BCD	BEF	ADF

A $1/2^p$ fraction of a 2^n design is designated as a 2^{n-p} fractional factorial. In general a $1/2^p$ fraction of the 2^n factorial will require p design generators with

$2^p - p - 1$ generalized interactions. Each effect will have $2^p - 1$ aliases. The $1/2^2$ or quarter fraction 2_{IV}^{6-2} design in Table 12.7 required $p = 2$ design generators resulting in $2^2 - 2 - 1 = 1$ generalized interaction. There were $2^2 - 1 = 3$ aliases for each effect.

Choose Design Generators to Avoid a Bad Alias Structure

Care must be exercised when selecting design generators to avoid aliasing those effects of interest in the study with one another. Thus, it is important to look at the alias structure prior to assigning actual factors to the factor letters when setting up the experiment.

Notice that the 2_{IV}^{6-2} design in Table 12.7 consists of a complete replication of the 2^4 design in A, B, C, and D. It is also a complete replication of a 2^4 design in any four factors that do not combine to form the defining relationships for the design. Any combination of factors other than $ABDE$, $ABCF$, and $CDEF$ will produce a complete 2^4 factorial. Upon close inspection the design in Table 12.7 is seen to consist of two replications of a 2^{4-1} design for experiments, consisting of the four factors used in any one of the defining relationships considered separately.

The 2^{n-p} design will contain a complete factorial for any $n - p$ factors that do not make up a defining relationship. A fractional factorial can be made up from some subset of $n - p$ factors. If it is suspected at the outset that some of the factors will have negligible effects, then the original 2^{n-p} fractional factorial can be set up so that a full factorial will exist for the factors of most interest or those expected to have real effects.

12.6 Construction of 2^{n-p} Designs with Resolution III and IV

Resolution III Designs with N Experimental Units for $N - 1$ Factors

A design of Resolution III can be constructed to investigate $n = N - 1$ factors using N experimental units, where N is a multiple of 4. For example, the half fraction of a 2^3 factorial in Table 12.1 is a 2_{III}^{3-1} fractional factorial that has $N - 1 = 3$ factors investigated with $N = 4$ experimental units. Designs of Resolution III have no main effects aliased with each other but do have main effects aliased with two-factor interactions.

A 2^{n-p} design of Resolution III requires p design generators. Suppose a design of Resolution III is desired for $N - 1 = 7$ factors in $N = 8$ runs. The eight runs would be a (1/16) fraction of the $2^7 = 128$ runs required for a full factorial. Thus, the design is a 2_{III}^{7-4} fractional factorial requiring $p = 4$ design generators.

The design construction begins by writing the $+$ and $-$ coefficients for $n - p$ main effects in standard order. The $+$ and $-$ coefficients for $n - p = 3$ main effects, A, B, and C, in standard order for a complete $2^{7-4} = 2^3$ factorial are shown in Table 12.9.

Table 12.9 A 2_{III}^{7-4} fractional factorial with design generators ABD, ACE, BCF, and $ABCG$

Unit	A	B	C	D = AB	E = AC	F = BC	G = ABC	Treatment
1	−	−	−	+	+	+	−	def
2	+	−	−	−	−	+	+	afg
3	−	+	−	−	+	−	+	beg
4	+	+	−	+	−	−	−	abd
5	−	−	+	+	−	−	+	cdg
6	+	−	+	−	+	−	−	ace
7	−	+	+	−	−	+	−	bcf
8	+	+	+	+	+	+	+	$abcdefg$

The coefficients required for the remaining factors, D, E, F, and G, are associated with the coefficients of all the interaction columns of A, B, and C. Thus, the columns that indicate the levels for factors D, E, F, and G are generated from the equivalence relationships

$$D = AB \qquad E = AC \qquad F = BC \qquad \text{and} \qquad G = ABC$$

The $p = 4$ design generators are then

$$D \times AB = ABD$$
$$E \times AC = ACE$$
$$F \times BC = BCF$$

and

$$G \times ABC = ABCG$$

The defining relationship is found by multiplying the four generators in all possible ways. The complete defining relationship is

$$I = ABD = ACE = AFG = BCF = BEG = CDG = DEF = ABCG$$

$$= ABEF = ACDF = ADEG = BCDE = BDFG = CEFG$$

$$= ABCDEFG$$

The alias structure of the design for the main effects is found by multiplying each of the main effects by the interactions included in the defining relationship. The two-factor interactions aliased with the main effects are

$$A = BD = CE = FG$$

$$B = AD = CF = EF$$

$$C = AE = BF = DG$$

$$D = AB = CG = EF$$

$$E = AC = BG = DF$$

$$F = AG = BC = DE$$

$$G = AF = BE = CD$$

The design in Table 12.9 has all available contrasts of the $2^{n-p} = 2^3$ factorial associated with the main effects along with their aliases, and the design is referred to as a *saturated design*. The design has 7 degrees of freedom available to estimate each of the seven main effects and no degrees of freedom to estimate experimental error. Designs of Resolution III can be used in screening experiments with many factors to identify clearly dominant factors without major expenses of time and other resources.

The Plackett–Burman Designs

Resolution III saturated designs with $n = N - 1$ factors requiring $N = 2^{n-p}$ experimental units can be generated from the basic 2^{n-p} factorial arrangement, as shown in the preceding paragraphs. The result is a design with the number of experimental units equal to the power of 2—experiments with $4, 8, 16, \ldots, 2^{n-p}$ units.

A class of Resolution III designs developed by Plackett and Burman (1946) requiring a number of experimental units equal to a multiple of 4 have been used extensively for screening experiments in industrial research. They provide designs for intermediate values of N that are not a power of 2. Designs were provided for $N \leq 100$, except for 92, by Plackett and Burman (1946). A sample of the most practical and easily constructed designs is given in Table 12.10 for $N = 12, 16, 20, 24$, and 32 experimental runs; the designs can be generated from rows of $+$ and $-$ signs.

Each row of generators in Table 12.10 has $N - 1$ coefficients. These are the $+$ and $-$ coefficients required for the first run of the $N - 1$ factors. The factor coefficients for the second run are generated from the first run by taking the first coefficient in the first run and placing it in the last position of the second run and shifting the coefficients of the first run one position to the left. Succeeding runs are generated in the same manner. Upon completing the cycle of coefficient replacements for $N - 1$ runs a final run is added with $-$ coefficients for all $N - 1$ factors. A design for $N - 1 = 11$ factors in $N = 12$ runs is shown in Table 12.11.

Table 12.10 Generators for Plackett–Burman designs with $N = 12, 16, 20, 24,$ and 32 runs

N	
12	$+ + - + + + - - - + -$
16	$+ - - - + - - + + - + - + + +$
20	$+ + - - + + + + - + - + - - - - - + + -$
24	$+ + + + + - + - + + - - + + - - + - + - - - -$
32	$- - - - + - + - + + + - + + - - - - + + + + + - - + + - + - - +$

Table 12.11 A Plackett–Burman design for $N - 1 = 11$ factors in $N = 12$ runs

					Factor						
Run	1	2	3	4	5	6	7	8	9	10	11
1	+	+	−	+	+	+	−	−	−	+	−
2	+	−	+	+	+	−	−	−	+	−	+
3	−	+	+	+	−	−	−	+	−	+	+
4	+	+	+	−	−	−	+	−	+	+	−
5	+	+	−	−	−	+	−	+	+	−	+
6	+	−	−	−	+	−	+	+	−	+	+
7	−	−	−	+	−	+	+	−	+	+	+
8	−	−	+	−	+	+	−	+	+	+	−
9	−	+	−	+	+	−	+	+	+	−	−
10	+	−	+	+	−	+	+	+	−	−	−
11	−	+	+	−	+	+	+	−	−	−	+
12	−	−	−	−	−	−	−	−	−	−	−

Although this 12-run design could be used to screen 11 factors, it has been recommended that the designs be used for 6 less factors than there are runs (Mason, Gunst, & Hess, 1989). Upon deleting 6 factors from the design, 5 degrees of freedom remain for an estimate of experimental error variance after the main effects are estimated from the design. Of course, this practice still assumes the main effects are the dominant effects in the experiment. For example, a 12-run design for 6 factors is obtained by using only the first six columns of the design in Table 12.11. There are a total of 11 degrees of freedom in the design, of which 6 can be used to estimate the main effects and 5 used for an estimate of experimental error variance.

Resolution IV Designs Require $2N$ Experimental Units for N Factors

The 2^{n-p} designs of Resolution IV have no main effects aliased with other main effects or two-factor interactions, and main effects can be estimated free of confounding with two-factor interactions. Such was not the case with Resolution III designs. A 2_{IV}^{n-p} design with $N = n$ must have at least $2n$ experimental units or runs.

A design of Resolution IV with n factors may be generated from a design of Resolution III with $n - 1$ factors by a *fold-over* technique. A second fraction is added to a Resolution III design for $n - 1$ factors in n runs. All signs in the second fraction are reversed from the first fraction. The I column in the first fraction has all $+$ coefficients, whereas in the second fraction the I column has all $-$ coefficients. This column of coefficients is used for an added factor. The design now has n factors with $2n$ runs.

The procedure is illustrated in Table 12.12 with the generation of a 2_{IV}^{4-1} design from a 2_{III}^{3-1} design. The first fraction of the design is a 2^{3-1} design of Resolution III, in which the coefficients for factor C are derived from the coefficients for AB interaction. The second fraction is derived from the first with opposite signs for the coefficients. The column originally labeled I is used for the fourth factor, D, to complete the 2^{4-1} design. The design can be seen to be of Resolution IV since it is a half fraction of a 2^4 design requiring one defining contrast. It is a complete 2^3 factorial in A, B, and D with the coefficients of C derived from $C = ABD$, so that the defining contrast is $C \times ABD = ABCD$. The resolution of the design is equal to the number of letters in the defining contrast $ABCD$, or Resolution IV.

Table 12.12 Generation of a 2^{4-1} Resolution IV design by folding over a 2^{3-1} Resolution III design

$I = D$	A	B	$C = AB$	
$+$	$-$	$-$	$+$	First fraction is a 2_{III}^{3-1} design
$+$	$+$	$-$	$-$	
$+$	$-$	$+$	$-$	
$+$	$+$	$+$	$+$	
$-$	$+$	$+$	$-$	Second fraction with opposite signs
$-$	$-$	$+$	$+$	
$-$	$+$	$-$	$+$	
$-$	$-$	$-$	$-$	

12.7 Genichi Taguchi and Quality Improvement

Finally, we note one special application of fractional factorial designs because of recent developments in the application of statistics to manufacturing, part of which deals directly with the issue of statistical designs.

Fractional factorial designs are used extensively in off-line experiments for product quality improvement. Off-line investigations integrate engineering design and statistical design principles to improve the quality of products and increase productivity. In particular, the Taguchi methodology (Taguchi, 1986) has had a major impact on product and process improvement design in manufacturing.

Robust parameter design is the one part of the complete Taguchi methodology that involves factorial treatment designs. The design consists of both factors that are controllable in the manufacturing process and those that are not controllable. The controllable factors are called *parameters* in Taguchi's terminology; those not controllable are called *noise* factors, or variables. The noise variables or factors are those most likely to be sensitive to changes in environmental conditions during production and thus transmit variability to the responses of interest in the process.

Both types of factors, controllable and noise, can be controlled in off-line experimentation, and that is an important consideration in the Taguchi methodology. One objective is to determine which combination of controllable factors is least sensitive to changes in the noise variables. This is the concept from which the term *robust parameter design* is derived. The best choice of controllable factor levels leads to a manufacturing process that results in the desired product and is *robust* to any fluctuations in the uncontrollable noise factors.

The design concept is illustrated with a hypothetical off-line experiment with three controllable factors, A, B, and C, each at two levels, and one noise factor, D, with three levels. An orthogonal array for the control factors is crossed with an orthogonal array for the noise factors. For our example with four factors, the 2^3 factorial for factors A, B, and C would constitute the *inner array* in Taguchi terminology, and the 3^1 for factor D would constitute the *outer array*.

The three control factors are configured in Table 12.13 in a $\frac{1}{2}$ fraction of a 2^3 factorial with 4 runs, or a 2^{3-1}_{III} fractional factorial, in the inner array. Three levels of the noise factor, D, in the outer array are used for each run in the 2^{3-1}_{III} design, resulting in 12 runs for the experiment.

Table 12.13   Hypothetical Taguchi experiment

A	B	C	D_0	D_1	D_2	$\bar{y}$	s
$-$	$-$	$+$	54	56	52	54	2.0
$+$	$-$	$-$	69	70	71	70	1.0
$-$	$+$	$-$	58	55	49	54	4.6
$+$	$+$	$+$	58	65	69	64	5.6

Suppose the manufactured product has a desired response target of 54. The mean, $\bar{y}$, and standard deviation, s, of the observations from the runs on the three levels of the noise factor, D, for each of the four runs on the control factors are shown in Table 12.13.

Two of the treatments for the control factors, $(-, -, +)$ and $(-, +, -)$, meet the desired product response of 54. However, the $(-, -, +)$ treatment, with a standard deviation of 2.0, is less sensitive to changes in levels of the noise factor (D) than the $(-, +, -)$ treatment with a standard deviation of 4.6. Thus, the product from the $(-, -, +)$ treatment would be considered the *more robust* product since fluctuations on the noise factor (D) transmit less variability to the product's response.

The Taguchi analysis method concentrates on the maximization of a signal-to-noise ratio (SNR) specific to the process goals. The primary choices for goals are

(1) minimize the response, (2) maximize the response, and (3) achieve a specified response target value other than minimum or maximum.

The method proceeds rather simply with an analysis of variance for the SNR that determines which control factors affect the SNR. Then the method continues with an analysis of variance for the $\bar{y}$ that ascertains which control factors impact the mean response and thus which control factors' levels can be set to achieve a process target response. Myers and Montgomery (1995) illustrate other details of the analysis for the interested reader.

Pignatiello and Ramberg (1991) discuss major advantages and disadvantages of the Taguchi approach. Myers and Montgomery (1995), Montgomery (1997), and Vining (1998) present more detailed descriptions, examples, and critiques of the Taguchi methodology. A major contribution of the Taguchi method was the validation of experimentation and statistical design as part of the quality improvement process. However, the Taguchi designs have evoked considerable controversy over the particular designs used and their implementation.

Although many standard 2^{n-p} fractional factorial designs are recommended in the Taguchi method, many are saturated or nearly saturated Plackett–Burman designs or fractional designs with three levels to detect response curvature. An assumption of no interaction among the control factors often is necessary for successful use of these designs. The crossing of orthogonal design for the control factors with that for the noise factors provides ample ability to estimate interactions between control and noise factors, but the ability to estimate important interactions among the control factors has been sacrificed to do so. Myers and Montgomery (1995) suggest several design strategies that could be more economical of experimental runs and yet provide the means to estimate interactions of interest to the investigator. Hunter (1985) presents a detailed discussion of the Taguchi three-level fractional designs based on Latin squares.

Better sequences of design strategies at the onset also have been suggested by Hunter (1989) and Pignatiello and Ramberg (1991). As one example, the sequence could start with 2^{n-p} fractional factorials to detect interactions with factor levels added to the center of the design to detect curvature. Additional suggested refinements as the process is characterized include the use of complete factorials and response surface designs (see Chapter 13).

As a final point, randomization is not advocated in the Taguchi approach; it can lead to important effects confounded with external variations being unaccounted for by the design.

12.8 Concluding Remarks

As we have seen in Chapters 11 and 12, factorials are versatile treatment designs that can be adapted to a variety of experimental conditions, including single replications, fractional replications, and experiments requiring incomplete block designs.

Only the rudimentary principles for the construction and analysis of these designs have been presented here. An extensive list of literature exists on fractional factorial designs for experimental work in industry. Extensive discussions on the design and analysis of fractional factorials in various fields of application may be found in Daniel (1976), Box, Hunter, and Hunter (1978), Diamond (1989), Haaland (1989), and Mason, Gunst, and Hess (1989). Some of the cited references contain tables of specific designs derived from the basic principles presented in this chapter. Diamond (1989) provides an extensive bibliography of published literature relating to fractional factorials.

An abundant number of commercial software products have been developed to produce designs and analyses for any experiments that require fractional factorial designs.

Fractional Designs for 3^n Factorials

Fractional replications may be constructed for 3^n factorials and other factorial series using the same principles as those for the 2^n series. One-third and one-ninth fractional factorials are possible with the 3^n series. Aliases for factorial effects are determined on the basis of symbolic products with the defining contrasts and squares of defining contrasts. An extensive selection of fractional factorial designs for the 3^n series is available in Conar and Zelen (1959). Details on the construction of fractional factorials in other than the 3^n series can be found in Kempthorne (1952) and John (1987). Examples of some designs can be found in Cochran and Cox (1957), Johnson and Leone (1977), and Montgomery (1997).

EXERCISES FOR CHAPTER 12

---

1. A list of designs is shown below.

i.	2^{5-1}	v.	2^{7-2}	ix.	2^{9-5}
ii.	2^{6-1}	vi.	2^{7-3}	x.	2^{10-6}
iii.	2^{6-2}	vii.	2^{7-4}	xi.	2^{11-6}
iv.	2^{6-3}	viii.	2^{8-4}	xii.	2^{15-11}

Indicate the following quantities for each design:
a. the fraction of the full design
b. the number of generators required
c. the number of generalized interactions
d. the number of aliases for each effect
e. the number of experimental units or runs required

2. Construct a 2^{5-1} fractional factorial design. Use $I = ABCDE$ as the defining relationship.
a. What is the resolution of the design?
b. Show the alias structure for main effects and two-factor interactions.

3. Construct a 2^{6-2} fractional factorial design. Use $ABCD$ and $CDEF$ as the design generators.
  a. What is the generalized interaction?
  b. What is the resolution of the design?
  c. Show the alias structure for all main effects and two-factor interactions.

4. Construct a Resolution IV design that is a $\frac{1}{8}$ fraction of a 2^7 factorial.

5. If you fold over a 2^{5-2}_{III} fractional factorial, is the resulting design a 2^{6-2}_{IV} fractional factorial? Explain.

6. A 2^{5-2} fractional factorial is proposed with two possible generating relations:

        i. $I = ABCD = BCE$

       ii. $I = ABCDE = ABCD$

  a. What fraction of a 2^5 design will it be?
  b. Which defining relationship is preferred for the design? Explain.

7. A 2^{5-2}_{III} fractional factorial is designed with defining relationship $I = ACE = BCDE$. It is known that factors A, B, and C do not interact with one another and factors C, D, and E do not interact with one another.
  a. Which effects can be estimated ignoring three-factor and larger interactions?
  b. Is it possible to have a better 2^{5-2}_{III} design for this situation? Explain.

8. A researcher conducts an experiment with eight runs using a 2^{5-2} fractional factorial constructed with the equivalencies $D = AB$ and $E = ABC$. Then, the researcher conducts another eight runs with the same five factors but reverses the signs used in the original eight runs for each of the factors.
  a. What is the resolution of the design with the original eight runs?
  b. What is the resolution of the 16-run design?
  c. Suppose it is known at the start that 16 runs are going to be used for the experiment. Can a better design of 16 runs be constructed? If so, construct the design and give its properties.

9. A 2^{6-3}_{III} fractional factorial design is generated with the equivalencies $D = AB$, $E = AC$, and $F = ABC$. What is the resulting design if the factors D and E are dropped from the design?

10. A process under study involved the heat treatment of leaf springs for trucks. The assembled leaf spring was put through a high-temperature furnace, transferred to a forming machine where the curvature of the spring was formed, and then submersed in an oil quench. The measurement of importance was the free height of the spring in the unloaded condition. The target value of the free height was 8 inches. Deviations from this value were considered undesirable. An experiment was conducted to evaluate the effect of four factors on the deviation of free height from the target value in the manufacturing process. The factors considered were (A) furnace temperature; (B) heating time; (C) transfer time, or the length of time to transfer the spring assembly from furnace to curvature former; and (D) hold-down time, or the time the curvature former is closed on the hot spring.

In addition, the engineers were interested in the two-factor interactions AB, AC, and BC as well as the main effects of each factor. A 2^{4-1} fractional factorial was used for the experiment with the defining relationship $I = ABCD$. The eight treatment combinations follow with the signal-to-noise ratio (Z) for six springs constructed with each treatment. The signal-to-noise ratio is a measure of the deviation of the springs from the target value.

Run	A	B	C	D	Z
1	+	−	−	+	29.46
2	−	−	+	+	28.11
3	−	−	−	−	28.00
4	+	−	+	−	30.59
5	+	+	+	+	35.31
6	+	+	−	−	38.68
7	−	+	+	−	31.55
8	−	+	−	+	47.70

Source: J. J. Pignatiello and J. S. Ramberg (1985), *Journal of Quality Technology* 17, 198–206. Discussion of article by R. N. Kackar (1985). Off-line quality control, parameter design, and the Taguchi method. *Journal of Quality Technology* 17, 176–188.

a.   What is the resolution of this design?
b.   Show how the + and − signs were determined for factor D.
c.   Show the alias structure for this design.
d.   What must be assumed to obtain estimates of the two-factor interactions AB, AC, and BC?
e.   Estimate the main effects and two-factor interactions of interest and their standard errors.
f.   Interpret the results.

11.   The manufacturer of instant soup products wanted to produce dry soup mix packages with minimum weight variation among packages. Five factors were identified that might influence variation in the filling process. A 2^{5-1} fractional experiment with the defining relationship $I = -ABCDE$ was conducted to evaluate the effects of the factors and their interactions. The factors and levels were (A) the number of mixer ports through which the vegetable oil was added (1 and 3), (B) the temperature surrounding the mixer (− = cooled or + = ambient temperature), (C) the mixing time (60 and 80 sec), (D) the batch weight (1500 and 2000 lb), and (E) the days of delay between mixing and packaging (1 and 7). Between 125 and 150 packages of soup mix were sampled over an eight-hour production run for each treatment combination. The standard deviation for the weight of the packages was computed as a measure of variation in the filling process and used as the response variable y. The factor levels and the process standard deviation for each of the 16 treatment combinations follow.

Run	A	B	C	D	E	y
1	-1	-1	-1	1	1	0.78
2	1	-1	1	1	1	1.10
3	1	1	-1	-1	-1	1.70
4	1	-1	1	-1	-1	1.28
5	-1	1	-1	-1	1	0.97
6	-1	-1	1	-1	1	1.47
7	-1	1	-1	1	-1	1.85
8	1	1	1	1	-1	2.10
9	-1	1	1	1	1	0.76
10	1	1	-1	1	1	0.62
11	-1	-1	1	1	-1	1.09
12	-1	-1	-1	-1	-1	1.13
13	1	-1	-1	-1	1	1.25
14	1	1	1	-1	1	0.98
15	1	-1	-1	1	-1	1.36
16	-1	1	1	-1	-1	1.18

Source: L. B. Hare (1988), In the soup: A case study to identify contributors to filling variability. *Journal of Quality Technology* 20, 36–43.

a. What is the resolution of this design?
b. Show how the $+$ and $-$ signs for levels of factor E were determined for the design.
c. Show the alias structure for the design.
d. What assumptions must be made to estimate main effects and two-factor interactions free of any other effects?
e. Estimate the main effects and two-factor interactions of interest and their standard errors.
f. Interpret the results.
g. Construct a quarter fraction design from the present design.

12. A 2^{7-4} fractional factorial, or eight-run, Plackett–Burman design was used to study the effects of seven factors on oil consumption in diesel engines. Good oil control consumption is considered to be 0.195 grams/horsepower hour (g/hp hr). The seven engine factors each used at two levels were (A) top ring fit, (B) top ring twist, (C) intermediate ring face, (D) intermediate ring type, (E) piston crown clearance, (F) oil ring width, and (G) use of liner tabs. The oil consumption in grams/horsepower hour (y) is shown for each of the eight runs.

Run	A	B	C	D	E	F	G	y
1	$-$	$-$	$-$	$+$	$+$	$-$	$+$	0.204
2	$-$	$-$	$+$	$+$	$-$	$+$	$-$	0.662
3	$-$	$+$	$-$	$-$	$+$	$+$	$-$	1.075
4	$-$	$+$	$+$	$-$	$-$	$-$	$+$	0.404
5	$+$	$-$	$-$	$-$	$-$	$+$	$+$	0.445
6	$+$	$-$	$+$	$-$	$+$	$-$	$-$	1.297
7	$+$	$+$	$-$	$+$	$-$	$-$	$-$	0.386
8	$+$	$+$	$+$	$+$	$+$	$+$	$+$	1.157

Source: P. R. Shepler (1975), Fractional factorial plans for diesel engine oil control and for seals. *Proceedings of the 31st National Conference on Fluid Power*, 204–234.

a. What is the resolution of this design? What does the resolution of this design imply with regard to the estimation of factorial effects?
b. What fraction of a 2^7 factorial is this design?
c. Determine the design generators.
d. Show the alias structure for the main effects of the design.
e. Estimate the main effects.
f. Interpret the results.

The objective was to find the best configuration of the seven factors to reduce oil consumption. The engineers determined that the best test was obtained with run 1 (0.204 g/hp hr). They also considered a ratio of the average response of the − levels and + levels (largest average/smallest average) for each factor and called it the "favor ratio." If the largest average was from the − levels, the favor ratio was given a − sign; and if the largest average was from the + levels, the favor ratio was given a + sign. The favor ratio signs were the same as that for run 1 except for the sign of factor E, piston crown clearance. To check the results of their previous tests they conducted three additional runs. The first run, labeled "check," included factor levels the same as those for the favor ratio except for two factors A and G. The second run, labeled "proof," included all the same factor levels as those for the favor ratio. The third run was a repeat of run 1 with factor E at the + level. Their past experience indicated that large piston crown clearance (+ level) almost always had been better for oil control. The results of the three additional runs are shown along with the favor ratio.

Run	A	B	C	D	E	F	G	y
Check	+	−	−	+	−	−	−	0.435
Proof	−	−	+	+	−	−	+	0.662
Repeat 1	−	+	−	−	+	+	−	1.075
Favor ratio	−	−	−	+	−	−	−	

g. Were the engineers' conclusions based on the favor ratio consistent with the estimates of the effects from the test runs? Explain. Under what circumstances would the favor ratio be consistent with main effect estimates?
h. Were the engineers' suspicions about factor E correct? Explain.

12A Appendix: Fractional Factorial Design Plans

Table 12A.1 Selected 2^{n-p} fractional factorial designs

Number of Factors	Experimental Units	Fraction	Design Resolution	Design Generator*
3	4	$\frac{1}{2}$	III	$C = AB$
4	8	$\frac{1}{2}$	IV	$D = ABC$
5	8	$\frac{1}{4}$	III	$D = AB$
				$E = AC$
	16	$\frac{1}{2}$	V	$E = ABCD$
6	16	$\frac{1}{4}$	IV	$E = ABC$
				$F = BCD$
	8	$\frac{1}{8}$	III	$D = AB$
				$E = AC$
				$F = BC$
7	32	$\frac{1}{4}$	IV	$F = ABCD$
				$G = ABDE$
	16	$\frac{1}{8}$	IV	$E = ABC$
				$F = BCD$
				$G = ACD$
	8	$\frac{1}{16}$	III	$D = AB$
				$E = AC$
				$F = BC$
				$G = ABC$
8	64	$\frac{1}{4}$	V	$G = ABCD$
				$H = ABEF$
	32	$\frac{1}{8}$	IV	$F = ABC$
				$G = ABD$
				$H = BCDE$
	16	$\frac{1}{16}$	IV	$E = BCD$
				$F = ACD$
				$G = ABC$
				$H = ABD$

Table 12A.1 (*continued*)

Number of Factors	Experimental Units	Fraction	Design Resolution	Design Generator*
9	16	$\frac{1}{32}$	III	$E = ABC$
				$F = BCD$
				$G = ACD$
				$H = ABD$
				$J = ABCD$
	64	$\frac{1}{8}$	IV	$G = ABCD$
				$H = ACEF$
				$J = CDEF$
	32	$\frac{1}{16}$	IV	$F = BCDE$
				$G = ACDE$
				$H = ABDE$
				$J = ABCE$
10	16	$\frac{1}{64}$	III	$E = ABC$
				$F = BCD$
				$G = ACD$
				$H = ABD$
				$J = ABCD$
				$K = AB$
	32	$\frac{1}{32}$	IV	$F = ABCD$
				$G = ABCE$
				$H = ABDE$
				$J = ACDE$
				$K = BCDE$
	64	$\frac{1}{16}$	V	$G = BCDF$
				$H = AGDB$
				$J = ABDE$
				$K = ABCE$
	128	$\frac{1}{8}$	V	$H = ABCG$
				$J = BCDE$
				$K = ACDF$

*Either the positive or negative half of the design generators may be used to construct the fractional design.

13   Response Surface Designs

The central topic in this chapter is constructing designs for efficiently estimating response surfaces from factorial treatment designs with quantitative factors. The nature of linear and quadratic response surfaces is discussed, and designs developed specifically for response surface experiments are described. The discussions include estimating response surface equations and methods to explore the surfaces. Special designs are presented for experiments with factors that are ingredients of mixtures.

13.1   Describe Responses with Equations and Graphs

The objective of all experiments includes describing the response to treatment factors. Throughout this text when treatment factors had quantitative levels we have characterized the response (y) to the factor levels (x) with the polynomial regression equation. For example, in Chapter 3 a polynomial equation was used to estimate the relationship between seed production of plants, y, and density of plants in the plot, x. The estimated quadratic regression equation was graphed as a curve, and we were able to visualize the response of seed production to plant density throughout the range of plant densities included in the experiment. One of the main advantages of the response curve includes the ability to visualize the responses throughout the range of factor levels included in the experiment.

Response Surface Graphs for Two Treatment Factors

The response equation can be displayed as a surface when experiments investigate the effect of two quantitative factors such as the effect of temperature and pressure on the rate of a chemical reaction. In Chapter 6 a quadratic polynomial for two quantitative factors, salinity of media and number of days, was estimated to characterize plant response.

The quadratic response equation is represented as a solid surface in the three-dimensional display of Figure 13.1a. The equation is displayed in Figure 13.1b as a contour plot with lines of equal response values similar to contours of equal elevation levels shown on topographic maps.

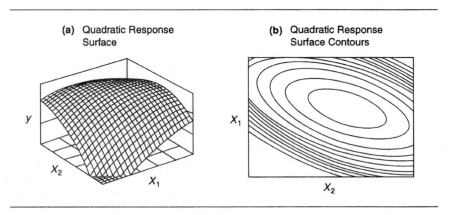

(a) Quadratic Response Surface

(b) Quadratic Response Surface Contours

Figure 13.1 The response equation for two factors displayed as (a) a response surface and (b) contours of equal response

The response surface enables the investigator to visually inspect the response over a region of interesting factor levels and to evaluate the sensitivity of the response to the treatment factors. In certain industrial applications the response surfaces are explored to determine the combination of factor levels that provides an optimum operating condition, such as the combination of temperature and time to maximize the yield of chemical production. In other applications the surfaces are explored to find factor-level combinations that economically improve the responses over current operating conditions if it is too expensive to attain optimum conditions.

Response surfaces also can be used for analytical studies of fundamental processes. For example, they are used frequently in biological sciences to investigate the interplay of factors on the response variable, such as the interaction between nitrogen and phosphorus on the growth of plants.

Polynomial Models Approximate the True Response

Response surface design and analysis strategy assumes the mean of the response variable μ_y is a function of the quantitative factor levels represented by the variables $x_1, x_2, \ldots, x_k$. Polynomial models are used as practical approximations to the true response function. The true function commonly is unknown, and the polynomial functions most often provide good approximations in relatively small regions of the quantitative factor levels.

The most common polynomial models used for response surface analysis are the linear, or *first-order,* model and the quadratic, or *second-order,* model. The first-order model for two factors is

$$\mu_y = \beta_o + \beta_1 x_1 + \beta_2 x_2 \tag{13.1}$$

and the second-order model is

$$\mu_y = \beta_o + \beta_1 x_1 + \beta_2 x_2 + \beta_{11} x_1^2 + \beta_{22} x_2^2 + \beta_{12} x_1 x_2 \tag{13.2}$$

The contour plots for first-order models have series of parallel lines representing coordinates of the factor levels that produce equal response values. The contour plots for quadratic models are more complex with a variety of possible contour patterns. One such pattern shown in Figure 13.1b is a symmetrical mound shape contour with a maximum response occurring within the central contour. Four other patterns are shown in Figure 13.2. Figure 13.2a demonstrates a contour plot with a minimum response occurring within the center contour, indicating a symmetrical surface with a depression in the center. Figure 13.2b depicts a rising ridge with the maximum occurring outside the experimental region. Figure 13.2c shows a stationary ridge in the center of the plot with a decreasing response to the right or left of the center line of maximum response. A saddle contour plot or minimax is portrayed in Figure 13.2d in which the response can increase or decrease from the center of the region, depending on the direction of movement from the center.

Sequential Experiments for Response Surface Analysis

Box and Wilson (1951) accelerated the promotion of response surface analysis for industrial applications. Their primary theme was the use of sequential experimentation with the purpose of determining the optimum operating conditions for an industrial process.

The general approach begins with 2^n factorial treatment designs to identify factors that influence the process. Subsequent experiments use factor treatment combinations to locate an area in the factor space that most likely produces optimum responses. Ultimately, a 2^n factorial arrangement in this region is augmented with treatment combinations to characterize the response surface with quadratic polynomials.

Time Scales Can Prevent Effective Sequential Experiments

In some fields of application the time scale for completion of experiments prohibits sequential experimentation. Many biological studies may require months to complete a single experiment. Often accumulated information from previous biological studies enables the investigator to identify the regions of optimum response, and experiments can be designed to explore the response surface in those regions.

The objective of this chapter is to present some of the basic designs and methods of analysis used for identifying optimum conditions in sequential

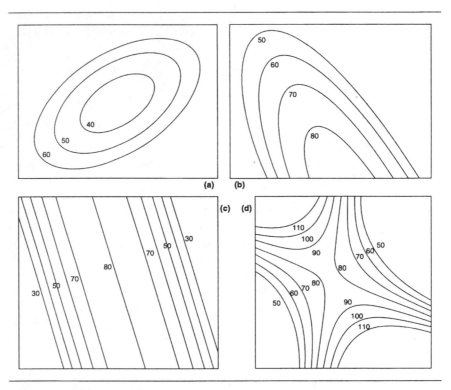

`Figure 13.2` Contour plots for a (a) minimum surface, (b) rising ridge, (c) stationary ridge, and (d) saddle

experimentation and to present designs for efficiently estimating the response surface equations when the region of optimum response is identified.

13.2   Identify Important Factors with 2^n Factorials

Complete or fractional factorial experiments are used for the initial experiments conducted in the study of response surfaces. When the region of optimum response is unknown, the 2^n factorials or fractions of the 2^n factorials are used to identify factors that affect the response variable. (These designs were discussed in Chapters 11 and 12.)

Estimate Linear Responses to Factors

The 2^n factorials are suitable designs to estimate the mean responses for the *linear,* or *first-order*, model in Equation (13.1). The inclusion of two or more observations at the middle level of all factors is the usual recommended procedure to estimate experimental error and to provide a means to evaluate the adequacy of the linear response surface model.

An extensive coverage of methods to identify important factors and factor levels in the region of optimum response conditions can be found in Box and Draper (1987) and in Myers and Montgomery (1995). A brief introduction to the identification of important factors is illustrated with an industrial chemistry experiment to evaluate the factors that affect the vinylation of methyl glucoside.

Example 13.1 Vinylation of Methyl Glucoside

Vinylation of methyl glucoside occurs when it is added to acetylene under high pressure and high temperature in the presence of a base to produce monovinyl ethers. The monovinyl ether products are useful for various industrial synthesis processes. The results of a study on vinylation of methyl glucoside by Marvel et al. (1969) are used to illustrate the methods to identify and evaluate important factors for response surface characterization. The ultimate goal of the project was to determine which conditions produced maximum conversion of methyl glucoside to each of several monovinyl isomers.

Some methods to identify important factors with first-order response surfaces are illustrated with two factors, pressure and temperature. The treatment design was a 2^2 factorial with "temperature" at 130° and 160° C and "pressure" at 325 and 475 psi as factors. Four replications were conducted in the center of the experimental region at a temperature of 145° C and a pressure of 400 psi to provide an estimate of experimental error variance and to evaluate the adequacy of the linear response model. The treatment combinations and percent conversion of methyl glucoside are shown in Table 13.1.

Table 13.1 Vinylation of methyl glucoside in a 2^2 factorial plus four replications at the design center with Temperature and Pressure as factors

Original Factors		Coded Factors		
Temperature	Pressure	x_1	x_2	%Conversion
130	325	-1	-1	8
160	325	$+1$	-1	24
130	475	-1	$+1$	16
160	475	$+1$	$+1$	32
145	400	0	0	21
145	400	0	0	23
145	400	0	0	20
145	400	0	0	24

Coded Factor Levels for Convenience

Coded factor levels provide a uniform framework to investigate factor effects in any experimental context since the actual values of factor levels depend on the particular factors in the study. Coded levels for the 2^n factorial design factors are

$$x_i = \frac{(A_i - \overline{A})}{D} \tag{13.3}$$

where A_i is the ith level of factor A, $\overline{A}$ is the average level for factor A; and D is $\frac{1}{2}(A_2 - A_1)$. Coded levels of temperature (T) and pressure (P) in Table 13.1 are

$$x_1 = \frac{T - 145}{15} \quad \text{and} \quad x_2 = \frac{P - 400}{75} \tag{13.4}$$

Estimates of the Linear Responses

The estimates of the coefficients for the first-order model in Equation (13.1) a

$$\hat{\beta}_0 = \bar{y} = \frac{1}{4}(8 + 24 + 16 + 32) = 20$$

$$\hat{\beta}_1 = \frac{1}{2}T = \frac{1}{4}(-8 + 24 - 16 + 32) = 8$$

$$\hat{\beta}_2 = \frac{1}{2}P = \frac{1}{4}(-8 - 24 + 16 + 32) = 4$$

The estimates of the linear coefficients, β_1 and β_2, are one-half of the factorial treatment effect estimates for a 2^2 factorial (see Chapter 11).

The variance of the four observations at the design center is $s^2 = 3.33$, and an estimate of the standard error for the coefficient estimates is

$$s_{\hat{\beta}} = \sqrt{\frac{4}{16}(3.33)} = 0.91$$

Whether the experimental error variance is adequately estimated with replication only at the center of the design factor levels can matter. If the variance of the response in any way depends on the factor level, then replication of the design at the high- and low-factor-level combinations is recommended to detect any heterogeneous variability among the treatment combinations.

The estimates of the regression coefficients indicate that increases in Temperature or Pressure will increase the vinylation of methyl glucoside. The estimated first-order model equation is

$$\hat{y} = 20 + 8x_1 + 4x_2$$

The temperature and pressure interaction TP measures lack of fit to the linear model and is represented by the term $\beta_{12}x_1x_2$ in the quadratic model in Equation (13.2). The estimate of the coefficient β_{12} is one-half of the TP interaction, or

$$\widehat{\beta}_{12} = \frac{1}{2}TP = \frac{1}{4}(8 - 24 - 16 + 32) = 0$$

The standard error of $\widehat{\beta}_{12}$ is 0.91, the same as that for the linear coefficients. The estimated interaction component of 0 indicates that Temperature and Pressure are acting independently on the conversion.

Center Design Points to Evaluate Surface Curvature

Replicate observations at the design center not only provide an estimate of experimental error, but they also provide a means to measure the degree of curvature in the experimental region. Let $\overline{y}_f$ be the mean of the four treatment combinations for the 2^2 factorial and $\overline{y}_c$ be the mean of the center points. There is some evidence for curvature on the response surface if the average response in the center of the design coordinates, $\overline{y}_c$, is larger or smaller than the average response at the extreme levels of the factors, $\overline{y}_f$. The difference $(\overline{y}_f - \overline{y}_c)$ is an estimate of $\beta_{11} + \beta_{22}$, where β_{11} and β_{22} are the quadratic regression coefficients in Equation (13.2). The observed means are $\overline{y}_f = 20$ and $\overline{y}_c = 22$, with a difference of $\overline{y}_f - \overline{y}_c = -2$. The standard error of the difference is estimated as $\sqrt{3.33(\frac{1}{4} + \frac{1}{4})} = 1.29$; the linear response appears to adequately describe the surface in this region.

The contour plot for the estimated linear response equation is shown in Figure 13.3. The values of the contours ascend as the levels of Temperature and Pressure increase. Ascending contours indicate that a combination of Temperature and Pressure to maximize conversion may exist in a direction perpendicular to the contours.

Path of Steepest Ascent to an Optimum Response

Ultimately, the investigator will want to characterize the region of optimum response. To do so requires the investigator to locate the region of factor levels that produces optimum conditions. The **method of steepest ascent** is a procedure developed to move the experimental region in the response variable in a direction of maximum change toward the optimum.

Based on the estimated linear equation $\widehat{y} = 20 + 8x_1 + 4x_2$, the path of steepest ascent perpendicular to the contours of equal response moves 4 units in the x_2 direction for every 8 units in the x_1 direction. Equivalently, the path has a movement of $4/8 = 0.5$ unit in x_2 for every 1 unit movement in x_1.

The path of steepest ascent is started at the center of the design with $(x_1, x_2) = (0, 0)$. The center of the design for values of temperature and pressure is $(T, P) = (145, 400)$ in Figure 13.3. A change of $\Delta x_1 = 1$ unit in the x_1 direction is a 15°C change in Temperature and a $\Delta x_2 = 0.5$ unit in the x_2 direction is a 37.5 psi change in Pressure.

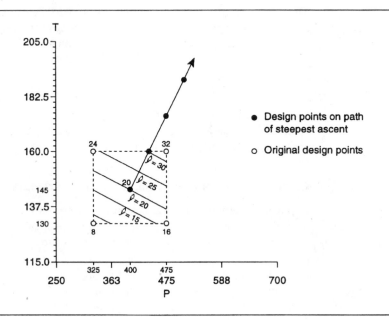

Figure 13.3 Contour plot for the linear response of methyl glucoside vinylation, % conversion, to temperature (T) and pressure (P)

The objective is to move along the path of steepest ascent until a maximum response is observed. The chemist will perform experiments at combinations of Temperature and Pressure along the path of steepest ascent. Suppose the chemist wants to make changes relative to a 1-unit change in x_1. The levels of Temperature and Pressure along the path beginning at $(T, P) = (145, 400)$, the design center, with 1-unit changes in x_1 and one-half unit changes in x_2 are shown in Table 13.2.

Table 13.2 Path of steepest ascent to search for region of maximum response in vinylation of methyl glucoside

Step	x_1	x_2	T	P
0	0	0	145	400.0
1	1	0.5	160	437.5
2	2	1.0	175	475.0
3	3	1.5	190	512.5
4	4	2.0	205	550.0
$\vdots$	$\vdots$	$\vdots$	$\vdots$	$\vdots$

Eventually, as the chemist advances along the path of steepest ascent the increases in the response become smaller until an actual decrease is observed in the response. The decrease in response should indicate that the region of maximum response is in the neighborhood of the current temperature and pressure conditions.

At that point in the process an experiment can be designed to estimate a quadratic polynomial equation that approximates the response surface.

13.3 Designs to Estimate Second-Order Response Surfaces

A new experiment has to be designed to characterize the response surface once the region of optimum response is identified. The surface usually is approximated by a quadratic equation to characterize any curvature in the surface.

The 2^n factorials or fractions thereof are useful designs to identify the important factors and regions of optimum response. However, in the region of optimum response, these designs provide insufficient information to estimate quadratic response equations. At least three levels are required for each factor, and the design must have $1 + 2n + n(n - 1)/2$ distinct design points to estimate the parameters in a quadratic regression model for approximation to the curved surface.

Desirable properties for experimental designs for response surface estimation include the ability to estimate experimental error variance and allow for a test of lack of fit to the model. Designs should also efficiently estimate the model coefficients and predict responses.

Several classes of designs with these desirable properties that have been developed for second-order response surface approximation are discussed in this section.

3^n Factorials for Quadratic Surface Estimation

The 3^n factorials can be used to estimate the quadratic polynomial equations. However, the number of treatment combinations required by the 3^n factorials leads to an impractical experiment size. While a 3^n design with two factors only requires 9 treatment combinations, a design with three factors requires 27 and one with four factors requires 81 treatment combinations.

Central Composite Designs Are an Alternative to 3^n Factorials

Box and Wilson (1951) introduced **central composite designs** requiring fewer treatment combinations than 3^n factorials to estimate quadratic response surface equations. The central composite designs are 2^n factorial treatment designs with $2n$ additional treatment combinations called *axial points* along the coordinate axes of the coded factor levels. The coordinates for the axial points on the coded factor axes are $(\pm \alpha, 0, 0, \ldots, 0)$, $(0, \pm \alpha, 0, \ldots, 0)$, $\ldots$, $(0, 0, 0, \ldots, \pm \alpha)$. Generally, m replications are added to the center of the design at coordinate $(0, 0, \ldots, 0)$.

The central composite designs are used to advantage in sequential experimentation. The first step of the sequence consists of a series of tests conducted along a path of steepest ascent, such as that illustrated in Table 13.2. Eventually, the tests lead to a set of factor levels that provide an apparent maximum on the path. For example, suppose the responses on the path of steepest ascent are those shown in Figure 13.4, with a maximum response of 36 observed at $T = 190°C$ and $P = 512.5$ psi on the path.

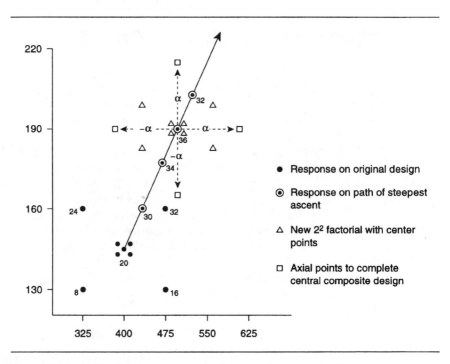

Figure 13.4   Path of steepest ascent and a central composite design

As a second step in the sequence the chemist can conduct a new 2^2 factorial experiment with several replications at the design center of $(T, P) = (190, 512.5)$.

Suppose the difference $(\bar{y}_f - \bar{y}_c)$ computed from the new experiment indicates a high degree of curvature on the surface. The third step in the sequential experiment consists of additional runs of the experiment at the axial points $(\pm \alpha, 0)$ and $(0, \pm \alpha)$ shown in boxes in Figure 13.4. This last set of treatment combinations on the axes, along with the 2^2 factorial and center points, constitutes a central composite design as a result of the sequential experimentation.

One replication of a central composite design consists of $N_f = 2^n$ treatment combinations from the 2^n factorial, $N_a = 2n$ treatment combinations at the axial points in the design, and m replications at the center for a total of $N = N_f + N_a + m$ design observations.

The coordinates on the coded x_1 and x_2 axes for the central composite design with two factors are shown in Display 13.1. A graphic display of the coordinate locations for the coded factor levels of two- and three-factor central composite designs are depicted in Figure 13.5. A quadratic equation can be estimated from this design because each factor has five levels. In addition, as we shall see in the

Display 13.1   Central Composite Design Coordinates

2^2 Design		Axial		Center*	
x_1	x_2	x_1	x_2	x_1	x_2
-1	-1	$-\alpha$	0	0	0
$+1$	-1	$+\alpha$	0		
-1	$+1$	0	$-\alpha$		
$+1$	$+1$	0	$+\alpha$		

*m replications

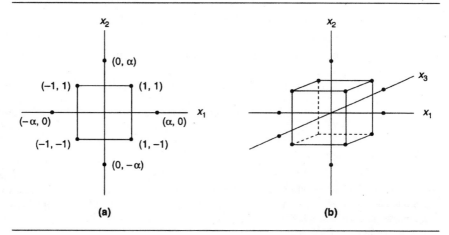

Figure 13.5   Central composite designs for (a) two factors and (b) three factors

next section, any significant deviations from the quadratic approximation can be evaluated.

The $N = 2^n + 2n + m$ experimental units required for the central composite design with n factors are fewer than those required by 3^n factorials with three or more factors. Thus, the central composite designs are more economical in the use of experimental resources and provide the ability to estimate quadratic response equations. Fractions of the 2^n designs with high-order interactions aliased can be used as the 2^n design base when there are many factors in the study.

Rotatable Designs to Improve Response Surface Explorations

Equal precision for all estimates of means is a desirable property in any experimental setting. However, the precision of the estimated values on the response surface based on the estimated regression equation will not be constant over the entire experimental region. A property of *rotatability* developed for central composite designs requires that the variance of estimated values be constant at points equally distant from the center of the design which is coded coordinates $(0, 0, \ldots, 0)$.

Rotatability of a design becomes important in the exploration of a response surface because the precision of the estimated surface does not depend on the orientation of the design with respect to the true response surface or the direction of the search for optimum conditions. The 2^n factorials used as first-order designs to implement the method of steepest ascent searches for regions of optimal responses are rotatable designs. Thus, the orientation of the design does not hinder the method of steepest ascent search because some responses are estimated with less precision than others.

The central composite design can be made rotatable by setting the axial point values as $\alpha = (2^n)^{1/4}$. The value of α for a two-factor design is $\alpha = (4)^{1/4} = \sqrt{2} = 1.414$, and for a three-factor design $\alpha = (8)^{1/4} = 1.682$. If there are r_f replications of the 2^n factorial and r_a replications of the axial treatment combinations a more general form for α is $\alpha = (r_f 2^n / r_a)^{1/4}$. If a 2^{n-p} fractional factorial is used as the basis for the central composite design, then $\alpha = (r_f 2^{n-p} / r_a)^{1/4}$.

Example 13. 2   Rotatable Design for Vinylation of Methyl Glucoside

Suppose the path of steepest ascent for the methyl glucoside study in Table 13.2 provided a maximum response at $T = 190°C$ and $P = 512.5$ psi and a rotatable central composite design is to be constructed with the design center at $(T, P) = (190, 512.5)$. Also, the relationship between the design coordinates (x_1, x_2) and temperature and pressure levels (T, P) remain as before where a change of one unit in x_1 is 15°C and a change of one unit in x_2 is 75 psi. With $\alpha = \sqrt{2}$, the design coordinates and the required temperature and pressure settings will be

	Axial				Center	2^n Design			
x_1	$-\sqrt{2}$	$+\sqrt{2}$	0	0	0	-1	$+1$	-1	$+1$
x_2	0	0	$-\sqrt{2}$	$+\sqrt{2}$	0	-1	-1	$+1$	$+1$
T	169*	211	190	190	190	175	205	175	205
P	512.5	512.5	406.4	618.6	512.5	437.5	437.5	587.5	587.5

*Example calculation, $169° = 190° - \sqrt{2}(15°)$

Designs for Uniform Precision in the Center of the Design

As previously stated, the variance of the estimated surface is not constant over the entire surface. Box and Hunter (1957) showed that the number of center points in the rotatable central composite designs could be chosen to provide a design with uniform precision for the estimated surface within one unit of the design center coordinates on the coded scale. They reasoned that the investigator is most interested in the response surface near the center of the design when a stationary point of the surface is located near the center of the design. The stationary point is a point of maximum or minimum response or a saddle point as shown in Figure 13.2d.

Some central composite rotatable designs with uniform precision are shown in Table 13.3.

Table 13.3   Uniform precision rotatable central composite designs

Number of Factors	2	3	4	5	5	6	6
Fraction of 2^n	1	1	1	1	$\frac{1}{2}$	1	$\frac{1}{2}$
α	1.414	1.682	2.000	2.378	2.000	2.828	2.378
N_f	4	8	16	32	16	64	32
N_a	4	6	8	10	10	12	12
m	5	6	7	10	6	15	9
N	13	20	31	52	32	91	53

The central composite designs require five levels of each factor coded as $-\alpha, -1, 0, 1, \alpha$. In some instances the preparation of five levels for some factors may be too difficult, expensive, or time-consuming. The *face-centered cube design* is a variation of the central composite design with $\alpha = 1$ that requires only three levels of each factor. Substituting $\alpha = 1$ into Display 13.1, the design for two factors becomes a 3^2 factorial. The design is most attractive when the region of interest is a cuboidal region produced by the design rather than the spherical region produced by the central composite design.

The design is not rotatable but the absence of this desirable property may be offset by the desire to have a cuboidal inference region and also by the savings in experimental resources. The face-centered cube design requires fewer runs at the center point of the design than does the central composite design to achieve a stable variance of estimated values throughout the design region. However, it should be remembered that replicate runs are needed at some design point or points to estimate experimental error variance. A face-centered cube design for three or more factors requires fewer treatment combinations than the 3^n factorials; thus, it is another alternative to the 3^n design, requiring fewer experimental units.

Box–Behnken Designs Another Alternative to 3^n Factorials

A class of three-level designs to estimate second-order response surfaces was proposed by Box and Behnken (1960). The designs are rotatable, or nearly so, with a reduction in the number of experimental units compared to the 3^n designs. The designs are formed by combining 2^n designs with incomplete block designs. Details for construction can be found in Box and Draper (1987). The coded factor levels for the treatment combinations required in a design for three factors are shown in Display 13.2. A complete set of treatment combinations for a 2^2 factorial occurs for each pair of factors accompanied by the 0 level of the remaining factors. Several replications of the design center $(0, 0, \dots, 0)$ are included.

> **Display 13.2   Box–Behnken Design Coordinates for a Three-Factor Design**
>
> | Factor | A | B | C |
> |---|---|---|---|
> | Coded Level | x_1 | x_2 | x_3 |
> | | -1 | -1 | 0 |
> | 2^2 factorial | $+1$ | -1 | 0 |
> | for A and B | -1 | $+1$ | 0 |
> | | $+1$ | $+1$ | 0 |
> | | -1 | 0 | -1 |
> | 2^2 factorial | $+1$ | 0 | -1 |
> | for A and C | -1 | 0 | $+1$ |
> | | $+1$ | 0 | $+1$ |
> | | 0 | -1 | -1 |
> | 2^2 factorial | 0 | $+1$ | -1 |
> | for B and C | 0 | -1 | $+1$ |
> | | 0 | $+1$ | $+1$ |
> | | 0 | 0 | 0 |
> | Design center | 0 | 0 | 0 |
> | | 0 | 0 | 0 |

These designs are spherical rather than cuboidal since the design points fall on the edges of a cube rather than on the corners like those of the face-centered cube design. The Box–Behnken design should only be used if one is *not* interested in predicting responses at the corners of the cuboidal region.

Incomplete Block Designs to Increase Precision

Incomplete block designs are useful to reduce experimental error variance when the number of treatments is large or when the experimental conditions preclude the conduct of complete replications at one time or under the same conditions.

Box and Hunter (1957) gave the conditions for blocking second-order response surface designs so that the block effects do not affect the estimates of the parameters for the response surface equation. They showed that two conditions must be satisfied for the blocks to be orthogonal to the parameter estimates of the response surface equation. Let n_b be the number of treatments in the bth block. The two conditions necessary are

1.   Each block must be a first-order orthogonal design. For each block the following relationship must hold for each pair of design variables x_i and x_j:

$$\sum_{k=1}^{n_b} x_{ik} x_{jk} = 0 \qquad i \ne j = 0, 1, 2, \ldots, n \qquad (13.5)$$

2. The fraction of the total sum of squares for each design variable contributed by every block must be equal to the fraction of the total observations placed in the block. Thus, the following relationship must hold between the design variables and the number of observations for every block:

$$\frac{\displaystyle\sum_{k=1}^{n_b} x_{ik}^2}{\displaystyle\sum_{k=1}^{N} x_{ik}^2} = \frac{n_b}{N} \qquad i = 1, 2, \ldots, n \qquad (13.6)$$

A suggested strategy for blocking the central composite design places the N_f treatments for the 2^n design and m_f design center points in one block and the N_a axial treatments with m_a design center points in a second block. This blocking arrangement satisfies the first condition, Equation (13.5).

The central composite rotatable design for two factors arranged into two blocks is shown in Display 13.3. The first block is composed of $N_f = 4$ treatment combinations of the 2^2 factorial plus $m_f = 2$ design center points, and the second block consists of $N_a = 4$ axial treatment combinations plus $m_a = 2$ design center points. The computations required to evaluate the first condition for an orthogonal block design are the sums of crossproducts between x_1 and x_2 in each block. It is easy to verify that $\Sigma x_1 x_2 = 0$ in both blocks.

Display 13.3	**Central Composite Rotatable Design for Two Factors in Two Incomplete Blocks**		
	Factor	*A*	*B*
	Coded Level	x_1	x_2
		− 1	− 1
		+ 1	− 1
Block 1		− 1	+ 1
		+ 1	+ 1
		0	0
		0	0
		1.414	0
		− 1.414	0
Block 2		0	1.414
		0	− 1.414
		0	0
		0	0

For the complete design

$$\sum_{k=1}^{12} x_{1k}^2 = \sum_{k=1}^{12} x_{2k}^2 = 8$$

and for both block 1 and block 2

$$\sum_{k=1}^{6} x_{1k}^2 = \sum_{k=1}^{6} x_{2k}^2 = 4$$

The number of treatment observations in blocks 1 and 2 are $n_1 = n_2 = 6$, and the total number of observations is $N = 12$ with a ratio $n_i/N = 6/12 = 1/2$. The second condition, Equation (13.6), requires that the ratio of the sums of squares of x_1 and x_2 in every block to that for the entire experiment be equal to n_i/N. For both block 1 and block 2 the sum of squares ratio is $4/8 = 1/2$, which is equivalent to the ratio for n_i/N; the design is orthogonal.

For the second condition to be satisfied the following relationship must hold:

$$\alpha^2 = n \left[\frac{1 + p_a}{1 + p_f} \right] \qquad (13.7)$$

where $p_a = m_a/N_a$ and $p_f = m_f/N_f$. For the design to satisfy the two conditions and be rotatable $\alpha = (2^n r_f/r_a)^{1/4}$. It is not always possible to find a design that exactly satisfies Equation (13.7) with $\alpha = (2^n r_f/r_a)^{1/4}$, but in practice values of the design observation numbers can be determined to provide designs with near orthogonal blocking and rotatability. Box and Draper (1987) provide relative proportions of r_f and r_a required for rotatability and orthogonal blocking when $p_a = p_f$.

For the design in Display 13.3 the fraction of design center observations in each block is $p_a = p_f = 1/2$ and $\alpha = \sqrt{2}$.

Evaluating the condition for rotatability and orthogonality in Equation (13.7) we have

$$\alpha^2 = n \left[\frac{1 + p_a}{1 + p_f} \right] = 2 \left[\frac{1 + 1/2}{1 + 1/2} \right] = 2$$

and $\alpha = \sqrt{2}$ as required for rotatability.

The central composite rotatable designs listed in Table 13.3 can be placed in useful incomplete block designs for near rotatable and orthogonal central composite designs. The 2^n factorial or 2^{n-p} fractional factorial is placed in one or more incomplete blocks and the axial treatment combinations are placed in one separate block. For the designs listed in Table 13.3 the number of blocks for the 2^n factorial or fractional factorial, and a suggested number of design center points in each block are shown in Table 13.4.

Table 13.4 Incomplete block designs for near rotatable and orthogonal central composite designs

Number of Factors	2	3	4	5	5	6	6
Fraction of 2^n	1	1	1	1	$\frac{1}{2}$	1	$\frac{1}{2}$
N_f	4	8	16	32	16	64	32
m_f	2	2	2	4	2	2	2
Number of Blocks*	1	2	2	4	1	8	2
α	1.414	1.682	2.000	2.378	2.000	2.828	2.378
N_a	4	6	8	10	10	12	12
m_a	2	2	1	1	4	1	4

*See Appendix 11A for defining contrasts to block the 2^n factorial or fractional factorial design portion.

Reducing the Number of Design Points

The expense, difficulty, or time consumption with certain types of experiments may necessitate reducing the experiment size. The amount of reduction is limited by the statistical model to estimate the response surface. The second-order response surface equation for n factors has a constant term, n linear terms, n quadratic terms, and $n(n-1)/2$ interaction terms for a total of $(n+1)(n+2)/2$ terms. Thus, the minimum number of points a design could have to estimate the second-order response surface is $(n+1)(n+2)/2$.

Designs have been developed to have as close to the minimum number of points as possible to estimate the second-order response surface. Tables of these designs or methods to construct them can be found in Box and Draper (1974), Roquemore (1976), Notz (1982), Draper (1985), Draper and Lin (1990), and Myers and Montgomery (1995).

Most of the designs are based on 2^{n-p} fractional factorials augmented with design points to estimate second-order response surface models. In most cases the designs are saturated with few or no replicated design points. An independent estimate of experimental error is required to test the efficacy of the response surface model, unless the design is replicated. In addition, the saturated designs do not allow a test for lack of fit of the hypothesized second-order response surface model.

An Evaluation of Response Surface Designs

Myers et al. (1992) used the prediction variance of the second-order response surface equation to evaluate many of the popular second-order response surface designs. A design was considered superior if the variance of predicted values was smaller than that of other designs.

The central composite designs were found superior in general over spherical regions covered by the design points (see Figure 13.5). When designs were restricted to the cuboidal regions ($\alpha = 1$ in Figure 13.5), the resulting face-centered cube design formed by the central composite design was found generally superior to the Box–Behnken design in the cuboidal region.

Among the saturated designs, the designs by Roquemore (1976), Notz (1982), and Box and Draper (1974) were found quite efficient relative to others that were evaluated.

Myers and Montgomery (1995) presented design efficiencies for estimating model coefficients and for prediction variances in a spherical region. Their general conclusions were that central composite and Box–Behnken designs were quite efficient as were some of the saturated designs from Roquemore (1976).

13.4 Quadratic Response Surface Estimation

When the supposed region of optimum response has been identified by the method of steepest ascent or other methods of experimentation it is often necessary to characterize the response surface in that region of the factors. Using the designs described in the previous section, experiments can be conducted to obtain data for estimating a quadratic approximation to the response surface.

The estimated response equation will enable the researcher to locate a stationary response point that could be a maximum, a minimum, or a saddle point on the surface. An examination of the contour plot will indicate how sensitive the response variable is to each of the factors and to what degree the factors interplay as they affect the response variable.

Example 13.3   Tool Life Response to Lathe Velocity and Cutting Depth

A new cutting tool available from a vendor was going to be used by a company. The vendor claimed the new model tool would reduce production costs because it would last longer than the old model; thus, tool replacement cost would be reduced. The life of a metal cutting tool is dependent on several operating conditions, including the speed of the lathe and the depth of the cut made by the tool.

The plant engineer had determined from previous studies that maximum tool life was achieved for the current tool with a lathe velocity setting of 400 and a cutting depth setting of 0.075. The engineer wanted to determine the optimum settings required for the new tool. A central composite design was used for an experiment to characterize the life of the new tool under varying lathe speeds and cutting depths within the region of current optimum operating conditions for maximum tool life. The data from the experiment are shown in Table 13.5.

Table 13.5 Observed tool life from a factorial experiment with lathe speed and cutting depth treatment factors in a central composite design

| Original Factors | | Coded Factors | | |
Lathe Speed	Cutting Depth	x_1	x_2	Tool Life
600	0.100	$+1$	$+1$	154
600	0.050	$+1$	-1	132
200	0.100	-1	$+1$	166
200	0.050	-1	-1	83
683	0.075	$\sqrt{2}$	0	156
117	0.075	$-\sqrt{2}$	0	144
400	0.110	0	$\sqrt{2}$	166
400	0.040	0	$-\sqrt{2}$	91
400	0.075	0	0	167
400	0.075	0	0	175
400	0.075	0	0	170
400	0.075	0	0	176
400	0.075	0	0	156
400	0.075	0	0	170

The Estimated Response Surface Equation

The second-order response surface model of Equation (13.2) is fit to the data by least squares regression procedures. The equation can be estimated by any computer program for regression analysis. A brief account of least squares estimation for regression models is given in Appendix 13A.1. A detailed presentation of regression analysis can be found in Rawlings (1988). The estimated second-order response surface equation for tool life from the data in Table 13.5 is

$$\hat{y} = 169 + 6.747x_1 + 26.385x_2 - 10.875x_1^2 - 21.625x_2^2 - 15.250x_1x_2$$

Sum of Squares Partitions for the Regression Analysis

The sum of squares partitions in the analysis of variance for the regression model are shown in Table 13.6. The sum of squares for the full second-order model is

$$SSR(x_1, x_2, x_1^2, x_2^2, x_1x_2) = 10{,}946$$

The regression sum of squares is partitioned into reductions for linear and quadratic components of the model using the principle of reduced model and full model sum of squares partitions.

The partition for the linear components of the model, x_1 and x_2, or

$$SSR(x_1, x_2) = 5{,}933$$

Table 13.6 Analysis of variance for quadratic response surface model

Source of Variation	Degrees of Freedom	Sum of Squares	Mean Square
Total	13	11,317	
Regression	5	10,946	2,189.2
Linear (x_1, x_2)	2	5,933	2,966.5
Quadratic $(x_1^2, x_2^2, x_1 x_2)$	3	5,013	1,671.0
Error	8	371	46.4
Lack of Fit	3	111	37.0
Pure Error	5	260	52.0

is the regression sum of squares for the reduced first-order model $y = \beta_o + \beta_1 x_1 + \beta_2 x_2 + e$. The partition for the quadratic components is the difference between the regression sum of squares for the full model and the reduced model, or

$$SSR(x_1^2, x_2^2, x_1 x_2 \mid x_1, x_2) = 10{,}946 - 5{,}933 = 5{,}013$$

The sum of squares for error, $SSE = 371$, is partitioned into two parts. The sum of squares for pure experimental error, $SSE(\text{pure error}) = 260$, with 5 degrees of freedom is computed from the six replicate observations at the center of the design with factor coordinates $(V, D) = (400, 0.075)$. The remaining 3 degrees of freedom sum of squares for error, $SSE(\text{Lack of fit}) = 111$, can be attributed to error in specification of the response surface model, referred to as lack of fit, or it can be attributed to experimental error. Since the six center points of the design provide an estimate of pure experimental error, the sum of squares designated as lack of fit can be used to test the significance of lack of fit to the quadratic model.

Tests of Hypotheses About the Second-Order Model

The hypotheses of interest in the analysis are

- Significance of the complete second-order model:

$$H_0: \beta_1 = \beta_2 = \beta_{11} = \beta_{22} = \beta_{12} = 0$$

$$F_0 = \frac{2{,}189.2}{52} = 42.1 \qquad \text{Reject } H_0 \text{ since } F_0 > F_{.05,5,5} = 5.05$$

- Significance of linear components for the model:

$$H_0: \beta_1 = \beta_2 = 0$$

$$F_0 = \frac{2{,}966.5}{52} = 57.0 \qquad \text{Reject } H_0 \text{ since } F_0 > F_{.05,2,5} = 5.79$$

- Significance of quadratic deviations from the linear model:

$$H_0: \beta_{11} = \beta_{22} = \beta_{12} = 0$$

$$F_0 = \frac{1{,}671}{52} = 32.1 \qquad \text{Reject } H_0 \text{ since } F_0 > F_{.05,3,5} = 5.41$$

- Significance of lack of fit to the quadratic model:

$$F_0 = \frac{37}{52} = 0.71 \qquad \text{Do not reject } H_0 \text{ since } F_0 < F_{.05,3,5} = 5.41$$

The complete quadratic regression model is significant, and the lack of fit to the quadratic model is not significant; thus, we can conclude the second-order model is an adequate approximation to the true response surface. A contour plot of the quadratic response surface model depicted in Figure 13.6 shows a maximum surface with a maximum tool life in the middle of the center contour.

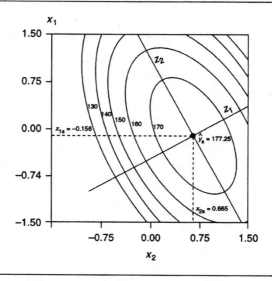

Figure 13.6 Response surface contour plot for the tool life experiment response equation $\hat{y} = 169 + 6.747x_1 + 26.385x_2 - 10.875x_1^2 - 21.625x_2^2 - 15.250x_1x_2$

The coordinates of the contour plot are displayed for the coded values of the two factors. The orientation of the contours indicates some interaction between lathe speed x_1 and cutting depth x_2. For example, a constant cutting tool life of 150 can be maintained for faster lathe speeds, increasing x_1, by decreasing the cutting depth, decreasing x_2.

The contours also indicate the relative sensitivity of tool life to the coded factor levels x_1 and x_2. The tool life contours increase more rapidly toward the maximum on the coded cutting depth axis x_2 than they do on the coded lathe speed axis x_1.

13.5 Response Surface Exploration

The significant quadratic equation and the contour plot of the equation have given us a general picture of the relationship between tool life and the two design factors, lathe speed and cutting depth.

Estimates of the coordinates for the stationary point on the surface and an estimate of the response at the stationary point provide a more specific characterization of the response surface. Sometimes, it is useful to know the direction and amount of change to make in one or more of the design factor levels to achieve the maximum change in response.

A more specific characterization of the sensitivity of response to the design factors can be achieved with the canonical form of the response equation. The location of coordinates for the stationary point and derivation of the canonical form of the response equation requires some knowledge of calculus and matrix algebra. However, the results of the computations are understandable when they are displayed in graphic form on the contour plot of Figure 13.6.

Locating Coordinates for the Stationary Point of the Response Surface

The x_1 and x_2 coordinates for the stationary point are obtained from the partial derivatives of the estimated response function with respect to x_1 and x_2. The estimated response for tool life is

$$\hat{y} = 169 + 6.747x_1 + 26.385x_2 - 10.875x_1^2 - 21.625x_2^2 - 15.250x_1x_2 \quad \textbf{(13.8)}$$

The partial derivatives are set equal to 0:

$$\frac{\partial \hat{y}}{\partial x_1} = 0 \qquad \frac{\partial \hat{y}}{\partial x_2} = 0$$

to produce the equations

$$2(-10.875)x_1 + (-15.250)x_2 = -6.747$$

$$(-15.250)x_1 + 2(-21.625)x_2 = -26.385$$

The solutions of the equations for x_1 and x_2 are $\hat{x}_{1s} = -0.156$ and $\hat{x}_{2s} = 0.665$. These values are the coordinates for the maximum response on the surface at the stationary point indicated on Figure 13.6.

The estimated response at the stationary point is found by substituting $\hat{x}_{1s} = -0.156$ and $\hat{x}_{2s} = 0.665$ into Equation (13.8); the estimated stationary point response is

$$\hat{y}_s = 169 + 6.747(-0.156) + 26.385(0.665) - 10.875(-0.156)^2$$

$$- 21.625(0.665)^2 - 15.250(-0.156)(0.665) = 177.25$$

Given $x_1 = (V - 400)/200$ and $x_2 = (D - 0.075)/0.025$, the values of lathe speed (V) and cutting depth (D) at the stationary point are

$$V = -0.156(200) + 400 = 368.8$$

and

$$D = 0.665(0.025) + 0.075 = 0.092$$

The general solution for a stationary point with any number of x_i variables in the response equation is given in Appendix 13A.2.

The Canonical Analysis to Simplify the Quadratic Equation

The canonical form of a quadratic equation is an effective aid to visualize the surface and to determine the relative sensitivity of the response variable to each of the factors. It is difficult to visualize the surface by examining the estimated coefficients for the normal form of the quadratic response equation. Likewise, it is difficult to determine the changes in factor levels necessary to produce a specified change in the response.

The canonical analysis rotates the axes of the x_i variables to a new coordinate system, and the center of the new coordinate system is placed at the stationary response point of the surface. The canonical form of the equation for two variables is

$$\hat{y} = \hat{y}_s + \lambda_1 Z_1^2 + \lambda_2 Z_2^2 \tag{13.9}$$

where Z_1 and Z_2 are the rotated axes' variables. Notice only the quadratic terms of the canonical variables Z_1 and Z_2 are included in the canonical form of the response equation. An outline of the computations required to obtain the canonical form for the tool life response equation is given in Appendix 13A.3.

The canonical form for the tool life response equation is

$$\hat{y} = 177.25 - 25.58Z_1^2 - 6.92Z_2^2 \tag{13.10}$$

where the center of the new coordinate system is located at $x_1 = -0.156$ and $x_2 = 0.665$ in the original coordinate system shown in Figure 13.6. The relationship between the two coordinate systems was determined (Appendix 13A.3) to be

$$Z_1 = 0.4603x_1 + 0.8877x_2 - 0.5185$$

$$Z_2 = 0.8877x_1 - 0.4603x_2 + 0.4446 \tag{13.11}$$

Notice the Z_1 and Z_2 canonical axes are oriented with the contours of the surface. The sizes and signs of the λ_i indicate the type of quadratic response surface that has been estimated.

The λ_i coefficients for the tool life surface are $\lambda_1 = -25.58$ and $\lambda_2 = -6.92$. Examination of the surface in Figure 13.6 reveals that any movement away from the center of the Z_1, Z_2 coordinate system results in a response decrease. Thus, when all λ_i coefficients are negative the surface is a maximum surface such as that for the tool life surface in Figure 13.6.

If the λ_i coefficients are positive, then any movement from the center of the Z_1, Z_2 coordinate system results in a response increase and the surface is a minimum surface as shown in Figure 13.2a. If one coefficient is positive and the other negative, say $\lambda_1 > 0$ and $\lambda_2 < 0$, then movement away from $(0, 0)$ along the Z_1 axis results in an increased response and movement along the Z_2 axis results in a decrease. Thus, the surface is a saddle or minimax at the stationary point as shown in Figure 13.2d. If one of the $\lambda_i = 0$, the surface is a stationary ridge (Figure 13.2c) because the response will not change along the Z_i axis.

The lengths of the principal axes of the ellipses formed by the contours are proportional to $|\lambda_i|^{-1/2}$. For the tool life surface $|-25.58|^{-1/2} = 0.20$ and $|-6.92|^{-1/2} = 0.38$, and the fitted surface is attenuated along the Z_2 axis as seen in Figure 13.6.

Suppose, for illustration, that lathe speed and cutting depth for maximum tool life at coordinates $x_1 = -0.156$ and $x_2 = 0.665$ were impractical. The least change in tool life as lathe speed and cutting depth change is exhibited on the surface along the Z_2 axis direction when $Z_1 = 0$. The x_1 and x_2 coordinates along the Z_2 axis when $Z_1 = 0$ can be obtained from the first equation in Equations (13.11). The least loss in tool life can be found on settings corresponding to values of x_1 and x_2 that satisfy $0.4603x_1 + 0.8877x_2 - 0.5185 = 0$.

The coefficients of the x_i in Equations (13.11) can provide information about the relationships of lathe speed and cutting depth to tool life. Consider the coefficients for the second equation relating Z_2 to x_1 and x_2, $Z_2 = 0.8877x_1 - 0.4603x_2 + 0.4446$. The pair of coefficients $(0.8877, -0.4603)$ indicate a compensation between lathe speed and cutting depth on tool life. An increase in lathe speed, to some extent, can be compensated for by a decrease in cutting depth along the elongated Z_2 axis.

The estimated response equation in its original form or in the canonical form is only valid for the region of factor levels included in the experiment. Any attempt to estimate the tool life outside of the limits bounded by lathe speeds of 117 and 683 and cutting depths of 0.04 and 0.11 could be quite misleading. An entirely different response model may be necessary to describe tool life outside of the region used by the current study.

13.6 Designs for Mixtures of Ingredients

Some treatment designs involve two or more factors that are ingredients of a mixture in which the percentages of the ingredients must sum to 100% of the mixture. Therefore, the levels of one factor are not independent of other factor levels.

Many food products, construction materials, and other commercial products are formed from mixtures of two or more ingredients in a recipe. Some examples are

- fabrics with a blend of cotton and polyester fiber
- fruit juice blends of orange, pineapple, and apple juices with water
- concrete formed from water, aggregate, and cement
- fertilizer formulations of nitrogen, phosphorus, and potassium

This section includes a brief introduction to selecting designs and estimating response surface equations for mixture experiments. Cornell (1990) provides an indepth coverage of design and analysis of mixture experiments.

Factor Levels Are Proportions of Ingredients

Variation in the proportions of the ingredients in mixtures can affect the properties of the end product. Investigations with mixture experiments concentrate on the relationship of the measured response variable to the relative proportions of the separate ingredients present in the product rather than on the total amounts of the factors.

If $x_1, x_2, \ldots, x_k$ are the variables representing the proportions of the k ingredients or components of the mixture, the values of the x_i are constrained such that

$$0 \leq x_i \leq 1 \quad i = 1, 2, \ldots, k \tag{13.12}$$

and the proportions of the k ingredients in the mixture sum to unity, or

$$\sum_{i=1}^{k} x_i = x_1 + x_2 + \cdots + x_k = 1 \tag{13.13}$$

If the proportion of one ingredient is $x_i = 1$, then the other ingredients are absent from the mixture and the product is a pure or single-component mixture. For example, a two-component mixture experiment with cotton and polyester fabric blends represented by the proportions x_1 and x_2 may have pure cotton fabric, in which case $x_1 = 1$ and $x_2 = 0$, or a pure polyester fabric, where $x_1 = 0$ and

$x_2 = 1$. The allowable values of x_1 and x_2 for a two-component mixture design are coordinate values along the line $x_1 + x_2 = 1$ (shown in Figure 13.7).

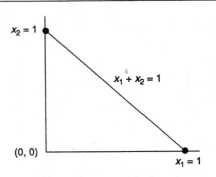

Figure 13.7   Factor space for a two-component mixture, $x_1 + x_2 = 1$

Depict the Factor Space with a Simplex Coordinate System

The coordinate values for a three-component mixture design are the coordinate values found on the plane defined by $x_1 + x_2 + x_3 = 1$ in Figure 13.8a. The geometric description of the factor space for k components is that of a simplex in $(k - 1)$ dimensions. The two-dimensional simplex coordinate system for the three-component mixture design is shown in Figure 13.8b as an equilateral triangle.

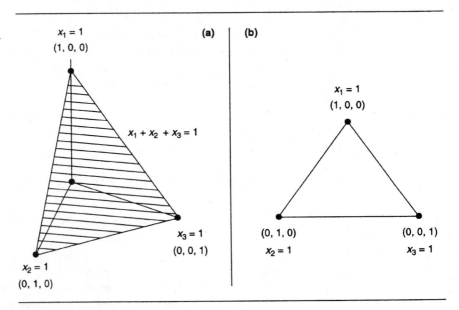

Figure 13.8   Factor space for (a) a three-component mixture, $x_1 + x_2 + x_3 = 1$ and (b) a three-component simplex coordinate system

The vertices of the triangle represent single-component mixtures with one $x_i = 1$ and all others equal to 0. The sides of the triangle represent design coordinates for two-component mixtures with one $x_i = 0$. Design coordinates in the triangle interior represent three-component mixtures with $x_1 > 0$, $x_2 > 0$, and $x_3 > 0$. Any combination of component proportions for a mixture experiment must be on the boundaries or inside the triangle of coordinates in Figure 13.8b.

The axes of the x_i variables in the three-variable simplex coordinate system are shown in Figure 13.9. The axis for component i is the line from the base point $x_i = 0$ and $x_j = 1/(k - 1)$ for all other components $j \neq i$ to the vertex, where $x_i = 1$ and $x_j = 0$ for $j \neq i$. For example, with the three-component design in Figure 13.9, the x_1 axis extends from the base coordinate $(0, \frac{1}{2}, \frac{1}{2})$ to the vertex coordinate $(1, 0, 0)$.

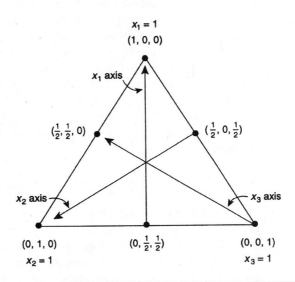

Figure 13.9 Simplex coordinate axes for a three-component mixture with design coordinates for a $\{3, 2\}$ simplex-lattice design

Treatment Designs for Mixtures

Simplex-Lattice Designs

The array made up of a uniform distribution of design coordinates on the simplex coordinate system is known as a *lattice*; see Figure 13.9. The simplex-lattice designs consist of a lattice of design coordinates constructed to enable the estimation of polynomial response surface equations.

The designation $\{k, m\}$ is used for a simplex-lattice design with k components to estimate a polynomial response surface equation of degree m. For example, a

{3, 2} simplex-lattice design has three components in the mixture design to estimate a quadratic response surface equation.

The proportions of each component included in a {k, m} simplex-lattice design are

$$x_i = 0, \frac{1}{m}, \frac{2}{m}, \cdots, 1 \tag{13.14}$$

The design consists of all possible combinations of those levels of the x_i, where $\Sigma\, x_i = 1$ for any combination of proportions.

The combinations of mixture proportions shown at each • in Figure 13.9 are the coordinate values for a {3, 2} simplex-lattice. The proportions for each x_i with $m = 2$ are $x_i = 0$, $\frac{1}{2}$, and 1. It can be seen in Figure 13.9 that $\Sigma\, x_i = 1$ for each design point.

The {k, 2} simplex-lattice design to estimate quadratic response surface equations only has mixtures on the boundaries of the coordinate system with one or more of the components absent from the mixture. The general {k, m} simplex-lattice will consist of single-component mixtures, two-component mixtures, and so forth up to mixtures consisting of at most m components. If $m = k$, there will be one mixture at the centroid of the coordinate system in the experiment that contains all mixture components. For example, the {3, 3} simplex-lattice would include the mixture with component proportions $(\frac{1}{3}, \frac{1}{3}, \frac{1}{3})$ as well as the single-component mixtures and the two-component mixtures with proportions $\frac{1}{3}$ and $\frac{2}{3}$ for the two components.

Simplex-Centroid Designs

The simplex-centroid design is a design on the simplex coordinate system that consists of mixtures containing $1, 2, 3, \ldots,$ or k components, each in equal proportions. Consequently, there are k single-component mixtures, all possible two-component mixtures with proportion $\frac{1}{2}$ for each component, all possible three-component mixtures with proportion $\frac{1}{3}$ for each component, and so forth up to one k-component mixture with proportion $\frac{1}{k}$ for each component. The mixtures for the simplex-centroid design are contrasted with the mixtures for the {3, 2} and {3, 3} simplex-lattice designs in Table 13.7.

Augmented Simplex-Centroid Designs

The mixture combinations for the simplex-lattice and simplex-centroid designs lie on the edges of the simplex factor space with the exception of one centroid point that contains all mixture components. More complete mixtures are possible by augmenting the simplex-centroid design with mixtures on the axes of the simplex factor space.

The design points are positioned on each axis equidistant from the centroid toward the vertices. A k-component design will have k additional design points with coordinates

x_1	x_2	$\cdots$	x_k
$\dfrac{(k+1)}{2k}$	$\dfrac{1}{2k}$	$\cdots$	$\dfrac{1}{2k}$
$\dfrac{1}{2k}$	$\dfrac{(k+1)}{2k}$	$\cdots$	$\dfrac{1}{2k}$
$\vdots$	$\vdots$	$\vdots$	$\vdots$
$\dfrac{1}{2k}$	$\dfrac{1}{2k}$	$\cdots$	$\dfrac{(k+1)}{2k}$

The addition of the axial points will provide a better distribution of information throughout the experimental region. The three additional design points required by augmenting the simplex-centroid design for three components are $(\frac{4}{6}, \frac{1}{6}, \frac{1}{6})$, $(\frac{1}{6}, \frac{4}{6}, \frac{1}{6})$, and $(\frac{1}{6}, \frac{1}{6}, \frac{4}{6})$. The complete design is depicted in Figure 13.10.

Table 13.7 Simplex-lattice designs and a simplex-centroid design for a three-component mixture

\{3,2\} Lattice			\{3,3\} Lattice			Centroid		
x_1	x_2	x_3	x_1	x_2	x_3	x_1	x_2	x_3
1	0	0	1	0	0	1	0	0
0	1	0	0	1	0	0	1	0
0	0	1	0	0	1	0	0	1
$\frac{1}{2}$	$\frac{1}{2}$	0	$\frac{2}{3}$	$\frac{1}{3}$	0	$\frac{1}{2}$	$\frac{1}{2}$	0
$\frac{1}{2}$	0	$\frac{1}{2}$	$\frac{2}{3}$	0	$\frac{1}{3}$	$\frac{1}{2}$	0	$\frac{1}{2}$
0	$\frac{1}{2}$	$\frac{1}{2}$	0	$\frac{2}{3}$	$\frac{1}{3}$	0	$\frac{1}{2}$	$\frac{1}{2}$
			$\frac{1}{3}$	$\frac{2}{3}$	0	$\frac{1}{3}$	$\frac{1}{3}$	$\frac{1}{3}$
			$\frac{1}{3}$	0	$\frac{2}{3}$			
			0	$\frac{1}{3}$	$\frac{2}{3}$			
			$\frac{1}{3}$	$\frac{1}{3}$	$\frac{1}{3}$			

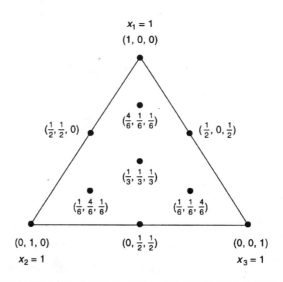

Figure 13.10 Augmented simplex-centroid design for a three-component mixture experiment

Pseudocomponents for Ingredients with Lower Bounds

Many mixtures require all components to be present in at least some minimum proportions. Clearly, concrete requires some minimum proportions of water, cement, and aggregate. Lower bounds, L_i, on component proportions impose the constraint

$$0 \leq L_i \leq x_i \leq 1 \tag{13.15}$$

on the component proportions. Suppose the lower bounds for cement (x_1), water (x_2), and aggregate (x_3) are

$$0.10 \leq x_1 \qquad 0.20 \leq x_2 \qquad 0.30 \leq x_3$$

and a $\{3,2\}$ simplex-lattice design is going to be used for the experiment. The lower bounds on the component proportions limit the design to a subregion of the original factor space on the simplex shown in Figure 13.9 or Table 13.7.

To simplify the construction of the design coordinates a set of pseudocomponents are constructed by coding the original component variables to a simplex coordinate system for the pseudocomponent variables $\tilde{x}_i$ with constraints $0 \leq \tilde{x}_i \leq 1$. If the lower bound for component i is L_i and $L = \Sigma L_i$, then the pseudocomponent $\tilde{x}_i$ is computed as

$$\tilde{x}_i = \frac{x_i - L_i}{1 - L}$$

A design in the original components can be constructed on the basis of coordinates for the pseudocomponents that are set up in a regular simplex with $\Sigma \tilde{x}_i = 1$. The proportions of the original components required for mixtures in the experiment can be derived by the reverse transformation

$$x_i = L_i + \tilde{x}_i(1 - L)$$

For the concrete example the lower bounds were $L_1 = 0.10$, $L_2 = 0.20$, and $L_3 = 0.30$ with sum $L = 0.10 + 0.20 + 0.30 = 0.60$. The pseudocomponents are

$$\tilde{x}_1 = \frac{x_1 - 0.10}{0.40} \qquad \tilde{x}_2 = \frac{x_2 - 0.20}{0.40} \qquad \tilde{x}_3 = \frac{x_3 - 0.30}{0.40}$$

and the transformations back to original component proportions from the pseudo-components are

$$x_1 = 0.10 + 0.40\tilde{x}_1 \qquad x_2 = 0.20 + 0.40\tilde{x}_2 \qquad x_3 = 0.30 + 0.40\tilde{x}_3$$

The complete design for the concrete mixture experiment is shown in Table 13.8 with values for the pseudocomponent coordinates and the original components on the subregion of the original simplex.

Table 13.8 Pseudocomponent and original component coordinates in a $\{3, 2\}$ simplex-lattice design for the concrete mixture experiment

Pseudocomponents			*Original Components*		
$\tilde{x}_1$	$\tilde{x}_2$	$\tilde{x}_3$	*Cement*	*Water*	*Aggregate*
1	0	0	0.50	0.20	0.30
0	1	0	0.10	0.60	0.30
0	0	1	0.10	0.20	0.70
$\frac{1}{2}$	$\frac{1}{2}$	0	0.30	0.40	0.30
$\frac{1}{2}$	0	$\frac{1}{2}$	0.30	0.20	0.50
0	$\frac{1}{2}$	$\frac{1}{2}$	0.10	0.40	0.50

13.7 Analysis of Mixture Experiments

Canonical Polynomials to Approximate Surfaces

The general form of the polynomial function used to approximate linear response surfaces is

$$\mu_y = \beta_0 + \beta_1 x_1 + \cdots + \beta_k x_k \tag{13.16}$$

The restriction on mixture components, $x_1 + x_2 + \cdots + x_k = 1$, creates a dependency among the x_i in the linear function. Multiplying β_0 by $(x_1 + x_2 + \cdots + x_k) = 1$ provides a reexpression of the model as

$$\mu_y = \beta_0 \left(\sum_{i=1}^{k} x_i \right) + \sum_{i=1}^{k} \beta_i x_i = \beta_1^* x_1 + \beta_2^* x_2 + \cdots + \beta_k^* x_k \qquad (13.17)$$

where $\beta_i^* = \beta_0 + \beta_i$, $i = 1, 2, \ldots, k$. The reexpressed equation with parameters β_i^* is known as a **canonical polynomial**. The canonical polynomial and the original polynomial are equivalent because one is derived from the other and the degree of the polynomial and the number of components are unchanged upon reexpression.

The quadratic polynomial function used to approximate response surfaces is

$$\mu_y = \beta_0 + \sum_{i=1}^{k} \beta_i x_i + \sum_{i=1}^{k} \beta_{ii} x_i^2 + \sum \sum_{i < j} \beta_{ij} x_i x_j \qquad (13.18)$$

The quadratic canonical polynomial produced by enacting the restriction $\Sigma x_i = 1$ is

$$\mu_y = \sum_{i=1}^{k} \beta_i^* x_i + \sum \sum_{i < j} \beta_{ij}^* x_i x_j \qquad (13.19)$$

where $\beta_i^* = \beta_0 + \beta_i + \beta_{ii}$ and $\beta_{ij}^* = \beta_{ij} - \beta_{ii} - \beta_{jj}$. The new parameters of the quadratic canonical polynomial for three-mixture components expressed in terms of the original polynomial parameters are

$$\beta_1^* = \beta_0 + \beta_1 + \beta_{11} \qquad \beta_2^* = \beta_0 + \beta_2 + \beta_{22} \qquad \beta_3^* = \beta_0 + \beta_3 + \beta_{33}$$

$$\beta_{12}^* = \beta_{12} - \beta_{11} - \beta_{22} \qquad \beta_{13}^* = \beta_{13} - \beta_{11} - \beta_{33} \qquad \beta_{23}^* = \beta_{23} - \beta_{22} - \beta_{33}$$

The interpretation of the canonical polynomials is illustrated with a mixture experiment on gasoline component blends.

Example 13.4   A Gasoline-Blending Mixture Experiment

The octane of a gasoline blend depends upon the proportions of the various petroleum components blended to produce the fuel. The objective of most gasoline-blending studies is to develop a linear blending model to determine the most profitable blend of gasoline components. Coefficients in the linear blending model, referred to as *blending values*, describe the blending behavior of a given fuel component. However, the blending composition depends on a number of factors including the quality of the components. Thus, the linearity of the blending components is lost, and a more complex quadratic or blending component interaction model must be considered.

The analysis of a mixture experiment is illustrated with a gasoline-blending mixture experiment. To evaluate the need for an interaction blending model the experiment was designed to estimate the quadratic canonical polynomial.

A mixture experiment was set up to evaluate the effect of three components on the octane ratings of gasoline. The components alkylate (A), light straight run (B), and reformate (C), were used in a simplex-centroid design with seven mixtures. The octane ratings were determined for two replicate runs of each mixture. The octane ratings for each of the mixtures are shown in Table 13.9.

Table 13.9 Octane ratings from a mixture experiment on gasoline blends

	Components*			
x_1	x_2	x_3	y_{ij}	$\bar{y}_{i.}$
1	0	0	106.6, 105.0	105.80
0	1	0	83.3, 81.4	82.35
0	0	1	99.4, 91.4	95.40
$\frac{1}{2}$	$\frac{1}{2}$	0	94.1, 91.4	92.75
$\frac{1}{2}$	0	$\frac{1}{2}$	101.9, 98.0	99.95
0	$\frac{1}{2}$	$\frac{1}{2}$	92.3, 86.5	89.40
$\frac{1}{3}$	$\frac{1}{3}$	$\frac{1}{3}$	96.3, 91.7	94.00

*x_1 = alkylate, x_2 = light straight run, x_3 = reformate
Source: R. D. Snee (1981), Developing blending models for gasoline and other mixtures. *Technometrics* 23, 119–130.

Estimating the Quadratic Canonical Polynomial Response Surface Model

For convenience the asterisk will be dropped from the coefficients in the canonical polynomial equations. The full quadratic canonical polynomial model for the gasoline-blending experiment is

$$y_{ij} = \beta_1 x_{1j} + \beta_2 x_{2j} + \beta_3 x_{3j} + \beta_{12} x_{1j} x_{2j} + \beta_{13} x_{1j} x_{3j} + \beta_{23} x_{2j} x_{3j} + e_{ij}$$

$$i = 1, 2, \ldots, t \quad j = 1, 2, \ldots, r \tag{13.20}$$

where the experimental errors e_{ij} are assumed to be independent and normally distributed with mean 0 and variance σ^2. Also, $t = 7$ mixture treatments and $r = 2$ replications per mixture produce a total of $N = rt = 14$ observations.

The hypothesis of initial interest is whether the response depends on the mixture components according to the quadratic model. When the null hypothesis is true the mean response is adequately described by the reduced model $y_{ij} = \beta_0 + e_{ij}$, where $\beta_1 = \beta_2 = \beta_3 = \beta_0$ and $\beta_{12} = \beta_{13} = \beta_{23} = 0$.

An analysis of variance of the data in Table 13.9 for the mixture treatments in a completely randomized design will provide an estimate of pure experimental error. The sum of squares among the mixtures is

$$SST = r \sum_{i=1}^{t} (\bar{y}_{i.} - \bar{y}_{..})^2 = 669.32$$

with $(t-1) = 6$ degrees of freedom. The sum of squares for experimental error is

$$SSE = \sum_{i=1}^{t} \sum_{j=1}^{r} (y_{ij} - \bar{y}_{i.})^2 = 73.74$$

with $(N-t) = 7$ degrees of freedom. Thus, the mean square for experimental error is $MSE = 10.53$ with 7 degrees of freedom.

The least squares estimates of the parameters for the canonical polynomials require a fit to a regression model without the usual intercept term β_0. Many computer regression programs have the capability to fit the regression model without the intercept term, and they will provide the correct least squares estimates of the parameters for the canonical polynomials.

The estimated full quadratic equation is

$$\hat{y} = 105.8x_1 + 82.3x_2 + 95.4x_3 - 5.1x_1x_2 - 2.4x_1x_3 + 2.3x_2x_3 \quad \textbf{(13.21)}$$

with experimental error sum of squares $SSE_f = 73.76$ with $14 - 6 = 8$ degrees of freedom.

Tests of Hypotheses About the Model

A Test for the Complete Model

If the response does not depend on the mixture components, the fully reduced model is $y_{ij} = \beta_0 + e_{ij}$ and the surface has a constant height. The experimental error sum of squares for this reduced model is

$$SSE_r = \sum_{i=1}^{t} \sum_{j=1}^{r} (y_{ij} - \bar{y})^2 = 743.05$$

with $N - 1 = 13$ degrees of freedom.

The sum of squares reduction for the full quadratic response surface model is

$$SSR = SSE_r - SSE_f = 743.05 - 73.76 = 669.29$$

with $13 - 8 = 5$ degrees of freedom. The sum of squares for the quadratic model accounts for 5 of the 6 degrees of freedom for treatments with 1 degree of freedom remaining for lack of fit to the quadratic model. The analysis of variance is summarized in Table 13.10.

The null hypothesis for the quadratic response equation is H_0: $\beta_1 = \beta_2 = \beta_3 = \beta_0$ and $\beta_{12} = \beta_{13} = \beta_{23} = 0$. The test statistic $F_0 = MSR/MSE = 133.86/10.53 = 12.71$ exceeds the critical value of $F_{.05,5,7} = 3.97$, and the null hypothesis is rejected.

Table 13.10   Analysis of variance for mixture experiment with gasoline blends

Source of Variation	Degrees of Freedom	Sum of Squares	Mean Square
Treatments	6	669.32	111.55
Regression	5	669.29	133.86
Lack of fit	1	0.03	0.03
Error	7	73.74	10.53

A Test for the Quadratic Terms

The investigator would want to know whether the full quadratic model is necessary to approximate the response surface or whether the linear surface is adequate to explain the relationship between octane rating and the component mixtures. A test of the null hypothesis $H_0: \beta_{12} = \beta_{13} = \beta_{23} = 0$ will determine whether the quadratic components of the model are necessary. The full model and reduced model principle can be used to determine the significance of the quadratic components. The sums of squares for experimental error are required from the full quadratic model and the reduced linear model $y_{ij} = \beta_1 x_{1j} + \beta_2 x_{2j} + \beta_3 x_{3j} + e_{ij}$. The estimated reduced model linear equation is

$$\widehat{y} = 105.1x_1 + 82.1x_2 + 95.5x_3 \tag{13.22}$$

with experimental error sum of squares $SSE_{lin} = 77.37$ with $14 - 3 = 11$ degrees of freedom. The sum of squares reduction due to the quadratic terms after the linear terms are fit is

$$SSR(\text{quadratic}) = SSE_{lin} - SSE_f = 77.37 - 73.76 = 3.61$$

with $11 - 8 = 3$ degrees of freedom with $MSR(\text{quadratic}) = 3.61/3 = 1.20$. The test statistic $F_0 = MSR(\text{quadratic})/MSE = 1.20/10.53 = 0.11$ with critical region $F_0 > F_{.05,3,7} = 4.35$ is not significant, and the quadratic terms do not improve the approximation of the model to the mixture response surface. Thus, a linear blending model is adequate for this set of blending components.

A Test for the Linear Terms

Since the quadratic components of the model account for very little of the regression sums of squares it is fairly obvious the linear terms account for most of the sum of squares for the regression model. A formal test of the null hypothesis $H_0:$ $\beta_1 = \beta_2 = \beta_3 = \beta_0$ is derived from the sum of squares reduction for the linear model as

$$SSR(\text{linear}) = SSE_r - SSE_{lin} = 743.05 - 77.37 = 665.68$$

with $13 - 11 = 2$ degrees of freedom with MSR(linear) $= 665.68/2 = 332.84$. The test statistic is $F_0 = MSR$(linear)$/MSE = 332.84/10.53 = 31.61$ with critical region $F_0 > F_{.05,2,7} = 4.74$, and the null hypothesis is rejected.

Interpretations for the Estimated Response Equation

The estimated linear canonical polynomial, $\widehat{y} = 105.1x_1 + 82.1x_2 + 95.5x_3$ in Equation (13.22), provides a significant and adequate fit to the mixture response surface. The estimated standard error for each of the $\widehat{\beta}_i$ determined from the regression program is $s_{\widehat{\beta}_i} = 1.89$, and the individual coefficient estimates are significant by the Student t test, $t_0 = \widehat{\beta}_i/s_{\widehat{\beta}_i}$, with 7 degrees of freedom.

The coefficient $\widehat{\beta}_i$ is the estimated response at the vertex of the simplex design representing the mixture with 100% of that component or the single-component mixture. Alternatively, it represents the estimated response at the maximum value for that component. The variables for Example 13.4 were the proportions $x_1 =$ alkylate, $x_2 =$ light straight run, and $x_3 =$ reformate. For example, with 100% alkylate and 0% each of light straight run and reformate the estimated octane rating is $\widehat{y} = 105.1$. Likewise, with 100% light straight run the estimated octane is $\widehat{y} = 82.1$, and with 100% reformate the estimated octane is $\widehat{y} = 95.5$. The estimated linear surface is depicted in Figure 13.11.

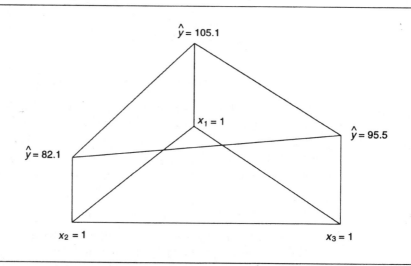

Figure 13.11   Estimated linear surface for the gasoline-blending mixture experiment

The β_{ij} describe the departures from the linear response surface. An illustration of a quadratic response for a two-component system is depicted in Figure 13.12. If the two components are additive, with a linear blending of the two components the mean response is $\mu_l = \beta_1 x_1 + \beta_2 x_2$ (shown as the straight line blending in Figure

(13.12). A nonlinear quadratic blending of the two components with $\beta_{12} > 0$ is shown by the response curve $\mu_q = \beta_1 x_1 + \beta_2 x_2 + \beta_{12} x_1 x_2$. The coefficient β_1 represents the height of the curve when $x_1 = 1$ and $x_2 = 0$, and β_2 represents the height of the curve when $x_1 = 0$ and $x_2 = 1$. The $\beta_{12} x_1 x_2$ term contributes to the response whenever $x_1 > 0$ and $x_2 > 0$. The maximum departure from the linear blending occurs at $x_1 = x_2 = \frac{1}{2}$ when $\beta_{12} x_1 x_2 = \frac{1}{4}\beta_{12}$. The blending of the two components in Figure 13.12 is said to be synergistic because the response for the 1:1 mixture at $x_1 = x_2 = \frac{1}{2}$ exceeds the simple average of the pure mixtures depicted by the linear blending line. If the coefficient β_{12} were negative the nonlinear blending line would fall below that of the linear blending line and the components would be antagonistic to one another.

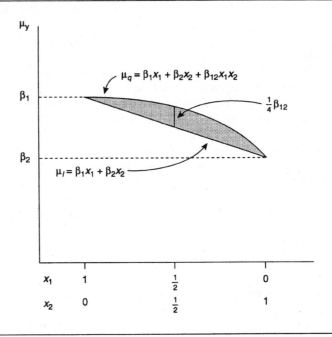

Figure 13.12 Comparison of linear and nonlinear blending for a two-component mixture system

EXERCISES FOR CHAPTER 13

1. A 2^2 factorial experiment was conducted to determine whether the volume of two reagents had an effect on the ability of an assay method to measure levels of a specific drug in serum. Two replications of the treatments were tested in a completely randomized design with two additional

replications at the center of the design. The serum for each test was sampled from a serum pool spiked with a single dose of the drug. The data are shown next with reagent volumes in μl.

Reagent		% Drug Recovered
A	B	
10	20	32, 35
40	20	44, 47
10	50	51, 53
40	50	68, 72
25	35	48, 53

a.   Estimate the experimental error variance.

b.   Estimate the linear response equation and standard errors of the coefficient estimates. Are the linear effects of the reagents significant?

c.   Estimate β_{12} for AB interaction and standard error of the estimate. Is there significant interaction?

d.   Estimate the departure from a linear surface, $\beta_{11} + \beta_{22}$, and standard error of the estimate. Is there a significant departure from a linear surface?

e.   Determine the first five steps on the path of steepest ascent from the center of the design with steps of one unit in x_1 for reagent A. Show the levels of both factors at each step.

2.   Consider the experiment on vinylation of methyl glucoside in Example 13.1. Suppose the chemist has observed a maximum response at $T = 175°C$ and $P = 475.5$ psi via the path of steepest ascent in Table 13.2. Use those levels as the average factor levels in Equation (13.4), and design a central composite rotatable design to estimate the quadratic response surface equation. Show the actual levels of T and P required for each treatment combination in the design.

3.   Show the coded design coordinates required for a uniform precision rotatable central composite design with four factors.

4.   Describe the response surface for the following canonical forms given for quadratic response surfaces:

a. $\hat{y} = 100 - Z_1^2 - 2Z_2^2$

b. $\hat{y} = 50 + 2Z_1^2$

c. $\hat{y} = 75 + Z_1^2 - 2Z_2^2$

5.   An animal scientist studied the relationship between metabolism of methionine, a sulfur amino acid, and carotene, vitamin A, as they affect the growth of chickens. The optimum levels of methionine and carotene were thought to be 0.9% methionine in the diet and 50 micrograms carotene per day. A central composite rotatable design was used for the experiment. Eight chicks were randomly assigned to each of the treatment diets, and their weight gains were recorded after 38 days. The average weight gains for the treatments follow.

Original Factors		Coded Factors		Weight Gain
Methionine	Carotene	x_1	x_2	
1.183	85.36	+1	+1	445
1.183	14.64	+1	−1	331
0.617	85.36	−1	+1	443
0.617	14.64	−1	−1	336
1.183	50.00	$\sqrt{2}$	0	414
0.500	50.00	$-\sqrt{2}$	0	389
0.900	100.00	0	$\sqrt{2}$	435
0.900	0.00	0	$-\sqrt{2}$	225
0.900	50.00	0	0	442
0.900	50.00	0	0	412
0.900	50.00	0	0	418
0.900	50.00	0	0	440
0.900	50.00	0	0	441

a. Estimate the quadratic response surface equation for weight gain, and summarize the sum of squares partitions in an analysis of variance table.

b. Test the significance of the complete quadratic model, the quadratic deviations from the linear model, the significance of the linear components of the model, and the lack of fit to the quadratic model. What are your conclusions?

c. The response surface has a maximum within the design coordinates. Determine the levels of methionine and carotene that produce the maximum response, and estimate the maximum response.

d. Compute the canonical equation (see Appendix 13A.3), and describe the response surface. Based on the canonical equation, what is the relationship between methionine and carotene? Can one be used to compensate for the other in the animal's diet?

6. The experiment on vinylation of methyl glucoside used in Example 13.1 (Marvel et al., 1969) included four factors in a central composite rotatable design placed in an incomplete block design. The percent conversion of methyl glucoside to a vinylation product was the response variable of interest. The actual and coded levels of the four factors used in the experiment were

	Coded Level	− 2	− 1	0	1	2
x_1	Time, hours	1	3	5	7	9
x_2	Temperature, °C	115	130	145	160	175
x_3	Pressure, psi	250	325	400	475	550
x_4	Solvent ratio (water/dioxane)	95	80	65	50	35

The percent conversion of methyl glucoside to a vinylation product for each treatment combination in the experiment is shown in the table that follows:

		Block 1					Block 2					Block 3		
y	x_1	x_2	x_3	x_4	y	x_1	x_2	x_3	x_4	y	x_1	x_2	x_3	x_4
10	−1	−1	−1	−1	56	−1	1	1	1	13	−2	0	0	0
21	1	1	−1	−1	7	−1	−1	−1	1	43	2	0	0	0
5	−1	−1	1	1	35	1	1	1	−1	2	0	−2	0	0
46	1	1	1	1	27	1	−1	1	1	52	0	2	0	0
6	1	−1	−1	1	19	1	−1	−1	−1	18	0	0	−2	0
16	1	−1	1	−1	52	1	1	−1	1	52	0	0	2	0
22	−1	1	1	−1	17	−1	−1	1	−1	15	0	0	0	−2
18	−1	1	−1	1	24	−1	1	−1	−1	58	0	0	0	2
21	0	0	0	0	33	0	0	0	0	34	0	0	0	0
31	0	0	0	0	30	0	0	0	0	39	0	0	0	0

a. Estimate the quadratic response surface equation for percent conversion, and summarize the sum of squares partitions in an analysis of variance table.

b. Test the significance of the complete quadratic model, the quadratic deviations from the linear model, the significance of the linear components of the model, and the lack of fit to the quadratic model. What are your conclusions?

c. Determine the factor levels that produce the maximum response with the quadratic model. Is the optimum within the current design factor levels?

d. Compute the canonical equation from the quadratic model estimates (see Appendix 13A.3), and describe the response surface. Based on the canonical equation, what type of surface has been estimated?

7. Construct an incomplete block design for a central composite design with three treatment factors such that the conditions in Equations (13.5) and (13.6) are satisfied. What is the value of α required for the relationship to hold in Equation (13.7)?

8. An experiment is planned to evaluate the flavor quality of a fruit juice containing orange, pineapple, lime, and papaya juices. The minimum allowable proportions of the four juices in the mix are orange > .15, pineapple > .10, lime > .10, and papaya > .20. Design a mixture experiment with a simplex-lattice design. List the design both as pseudocomponents and as original components.

9. A highway-paving mixture is formed by dispersing liquid sulfur into liquid asphalt to produce a sulfur–asphalt binder. The binder is then mixed with sand to produce the paving mixture. A mixture experiment is to be conducted with a minimum of 5% sulfur and a minimum of 10% asphalt.

a. Design the mixture experiment with a simplex-centroid design. Show the design as pseudocomponents and as original components.

b. Construct the design as an augmented simplex-centroid. Show the pseudocomponents and original components.

10. A mixture experiment was conducted with the sulfur–asphalt binder described in Exercise 13.9. The experiment was conducted with a minimum of 10% sulfur, a minimum of 20% asphalt, and a minimum of 50% sand in the mixtures. Two replications of each mixture were prepared, and a

specimen from each replication was subjected to a strength test. A simplex-lattice design was used for the experiment. The strength data with mixture proportions for pseudocomponents, where x_1 is the pseudocomponent for sulfur, x_2 for asphalt, and x_3 for sand, follow.

Pseudocomponents			Strength
x_1	x_2	x_3	y_{ij}
1	0	0	12.0, 13.7
0	1	0	2.4, 3.6
0	0	1	2.6, 4.3
0.5	0.5	0	18.9, 16.8
0.5	0	0.5	19.4, 17.1
0	0.5	0.5	4.6, 7.3

a. Given the minimum values for sulfur, asphalt, and sand, determine the actual proportions for the three components for each of the treatment mixtures in the experiment.
b. Estimate the experimental error variance.
c. Estimate the linear and quadratic response surface polynomial. Determine the significance of the linear terms and the significance of the quadratic addition to the model.
d. Interpret the coefficients in the model.

13A.1 Appendix: Least Squares Estimation of Regression Models

The least squares estimation of parameters for the regression model follows the procedures illustrated in previous chapters for various experiment designs. The difference to be noted for the regression model is the inclusion of the continuous value x_i variables in the model that were not seen previously in the experiment design models.

Estimation is illustrated for a model with two x_i variables. The multiple linear regression model is

$$y_j = \beta_0 x_{0j} + \beta_1 x_{1j} + \beta_2 x_{2j} + e_j \qquad j = 1, 2, \ldots, n$$

The intercept or constant term for the model written in the general form as $\beta_0 x_{0j}$ is most often identified in the model only as β_0 since the variable x_{0j} takes on a constant value $x_{0j} = 1$ for all observations.

The least squares estimators for the regression coefficients are found by differentiating

$$\sum_{j=1}^{n} e_j^2 = \sum_{j=1}^{n} (y_j - \beta_0 x_{0j} - \beta_1 x_{1j} - \beta_2 x_{2j})^2$$

with respect to each of the β_i and setting the result equal to 0. The partial derivatives set to 0 are

$$\frac{\partial}{\partial \beta_0} = -2 \sum_{j=1}^{n} x_{0j}(y_j - \beta_0 x_{0j} - \beta_1 x_{1j} - \beta_2 x_{2j}) = 0$$

$$\frac{\partial}{\partial \beta_1} = -2 \sum_{j=1}^{n} x_{1j}(y_j - \beta_0 x_{0j} - \beta_1 x_{1j} - \beta_2 x_{2j}) = 0$$

$$\frac{\partial}{\partial \beta_2} = -2 \sum_{j=1}^{n} x_{2j}(y_j - \beta_0 x_{0j} - \beta_1 x_{1j} - \beta_2 x_{2j}) = 0$$

The resulting normal equations are

$$\widehat{\beta}_0 \sum x_{0j}^2 + \widehat{\beta}_1 \sum x_{0j}x_{1j} + \widehat{\beta}_2 \sum x_{0j}x_{2j} = \sum x_{0j}y_j$$

$$\widehat{\beta}_0 \sum x_{0j}x_{1j} + \widehat{\beta}_1 \sum x_{1j}^2 + \widehat{\beta}_2 \sum x_{1j}x_{2j} = \sum x_{1j}y_j$$

$$\widehat{\beta}_0 \sum x_{0j}x_{2j} + \widehat{\beta}_1 \sum x_{1j}x_{2j} + \widehat{\beta}_2 \sum x_{2j}^2 = \sum x_{2j}y_j$$

The simultaneous solutions of equations for the $\widehat{\beta}_i$ result in the least squares estimators for the β_i. The estimates for a given problem can be obtained with any statistical computing package that includes a program for multiple linear regression.

The interested reader can find a detailed account of regression analysis methodology in Rawlings (1988). A brief outline of the model formulation, construction of the normal equations, and solutions to the normal equations are given here assuming some knowledge of matrix notation. The multiple linear regression model can be written in matrix form as

$$y = X\beta + e$$

where

$$y = \begin{bmatrix} y_1 \\ y_2 \\ \vdots \\ y_n \end{bmatrix} \qquad X = \begin{bmatrix} 1 & x_{11} & \cdots & x_{k1} \\ 1 & x_{12} & \cdots & x_{k2} \\ \vdots & \vdots & \ddots & \vdots \\ 1 & x_{1n} & \cdots & x_{kn} \end{bmatrix} \qquad \beta = \begin{bmatrix} \beta_o \\ \beta_1 \\ \vdots \\ \beta_k \end{bmatrix} \qquad e = \begin{bmatrix} e_1 \\ e_2 \\ \vdots \\ e_n \end{bmatrix}$$

Denoting X' as the transpose of the X matrix the normal equations in matrix form are

$$X'X\widehat{\beta} = X'y$$

where

$$\boldsymbol{X'X} = \begin{bmatrix} n & \Sigma x_1 & \Sigma x_2 & \cdots & \Sigma x_k \\ \Sigma x_1 & \Sigma x_1^2 & \Sigma x_1 x_2 & \cdots & \Sigma x_1 x_k \\ \Sigma x_2 & \Sigma x_1 x_2 & \Sigma x_2^2 & \cdots & \Sigma x_2 x_k \\ \vdots & \vdots & \vdots & \ddots & \vdots \\ \Sigma x_k & \Sigma x_1 x_k & \Sigma x_2 x_k & \cdots & \Sigma x_k^2 \end{bmatrix} \qquad \boldsymbol{X'y} = \begin{bmatrix} \Sigma y \\ \Sigma x_1 y \\ \Sigma x_2 y \\ \vdots \\ \Sigma x_k y \end{bmatrix}$$

The solutions to the normal equations for $\widehat{\beta}$ are found by multiplying both sides of the normal equations by the inverse matrix for $\boldsymbol{X'X}$ denoted $(\boldsymbol{X'X})^{-1}$. The solution is written as

$$\widehat{\beta} = (\boldsymbol{X'X})^{-1}\boldsymbol{X'y}$$

The computations are illustrated with data for two independent variables, x_1 and x_2, to estimate the coefficients for the first-order model $y = \beta_o + \beta_1 x_1 + \beta_2 x_2 + e$. The data matrices are

$$\boldsymbol{y'} = [\,41, \quad 52, \quad 54, \quad 73, \quad 66, \quad 67, \quad 79\,]$$

$$\boldsymbol{X'} = \begin{bmatrix} 1 & 1 & 1 & 1 & 1 & 1 & 1 \\ 1 & 2 & 3 & 4 & 5 & 6 & 7 \\ 1 & 2 & 1 & 3 & 2 & 2 & 3 \end{bmatrix}$$

$$\boldsymbol{X'X} = \begin{bmatrix} 7 & 28 & 14 \\ 28 & 140 & 63 \\ 14 & 63 & 32 \end{bmatrix} \qquad \boldsymbol{X'y} = \begin{bmatrix} 432 \\ 1{,}884 \\ 921 \end{bmatrix}$$

$$(\boldsymbol{X'X})^{-1} = \begin{bmatrix} 1.16 & -0.03 & -0.44 \\ -0.03 & 0.06 & -0.11 \\ -0.44 & -0.11 & 0.44 \end{bmatrix}$$

The solution to the normal equations is

$$\widehat{\beta} = (\boldsymbol{X'X})^{-1}\boldsymbol{X'y} = \begin{bmatrix} 1.16 & -0.03 & -0.44 \\ -0.03 & 0.06 & -0.11 \\ -0.44 & -0.11 & 0.44 \end{bmatrix} \begin{bmatrix} 432 \\ 1{,}884 \\ 921 \end{bmatrix} = \begin{bmatrix} 31.43 \\ 3.57 \\ 8.00 \end{bmatrix}$$

The estimated equation is $\widehat{y}_j = \widehat{\beta}_0 x_0 + \widehat{\beta}_1 x_{1j} + \cdots + \widehat{\beta} x_{kj}$, and the sum of squares for experimental error for the full regression model is calculated as

$$SSE_f = \sum_{j=1}^{n} (y_j - \widehat{y}_j)^2 = (\boldsymbol{y} - \boldsymbol{X}\widehat{\beta})'(\boldsymbol{y} - \boldsymbol{X}\widehat{\beta}) = \boldsymbol{y'y} - \widehat{\beta}'\boldsymbol{X'y}$$

$$= \sum_{j=1}^{n} y_j^2 - \widehat{\beta}_o \Sigma y_j - \widehat{\beta}_1 \Sigma x_{1j} y_j - \cdots - \widehat{\beta}_k \Sigma x_{kj} y_j$$

with $n - k - 1$ degrees of freedom.

The reduced model without the variables x_i, $i = 1, 2, \ldots, k$, is $y_j = \beta_0 x_{0j} + e_j = \mu + e_j$. The estimated regression equation with the reduced model is $\widehat{y}_j = \widehat{\beta}_0 x_0$. The sum of squares for experimental error for the reduced model is

$$SSE_r = \sum_{j=1}^{n} (y_j - \widehat{y}_j)^2 = \sum_{j=1}^{n} y_j^2 - \widehat{\beta}_0 \sum_{j=1}^{n} y_j$$

with $n - 1$ degrees of freedom.

For the reduced model $\widehat{\beta}_0 = \overline{y}$ and $SSE_r = \sum_{j=1}^{n} (y_j - \overline{y})^2$. The sum of squares for regression due to the inclusion of the independent variables x_i, $i = 1, 2, \ldots, k$, in the model is

$$SSR = SSE_r - SSE_f$$

with k degrees of freedom.

Given $\Sigma y_j^2 = 27{,}716$, the experimental error sum of squares for the full model of the example is

$$SSE_f = 27{,}716 - 31.43(432) - 3.57(1{,}884) - 8.0(921) = 44.4$$

with $n - k - 1 = 7 - 2 - 1 = 4$ degrees of freedom. The estimate of β_0 for the reduced model is $\overline{y} = 61.7$, and the experimental error sum of squares is $SSE_r = 1055.4$. The sum of squares for regression is

$$SSR = 1055.4 - 44.4 = 1011.0$$

with $k = 2$ degrees of freedom.

13A.2 Appendix: Location of Coordinates for the Stationary Point

The estimated quadratic model expressed in matrix form is

$$\widehat{y} = \widehat{\beta}_o + \boldsymbol{x}'\boldsymbol{b} + \boldsymbol{x}'\boldsymbol{B}\boldsymbol{x} \tag{13A.1}$$

where

$$\boldsymbol{x}' = \begin{bmatrix} x_1 & x_2 & \cdots & x_k \end{bmatrix} \qquad \boldsymbol{b}' = \begin{bmatrix} \widehat{\beta}_1 & \widehat{\beta}_2 & \cdots & \widehat{\beta}_k \end{bmatrix}$$

$$\boldsymbol{B} = \begin{bmatrix} \widehat{\beta}_{11} & \widehat{\beta}_{12}/2 & \widehat{\beta}_{13}/2 & \cdots & \widehat{\beta}_{1k}/2 \\ \widehat{\beta}_{12}/2 & \widehat{\beta}_{22} & \widehat{\beta}_{23}/2 & \cdots & \widehat{\beta}_{2k}/2 \\ \widehat{\beta}_{13}/2 & \widehat{\beta}_{23}/2 & \widehat{\beta}_{33} & \cdots & \widehat{\beta}_{3k}/2 \\ \vdots & \vdots & \vdots & \vdots & \vdots \\ \widehat{\beta}_{1k}/2 & \widehat{\beta}_{2k}/2 & \widehat{\beta}_{3k}/2 & \cdots & \widehat{\beta}_{kk} \end{bmatrix}$$

The stationary point is found by setting the derivative of $\widehat{y}$ with respect to the $\boldsymbol{x}$ vector equal to 0,

$$\frac{\partial \widehat{y}}{\partial \boldsymbol{x}} = \boldsymbol{b} + 2\boldsymbol{Bx} = 0 \tag{13A.2}$$

The vector of design coordinates for the stationary point is the solution to Equation (13A.2), or

$$\boldsymbol{x}_s = -\frac{1}{2}\boldsymbol{B}^{-1}\boldsymbol{b} \tag{13A.3}$$

Substituting the solution into Equation (13A.1), the estimated response at the stationary point is

$$\widehat{y}_s = \widehat{\beta}_0 + \frac{1}{2}\boldsymbol{x}'_s\boldsymbol{b} \tag{13A.4}$$

13A.3 Appendix: Canonical Form of the Quadratic Equation

The canonical form of the quadratic response equation is

$$\widehat{y} = \widehat{y}_s + \lambda_1 Z_1^2 + \lambda_2 Z_2^2 + \cdots + \lambda_k Z_k^2 \tag{13A.5}$$

where the λ_i are the eigenvalues of the matrix $\boldsymbol{B}$ in Equation (13A.1) and the Z_i are variables associated with the rotated axes that correspond to the axes of the contours of the response surface. The origin for the rotated coordinate system is the stationary point with all $Z_i = 0$ and response $\widehat{y}_s$.

The eigenvalues of $\boldsymbol{B}$ are the roots of the determinantal equation

$$|\boldsymbol{B} - \lambda\boldsymbol{I}| = 0 \tag{13A.6}$$

where $\boldsymbol{I}$ is the identity matrix. The relationship between the matrix $\boldsymbol{B}$ and the λ_i is

$$\boldsymbol{Bm}_i = \boldsymbol{m}_i \lambda_i \quad i = 1, 2, \ldots, k \tag{13A.7}$$

where the $\boldsymbol{m}_i$ are the eigenvectors corresponding to the λ_i. The $\boldsymbol{m}_i$ are normalized so that $\boldsymbol{m}'_i \boldsymbol{m}_i = 1$.

The relationship between the variables representing the coded factor levels $\boldsymbol{x}$ and the canonical equation variables $\boldsymbol{Z}$ is

$$Z = M'(x - x_s) \tag{13A.8}$$

where the columns of M are the normalized eigenvectors m_i.

The various methods required for the calculations can be found in standard matrix algebra books such as Graybill (1983). The matrix calculations can be performed with many of the common computer programs. The calculations are illustrated with the response equation for Example 13.3,

$$\hat{y} = 169 + 6.747x_1 + 26.385x_2 - 10.875x_1^2 - 21.625x_2^2 - 15.25x_1x_2$$

The determinantal equation is

$$\begin{vmatrix} -10.875 - \lambda & -7.625 \\ -7.625 & -21.625 - \lambda \end{vmatrix} = 0$$

with $\lambda^2 + 32.5\lambda + 177.03 = 0$. The roots of the quadratic equation are $\lambda_1 = -25.58$ and $\lambda_2 = -6.92$. Thus, the canonical equation with $\hat{y}_s = 177.25$ is

$$\hat{y} = 177.25 - 25.58Z_1^2 - 6.92Z_2^2$$

The matrix of normalized eigenvectors is

$$M = \begin{bmatrix} 0.4603 & 0.8877 \\ 0.8877 & -0.4603 \end{bmatrix}$$

The coordinate of the stationary point is $x_{1s} = -0.156$ and $x_{2s} = 0.665$, and the relationship between the canonical variables and the coded factor variables is

$$\begin{bmatrix} Z_1 \\ Z_2 \end{bmatrix} = \begin{bmatrix} 0.4603 & 0.8877 \\ 0.8877 & -0.4603 \end{bmatrix} \begin{bmatrix} (x_1 + 0.156) \\ (x_2 - 0.665) \end{bmatrix}$$

or

$$Z_1 = 0.4603(x_1 + 0.156) + 0.8877(x_2 - 0.665)$$

$$Z_2 = 0.8877(x_1 + 0.156) - 0.4603(x_2 - 0.665)$$

14   Split-Plot Designs

This chapter introduces the split-plot design for experiments with a factorial treatment design and describes some unique features of the design relative to its structure, composition of experimental errors, and analysis. The relative efficiency for split-plot designs is also discussed. Extensions and variations of the design include the split-split-plot and split-block designs.

14.1   Plots of Different Size in the Same Experiment

One factor sometimes requires more experimental material for its evaluation than a second factor in factorial experiments. In agronomic or horticultural field trials a factor such as cultural methods may require the use of equipment that is best-suited for large plots, whereas another factor in the experiment such as cultivar or fertility level may be applied easily to a much smaller plot of land. The larger cultural treatment plot, the *whole plot*, is split into smaller *subplots* to which the different cultivars or fertility treatments are applied. This is known as a *split-plot* design, and in this particular example there are two different sizes of experimental units.

The experiment used for the following example illustrates the creation of a split-plot design when a second factor was introduced to subdivisions of the existing experimental units for an experiment already in progress.

> ### Example 14.1   Nitrogen Fertilizer and Thatch Accumulation in Penncross Creeping Bent Grass
>
> The soil for most golf greens is almost pure sand and frequent irrigation and fertilization are required to maintain the turf. The sandy soil has little capacity to retain nitrogen, and after fertilization the nitrogen quickly leaches from the root zone after irrigation. Administering large initial doses of nitrogen to

retain nitrogen for a longer period of time is harmful to the turfgrass and soil microbes. Thus, a means to apply moderate rates of nitrogen and to retain nitrogen in the root zone would be very beneficial.

Nitrogen fertilizers are manufactured in various chemical configurations. Two commonly used fertilizers, ammonium sulphate and urea, were known to be fast-release nitrogen forms and were expected to leach out of the soil very quickly. Others such as isobutylidene diurea (IBDU) and sulphur-coated urea, urea(SC), although more costly, were known to be slow-release nitrogen forms and expected to stay in the soil for a longer period of time.

A second factor that could affect nitrogen retention was the turf thatch, or the buildup of dead grass. The thatch was removed frequently because it was thought plant diseases accumulated in the buildup. However, thatch provides a capacity to retain nitrogen fertilizer and could partially relieve some of the difficulty of nitrogen loss with the sandy soil.

Research Objective: A soil scientist wanted to investigate the effects of nitrogen supplied in different chemical forms and later evaluate those effects combined with those of thatch accumulation on the quality of an established turf.

Treatment Design: The four forms of nitrogen fertilizer used for the study were (1) urea, (2) ammonium sulphate, (3) isobutylidene diurea (IBDU), and (4) sulphur-coated urea, urea(SC). Each of the fertilizers was to be supplied to the turf at a rate of 1 pound of nitrogen per 1000 square feet of turf. Any differences in responses to the fertilizers then could be attributed to the mode of nitrogen release because an equivalent amount of nitrogen was supplied by each form.

Experiment Design: A golf green had been constructed and seeded with Penncross creeping bent grass on the experimental plots. The nitrogen treatment plots were arranged on the golf green in a randomized complete block design with two replications.

After two years the second treatment factor, years of thatch accumulation, was added to the experiment. Each of the eight experimental plots was split into three subplots to which levels of the second treatment factor were randomly assigned. The lengths of time the thatch was allowed to accumulate on the subplot green were two, five, or eight years.

The resulting split-plot design had the whole-plot treatment factor of nitrogen source in a randomized complete block design with years of thatch accumulation as a subplot treatment factor.

The field plan for one replication of the split-plot design for the turfgrass experiment after randomization is shown in Display 14.1. The nitrogen whole-plot treatment, factor A, has four levels, and the years of thatch accumulation subplot factor, factor B, has three levels.

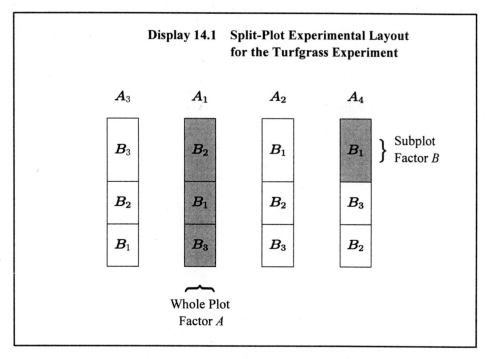

**Display 14.1 Split-Plot Experimental Layout
for the Turfgrass Experiment**

Source: Dr. I. Pepper, Department of Soil and Water Science, University of Arizona.

An Altered Randomization for Split-Plot Designs

The usual randomization of the factorial treatment combinations to the experimental units has been altered to accommodate the particular requirements of the experiment. For example, the split-plot design in Display 14.1 has one additional restriction in the randomization relative to that of the usual randomized design. In the usual randomized design the 12 treatment combinations are assigned randomly to the 12 subplots; however, in the split-plot design the three levels of factor B combined with a single form of nitrogen, factor A, are restricted to the same whole plot. Thus, the split-plot design is a product of specific changes in the random allocation of factorial treatment combinations to the experimental units.

More Possibilities for Two Unit Sizes

If one factor, such as temperature, humidity, or photoperiod, requires environmental control chambers, then each level of that factor requires a separate chamber. A second treatment factor such as culture media for plants and even a third treatment factor can be included within each of the chambers, thereby achieving an economical use of the chambers to study more than one factor at a time. The chambers represent whole plots, and the subplots are units to which the second factor levels are applied inside each chamber.

In education research a teaching method is applied as a whole-plot treatment to an entire classroom; however, subgroups of students within each classroom can be used as subplots to study an additional factor, such as the use of certain library materials or microcomputers.

Industrial research may require replicate batches of raw material mixtures. The whole-plot treatment, such as product mixtures, can be one batch of raw material, and a subplot treatment, such as curing time, can be applied to subbatches of the larger product mixture batches.

14.2   Two Experimental Errors for Two Plot Sizes

The statistical analysis must take into account the presence of two different types or sizes of experimental units in the experiment. Factor A effects are estimated from the whole plots for factor A. The factor B effects and the AB interaction effects are estimated from the subplots for factor B. Since the whole plots and subplots are experimental units of different sizes or types, they have different precision. The differing precision must be taken into account for making comparisons among treatment means.

The consideration of two separate errors is a consequence of the fact that observations from different subplots in the same whole plot may be positively correlated. The correlation reflects the nature of experimental units to respond similarly when they are adjacent to one another, such as neighboring field plots, students in a classroom, cultures in a growth chamber, or units from the same batch of raw materials in the industrial experiment.

We assume a correlation ρ between observations on any two subplots in the same whole plot and also assume observations from two different whole plots are uncorrelated. Given these assumptions, it can then be shown that the error variance for the main effects of A on a per-subplot basis is $\sigma^2[1 + (b - 1)\rho]$ if there are b subplots in each whole plot. Likewise, for comparisons among the main effects of B and AB interaction comparisons the error variance per subplot is $\sigma^2(1 - \rho)$.

As a consequence of these differences in the errors associated with the whole plot and subplot treatment comparisons, the partition of the sum of squares in the analysis of variance is altered somewhat from the usual partition for a two-factor factorial design. The partitions for factor effects and blocking factors remain the same as that for the usual factorial designs, but the experimental error is partitioned into two components. One component of experimental error is associated with the whole-plot treatment factor, and the other component is associated with the subplot treatment factor and interaction.

14.3 The Analysis for Split-Plot Designs

The Split-Plot Model

A mixed-model formulation is used for the split-plot design to reflect the separate experimental error variances for the subplot and whole-plot units. It includes separate random error effects for the whole-plot units and the subplot units. If the whole-plot treatment factor is placed in a randomized complete block design the linear model is

$$y_{ijk} = \mu + \alpha_i + \rho_k + d_{ik} + \beta_j + (\alpha\beta)_{ij} + e_{ijk} \qquad (14.1)$$

$$i = 1, 2, \ldots, a \qquad j = 1, 2, \ldots, b \qquad k = 1, 2, \ldots, r$$

where μ is the general mean, α_i is the effect of the ith level of factor A, ρ_k is the effect of the kth block, d_{ik} is the whole-plot random error, β_j is the effect of the jth level of factor B, $(\alpha\beta)_{ij}$ is the interaction effect between factors A and B, and e_{ijk} is the subplot random error.

The whole-plot and subplot errors are assumed to be independent, normally distributed random errors with mean 0 and variances σ_d^2 and σ_e^2, respectively. Randomization of treatments to the experimental units justifies the assumption of independence for the two random errors and the equal correlation between the errors for subplot units on the same whole-plot unit.

The Split-Plot Analysis of Variance

The expectation of the mean squares for the analysis of variance using the mixed model in Equation (14.1) is shown in Table 14.1 with fixed effects for factors A and B.

Table 14.1 Expected mean squares for the split-plot analysis of variance

Source of Variation	Degrees of Freedom	Mean Square	Expected Mean Square
Blocks	$r - 1$	MS Blocks	
A	$a - 1$	MSA	$\sigma_e^2 + b\sigma_d^2 + rb\theta_a^2$
Error(1)	$(a - 1)(r - 1)$	$MSE(1)$	$\sigma_e^2 + b\sigma_d^2$
B	$b - 1$	MSB	$\sigma_e^2 + ra\theta_b^2$
AB	$(a - 1)(b - 1)$	$MS(AB)$	$\sigma_e^2 + r\theta_{ab}^2$
Error(2)	$a(r - 1)(b - 1)$	$MSE(2)$	σ_e^2

The expected mean squares for Error(1) and Error(2) based on the mixed model reflect the differences in the variability for the two different types of experimental units. The expected error variances for the whole plots are larger than those for the subplots.

Tests of Hypotheses About Factor Effects

The F statistics to test the null hypotheses for interaction and main effects are

- (interaction) $H_0: (\alpha\beta)_{ij} = 0$ versus $H_a: (\alpha\beta)_{ij} \neq 0$ for some i, j

$$F_0 = \frac{MS(\boldsymbol{AB})}{MSE(2)} \text{ with } (a-1)(b-1) \text{ and } a(r-1)(b-1) \text{ d.f.}$$

- (factor B main effects) $H_0: \overline{\mu}_{.1} = \cdots = \overline{\mu}_{.b}$ versus $H_a: \overline{\mu}_{.i} \neq \overline{\mu}_{.j}$ for some i, j

$$F_0 = \frac{MS\boldsymbol{B}}{MSE(2)} \text{ with } b-1 \text{ and } a(r-1)(b-1) \text{ d.f.}$$

- (factor A main effects) $H_0: \overline{\mu}_{1.} = \cdots = \overline{\mu}_{a.}$ versus $H_a: \overline{\mu}_{i.} \neq \overline{\mu}_{j.}$ for some i, j

$$F_0 = \frac{MS\boldsymbol{A}}{MSE(1)} \text{ with } a-1 \text{ and } (a-1)(r-1) \text{ d.f.}$$

Example 14.2   Data and Analysis

One of the measurements made on the turfgrass plots of Example 14.1 was the chlorophyll content of clippings (mg/g) sampled from each plot. The data are shown in Table 14.2, and the analysis of variance is shown in Table 14.3.

Table 14.2 Chlorophyll content (mg/gm) of grass clippings

| Nitrogen Source | Block | Years of Thatch Accumulation | | |
		2	5	8
Urea	1	3.8	5.3	5.9
	2	3.9	5.4	4.3
Ammonium sulphate	1	5.2	5.6	5.4
	2	6.0	6.1	6.2
IBDU	1	6.0	5.6	7.8
	2	7.0	6.4	7.8
Urea (SC)	1	6.8	8.6	8.5
	2	7.9	8.6	8.4

Table 14.3 Analysis of variance of chlorophyll content of Penncross creeping bent grass clippings

Source of Variation	Degrees of Freedom	Sum of Squares	Mean Square	F	Pr > F
Total	23	48.78			
Block	1	0.51	0.51		
Nitrogen (N)	3	37.32	12.44	29.62	0.010
Error (1)	3	1.26	0.42		
Thatch (T)	2	3.82	1.91	9.10	0.009
N × T	6	4.15	0.69	3.29	0.065
Error (2)	8	1.72	0.21		

Computational Notes

The analysis of variance for the split-plot design can be computed with many available statistical computing packages. The instructions for the analysis are equivalent to those for a two-factor factorial treatment design except for the necessity to compute two separate error terms for the analysis.

The sum of squares for Error(1) is numerically equivalent to the error sum of squares for the experiment design that was utilized for the whole plots. With completely randomized designs it is equivalent to the sum of squares for whole plots within factor A treatments. For randomized complete block designs it is equivalent to the computation of the sum of squares for Blocks $\times A$ interaction. The sum of squares for Error(2) is ordinarily the residual sum of squares computed automatically by the program.

Alternatively, the analysis can be conducted with statistical programs specifically written for mixed models, which have been incorporated into many statistical computing packages. They can be particularly useful for split-plot designs in blocked designs with random blocks. The statistical estimation in these programs is based on maximum likelihood methods, which are beyond the scope of this book.

The test for interaction between nitrogen and thatch is $F_0 = 0.69/0.21 = 3.29$, and the test is not significant with $Pr > F = .065$ (Table 14.3). The test for differences among the thatch means is $F_0 = 1.91/0.21 = 9.10$, and it is significant with $Pr > F = .009$. The test for differences among the whole-plot nitrogen treatment means is $F_0 = 12.44/0.42 = 29.62$, and it is significant with $Pr > F = .01$.

The cell means and the marginal means for source of nitrogen and years of thatch accumulation are shown in Table 14.4. Turf that received the sulphur-coated urea, urea(SC), had the highest chlorophyll content followed by IBDU and ammonium sulphate while urea resulted in the lowest amount of chlorophyll. This relative ranking of the chlorophyll content by source of nitrogen was the same for each year of thatch accumulation.

The graph in Figure 14.1 shows a plot of the mean chlorophyll content versus years of thatch accumulation separately for each of the nitrogen sources. The marginal means in Table 14.4 indicate an increase in chlorophyll content as the

Table 14.4 Mean chlorophyll content (mg/g) of Penncross creeping bent grass clippings

| Nitrogen Source | Years of Thatch Accumulation | | | Nitrogen Means |
	2	5	8	
Urea	3.85	5.35	5.10	4.77
Ammonium sulphate	5.60	5.85	5.80	5.75
IBDU	6.50	6.00	7.80	6.77
Urea(SC)	7.35	8.60	8.45	8.13
Thatch means	5.83	6.45	6.79	

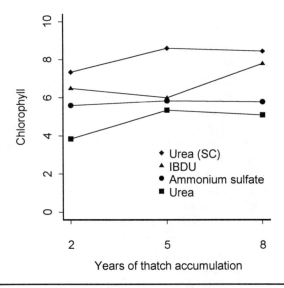

Figure 14.1 Mean chlorophyll content of Penncross creeping bent grass clippings versus nitrogen source for two, five, and eight years of thatch accumulation

years of thatch accumulation increase. However, observation of the cell means and their plotted values in Figure 14.1 shows that this general trend was not manifested for each of the separate sources of nitrogen.

This difference in the trends among the nitrogen sources indicates the possibility that some form of interaction may be present even though the global test for interaction was not significant. Partitions of the interaction sum of squares will sometimes reveal the presence of certain forms of interaction between the factors.

Interpretation with Regression Contrasts

The factorial treatment design is composed of one quantitative factor, years of thatch accumulation (T), and one qualitative factor, source of nitrogen (N). The treatment design lends itself to the use of the polynomial regression partitions in the analysis of variance for one quantitative factor with one qualitative factor (discussed in Chapter 6). Linear and quadratic regression sums of squares partitions can be computed for years of thatch accumulation (T) with corresponding partitions for the nitrogen $\times$ thatch interaction. These partitions are shown in the analysis of variance for the experiment in Table 14.5.

Table 14.5 Analysis of variance for chlorophyll content of Penncross creeping bent grass clippings with orthogonal polynomial regression partitions for thatch factor

Source of Variation	Degrees of Freedom		Sum of Squares	Mean Square	F	Pr > F
Total	23		48.78			
Blocks	1		0.51	0.51		
Nitrogen (N)	3		37.32	12.44	29.62	0.010
Error(1)	3		1.26	0.42		
Thatch (T)	2		3.82	1.91	9.10	0.009
T linear		1	3.71	3.71	17.67	0.003
T quadratic		1	0.11	0.11	0.52	0.494
$N \times T$	6		4.15	0.69	3.29	0.065
$N \times T$ linear		3	0.80	0.27	1.29	0.358
$N \times T$ quadratic		3	3.36	1.12	5.33	0.028
Error(2)	8		1.72	0.21		

The interaction for quadratic deviations of nitrogen $\times$ thatch is significant with $F_0 = 1.12/0.21 = 5.33$ and $Pr > F = .028$. The linear regression partition for thatch (T) is significant with $F_0 = 3.71/0.21 = 17.67$ and $Pr > F = .003$. Thus, there are significant quadratic deviations from the linear response to years of thatch accumulation that differ among the sources of nitrogen. The differing patterns were observed in Figure 14.1.

Several types of comparisons may be of interest at this point to further clarify the interpretation. Given there was a significant regression interaction component, it will be beneficial to make a comparison among the nitrogen means for each of the years of thatch accumulation to determine if the sulphur-coated urea always resulted in the highest chlorophyll content of the turf. Also a quadratic regression of chlorophyll content on years of thatch accumulation can be computed for each of the nitrogen sources to characterize the differences among the nitrogen sources relative to years of thatch accumulation.

14.4 Standard Errors for Treatment Factor Means

The standard errors shown in Table 14.6 are used for tests of hypotheses for comparisons among the estimated treatment means. The degrees of freedom associated with each of the standard errors are those for the mean square used in the standard error. The only exception is the final comparison shown in the table where the standard error is a weighted combination of the two mean squares for error. Consequently, the appropriate degrees of freedom can be approximated by the procedure introduced by Satterthwaite (1946) (Chapter 5). The approximation is

$$\text{d.f.} = \frac{[(b-1)MSE(2) + MSE(1)]^2}{\dfrac{[(b-1)MSE(2)]^2}{f_2} + \dfrac{[MSE(1)]^2}{f_1}} \tag{14.2}$$

where f_1 and f_2 are the degrees for $MSE(1)$ and $MSE(2)$, respectively, from the analysis of variance.

Table 14.6   Standard error estimators for the split-plot design (A = whole-plot factor, B = subplot factor)

Treatment Comparison	Standard Error Estimator
Difference between two A means $\bar{y}_{u..} - \bar{y}_{v..}$	$\sqrt{\dfrac{2MSE(1)}{rb}}$
Difference between two B means $\bar{y}_{.u.} - \bar{y}_{.v.}$	$\sqrt{\dfrac{2MSE(2)}{ra}}$
Difference between two B means at the same level of A $\bar{y}_{ju.} - \bar{y}_{ju.}$	$\sqrt{\dfrac{2MSE(2)}{r}}$
Difference between two A means at the same or different level of B $\bar{y}_{uk.} - \bar{y}_{vk.}$ or $\bar{y}_{uk.} - \bar{y}_{vm.}$	$\sqrt{\dfrac{2[(b-1)MSE(2) + MSE(1)]}{rb}}$

It should be noted that comparisons requiring standard errors based on the linear combination of mean squares, $(b-1)MSE(2) + MSE(1)$, only have approximate probability levels for tests of hypotheses and confidence intervals. The linear combination of mean squares no longer shares the same probability distribution properties of the individual mean squares.

Standard Errors for the Turfgrass Experiment

Referring to Table 14.6 the standard errors required for the analysis of Example 14.1 are differences between

- two nitrogen means

$$\sqrt{\frac{2MSE(1)}{rb}} = \sqrt{\frac{2(0.42)}{6}} = 0.37 \text{ with 3 degrees of freedom}$$

- two thatch means

$$\sqrt{\frac{2MSE(2)}{ra}} = \sqrt{\frac{2(0.21)}{8}} = 0.23 \text{ with 8 degrees of freedom}$$

- two thatch means for the same nitrogen

$$\sqrt{\frac{2MSE(2)}{r}} = \sqrt{\frac{2(0.21)}{2}} = 0.46 \text{ with 8 degrees of freedom}$$

- two nitrogen means at the same or different thatch levels

$$\sqrt{\frac{2[(b-1)MSE(2) + MSE(1)]}{rb}} = \sqrt{\frac{2[2(0.21) + 0.42]}{6}} = 0.53$$

where the degrees of freedom for the last standard error obtained from the Satterthwaite approximation are

$$\text{d.f.} = \frac{[2(0.21) + 0.42]^2}{\frac{[2(0.21)]^2}{8} + \frac{[0.42]^2}{3}} = 8.73 \text{ or } 9 \tag{14.3}$$

The experimental error variance for subplots, $MSE(2) = 0.21$, is one-half the magnitude of experimental error variance estimated for the whole plots, $MSE(1) = 0.42$. Thus, the comparisons among years of thatch accumulation means and interaction means will be more precise than comparisons among source of nitrogen means. The question of whether there has been an equitable trade-off between the advantages and disadvantages of splitting the plots will be partially answered by the relative efficiency computations in Section 14.6.

14.5 Features of the Split-Plot Design

Randomization for the split-plot design requires that levels of factor A be randomized to the whole units in accordance with the protocol for the experiment design in which the whole units are arranged—that is, completely randomized, randomized complete block, and so forth. Levels of factor B are assigned to the subunits *within* each of the whole units at random, separately for each of the whole units.

The design also can be described as a confounded factorial design (discussed in Chapter 11). The subunits are regarded as the experimental units, so that the levels of factor A are applied to groups or blocks of the subunits. Consequently, comparisons among the levels of A are confounded with the blocks of subunits. The split-plot design is often referred to as a *confounded factorial design* in which the *main effects* are confounded with blocks, whereas confounding was restricted to *interactions* for the designs in Chapter 11.

The Analysis of Variance for Common Experiment Designs

The sources of variation and degrees of freedom for the sum of squares partitions are shown in Table 14.7 for the split-plot design in which the whole plots are arranged in three of the most common experiment designs—the completely randomized, randomized complete block, and Latin square designs. Each design has r replications, a levels of A, and b levels of B.

Table 14.7 Sources of variation and degrees of freedom for the split-plot design analysis of variance

Completely Randomized		Randomized Complete Block		Latin Square	
		Whole Plots			
Source	d.f.	Source	d.f.	Source	d.f.
				Rows	$a - 1$
		Blocks	$r - 1$	Columns	$a - 1$
A	$a - 1$	A	$a - 1$	A	$a - 1$
Error(1)	$a(r - 1)$	Error(1)	$(r - 1)(a - 1)$	Error(1)	$(a - 1)(a - 2)$
		Subplots			
B	$b - 1$	B	$b - 1$	B	$b - 1$
AB	$(a - 1)(b - 1)$	AB	$(a - 1)(b - 1)$	AB	$(a - 1)(b - 1)$
Error(2)	$a(r - 1)(b - 1)$	Error(2)	$a(r - 1)(b - 1)$	Error(2)	$a(a - 1)(b - 1)$

Different Precision for Whole-Plot and Subplot Factor Means

The larger degrees of freedom associated with the estimates of experimental error for the factor B means and the AB interaction means and comparisons among them indicate that the B and AB effects are more precisely estimated than the A

effects measured on the whole plots. However, this can be misleading as both $MSE(1)$ and $MSE(2)$ for the split-plot design have fewer degrees of freedom than does the mean square for experimental error in the design that would occur without the split-plot feature for the two-factorial effects. For example, the same factorial design in a randomized complete block design without the restrictions of the split-plot design would have $(r - 1)(ab - 1)$ degrees of freedom for experimental error, which exceeds that for either $MSE(1)$ or $MSE(2)$ with $(r - 1)(a - 1)$ and $a(r - 1)(b - 1)$ degrees of freedom, respectively, for the split-plot design.

Practical experience with split-plot designs has shown that $MSE(2)$ often is smaller than $MSE(1)$ as anticipated from the expected mean squares. Consequently, there can be an increase in the precision of estimates of B and AB effects that offsets the loss in degrees of freedom. However, the average experimental error over all treatment effects is the same with or without the split-plot feature. Thus, there is no net gain from the split-plot design. Rather, if there is a gain in the precision on B and AB effects it is offset by a loss in the precision on A effects. Consequently, large and interesting A effects possibly may be judged nonsignificant. Evaluation of the relative efficiency of the split-plot design in this context is discussed in Section 14.6.

Advantages of the Split-Plot Experiment

The primary advantages of the split-plot have already been mentioned: when one factor requires considerably more experimental material than another factor, such as in the agronomic field studies, or when there is an opportunity to study responses to a second factor while efficiently utilizing resources, such as in the growth chamber studies. The experiment in Example 14.1 illustrated the creation of a split-plot design by introducing a second factor to subdivisions of the existing experimental units of an experiment already in progress.

Disadvantages of the Split-Plot Experiment

The primary disadvantages of the split-plot design include the potential loss in precision in treatment comparisons and an increase in complexity of the statistical analysis. The analysis of variance and the estimation of the standard errors for different types of treatment comparisons require additional computations.

14.6 Relative Efficiency of Subplot and Whole-Plot Comparisons

Ordinarily, the relative efficiency of an experiment design refers to the efficiency as a result of blocking by some factor relative to that ignoring the blocking factor. With split-plot designs it is informative to consider the relative efficiency of using the split-plot design in lieu of the same experiment design without the split-plot feature. For example, when the whole-plot treatment factor is arranged in a randomized complete block design it is possible to determine which of the designs is more efficient for the whole-plot treatment comparisons and which is more efficient for the subplot treatments and interaction comparisons. As discussed in Section 14.5 there is a compromise between the increase in precision on the subplot treatment means and the decrease in precision on the whole-plot treatment means.

Efficiency of Subplot Comparisons

Federer (1955) shows the efficiency of the split-plot design relative to the randomized complete block design for the subplot comparisons to be

$$RE = K\frac{a(b-1)MSE(2)+(a-1)MSE(1)}{(ab-1)MSE(2)} \qquad \textbf{(14.4)}$$

where

$$K = \frac{(f_1+1)(f_2+3)}{(f_1+3)(f_2+1)}$$

$f_1 = a(b-1)(r-1)$, the degrees of freedom for the subplot error $MSE(2)$, and $f_2 = (ab-1)(r-1)$, the degrees of freedom for the randomized complete block experimental error.

Efficiency of Whole-Plot Comparisons

The relative efficiency of the split-plot design to the randomized complete block design for the whole-plot comparisons is

$$RE = K\frac{a(b-1)MSE(2)+(a-1)MSE(1)}{(ab-1)MSE(1)} \qquad \textbf{(14.5)}$$

with $f_1 = (a-1)(r-1)$ for the whole-plot error $MSE(1)$.

The efficiency of the split-plot design for thatch and nitrogen × thatch subplot comparisons relative to that if all 12 different combinations had been randomly allocated to the 12 plots in each block with $a = 4$, $b = 3$, $r = 2$, $f_1 = 8$, and $f_2 = 11$ is

$$RE = \frac{(9)(14)}{(11)(12)} \times \frac{8(0.21)+3(0.42)}{11(0.21)} = 0.95(1.27) = 1.21$$

or a gain of 21%. For nitrogen whole-plot comparisons with $f_1 = 3$ and $f_2 = 11$ the relative efficiency is

$$RE = \frac{(4)(14)}{(6)(12)} \times \frac{8(0.21)+3(0.42)}{11(0.42)} = 0.78(0.64) = 0.49$$

or only 49%. Therefore, a gain of only 21% in efficiency was realized for the subplot comparisons, and a loss of 51% was realized for the whole-plot comparisons with the split-plot design relative to the standard randomized complete block design.

14.7 The Split-Split-Plot Design for Three Treatment Factors

The convenience of introducing a third factor into the treatment design requires a subdivision of the subplots such that all levels of the third factor are administered to these new subdivisions, referred to as sub-subplots. The design, commonly referred to as the **split-split-plot**, has three different sizes or types of experimental units. The analysis requires the computation of an additional sum of squares for experimental error associated with the sub-subplots. The partition of the sum of squares for factors A and B is that shown in Table 14.7. If the third factor, C, has c levels the additional sources of variation and associated degrees of freedom in the analysis of variance are

Source	Degrees of Freedom
C	$c - 1$
AC	$(a - 1)(c - 1)$
BC	$(b - 1)(c - 1)$
ABC	$(a - 1)(b - 1)(c - 1)$
Error(3)	$ab(r - 1)(c - 1)$

The sum of squares for Error(3) is the usual residual sum of squares produced by a computer program. Special instructions must be given for separate computation of the sum of squares for Error(2). For randomized complete block designs the sum of squares for Error(2) is computed as the pooled sum of squares for the Blocks $\times B$ and Blocks $\times AB$ interaction.

All of the standard errors applied to A and B effects shown in Table 14.6 are valid and need only have the value of c included as part of the divisor. For example, the standard error for the difference between two factor A means $\bar{y}_{u...} - \bar{y}_{v...}$ is $\sqrt{2MSE(1)/rbc}$. The standard errors required for comparisons involving the factor C effects are shown in Table 14.8.

14.8 The Split-Block Design

Numerous variations of the basic split-plot design have been used by investigators beyond further subdivisions of the experimental units. One common variation is the split-block design in which the subunit treatments occur in a strip across the whole-plot units. The design is sometimes called a *strip plot* design.

The split-block design can be useful in agricultural field studies when two treatment factors require the use of large field plots. The levels of one treatment factor A are randomly assigned to the plots in a randomized complete block design.

Table 14.8   Standard error estimators for the split-split-plot design

Treatment Comparison	Standard Error Estima
Two C means $(\bar{y}_{..u} - \bar{y}_{..v})$	$\sqrt{\dfrac{2MSE(}{rab}}$
Two C means at the same level of A $(\bar{y}_{i.u.} - \bar{y}_{i.v.})$	$\sqrt{\dfrac{2MSE(3}{rb}}$
Two C means at the same level of B $(\bar{y}_{.iu.} - \bar{y}_{.iv.})$	$\sqrt{\dfrac{2MSE(3}{ra}}$
Two C means at the same level of A and B $(\bar{y}_{iju.} - \bar{y}_{ijv.})$	$\sqrt{\dfrac{2MSE(3}{r}}$
Two B means at the same or different levels of C $(\bar{y}_{.ui.} - \bar{y}_{.vi.}$ or $\bar{y}_{.ui.} - \bar{y}_{.vj.})$	$\sqrt{\dfrac{2[(c-1)MSE(3)+MSE(2}{rac}}$
Two B means at the same level of A and C $(\bar{y}_{iuj.} - \bar{y}_{ivj.})$	$\sqrt{\dfrac{2[(c-1)MSE(3)+MSE(2}{rc}}$
Two A means at the same or different level of C $(\bar{y}_{u.i.} - \bar{y}_{v.i.}$ or $\bar{y}_{u.i.} - \bar{y}_{v.j.})$	$\sqrt{\dfrac{2[(c-1)MSE(3)+MSE(1}{rbc}}$
Two A means at the same or different levels of B and C (for example, $\bar{y}_{uij.} - \bar{y}_{vij.}$ or $\bar{y}_{uij.} - \bar{y}_{vlk.}$)	$\sqrt{\dfrac{2[b(c-1)MSE(3)+(b-1)MSE(2)+MSE(1}{rbc}}$

The plots for the second factor B are constructed in the same manner but are laid out perpendicular to the plots for factor A. The levels of the second factor B are then randomly assigned to this second array of plots across the same block. A sketch of one block for the split-block design is shown in Display 14.2, in which there are $a = 3$ levels of factor A and $b = 4$ levels of factor B.

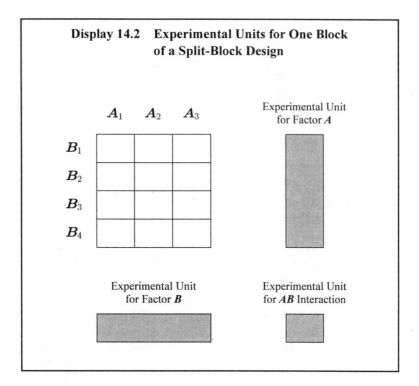

Display 14.2 Experimental Units for One Block of a Split-Block Design

Three Sizes of Units and Three Experimental Errors

The split-block design has three sizes of experimental units where the units for the main effects of A and B are equivalent to whole plots, each of different orientation. The unit for the AB interaction effect is a subplot where there is an intersection of the two whole plots for the respective levels of A and B. Consequently, a separate experimental error must be estimated from the analysis of variance for each of the three treatment effects.

The linear statistical model for the split-block design differs from that for the ordinary split-plot design in that it has separate random error terms for each of the three treatment effects. A blocking effect ρ_k is also included in the model

$$y_{ijk} = \mu + \rho_k + \alpha_i + d_{ik} + \beta_j + g_{jk} + (\alpha\beta)_{ij} + e_{ijk} \tag{14.6}$$

$$i = 1, 2, \ldots, a \quad j = 1, 2, \ldots, b \quad k = 1, 2, \ldots, r$$

The random error effects, d_{ik}, g_{jk}, and e_{ijk} are the experimental errors for the units associated with A, B, and AB effects, respectively, with variances σ_d^2, σ_g^2, and σ_e^2. The analysis of variance for this model along with expected mean squares for fixed treatment effects is shown in Table 14.9.

Table 14.9 Analysis of variance outline for the split-block design with two treatment factors

Source of Variation	Degrees of Freedom	Sum of Squares	Expected Mean Square
Blocks	$r - 1$	SS Blocks	
A	$a - 1$	SSA	$\sigma_e^2 + b\sigma_d^2 + rb\theta_a^2$
Error(1)	$(r - 1)(a - 1)$	$SSE(1)$	$\sigma_e^2 + b\sigma_d^2$
B	$b - 1$	SSB	$\sigma_e^2 + a\sigma_g^2 + ra\theta_b^2$
Error(2)	$(r - 1)(b - 1)$	$SSE(2)$	$\sigma_e^2 + a\sigma_g^2$
AB	$(a - 1)(b - 1)$	$SS(AB)$	$\sigma_e^2 + r\theta_{ab}^2$
Error(3)	$(r - 1)(a - 1)(b - 1)$	$SSE(3)$	σ_e^2

Standard Errors for Treatment Factor Means

The standard errors for the split-block are more complex than those for the split-plot design because of the altered randomization pattern of factor B treatment levels. The basic standard errors of the treatment means and treatment differences are shown in Table 14.10.

The degrees of freedom for the standard errors with a single mean square will be those associated with the mean square. For those standard errors with two or more mean squares it is necessary to obtain degrees of freedom values with the Satterthwaite approximation.

14.9 Additional Information About Split-Plot Designs

The basic split-plot design and several extensions of the design have been discussed in this chapter. Numerous other modifications and applications of the design have been presented elsewhere including subunit treatments in a Latin square design, confounding of comparisons among subunit treatments, and systematic arrangement of whole-unit treatments (Cochran & Cox, 1957; Petersen, 1985). Robinson (1967) gave a split-plot design and analysis for subplots in a balanced incomplete block arrangement and Coons et al. (1989) utilized a split-plot design with whole plots in a balanced incomplete block design. The statistical properties of split-plot designs with incomplete blocks were given by Mejza and Mejza (1996), and properties for the split-block designs with incomplete blocks were shown by Hering and Mejza (1997). Federer (1975) included an extensive array of modifications along with the connection to series of experiments repeated over locations or time. Little and Rubin (1987) and Jarrett (1978) provided detailed relevant discussions on the analysis of split-plot designs with missing data, including formulations to obtain standard errors for general treatment contrasts. Steel and Torrie (1980) included discussions on the analysis of repeated harvest data for perennial crops as split-plot

Table 14.10   Standard error estimators for the split-block design

Treatment Comparison	Standard Error Estimator
Two A means $(\bar{y}_{u..} - \bar{y}_{v..})$	$\sqrt{\dfrac{2MSE(1)}{rb}}$
Two B means $(\bar{y}_{.u.} - \bar{y}_{.v.})$	$\sqrt{\dfrac{2MSE(2)}{ra}}$
Two A means, same level of B $(\bar{y}_{uj.} - \bar{y}_{vj.})$	$\sqrt{\dfrac{2[(b-1)MSE(3) + MSE(1)]}{rb}}$
Two B means, same level of A $(\bar{y}_{ju.} - \bar{y}_{jv.})$	$\sqrt{\dfrac{2[(a-1)MSE(3) + MSE(2)]}{ra}}$

Any two means, different levels of A and B $(\bar{y}_{uj.} - \bar{y}_{vk.})$

$$\sqrt{\frac{2[aMSE(1) + bMSE(2) + (ab - a - b)MSE(3)]}{rab}}$$

designs. Since repeated harvests of the same plot fall under the general category of repeated measurements and longitudinal studies, a discussion of that topic is found in the next chapter.

EXERCISES FOR CHAPTER 14

1. A split-plot experiment was conducted on sorghum with two treatment factors, plant population density and hybrid. The whole plots were used for the four levels of plant population density—10, 15, 25, and 40 plants per meter of row. There were three hybrids randomly allocated to the subplots of each plot. The experiment was conducted in a randomized complete block design with four replications. The data that follow are the weights of the seed per plant in grams.
    a. Write the linear model for this experiment, explain the terms, and compute the analysis of variance for the data.
    b. Construct a summary table of cell means and marginal means for this experiment, and compute the estimated standard errors for the table of means.

Head Seed Weight (g) of a Sorghum Trial					
		Plants per Meter of Row			
Hybrid	*Block*	*10*	*15*	*25*	*40*
TAM 680	1	40.7	24.2	16.1	11.2
	2	37.8	44.4	17.6	12.7
	3	32.9	27.8	19.9	14.5
	4	43.1	34.1	20.1	15.4
RS 671	1	39.4	31.3	17.9	14.8
	2	47.8	34.5	30.5	17.3
	3	44.4	25.6	22.5	17.7
	4	49.0	50.4	25.2	18.7
Tx 399	1	68.7	26.2	20.5	18.9
× Tx 2536	2	56.2	48.1	28.2	26.2
	3	44.8	41.1	30.0	19.2
	4	59.3	46.0	24.7	22.0

Source: Dr. R. Voigt, Department of Plant Sciences, University of Arizona.

   c. Compute the estimated standard errors for the differences between two observed means:
     (i)   for hybrids
     (ii) for plant populations
     (iii) for two hybrids at the same plant population
   d. Test the hypotheses for interaction and main effects assuming fixed effects for hybrids and plant populations.
   e. Compute the relative efficiency of this split-plot design for subplot and whole-plot treatments relative to the ordinary randomized complete block design, and interpret it.
   f. Partition the sum of squares for plant population and the interaction sum of squares into the appropriate polynomial regression partitions, and interpret the results. Make a graph of the observed means to assist in the interpretation. The coefficients for linear, quadratic, and cubic partitions for the four levels of Plants per Meter are shown below.

Plants	10	15	25	40
Linear	− 0.546	− 0.327	0.109	0.764
Quadratic	0.513	− 0.171	− 0.741	0.399
Cubic	− 0.435	0.783	− 0.435	0.087

2.   A split-plot experiment was conducted in a completely randomized design with whole-plot treatments as a 2×2 factorial (factors A and B) and the subplot treatments as three levels of factor C. There were four replications of the experimental units. Assume all treatment effects were fixed.
    a. Write the linear model for the experiment. Identify each of the model components, and show the numerical ranges on the subscript.

b. Outline the analysis of variance table showing sources of variation, degrees of freedom, and expected mean squares.

3. Suppose the whole-plot treatments in Exercise 14.2 were arranged in a Latin square design. Repeat parts (a) and (b) in that case.

4. Suppose the subplot treatments in Exercise 14.2 were a 3×3 factorial arrangement of factors C and D while all other conditions given were the same. Repeat parts (a) and (b) in that case.

5. An investigator in food science wants to conduct an experiment to assess the effect of cold storage conditions on food quality. The two treatment factors to be used are storage temperature and container material. The food product will be placed in one of the containers and stored in a temperature control chamber for a fixed period of time after which a number of physical and subjective quality measures will be taken on the food product in each container. The investigator has three small temperature control chambers available for the experiment. The storage temperatures for the experiment are 2°C, 4°C, and 8°C. There are four container types for the study—sealed plastic, sealed cardboard covered with wax, sealed cardboard, and an open container as a control. There are four positions on the center shelf of the chamber in which the containers can be placed. Draw a diagram of a plan for the experiment so that the investigator may have three replications of the experiment. Use the following guidelines to construct your diagram.
    a. Use the labels I, II, and III to identify the three temperature chambers and the labels a, b, c, and d to identify the four positions within each chamber.
    b. Show the container type (C1, C2, C3, or C4) assigned to each position in each chamber as well as the temperature (2°, 4°, or 8°) assigned to the chamber for all three replications in your diagram.
    c. Show the randomization scheme you used to make the assignments in part (b).

6. A research specialist for a large seafood company investigated bacterial growth on oysters and mussels subjected to three different storage temperatures. Nine cold storage units were available. Three storage units were randomly selected to be used for each of the storage temperatures. Oysters and mussels were stored for two weeks in each of the cold storage units. A bacterial count was made from a sample of oysters and mussels at the end of two weeks. The logarithm of bacterial count for each sample is shown in the table at the end of the exercise.
    a. The investigator could have had three replications for the study by simply taking three random samples of each seafood from a single cold storage unit set at one temperature. In this way only three cold storage units would have been needed for the study, one for each temperature. Explain the potential difficulty with the study if it had been conducted in this way.
    b. Is there a significant increase in bacterial growth as temperature increases? Justify your answer.
    c. Is there a significant interaction between type of seafood and increase (if any) of bacterial growth with temperature? Justify your answer.
    d. Write the linear model for your analysis, state the assumptions, and explain the terms.
    e. Determine whether your assumptions about the linear model are correct for this data.

Storage Unit	Temperature (°C)	Seafood*	log (count)
1	0	1	3.6882
1	0	2	0.3565
2	0	1	1.8275
2	0	2	1.7023
3	0	1	5.2327
3	0	2	4.5780
4	5	1	7.1950
4	5	2	5.0169
5	5	1	9.3224
5	5	2	7.9519
6	5	1	7.4195
6	5	2	6.3861
7	10	1	9.7842
7	10	2	10.1352
8	10	1	6.4703
8	10	2	5.0482
9	10	1	9.4442
9	10	2	11.0329

*1 = oysters,   2 = mussels

7.   A split-plot experiment in a randomized complete block design evaluated the effects of nitrogen, water, and phosphorus rates on the water use efficiency in a drip irrigation culture of sweet corn. Two rates of phosphorus ($P_1 = 0$ and $P_2 = 245$ lb P_2O_5/acre) were randomized to whole plots in a randomized complete block design. The 3×3 factorial treatments of nitrogen (0, 130, and 260 lb N/acre) and water (16, 22, and 28 inches) were randomized to subplots within each of the main plots. The data shown are water use efficiency for each subplot.

    a.   Write a linear model for the experiment, explain the terms, and conduct the analysis of variance for the data.

    b.   Construct a summary table of cell means and marginal means for this experiment, and compute the estimated standard errors for the table of means.

    c.   Compute the estimated standard errors for the differences between two observed means:
       (i)   for phosphorus rates
       (ii)   for water levels
       (iii) for nitrogen rates
       (iv)   for water by nitrogen cell means

    d.   Test the hypotheses for all interactions and main effects, and interpret the results.

    e.   Partition the sum of squares for water and nitrogen into linear and quadratic polynomial regression partitions including interaction. Interpret the results utilizing a graph of the observed means to assist in the interpretation.

Water	Nitrogen	Block 1		Block 2	
		P_1	P_2	P_1	P_2
16	0	8.1	9.7	8.6	15.5
	130	36.0	34.2	34.5	33.1
	260	34.6	34.0	40.7	39.3
22	0	10.0	6.2	5.1	10.9
	130	21.5	19.7	19.9	21.9
	260	30.7	28.9	26.4	25.7
28	0	10.6	6.3	4.5	10.4
	130	19.4	19.7	21.7	19.9
	260	23.2	23.0	19.4	23.2

Source: Dr. T. Doerge, Department of Soil and Water Science, University of Arizona.

15 Repeated Measures Designs

The central topic of this chapter concerns repeated measurements of the response variable on each experimental unit. Experiments with observations made on successive occasions in time are emphasized. The statistical properties of the observations are discussed, and suitable methods for analyzing the data are demonstrated.

15.1 Studies of Time Trends

The time trend of individual responses to treatment is an important aspect of many experiments. Examples include experiments in which animals are weighed weekly to monitor growth under different nutrient conditions or field plots of perennial crops such as alfalfa are harvested several times in succession. **Repeated measures** occur frequently in clinical trials when patients are measured at regular intervals to monitor the response to medical treatment.

Repeated measures on each experimental unit provide information on the time trend of the response variable under different treatment conditions. Time trends can reveal how quickly the units respond to treatment or how long the treatment effects are manifested on the units of the study. Differences in these trends among the treatments also can be evaluated.

Repeated Observations Result in Increased Precision

Repeated observations on the same experimental unit over time are often a more efficient use of resources than the use of a different experimental unit for each observation time. Not only are fewer units required, thereby reducing costs, but the estimation of time trends will be more precise. The increased precision results because measurements on the same unit tend to be less variable than measurements on different units. Thus, the effect of repeated measures is similar to the effect of blocking.

Example 15.1   Early Detection of Phlebitis in Amiodarone Therapy

Treatment Design: An experiment described in detail for Example 2.2 was designed to explore mechanisms for early detection of phlebitis during Amiodarone therapy. Phlebitis is an inflammation of a vein that occurs upon intravenous administration of drugs. Three intravenous treatments were administered to test animals: (1) Amiodarone with a vehicle solution to carry the drug, (2) vehicle solution only, and (3) a saline solution.

Experiment Design: Rabbits, used as the test animals, were randomly assigned to the three treatment groups in a completely randomized design. A treatment solution was administered to the rabbit through an intravenous needle inserted in a vein of one ear. The temperature of both ears was monitored for several hours. An increase in the temperature of the treated ear was considered a possible early indicator of phlebitis. The difference in the temperatures of the two ears (treated minus untreated) was used as the response variable.

Repeated Measurements: The temperatures were observed every 30 minutes in each of the rabbits for the duration of the study. The observations, made at 0, 30, 60, and 90 minutes on the rabbits, are shown in Table 15.1.

A Profile Plot Reveals a Trend

The observed trends over time for the three treatments in the Amiodarone study are shown in Figure 15.1. The profile plots in Figure 15.1 show an increase in the observed temperature differences for the rabbits in the Amiodarone treatment. A less definite increase is observed with the vehicle treatment, and the profile for the saline treatment indicates only a fluctuating response over time with no definite trend.

The objective of the analysis for this study will be to determine whether there is a significant upward trend in the temperature for any of the treatments. If so, it will be important to determine whether the Amiodarone is responsible for any significant temperature increase rather than the vehicle solution or the intravenous procedure itself represented by the saline treatment. Thus, contrasts between the trends of the Amiodarone treatment and that of the two control treatments will be of utmost importance in the analysis.

Between- and Within-Subjects Designs

Repeated measures designs can be described in terms of the *between-subjects* design and the *within-subjects* design. The between-subjects design refers to the treatment design and the experiment design used for the experimental units. The within-subjects design refers to the repeated measures on each experimental unit. The

Table 15.1 Ear temperature differences (°C), treated minus untreated, of rabbits at 0, 30, 60, and 90 minutes after treatment

Treatment	Rabbit	*Time of Observation (minutes)*			
		0	*30*	*60*	*90*
Amiodarone	1	− 0.3	− 0.2	1.2	3.1
	2	− 0.5	2.2	3.3	3.7
	3	− 1.1	2.4	2.2	2.7
	4	1.0	1.7	2.1	2.5
	5	− 0.3	0.8	0.6	0.9
	Mean	− 0.24	1.38	1.88	2.58
Vehicle	6	− 1.1	− 2.2	0.2	0.3
	7	− 1.4	− 0.2	− 0.5	− 0.1
	8	− 0.1	− 0.1	− 0.5	− 0.3
	9	− 0.2	0.1	− 0.2	0.4
	10	− 0.1	− 0.2	0.7	− 0.3
	Mean	− 0.58	− 0.52	− 0.06	0.00
Saline	11	− 1.8	0.2	0.1	0.6
	12	− 0.5	0.0	1.0	0.5
	13	− 1.0	− 0.3	− 2.1	0.6
	14	0.4	0.4	− 0.7	− 0.3
	15	− 0.5	0.9	− 0.4	− 0.3
	Mean	− 0.68	0.24	− 0.42	0.22

Source: G. Ward, Department of Pharmaceutical Sciences, University of Arizona.

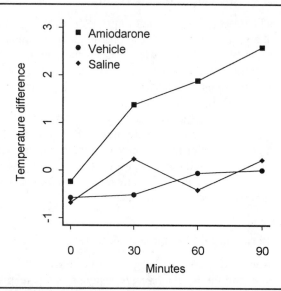

Figure 15.1 Profile plot of the means for each treatment at each time period for the Amiodarone study

design illustrated in Example 15.1 has three treatments in a completely randomized design for the between-subjects design. The within-subjects design consists of repeated measures on each rabbit.

Alternatively, two or more treatments can be administered to each of the subjects. Suppose athletes are used as subjects for a study on exercise physiology. After a training regimen is completed the athletes are given tests on a treadmill set in two different positions, horizontal and inclined. The objective is to determine whether differences exist between the results of horizontal and inclined treadmill tests. Each athlete is a block in a randomized complete block design if the treadmill tests are administered to the athletes in random order.

15.2 Relationships Among Repeated Measurements

The relationships among the observations govern the statistical methods required for the particular research design used in a study. The correspondence of the relationships to the method of analysis for repeated measures designs is explored in this chapter along with some useful strategies for the analysis.

Correlated Observations Among Repeated Measures

The time order of measurements at $0, 30, 60$, and 90 minutes on each rabbit in Example 15.1 cannot be randomized over time; thus, pairs of repeated measures on the same rabbit are likely to be correlated. Generally, pairs of observations adjacent in time are assumed to have a larger correlation than pairs of observations more separated in time. Observations at 0 and 30 minutes on any one of the rabbits in the Amiodarone study are assumed to have a larger correlation than observations at 0 and 90 minutes.

The correlation between two variables, say y_1 and y_2, is defined as

$$\rho_{12} = \frac{\sigma_{12}}{\sigma_1 \sigma_2} \tag{15.1}$$

where σ_1 and σ_2 are the standard deviations of y_1 and y_2 and σ_{12} is the covariance between y_1 and y_2. If the expected value or mean of the variable y is $E(y) = \mu$, the variance of y is $\sigma^2 = E(y - \mu)^2$. The covariance of two variables, y_1 and y_2, is $\sigma_{12} = E(y_1 - \mu_1)(y_2 - \mu_2)$. The covariance is a measure of how two variables will vary together. If one variable increases in value as the other increases in value the covariance is positive and the correlation between the variables is positive. The theoretical variances and covariances for repeated measures taken successively as y_1, y_2, y_3, and y_4 are illustrated in Display 15.1 as the 4×4 Σ matrix.

Analysis of Variance Assumptions

Equal variances for the treatment groups and independent, normally distributed observations are the usual assumptions required for a valid analysis of variance of the data. Independence of observations results in zero values for the covariances

	y_1	y_2	y_3	y_4
	Display 15.1	**Σ Matrix of Variances and Covariances for Four Repeated Measures**		
y_1	σ_1^2	σ_{12}	σ_{13}	σ_{14}
y_2	σ_{21}	σ_2^2	σ_{23}	σ_{24}
y_3	σ_{31}	σ_{32}	σ_3^2	σ_{34}
y_4	σ_{41}	σ_{42}	σ_{43}	σ_4^2

shown in Display 15.1. Under these assumptions σ^2 has the same value for all treatment groups and measurement times, and $\rho = 0$ or $\sigma_{ij} = \sigma_{ji} = 0$.

Compound Symmetry Means Equal Correlation Among Repeated Measures

A particular experiment with randomization of treatments to experimental units is only a random sample of all possible randomized experiments that could have been used. The act of randomization does not remove the correlation between observations on experimental units; however, the expected correlation between the experimental units is constant under all possible randomizations. If the variances and correlations are constant, the covariances will have the constant value $\sigma_{ij} = \rho\sigma^2$ from Equation (15.1). This condition is known as *compound symmetry*. The matrix of variances and covariances with compound symmetry is shown in Display 15.2.

	y_1	y_2	y_3	y_4
	Display 15.2	**Σ Matrix of Variances and Covariances of Four Repeated Measures Under Compound Symmetry**		
y_1	σ^2	$\rho\sigma^2$	$\rho\sigma^2$	$\rho\sigma^2$
y_2	$\rho\sigma^2$	σ^2	$\rho\sigma^2$	$\rho\sigma^2$
y_3	$\rho\sigma^2$	$\rho\sigma^2$	σ^2	$\rho\sigma^2$
y_4	$\rho\sigma^2$	$\rho\sigma^2$	$\rho\sigma^2$	σ^2

Split-Plot Treatments Are Randomized; Repeated Measures Are Not Randomized

The assumption of compound symmetry was used for the errors of observation in the split-plot experiment in Chapter 14 because treatments were randomly assigned to the subplots. The subject in the repeated measures design is equivalent to the whole-plot in the split-plot design, and the between-subjects treatment factor is equivalent to the whole-plot treatment factor in the split-plot design. The repeated measure on a subject is analogous to the subplot in the split-plot design. The difference between the subplot observations and the repeated measures is that treatments are randomized to the subplots in the split-plot design, whereas there is no randomization for the repeated measures. If all repeated measures on a subject are equally correlated, there is compound symmetry and the repeated measures design can be analyzed as a split-plot design with time of measurement as the subplot treatment factor. The split-plot analysis of variance was exhibited in Chapter 14 for various between-subject, or whole-plot, experiment designs.

The Huynh–Feldt Condition Less Stringent than Compound Symmetry

Huynh and Feldt (1970) showed that conditions required for the usual analysis of variance for repeated measures designs were less stringent than the compound symmetry condition. They showed the necessary condition is to have the same variance of the difference for all possible pairs of observations taken at different time periods, say y_i and y_j, or

$$\sigma^2_{(y_i - y_j)} = 2\lambda \ \text{ for } i \neq j \tag{15.2}$$

for some $\lambda > 0$. This condition also can be stated as

$$\sigma_{ij} = \frac{1}{2}(\sigma_i^2 + \sigma_j^2) - \lambda \ \text{ for } i \neq j \tag{15.3}$$

The matrix of variances and covariances satisfying this condition is known as the **Type H matrix**. The mean squares from an analysis of variance can be used to test hypotheses about the within-subjects treatments if the Huynh–Feldt condition is satisfied.

If Each Subject Receives All Treatments

The realities of many research studies from the standpoints of economy and control of experimental error require us to obtain more than one observation from each experimental unit. For example, the considerable cost of maintaining large animals makes it expedient to obtain as much information about the treatments as possible from the individual animals. Also, the variability of observations among animals tends to be much larger than that among multiple observations on the same animal. Thus, blocking on animals with all treatments administered to each animal increases the precision of treatment comparisons.

When each of the treatments is administered in random order to each subject, for example, B→D→A→C, then subjects are random blocks in a randomized complete block design. The expected mean squares for the randomized complete block design are shown in Table 15.2. It is a mixed model analysis with random blocks and fixed treatment effects. The statistic, $F_0 = MST/MSE$, tests the null hypothesis of no differences among the treatment means.

Table 15.2 Expected mean squares for the randomized complete block mixed-model analysis of variance

Source of Variation	Degrees of Freedom	Mean Square	Expected Mean Squares
Blocks	$r - 1$	MSB	$\sigma^2 + t\sigma_\rho^2$
Treatments	$t - 1$	MST	$\sigma^2 + r\theta_\tau^2$
Error	$(t - 1)(r - 1)$	MSE	σ^2

Carryover Effects

The effects of certain types of treatments carry over into the next treatment period when treatments are administered to subjects in sequence. These carryover effects can seriously bias the estimates of treatment means because treatments administered in previous periods influence the effects of the treatments in succeeding periods.

Carryover effects are particularly troublesome with animal and human subjects with successive administration of dietary or medical treatments that affect the physiology of the subject. A "washout," or rest period, between two successive treatments is often used to clear the effects of the most recent treatment before a second treatment is administered. Special crossover designs developed for these studies are the subject of Chapter 16.

15.3   A Test for the Huynh–Feldt Assumption

A univariate analysis of variance can be used under any of the three alternate sets of assumptions about the repeated measures discussed in Section 15.2. They were independence, compound symmetry, or the Huynh–Feldt condition for the repeated measures. The Huynh–Feldt condition with the Type H matrix for the variances and covariances of the repeated measures is the least restrictive of the three assumptions. The simpler univariate methods can be used for the analysis if we can assume the Huynh–Feldt condition holds for the repeated measures. The assumption of a Type H matrix can be evaluated with a test attributed to Mauchly (1940). The test is demonstrated in this section.

The Mauchly test of the Huynh–Feldt condition for repeated measures is illustrated with the Amiodarone study of Example 15.1. Recall the experiment had three treatments allocated to experimental units in a completely randomized design, and the experimental units were measured on four successive occasions.

The Mauchly W test (Mauchly, 1940), used to test the hypothesis of Type H form for Σ, is computed by many computer packages that have programs for the analysis of repeated measures designs. A brief outline on the details of the test statistic is given in Appendix 15A.1.

The result of the Mauchly W test for the Amiodarone study computed by a statistical program is shown in Table 15.3. The Mauchly test statistic, $W = 0.852$, is approximately distributed as a chi-square variable with $\nu = p(p-1)/2 - 1$ degrees of freedom, where p is the number of repeated measures. For the Amiodarone study $\nu = 4(3)/2 - 1 = 5$. The W statistic is not significant, with $P(\chi_5^2 > 1.72) = .886$—the probability of a chi-square variable with 5 degrees of freedom exceeding a chi-square value of 1.72. If the result of the Mauchly test is acceptable, then the F tests in the univariate analysis of variance are valid.

Table 15.3 Results of the Mauchly test for a Type H Σ matrix

Mauchly Sphericity Test
$W = .852$
Chi-square approx. $= 1.72$ with 5 D.F.
$P(\chi_5^2 > 1.72) = .886$

The Mauchly test for the variance–covariance matrix Σ tends to have low power unless sample sizes are very large. The ability to detect departures from the null hypothesis is not very good unless the experiments have a large number of replications. Boik (1981) indicated power of less than .20 for some specific cases of the Mauchly test when there were three treatment groups and as many as 12 subjects per group. Consequently, complete reliance on the test is not recommended. Given the uncertainty associated with the ability of the Mauchly test to detect departures from analysis of variance assumptions, our decision to use the univariate analysis of variance will have to be based on our experience with the specifics of our research material.

15.4 A Univariate Analysis of Variance for Repeated Measures

If we can reasonably assure ourselves that the analysis of variance assumptions are valid for repeated measures on each rabbit in the treatment groups, the split-plot analysis of variance mean squares can be used to test hypotheses about the treatment means and their interactions with time. The rabbits in the Amiodarone study are equivalent to whole plots for the three intravenous treatments, and repeated measures on the rabbits are equivalent to subplot treatments.

Use the Split-Plot Model for the Analysis

The linear model for the split-plot experiment is

$$y_{ijk} = \mu + \alpha_i + d_{ik} + \beta_j + (\alpha\beta)_{ij} + e_{ijk} \qquad (15.4)$$

$$i = 1, 2, \dots, t \qquad j = 1, 2, \dots, p \qquad k = 1, 2, \dots, r$$

where μ is the general mean, α_i is the effect of the ith treatment, d_{ik} is the random experimental error for rabbits within treatments with variance σ_d^2, β_j is the effect of the jth time, $(\alpha\beta)_{ij}$ is the interaction between treatments and time, and e_{ijk} is the normally distributed random experimental error on repeated measures with variance σ_e^2. The split-plot analysis of variance for the data from the Amiodarone study is shown in Table 15.4. The split-plot analysis of variance was exhibited in Chapter 14.

Table 15.4 Split-plot analysis of variance for repeated measures from the Amiodarone study in a completely randomized design

Source of Variation	Degrees of Freedom	Sum of Squares	Mean Square	F	Pr > F
Total	59	93.28			
Treatment (A)	2	35.38	17.69	19.44	0.000
Error(1)	12	10.94	0.91		
Time	3	16.08	5.36	9.24	0.000
A × Time	6	10.06	1.68	2.90	0.021
Error(2)	36	20.82	0.58		

The test for interaction between treatments and time, $F_0 = MS(A \times \text{Time})/MSE(2) = 1.68/0.58 = 2.90$, is significant with $Pr > F = .021$. The significant interaction between time and the intravenous treatments indicates the ear temperature responses over time are different among the three treatments.

Use Regression Contrasts on Repeated Measures to Study Time Trends

The global test for significance of the interaction between treatments and time indicates little about the specifics of interaction if it exists. The responses to individual treatments over time is an important component of the analysis on repeated measures. The interaction should be investigated as a difference in trend over time among the treatments.

The observed trends over time for the three treatments in the Amiodarone study were shown in Figure 15.1. The profile plots exhibited an increase in the observed temperature differences for the rabbits in the Amiodarone treatment. Less definite trends were observed with the vehicle and saline treatments.

The between-subject treatments and time constitute a factorial treatment design, with time as a quantitative factor and the between-subject treatments as a qualitative factor. The polynomial regression partitions for one qualitative factor in the analysis of variance have been illustrated several times in previous chapters (for example, see Example 6.3).

Linear, quadratic, and cubic regression sum of squares partitions can be computed for time with corresponding partitions for the treatment × time interaction. The sum of squares partitions for the Amiodarone study are shown in Table 15.5. The polynomial regression sum of squares partitions for the split-plot analysis of variance were exhibited in Example 14.1.

Table 15.5 Split-plot analysis of variance with polynomial contrasts for repeated measures from the Amiodarone study

Source of Variation	Degrees of Freedom		Sum of Squares	Mean Square	F	$Pr > F$
Total	59		93.28			
Treatment (A)	2		35.38	17.69	19.44	0.000
Error(1)	12		10.94	0.91		
Time (T)	3		16.08	5.36	9.24	0.000
T linear		1	14.52	14.52	25.03	0.000
T quadratic		1	0.60	0.60	1.03	0.315
T cubic		1	0.96	0.96	1.66	0.205
$A \times T$	6		10.06	1.68	2.90	0.021
$A \times T$ linear		2	7.80	3.90	6.72	0.003
$A \times T$ quadratic		2	0.56	0.28	0.48	0.622
$A \times T$ cubic		2	1.71	0.85	1.47	0.242
Error(2)	36		20.82	0.58		

It was determined previously that interaction between the treatments and time was significant, but the hypotheses tested with the mean squares in Table 15.5 provide more specific information about the form of the interaction. The interaction between treatments and linear regression on time, $A \times T$ linear, is significant with $F_0 = 3.90/0.58 = 6.72$ and $Pr > F = .003$. Neither the quadratic nor cubic regression on time has a significant interaction with time. Thus, the linear trends over time differ among the treatments.

The estimated linear contrasts for each of the treatment groups and their standard errors will indicate more specifically how the linear trends differ among the treatment groups. The linear contrasts, $a_{i(1)}$, are calculated for each of the treatment groups by computing the linear contrast among the time means for each treatment, where $a_{i(1)}$ is the linear contrast for treatment i. For example, the means for the Amiodarone treatment group at 0, 30, 60, and 90 minutes were -0.24, 1.38, 1.88, and 2.58 in Table 15.1. The linear contrast for the Amiodarone treatment group is

$$a_{1(1)} = \frac{\Sigma P_{1j}\bar{y}_{1j.}}{\Sigma P_{1j}^2} = \frac{(-3)(-0.24) + (-1)(1.38) + (1)(1.88) + (3)(2.58)}{(-3)^2 + (-1)^2 + 1^2 + 3^2}$$

$$= \frac{8.96}{20} = 0.45$$

with standard error

$$s_{c_1} = \sqrt{\frac{MSE(2)}{r(\Sigma P_{1j}^2)}} = \sqrt{\frac{0.58}{5(20)}} = 0.08$$

The linear contrasts for the vehicle and saline treatments are $a_{2(1)} = 0.11$ and $a_{3(1)} = 0.10$, respectively, each with standard error $s_c = 0.08$.

The 95% SCI for the linear contrasts require the Bonferroni $t_{.05,3,36} = 2.51$. The Bonferroni t has 36 degrees of freedom since the mean square for error, $MSE(2) = 0.58$ from Table 15.5, has 36 degrees of freedom. The 95% SCI are computed as $a_{i(1)} \pm 2.51(0.08)$.

The 95% SCI for the Amiodarone treatment is (0.25, 0.65), indicating a significant linear increase in the temperature over time with the Amiodarone treatment. The intervals for the vehicle and saline treatments are $(-0.09, 0.31)$ and $(-0.10, 0.30)$, respectively, indicating there is not a significant linear change in temperature for either the vehicle or saline treatments.

Since neither the saline nor vehicle control treatments resulted in a significant change in temperature differences we can conclude the significant increase in temperature for the Amiodarone treatment group was a function of the drug and not the vehicle or the manipulation of the intravenous injection.

15.5   Analysis When Univariate Analysis Assumptions Do Not Hold

When the Huynh–Feldt condition is not satisfied for the repeated measures the results from the univariate analysis illustrated in Section 15.4 are not valid. Several alternative analyses are suggested when the usual analysis of variance cannot be used.

Three Choices for the Analysis

A multivariate analysis is the most general method available, but general multivariate methods are beyond the intent of this book and their direct use is not considered here. A second alternative makes conservative adjustments to the usual F_0 statistics from the analysis of variance to better approximate the significance levels of the tests. These adjustments will be illustrated in this section. A third alternative analysis with attractive features, illustrated in this section, utilizes contrasts among the repeated measures. The analysis of contrasts uses features of multivariate analysis that can be applied in a straightforward manner to repeated measures. The study described in Example 15.2 resulted in repeated observations on the experimental units for which the Huynh–Feldt assumptions were not valid.

Example 15.2 Soil Moisture and Soil Microbe Activity

A productive agricultural soil requires a certain level of soil aeration to maintain active plant root growth and soil microbial activity. A soil scientist found that soil aeration levels had been affected in soils fertilized with the nutrient rich sludge by-product from a sewage treatment plant. The aeration level of the soil can be reduced by the high water content of the added sludge soil; through compaction by heavy machinery used to add the sludge; and, ironically, by the increased microbial activity caused by adding the high-organic sludge material to the soil.

Research Objective: One objective for this particular study was to determine moisture levels at which soil aeration became limiting to microbial activity in soils.

Treatment Design: The treatments included a control soil treatment with no sludge fertilizer and a moisture content of 0.24 kg water/kg soil. Three treatments of different moisture content were used for soil fertilized with sludge. The three moisture levels for the fertilized soil were 0.24, 0.26, and 0.28 kg water/kg soil.

Experiment Design: Samples of soil were randomly assigned to the four treatments in a completely randomized design. The treated soil samples were placed in airtight containers and incubated under conditions conducive to microbial activity. The soil was compacted in the containers to the degree experienced in the field.

Microbial activity, measured as CO_2 evolution, was used as a measure of the soil aeration level. The CO_2 evolution/kg soil/day was measured in each container on days 2, 4, 6, and 8 after the beginning of the incubation period. The microbial activity in each soil sample was recorded as the percent increase in CO_2 produced above atmospheric levels. The data are shown in Table 15.6.

The Univariate Analysis Assumption Is Not Valid

The Mauchly test of whether the Huynh–Feldt condition holds for the repeated CO_2 measurements is shown in Table 15.7. The Mauchly statistic is $W = 0.180$, and the chi-square distribution approximation to W has $\nu = p(p-1)/2 - 1 = 4(3)/2 - 1 = 5$ degrees of freedom. The significance level for the test is $P(\chi_5^2 > 11.52) = .042$, and the null hypothesis for the Type H matrix is rejected at the .05 significance level.

Table 15.6 Repeated measures on CO_2 evolution from microbial activity in soil under different moisture conditions

(Kg water/Kg soil) Moisture	Container	% CO_2 Evolution/kg Soil/Day			
		Day 2	Day 4	Day 6	Day 8
Control	1	0.22	0.56	0.66	0.89
	2	0.68	0.91	1.06	0.80
	3	0.68	0.45	0.72	0.89
	Mean	0.53	0.64	0.81	0.86
0.24	4	2.53	2.70	2.10	1.50
	5	2.59	1.43	1.35	0.74
	6	0.56	1.37	1.87	1.21
	Mean	1.89	1.83	1.77	1.15
0.26	7	0.22	0.22	0.20	0.11
	8	0.45	0.28	1.24	0.86
	9	0.22	0.33	0.34	0.20
	Mean	0.30	0.28	0.59	0.39
0.28	10	0.22	0.80	0.80	0.37
	11	0.22	0.62	0.89	0.95
	12	0.22	0.56	0.69	0.63
	Mean	0.22	0.66	0.79	0.65

Source: Dr. I. Pepper and J. Neilson, Department of Soil and Water Science, University of Arizona.

Table 15.7 Results of the Mauchly test for the Huynh–Feldt condition for CO_2 measurements on soil samples

Mauchly Sphericity Test
$W = 0.180$
Chi-square approx. = 11.52 with 5 D.F.
$P(\chi_5^2 > 11.52) = .042$

The Errors of Inference are Compromised

Given a significant Mauchly test we can assume results from the univariate analysis of variance may not be valid. Boik (1981) showed that very small departures from the Huynh–Feldt condition seriously affect the Type I errors and power of the univariate F tests for the repeated measure factor—the day of CO_2 measurement in the case of the current study. One of the earliest compromises to the univariate analysis of variance was to adjust the values of the computed F_0 statistics in the analysis.

Adjustments to Univariate Test Statistics

If the Huynh–Feldt condition does not hold for the repeated measures, then the F_0 statistic has only an approximate F distribution with reduced degrees of freedom (Box, 1954a, 1954b). Greenhouse and Geisser (1959) suggested an adjustment $\hat{\varepsilon}$ based on the work by Box, where the numerator and denominator degrees of freedom for the F_0 statistic are multiplied by $\hat{\varepsilon}$. More conservative tests result when the adjustment is used since $\hat{\varepsilon} \leq 1$ and the test requires a larger value of F_0 to be significant. The calculation of the $\hat{\varepsilon}$ adjustment is shown in Appendix 15A.2, and it is computed in most statistical programs for repeated measures analysis.

Huynh and Feldt (1976) suggested a less conservative adjustment than the Greenhouse–Geisser $\hat{\varepsilon}$ adjustment. Huynh (1978) reported the Huynh–Feldt $\tilde{\varepsilon}$ adjustment produced tests with probabilities of Type I errors closer to the chosen value of α than did the Greenhouse–Geisser adjustment. The Huynh–Feldt $\tilde{\varepsilon}$ adjustment is computed by most programs for repeated measures analysis, and the $\tilde{\varepsilon}$ adjustment computation is shown in Appendix 15A.2.

The adjustments are applied to the usual split-plot analysis of variance F_0 statistics for the repeated measures factor. No adjustments are necessary for tests about whole-plot, or between-subject, treatment factors since the treatments are randomly assigned to the experimental units. The split-plot analysis of variance for the repeated measures is shown in Table 15.8 with Moisture treatments as the whole-plot factor and Day as the subplot factor.

Table 15.8 Analysis of variance for CO_2 measurements on soil samples (SAS-GLM)

Source of Variation	Degrees of Freedom	Sum of Squares	Mean Square	F	$Pr > F$	G–G* $Pr > F$	H–F[†] $Pr > F$
Moisture	3	11.56	3.85	11.15	0.003		
Error(1)	8	2.77	0.35				
Day	3	0.49	0.16	1.22	0.324	0.317	0.324
Day × Moisture	9	1.55	0.17	1.28	0.296	0.330	0.304
Error(2)	24	3.21	0.13				

*Significance level after adjustment with Greenhouse–Geisser epsilon = 0.5245
[†]Significance level after adjustment with Huynh–Feldt epsilon = 0.8755

The Greenhouse–Geisser $\hat{\varepsilon}$ adjustment shown at the bottom of Table 15.8 is $\hat{\varepsilon} = .5245$ and the Huynh–Feldt $\tilde{\varepsilon}$ adjustment is $\tilde{\varepsilon} = .8755$. The Greenhouse–Geisser adjustment reduces the degrees of freedom about 48%, whereas the less conservative Huynh–Feldt adjustment reduces the degrees of freedom about 12%.

The Greenhouse–Geisser $\widehat{\varepsilon}$ adjustments to the numerator and denominator degrees of freedom for the test of interaction between moisture levels and days, Day × Moisture, are $9(.5245) = 4.7$ and $24(.5245) = 12.6$, respectively. With the adjusted degrees of freedom the probability of exceeding $F_0 = 1.28$ is $Pr > F = .330$, whereas the significance level was .296 without the adjustment. The net effect of the adjustment is to increase the P-Value for the F_0 statistic to a more conservative test which is less likely to reject the null hypothesis. The less conservative Huynh–Feldt adjustments result in $9(.8755) = 7.9$ and $24(.8755) = 21$ degrees of freedom for the test with $Pr > F = .304$.

Since the interaction between moisture levels and days was not significant, similar adjustments are made to F_0 degrees of freedom for a test of the main effects for days. The degrees of freedom adjustments have only a slight effect on $Pr > F$. In either case the main effects for days are not significant.

The F test results for main effects of moisture treatments is the usual F_0 for whole-plot main effects in the split-plot analysis of variance. The calculated value for soil moisture treatments is $F_0 = 11.15$ with $Pr > F = .003$, and the null hypothesis is rejected. The average CO_2 evolution differs among the soil moisture treatments, and the nonsignificant interaction indicates the levels did not change over the eight days of measurement.

Tests among the soil moisture treatment means require standard errors based on the whole-plot equivalent error mean square, $MSE(1) = 0.35$, in Table 15.8. For example, the standard error of the difference between two means is $\sqrt{2(0.35)/(3)(4)} = 0.24$, given $r = 3$ replications and $p = 4$ repeated measures for each main effect mean.

Contrasts on the Repeated Measures Provide Specific Inferences

The F tests based on the Greenhouse–Geisser or Huynh–Feldt adjustments are limited to global conclusions about the equality of treatment means, whereas questions of greater consequence usually involve interesting contrasts among the treatment means. An alternative analysis for repeated measures is based on important contrasts among the repeated measures. The analysis provides the appropriate test statistics from multivariate methods while using the familiar univariate analysis of variance.

The analysis requires the calculation of a contrast among the repeated measures for each experimental unit. The values of the contrast are used as if they were the original observations on the experimental units, and a univariate analysis of variance is computed from the observed contrast values.

Polynomial Regression Contrasts to Study Time Trends

The orthogonal polynomial regression contrasts are the most useful statistics for investigating the trend over time. For example, the linear contrast for CO_2 evolution over time in the soil moisture study is calculated for each soil sample as

$$z_{ij(1)} = (-3)(y_{ij1}) + (-1)(y_{ij2}) + (1)(y_{ij3}) + (3)(y_{ij4})$$

The linear model for a contrast c is

$$z_{ij(c)} = \mu_c + \tau_{i(c)} + e_{ij(c)} \tag{15.5}$$

where $z_{ij(c)}$ is the contrast value for the jth experimental unit on the ith treatment, μ_c is the general mean for the contrast, $\tau_{i(c)}$ is the treatment effect for the contrast, and $e_{ij(c)}$ is the normally distributed random experimental error for the contrast with variance σ_{ec}^2.

The contrasts of interest for the current study are the linear, quadratic, and cubic polynomial contrasts computed from the repeated observations on each of the soil containers found in Table 15.6. The resulting contrast values for each container are shown in Table 15.9. For example, the linear contrast for the first container in the control treatment is $z_{11(1)} = (-3)(0.22) + (-1)(0.56) + (1)(0.66) + (3)(0.89)$ $= 2.11$.

Table 15.9 Linear, quadratic, and cubic contrasts for three soil samples from each of four treatment groups in the CO_2 evolution study

Treatment	Unit	Contrasts		
		Linear	Quadratic	Cubic
Control	1	2.11	− 0.11	0.37
	2	0.51	− 0.49	− 0.33
	3	0.90	0.40	− 0.60
	Mean	1.17	− 0.07	− 0.19
0.24	4	− 3.69	− 0.77	0.77
	5	− 5.63	0.55	− 1.61
	6	2.45	− 1.47	− 0.85
	Mean	− 2.29	− 0.56	− 0.56
0.26	7	− 0.35	− 0.09	− 0.05
	8	2.19	− 0.21	− 2.47
	9	− 0.05	− 0.25	− 0.05
	Mean	0.60	− 0.18	− 0.86
0.28	10	0.45	− 1.01	0.15
	11	2.46	− 0.34	− 0.08
	12	1.36	− 0.40	0.02
	Mean	1.42	− 0.58	0.03

Linear(P_{1i}), $(-3, -1, 1, 3)$; quadratic(P_{2i}), $(1, -1, -1, 1)$; cubic(P_{3i})$(-1, 3, -3, 1)$

A Separate Analysis for Each Contrast

The analyses of variance computed for the linear, quadratic, and cubic contrasts are shown in Table 15.10. The analysis of variance follows the form used in Example 15.1. Polynomial contrasts were computed for the day factor along with their interaction contrasts with the moisture factor.

Table 15.10 Analyses of variance for polynomial contrasts on day of CO_2 measurements from soil samples

Source of Variation	Degrees of Freedom	Sum of Squares	Mean Square	F	Pr > F
Day linear	1	0.031	0.031	0.11	0.744
Day linear × Moisture	3	1.320	0.440	1.64	0.255
Error	8	2.143	0.268		
Day quadratic	1	0.366	0.366	4.19	0.075
Day quadratic × Moisture	3	0.156	0.052	0.60	0.635
Error	8	0.698	0.087		
Day cubic	1	0.093	0.093	2.02	0.193
Day cubic × Moisture	3	0.070	0.023	0.50	0.690
Error	8	0.369	0.046		

The primary difference from the analysis for Example 15.1 is the computation of separate error sums of squares for each of the contrasts and their interaction with moisture. The estimates of experimental error variance from the "Error" source of variation are $s_{e1}^2 = 0.268$, $s_{e2}^2 = 0.087$, and $s_{e3}^2 = 0.046$ for the linear, quadratic, and cubic contrasts, respectively. Many statistical computing packages can produce these analyses for repeated measures designs.

The Test for Interaction Between Treatments and Trends

The sum of squares for an interaction between a contrast and the Moisture treatments measures the variability in the regression contrast among the Moisture treatments. Therefore, it measures the interaction between the contrast and Moisture treatments. The F_0 statistic tests the hypothesis $H_0: \tau_{i(c)} = 0$ for all Moisture treatments.

The test statistic for no interaction between the linear contrast of time and Moisture treatments in the analysis of variance for Day linear, $F_0 = MS$(Day linear × Moisture)$/s_{e1}^2 = 0.440/0.268 = 1.64$, and with $Pr > F = .255$ we conclude there is no interaction. Likewise, the F_0 statistics for Day quadratic and Day cubic interaction with Moisture treatments test the hypotheses of no interaction between the Moisture treatments and the quadratic and cubic contrasts.

Since none of the interactions were significant we can test hypotheses concerning the existence of any trend in CO_2 evolution over time. The mean squares for each of the contrasts are used to test these hypotheses about the main effect of time. The statistic, $F_0 = MS(\text{Day linear})/s_{e1}^2 = 0.031/0.268 = 0.11$ tests the hypothesis that the linear contrast is zero for all treatments, and the hypothesis is not rejected with $Pr > F = .744$. Likewise, the F_0 statistics for Day quadratic and Day cubic test the hypotheses that the quadratic and cubic contrasts are equal to zero for all treatments, and neither contrast is significant.

We can conclude from these tests that there was no significant trend in CO_2 evolution from microbial activity during the first eight days of the incubation period. However, the analysis of variance in Table 15.8 with a significant Moisture main effect indicated the average levels of CO_2 differed among the soil treatments.

The Analysis of Individual Contrasts Are More Conservative

The analysis of the individual contrasts are not entirely free of drawbacks. The tests based on the analysis of the individual contrasts in Table 15.10 are more conservative and less powerful than those based on the usual split-plot analysis in Table 15.8. The F tests for the contrasts are based on error variances with 8 degrees of freedom in Table 15.10, whereas tests for the same contrasts from the split-plot analysis were based on an error variance with 24 degrees of freedom in Table 15.8. Even with the Greenhouse–Geisser or Huynh–Feldt adjustments in the univariate analysis the tests were less conservative than those from analyses of the individual contrasts.

Contrasts May Have Different Experimental Error Variances

The analysis based on individual contrasts illuminates a common occurrence in the analysis of data wherein the error variances associated with different contrasts can be quite disparate. The disparity in the error variances for polynomial contrasts in the soil microbe study can be seen in Table 15.10. The individual experimental error mean squares for the linear, quadratic, and cubic contrasts were $s_{e1}^2 = 0.268$, $s_{e2}^2 = 0.087$, and $s_{e3}^2 = 0.046$. Thus, a threefold difference exists between the error variances of the linear and quadratic contrasts and almost a sixfold difference between the linear and cubic contrasts. The error variance used in the univariate analysis of Table 15.8, $MSE = 0.13$, is the average error variance for the three contrasts. In certain settings the disparity between the error variances for the individual contrasts and their pooled value in the split-plot analysis of variance can lead to contradictory conclusions. Therefore, it is recommended that the potential disparity among error variances for a group of contrasts be evaluated in any particular study.

15.6   Other Experiments with Repeated Measures Properties

An analogy has been made in this chapter between the repeated measures on an experimental unit and the subplots in a split-plot experiment. Several features of the subplot in a traditional split-plot design distinguish it from a repeated measure. The subplots are usually distinct experimental units of a smaller size than the whole-plots. The subplot treatments are randomly assigned to the distinct subplot experimental units. These two properties of subplots allow a reasonable assumption that the variance and covariance structure of the observations are compatible with the requirements for the traditional analysis of variance.

The subplots in a traditional split-plot design ordinarily represent spatial variability as opposed to the time variability associated with repeated measures. The covariance relationships between the observations on the subplot treatments would be similar to those for repeated measures in time if the subunit treatments are not randomly assigned to the subplots. Randomization is not possible in certain types of split-plot experiments that have a spatial distribution of subunits.

Repeated Measures in Space

Consider a study for which an animal physiologist had hypothesized that size of tissue segment would affect the results of an assay for tyrosine concentration. Four segments of different size were taken from the same diaphragm muscle of an individual animal, and the tyrosine assay was conducted for each segment.

The four segments of different size from a single diaphragm muscle represent within-subject treatments where the individual diaphragm muscles are the subjects. The experiment design is a complete block design with diaphragm muscles as complete blocks and the muscle segment sizes as treatments.

The four segments can be taken from random locations of the diaphragm muscle but the spatial relationships among the segments are unknown. If there are different correlations between the segments of different sizes, then the usual assumptions for the randomized complete block design are not appropriate and a repeated measures analysis should be considered for the data.

Gradient Treatment Designs

Experiments with a gradient treatment design for the subplots illustrate a second type of a split-plot experiment without randomization of treatments to the subplots. Examples include experiments with sprinkler irrigation for agronomic crops used to create a gradient treatment design. Typically, the objective of the experiment is to ascertain the drought tolerance properties of several crop cultivars. The cultivars are randomly assigned to field plots in a randomized complete block design. The design is illustrated in Figure 15.2 for an experiment with five cultivars in each of two complete blocks.

A line sprinkler irrigation system is placed between the two blocks of cultivar plots. The sprinklers on the system can be adjusted to emit a high amount of water

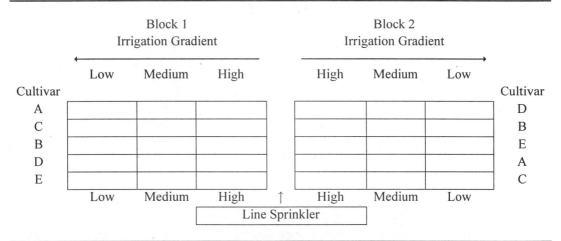

Figure 15.2   Illustration of a gradient treatment design for crop cultivars

close to the sprinkler line and a lesser amount of water in a gradient away from the sprinkler line. Consequently, a water gradient treatment is established along the length of each cultivar plot. Crop yields and other measurements are taken from subplots established on the irrigation gradient of each cultivar plot.

Three subplots are shown for each cultivar whole plot in Figure 15.2. The sub-plot treatments shown as "high," "medium," and "low" are the amounts of water applied to the subplots. The water levels are not randomly assigned to the subplots.

The usual split-plot design assumptions may not be appropriate for the observations on the subplots in the absence of randomization. It is more appropriate to consider the observations as repeated measures observations and proceed with an analysis of the data using the methods outlined in previous sections of this chapter.

The F test for the main effects of the cultivars will be valid in the whole-plot analysis of variance provided the cultivars are randomly assigned to the whole-plots in each block. The research question of initial interest in these studies relates to the differential performance of the cultivars over the water gradient treatment. The statistical test of cultivar $\times$ water-level interaction from the within-subjects subplot analysis will address that hypothesis.

15.7   Other Models for Correlation Among Repeated Measures

The most general structure for correlation among repeated measures is that shown in Display 15.1 in which all variances and covariances, with $\sigma_{ij} = \sigma_{ji}$, have the potential to be unique. Two versions based on assumptions that simplify the structure are either compound symmetry (Display 15.2), or the Huynh–Feldt conditions for equality of variances between all differences on repeated measures (Equation 15.2). Under these conditions a straightforward split-plot analysis of variance will provide valid inference.

Derived Variables Analysis When Analysis of Variance Assumptions Do Not Hold

Highly structured laboratory or field experiments such as the Amiodarone study in Example 15.1 or the soil microbe study in Example 15.2 are often effectively analyzed with the analysis used for the soil microbe study. That type of analysis may be considered an analysis of **derived variables**, which in Example 15.2 were orthogonal polynomial contrasts derived from the repeated measures over time.

Other variables can be derived from the repeated observations on each unit, depending on the nature of the study. For example, if the repeated measures reflect growth or wear response it may be quite reasonable to derive variables based on a nonlinear growth or wear model. In those cases, the derived variables would be parameters such as slope, asymptote, point of inflection, and so forth.

The advantage of this approach is its simplicity through inference by standard univariate analysis of variance. One of the disadvantages is that the derived variables are not necessarily independent; thus, the inferences for the derived variables are not independent of one another. The derived variable approach breaks down in studies with incomplete data because of missing values at some points in time, or in less structured studies in which subjects may be measured at different points in time. Under these circumstances the common variance assumption for standard analysis of variance methods is no longer valid.

Models with Reduced Number of Correlation Parameters

Multivariate models can be used for inference with the general correlation structure shown in Display 15.1. However, the model may include an unneccessarily large number of variances and covariances for estimation. Simpler structures have been proposed that have fewer covariance parameters to estimate. They also are not as restrictive as compound symmetry or the Huynh–Feldt condition and can be more representative of the correlation behavior.

Two such models are the **serial correlation** and the **random coefficients** models. The serial correlation model has errors correlated within subjects or units defined by the relationship

$$\rho_{(t_i - t_j)} = \frac{\sigma_{e_i e_j}}{\sigma^2}$$

where $\rho_{(t_i - t_j)}$ is the correlation between errors at times i and j and $\sigma_{e_i e_j}$ is the covariance between errors at times i and j on the same subject.

Correlation arises among the repeated measures in the random coefficients model as a consequence of the assumption that the treatment effects or regression coefficients vary across the subjects or experimental units. The simplest example is when the intercept of the time response profile varies between the units because some units are intrinsically low responders while others are high responders.

The model has subject- or unit-specific treatment effects as a result of this assumption. The random coefficient models are most useful when it is desired to make inferences about individuals rather than population averages, since the model coefficients for each individual can be estimated for prediction purposes.

Statistical estimation methods for the alternative correlation structures are based primarily on maximum likelihood methods and are beyond the scope of this book. Diggle, Liang, and Zeger (1994) provide detailed discussions of the models and estimation methods available for studies with repeated measurements.

EXERCISES FOR CHAPTER 15

1. A study was conducted on human subjects to measure the effects of three foods on serum glucose levels. Each of the three foods was randomly assigned to four subjects. The serum glucose was measured for each of the subjects at 15, 30, and 45 minutes after the food was ingested. The data are shown in the table.

		Time (minutes)		
Diet	Subject	15	30	45
1	1	28	34	32
	2	15	29	27
	3	12	33	28
	4	21	44	39
2	5	22	18	12
	6	23	22	10
	7	18	16	9
	8	25	24	15
3	9	31	30	39
	10	28	27	36
	11	24	26	36
	12	21	26	32

a. Describe the study in terms of the between-subjects and within-subjects designs.
b. Compute the mean of the observations for each diet at each time of measurement, and make a profile plot of the results for each treatment.
c. Write the linear model for a split-plot analysis of variance, identify the terms, and indicate the assumptions necessary for an analysis of the data.
d. Conduct the split-plot analysis for the data, test the necessary hypotheses, and compute treatment means and their standard errors. What are your conclusions?
e. Obtain the residual plots from the split-plot analysis, and interpret them.
f. Compute the sum of squares partitions for the linear and quadratic contrasts on time and their interactions with diet, test the null hypotheses, and interpret the results.
g. If a repeated measures analysis computer program is available, test the hypothesis that the Huynh–Feldt condition can be assumed for the Σ matrix of the experimental errors for within subject variances and covariances. What is your conclusion?

h. Suppose it is necessary to conduct a separate analysis for the linear and quadratic regression contrasts. Write the linear model for one of the contrasts, identify the terms, and indicate the necessary assumptions for the model. Compute the analyses for the contrasts, and interpret the results. Are the experimental error variances comparable for the contrasts? Do the results differ from those in part (f)? Explain.

2. A soil scientist conducted an experiment to evaluate the effects of soil compaction and soil moisture on the activity of soil microbes. Reduced levels of microbe activity will occur in poorly aerated soils. The aeration levels can be restricted in highly saturated or compacted soils. Treated soil samples were placed in airtight containers and incubated under conditions conducive to microbial activity. The microbe activity in each soil sample was measured as the percent increase in CO_2 produced above atmospheric levels.

The treatment design was a 3×3 factorial with three levels of soil compaction (bulk density = mg soil/m^3) and three levels of soil moisture (kg water/kg soil). There were two replicate soil container units prepared for each treatment.

The CO_2 evolution/kg soil/day was recorded on three successive days. The data for each soil container unit are shown in the table.

			Day		
Density	Moisture	Unit	1	2	3
1.1	0.10	1	2.70	0.34	0.11
		2	2.90	1.57	1.25
	0.20	3	5.20	5.04	3.70
		4	3.60	3.92	2.69
	0.24	5	4.00	3.47	3.47
		6	4.10	3.47	2.46
1.4	0.10	7	2.60	1.12	0.90
		8	2.20	0.78	0.34
	0.20	9	4.30	3.36	3.02
		10	3.90	2.91	2.35
	0.24	11	1.90	3.02	2.58
		12	3.00	3.81	2.69
1.6	0.10	13	2.00	0.67	0.22
		14	3.00	0.78	0.22
	0.20	15	3.80	2.80	2.02
		16	2.60	3.14	2.46
	0.24	17	1.30	2.69	2.46
		18	0.50	0.34	0.00

Source: Dr. I. Pepper and J. Neilson, Department of Soil and Water Science, University of Arizona.

a. Describe the study in terms of the between-subjects and within-subjects designs.

b. Compute the mean of the observations for each soil bulk density and soil moisture level at each time of measurement, and make a profile plot of the results for each treatment.

c. Write the linear model for a split-plot analysis of variance, identify the terms, and indicate the assumptions necessary for the analysis with this model.

d. Conduct the split-plot analysis for the data, test the necessary hypotheses, and compute treatment means and their standard errors. What are your conclusions?

e. Obtain the residual plots from the split-plot analysis, and interpret them.

f. Compute the sum of squares partitions for the linear, quadratic, and cubic contrasts on time and their interactions with density and moisture treatments; test the null hypotheses, and interpret the results.

g. If a repeated measures analysis computer program is available, test the hypothesis that the Huynh–Feldt condition can be assumed for the Σ matrix of the experimental errors for within-subject variances and covariances. What is your conclusion?

h. Apply the Greenhouse–Geisser adjustment to the F tests in part (d). Do the conclusions differ? Explain.

i. Conduct a separate analysis for the linear and quadratic regression contrasts. Write the linear model for one of the contrasts, identify the terms, and indicate the necessary assumptions for the model. Compute the analyses for the contrasts, and interpret the results. Are the experimental error variances comparable for the contrasts? Do the results differ from those in part (f)? Explain.

3. The fabric of athletic clothing may change the skin's hydration state because the fabric serves as a barrier to the dissipation of body-generated water. A textile scientist conducted a study to evaluate the effect of fiber type and fabric moisture content on evaporative water loss from the skin.

   Five male subjects were used for the study. Each subject was used to evaluate five fabric treatments. The test fabrics were cotton and polyester fabrics commonly used in athletic clothing. The test was conducted by placing a piece of fabric directly on the subject's forearm skin surface. An instrument was used to measure the amount of water that evaporated from the subject's skin surface. The test was conducted in a controlled environment room at 70° F and 65% relative humidity.

   The fabric treatments were (1) cotton at equilibrium, (2) cotton at saturation, (3) stiff polyester at equilibrium, (4) stiff polyester at saturation, and (5) soft polyester at saturation.

   Two of the treatments listed indicate the fabric was at equilibrium. The moisture content of the fabrics in this case had been allowed to come to equilibrium with the room moisture level prior to test.

   The evaporative water loss was measured by the instrument after 60 minutes of fabric application to the skin. The data for all subjects on each of the treatments are shown in the table.

Subject	Treatment				
	1	2	3	4	5
1	4.04	6.50	4.01	10.71	10.66
2	2.25	18.23	1.94	8.39	7.42
3	3.55	15.01	1.58	8.63	13.86
4	3.02	15.15	4.15	4.09	5.15
5	1.94	9.59	12.14	6.30	12.79

Source: Dr. K. Hatch, Family and Consumer Resources, University of Arizona.

    a. Describe the design.

    b. Write the analysis of variance mixed model for this study, identify the terms, and state the assumptions necessary for an analysis of the data.

    c. The textile scientist was interested in four particular contrasts among the treatments. The contrasts of interest were (i) 1 vs. 2, (ii) 2 vs. 4, (iii) 2 vs. 5, and (iv) 3 vs. 4. Write out the table of contrasts. Are the contrasts orthogonal? Explain.

    d. Conduct the mixed-model analysis of variance for the data. Include an analysis of the contrasts listed in part (c). Interpret the results.

    e. Obtain the residual plots for the analysis, and interpret them.

    f. Conduct an analysis for each of the contrasts listed in part (c). Write the linear model for one of the contrasts, identify the terms, and indicate the necessary assumptions for the model. Compute the analyses for the contrasts, and interpret the results. Are the experimental error variances comparable for the contrasts? Do the conclusions differ from those in part (d)? Explain.

4. An agronomist conducted a yield trial with five alfalfa cultivars in a randomized complete block design with three replications. Each plot was harvested four times in each of two years. The plot yields (lb/plot) from two harvests from each plot in each of two years are shown in the table.

		1986		1987	
Cultivar	Block	April	May	April	May
1	1	20.4	23.2	14.8	22.9
	2	21.5	23.7	18.8	22.6
	3	21.1	23.4	14.3	22.1
2	1	19.1	22.4	14.5	19.2
	2	20.8	22.1	10.1	22.0
	3	20.5	23.5	12.0	21.5
3	1	19.3	22.1	14.5	19.5
	2	19.8	25.4	16.9	23.1
	3	20.5	24.8	16.7	20.1
4	1	23.2	25.6	14.9	19.5
	2	21.8	24.4	16.0	18.1
	3	22.2	26.8	16.7	21.0
5	1	21.4	24.5	14.1	21.3
	2	20.7	22.9	12.6	20.0
	3	18.7	21.8	14.3	21.4

Source: Dr. M. Ottman, Department of Plant Sciences, University of Arizona.

    a. Describe the study in terms of the between-subjects and within-subjects designs.

    b. Compute the mean of the observations for each cultivar at each harvest of the two years, and make a profile plot of the results for each cultivar.

    c. Write the linear model for a split-plot analysis variance, identify the terms, and indicate the assumptions necessary for an analysis of the data.

d. If a repeated measures analysis computer program is available, obtain the Σ matrix of experimental errors for within-subject variances and covariances from the four measurements over months and years. Test the hypothesis that the Huynh–Feldt condition can be assumed for the Σ matrix. What is your conclusion?

e. The agronomist wants to compare the yields of the cultivars. The significance of interaction between cultivars, years, and months of harvest must be determined before the agronomist can compare the cultivar averages over months and years. The Huynh–Feldt assumption for the Σ matrix of experimental errors should be tested for the analysis of months $\times$ cultivars and for the analysis of years $\times$ months $\times$ cultivars if your program has the capability. Conduct the tests, compute the split-plot analysis for the data, and test the hypotheses for cultivar interactions with months and years. Use the G–G or H–F epsilon adjustments for the F tests if necessary. What are your conclusions from this analysis?

f. Obtain the residual plots from the split-plot analysis, and interpret them.

5. An agronomist conducted an experiment to evaluate the drought tolerance of four barley cultivars. He used a line source sprinkler system to create a water gradient treatment design on each cultivar plot. A description of the design was presented in Section 15.6. The four cultivars were randomly assigned to the plots in a randomized complete block design. The amount of water applied to each of the plots decreased with distance from the sprinkler line. The grain yield was measured on 12 sq ft subplots on each cultivar plot at four equally spaced distances from the line sprinkler. The data (grams of barley grain per 12 sq ft) for the subplots are shown in the table.

		Distance from Sprinkler*			
Variety	Block	1	2	3	4
1	1	416.7	376.1	328.9	178.1
	2	490.2	513.7	438.4	348.1
	3	341.2	452.0	541.5	458.8
2	1	644.7	555.4	587.8	413.7
	2	526.8	481.4	490.3	468.1
	3	540.6	504.3	495.9	523.3
3	1	388.9	491.8	355.0	222.2
	2	298.8	407.3	500.0	320.3
	3	386.7	388.4	492.4	438.2
4	1	512.0	598.9	442.1	186.0
	2	484.8	542.5	463.1	383.2
	3	368.5	547.8	702.9	445.3

*1 = closest to sprinkler, 4 = greatest distance from sprinkler
Source: Dr. M. Ottman, Department of Plant Sciences, University of Arizona.

a. Describe the study in terms of the between-subjects and within-subjects designs.

b. Compute the mean of the observations for each cultivar at each distance, and make a profile plot of the results.

c. Write the linear model for a split-plot analysis of variance of the data, identify the terms, and indicate the assumptions necessary for an analysis of the data.

d.   Conduct the split-plot analysis for the data, test the necessary hypotheses, and compute treatment means and their standard errors. What are your conclusions?

e.   Compute the sum of squares partitions for the linear, quadratic, and cubic contrasts on distance from the sprinkler; test the null hypotheses and interpret the results.

f.   If a repeated measures analysis computer program is available test the hypothesis that the Huynh–Feldt condition can be assumed for the Σ matrix of the experimental errors for within-subject variances and covariances. What is your conclusion?

g.   Apply the Greenhouse–Geisser adjustment to the F tests in part (d). Do the conclusions differ? Explain.

h.   Conduct a separate analysis of the linear, quadratic, and cubic regression contrasts. Write the linear model for one of the contrasts, identify the terms, and indicate the necessary assumptions for the model. Compute the analyses for the contrasts, and interpret the results. Are the experimental error variances comparable for the contrasts? Do the results differ from those in part (e)? Explain.

15A.1   Appendix: The Mauchly Test for Sphericity

The Huynh–Feldt condition for the matrix of variances and covariances of the p repeated measures of subjects requires $(p-1)$ normalized orthogonal contrasts for the repeated measures to be uncorrelated with equal variances. Let Σ be the covariance matrix of the repeated measures. Let the matrix C be a $(p-1) \times p$ matrix, where the rows are normalized orthogonal contrasts on the p repeated measures. The required Huynh–Feldt condition for the covariance of the contrasts is $C \Sigma C' = \lambda I$, where I is the identity matrix and C' is the transpose of C. If the condition is satisfied, the covariance matrix λI is said to be spherical.

Let s_{ij} be the element in the ith row and jth column of the pooled $p \times p$ covariance matrix for the within-subject experimental errors S, with ν degrees of freedom. Choose $(p-1)$ normalized orthogonal contrasts on the p repeated measures. Let the matrix C be the $(p-1) \times p$ matrix, where the rows are normalized orthogonal contrasts on the p repeated measures. Compute the $(p-1) \times (p-1)$ matrix CSC'. The test statistic (Mauchly, 1940) for the null hypothesis $H_0: C \Sigma C' = \lambda I$ is

$$W = \frac{(p-1)^{p-1} \mid CSC' \mid}{(\text{tr}\, CSC')^{p-1}} \qquad (15A.1)$$

where $\text{tr}\, CSC'$ is the trace of the matrix. The trace of a matrix is the sum of its diagonal elements. The test statistic is scaled to improve the accuracy of its approximation by the chi-square distribution. The scale factor for the chi-square approximation with $f = \frac{1}{2}p(p-1) - 1$ degrees of freedom is

$$\gamma = \nu - \frac{2p^2 - 3p + 3}{6(p-1)} \qquad (15A.2)$$

The null hypothesis is rejected at the α level of significance if $-\gamma \ln W > \chi^2_{\alpha, f}$.

15A.2   Appendix: Degrees of Freedom Adjustments for Repeated Measures Analysis of Variance

The Greenhouse–Geisser $\widehat{\varepsilon}$ (Greenhouse & Geisser, 1959), and the Huynh–Feldt $\widetilde{\varepsilon}$ (Huynh & Feldt, 1976) were proposed as degrees of freedom adjustments for F tests in the analysis of within-subjects treatment factors. The adjustments, based on work by Box (1954a, 1954b), were developed for designs with only one within-subject treatment factor. The Huynh–Feldt $\widetilde{\varepsilon}$ is a simple function of the Greenhouse–Geisser $\widehat{\varepsilon}$. The computations are outlined in the following paragraphs beginning with computations for $\widehat{\varepsilon}$.

Let s_{ij} be the element in the ith row and jth column of the pooled $p \times p$ covariance matrix for the within-subject experimental errors, $\boldsymbol{S}$. Choose $q = (p-1)$ normalized orthogonal contrasts on the p repeated measures. Let the matrix $\boldsymbol{C}$ be a $q \times p$ matrix, where the rows are normalized orthogonal contrasts on the p repeated measures. Compute the $q \times q$ matrix $\boldsymbol{A} = \boldsymbol{CSC}'$ with elements $\{a_{ij}\}$. The Greenhouse–Geisser $\widehat{\varepsilon}$ adjustment is computed as

$$\widehat{\varepsilon} = \frac{\left(\sum_{i=1}^{q} a_{ii}\right)^2}{q\sum_{i=1}^{q}\sum_{j=1}^{q} a_{ij}^2} \tag{15A.3}$$

The Huynh–Feldt $\widetilde{\varepsilon}$ adjustment is computed as

$$\widetilde{\varepsilon} = \frac{(Nq\widehat{\varepsilon} - 2)}{q(\nu - q\widehat{\varepsilon})} \tag{15A.4}$$

where N is the number of subjects and ν is the error degrees of freedom for the experimental error from the between-subjects analysis of variance. If the experiment consists of N subjects each with p treatments, then $\nu = N - 1$. With t between-subjects treatments in a completely randomized design, then $\nu = N - t$. With r replications of t between-subjects treatments in a randomized complete blocks design, then $\nu = (t-1)(r-1)$.

Given a within-subjects F_0 statistic based on mean squares with ν_1 and ν_2 degrees of freedom the adjusted degrees of freedom for the test are $\varepsilon\nu_1$ and $\varepsilon\nu_2$, where $\widehat{\varepsilon}$ is used for the Greenhouse–Geisser adjustment and $\widetilde{\varepsilon}$ is used for the Huynh–Feldt adjustment. The Huynh–Feldt $\widetilde{\varepsilon}$ is not used if $\widetilde{\varepsilon} \geq 1$.

16 Crossover Designs

The principal topic of this chapter is the design and analysis of experiments with different treatments administered in successive periods of time to the experimental units. The analysis of the experiments includes an evaluation of the effect of a treatment that may carry over to affect the response of a treatment in the following period. Designs specifically constructed to efficiently estimate the direct and carry-over effects of treatments are presented in this chapter.

16.1 Administer All Treatments to Each Experimental Unit

The **crossover study** describes experiments with treatments administered in sequence to each experimental unit. A treatment is administered to an experimental unit for a specific period of time after which another treatment is administered to the same unit. The treatments are successively administered to the unit until it has received all treatments.

Crossover Designs Can Increase Precision and Reduce Costs

When treatments are compared on the same experimental unit, the between-unit variation is removed from the experimental error. Thus, the individual experimental units are used as blocks in the crossover design to decrease the experimental error and increase efficiency of the experiment. The treatment comparisons in blocked designs are generally more precise than those in unblocked designs because the experimental unit or block variation is removed from comparisons between treatments on the same experimental unit. Thus, the primary advantage of crossover studies is the increased precision of treatment comparisons.

Crossover designs provide an economy of resources when a limited number of units are available for the study. Most commonly, crossover designs are used with human and animal subjects. The expense of maintaining large animals and the difficulties in recruiting adequate numbers of human subjects to achieve sufficient replication make crossover designs more attractive since they require fewer units for an equal amount of treatment replication. The following example of an experiment used a crossover design with beef steers in a feeding trial.

Example 16.1 Digestibility of Feedstuffs in Beef Cattle

Associative effects occur in animal diets when feedstuffs are combined and diet utilization or animal performance is different from that predicted from a sum of the individual ingredients. The addition of roughage to the diets of ruminant animals had been shown to influence various diet utilization factors such as ruminal retention time. However, information about the relative associative effects of different roughage was scarce, especially in mixed feedlot diets.

Research Hypothesis: An animal scientist hypothesized roughage source could influence utilization of mixed diets of beef steers by altering ruminal digestion of other diet ingredients.

Treatment Design: The basic mixed diet for the beef steers was a 65% concentrate based on steam-flaked milo and 35% roughage. Three roughage treatments were used with (A) 35% alfalfa hay as a control treatment, (B) 17.5% wheat straw and 17.5% alfalfa, and (C) 17.5% cottonseed hulls and 17.5% alfalfa.

Experiment Design: Twelve beef steers were available for the study. Each of the three roughage diets was fed to the steers in one of six possible sequences. Each diet in each sequence was fed to two steers for 30 days. The steers were allowed a period of 21 days to adapt to a diet change before any data were collected.

The Neutral Detergent Fiber (NDF) digestion coefficient calculated for each steer on each diet is shown in Table 16.1. The NDF digestion coefficient indicates the percent of dietary fiber digested by the steer.

Table 16.1 NDF digestion coefficients for two steers in each sequence of three roughage diets in a crossover design

		Sequence													
		1			*2*			*3*			*4*			*5*			*6*	
Steer:		*1*	*2*		*3*	*4*		*5*	*6*		*7*	*8*		*9*	*10*		*11*	*12*
Period I	(A)	50	55	(B)	44	51	(C)	35	41	(A)	54	58	(B)	50	55	(C)	41	46
Period II	(B)	61	63	(C)	42	45	(A)	55	56	(C)	48	51	(A)	57	59	(B)	56	58
Period III	(C)	53	57	(A)	57	59	(B)	47	50	(B)	51	54	(C)	51	55	(A)	58	61

Source: J. Moore, Department of Animal Science, University of Arizona.

Other Examples: A new drug and a standard drug are tested in a crossover design using patients afflicted with acute bronchial asthma to determine whether the new drug improves the patients' breathing over the standard drug. The new drug is administered to a patient during the first week, and the standard drug is administered to the same patient during the second week.

Three front panel designs developed to operate a laboratory instrument are tested in a crossover design with laboratory technicians as operators. The technician operates the instrument with each of the panel designs. Each technician tests the three panel designs on successive days, a different panel each day.

Design to Avoid Confounding Time Period Effects with Treatments

A comparison between two treatments on the same experimental unit is also a comparison between two time periods. Treatments and periods both can contribute to any observed differences. Crossover trials are designed to avoid confounding of period and treatment effects. For example, one group of experimental units will receive the sequence A→B, and a second group of units will receive the sequence B→A. Both treatments are administered in each period, and the treatment comparisons are independent of comparisons between periods.

A Carryover Effect Can Persist After the Treatment Period

A disadvantage of the crossover design is the possibility that a treatment given in one period will influence the response in the following treatment period. Effects of a treatment that continue into the next treatment period are *carryover effects*.

Typically, the treatments are administered in crossover designs for a length of time sufficient to allow the effect of the treatment to be manifested on the subjects. The subject is removed from the treatment for a resting, or washout, period between two treatment periods to bring the subject back to its original physiological or psychological state.

For example, after a drug is administered to a patient the rest period is intended to allow any residual of the drug to "wash out" of the patient's system so that it will not be present in the succeeding treatment period. Although sufficient resting time is allowed for the initial drug to disappear from the system, the physiological state may have been altered sufficiently to have some effect on the responses in the succeeding treatment period. The potential for a carryover effect cannot be ignored in crossover studies.

Since carryover effects are assumed to be present in studies, designs constructed specifically to measure carryover effects will be discussed in this chapter. A variety of designs have been developed for crossover studies to meet specific needs of different research problems. The discussions in this chapter are restricted to a few basic crossover designs to illustrate basic principles for design and analysis of crossover studies to meet specific goals of the research study.

The basic crossover design in Example 16.1 will be used to illustrate the relationship between carryover effects of treatments and other effects present in the experiment as well as to provide the foundation for the basic statistical model and analysis of crossover studies.

A Balanced Row–Column Design for the Digestibility Study

The crossover design for the roughage diets is a balanced row–column design. The periods and steers are the rows and columns of the design. Each of the roughage diets occurs one time in each steer and four times in each period of the design. The design was referred to as a Latin rectangle design in Chapter 8.

The six sequences of diet treatments are shown in Display 16.1. The six sequences of roughage diets were randomly assigned to the 12 steers. However, the order of diet administration to each of the steers was not randomized. Each of the six sequences of the three treatments must be present with equal frequency to avoid confounding the period effects with the treatment effects and to have a design *balanced* for carryover effects.

Display 16.1   A Crossover Design with Six Treatment Sequences for Three Treatments in Three Periods

	Sequence					
Period	*1*	*2*	*3*	*4*	*5*	*6*
I	A	B	C	A	B	C
	↓	↓	↓	↓	↓	↓
II	B	C	A	C	A	B
	↓	↓	↓	↓	↓	↓
III	C	A	B	B	C	A

Designs to Balance the Carryover Effects

A crossover design is balanced for carryover effects when each treatment follows each of the other treatments an equal number of times. Each treatment occurs equally frequently in each period, and it occurs once with each subject.

The crossover design for the roughage diet study shown in Display 16.1 is a balanced design. Diet A follows diet B twice, once each in sequences 5 and 6; and it follows diet C twice, once each in sequences 2 and 3. Likewise, diets B and C follow each of the other two treatments two times. The balance applies only to first-order carryover effects that alter the response in the first period following administration of the treatment. A second-order carryover effect alters the response in the second period following administration of the treatment.

The 21-day adaptation period in the roughage diet study is equivalent to the washout, or rest period, used in other studies to avoid a carryover effect. In the absence of carryover effects the effects of any previous diet are not manifested in the digestion physiology of the animal in the current period, and the digestion measurements reflect only the direct effects of the current diet.

The control diet treatment with 35% chopped alfalfa hay was the standard growing diet for feedlot steers. The treatments were designed to address the hypothesis that alternate roughage sources influenced utilization of mixed diets by altering rumen digestion. Diets B and C provided alternate roughage sources of wheat straw or cottonseed hulls mixed with the alfalfa. The remaining parts of the diet including grain and minerals were the same for all diets. If the research hypothesis is true the digestion of dietary fiber will differ among the three treatments.

The three treatment means calculated from Table 16.1 data are, $\bar{y}_A = 56.6$, $\bar{y}_B = 53.3$, and $\bar{y}_C = 47.2$. The observed mean measures the direct effect of a treatment in the periods it was active on the subjects plus the carryover effects of the other two treatments. The direct effects of diet A are not confounded with periods or steers because they were measured in each of the periods and on each of the steers. In the absence of carryover effects $\bar{y}_A = 56.6$ is an unbiased estimate of μ_A in the balanced design. Since diet A followed diet B in two sequences and diet C in two other sequences the response to diet A could be increased or decreased by any carryover effects of diets B or C administered to the steers in the previous periods.

16.2   Analysis of Crossover Designs

The Linear Model for Crossover Designs

The crossover design has n treatment sequence groups and r_i subjects in the ith group. There are t treatments and each group of subjects receives treatments in a different order for p treatment periods. Let y_{ijk} be the observation of the jth subject of the ith treatment sequence in the kth period.

The linear model for a crossover design is

$$y_{ijk} = \mu + \alpha_i + b_{ij} + \gamma_k + \tau_{d(i,k)} + \lambda_{c(i,k-1)} + e_{ijk} \qquad (16.1)$$

$$i = 1, 2, \ldots, n \quad j = 1, 2, \ldots, r_i \quad k = 1, 2, \ldots, p \quad d, c = 1, 2, \ldots, t$$

where μ is the general mean, α_i is the effect of the ith treatment sequence, b_{ij} is the random effect with variance σ_b^2 for the jth subject of the ith treatment sequence, γ_k is the period effect, and e_{ijk} is the random error with variance σ^2 for the subject in period k. The direct effect of the treatment administered in period k of sequence group i is $\tau_{d(i,k)}$, and $\lambda_{c(i,k-1)}$ is the carryover effect of the treatment administered in period $k-1$ of sequence group i. The value of the carryover effect for the observed response in the first period is $\lambda_{c(i,0)} = 0$ since there is no carryover effect in the first period.

For simplicity, the direct and carryover effects of the treatments are identified as $\tau_1, \tau_2, \ldots, \tau_t$ and $\lambda_1, \lambda_2, \ldots, \lambda_t$, respectively. The expected values for observations in the first and second treatment sequences of the roughage diet study are with y_{ik} representing sequence i in period k:

Sequence 1 (A→ B→ C)

$$E(y_{11}) = \mu + \alpha_1 + \gamma_1 + \tau_1$$
$$E(y_{12}) = \mu + \alpha_1 + \gamma_2 + \tau_2 + \lambda_1$$
$$E(y_{13}) = \mu + \alpha_1 + \gamma_3 + \tau_3 + \lambda_2$$

Sequence 2 (B→ C→ A)

$$E(y_{21}) = \mu + \alpha_2 + \gamma_1 + \tau_2$$
$$E(y_{22}) = \mu + \alpha_2 + \gamma_2 + \tau_3 + \lambda_2$$
$$E(y_{23}) = \mu + \alpha_2 + \gamma_3 + \tau_1 + \lambda_3$$

Carryover effects occur only in observations from the second and third period. E.g., $\lambda_{c(1,0)} = 0$ in sequence 1 because no treatment precedes diet A. The observations in periods 2 and 3 contain the carryover effects of diet A and diet B, λ_1 and λ_2, respectively. Likewise, for sequence 2 the carryover effects of diets B and C, λ_2 and λ_3, occur in the observations from the second and third periods, respectively.

Assume the Univariate Model Assumption Is Satisfactory

The observations on each experimental unit are repeated measures in time under different treatment conditions. They represent a multivariate observation on the experimental unit. The assumptions required for the univariate analysis of variance were discussed in Chapter 15, "Repeated Measures Designs." Kenward & Jones (1989) and Diggle et al. (1994) discuss details of repeated measures models for crossover designs beyond the scope of this book.

Whether the univariate analysis of variance can be used for the crossover designs depends on the relationships among variances and covariances of the experimental errors for the repeated measures. The univariate analysis can be used if any of the assumptions for independence, compound symmetry, or the Huynh–Feldt condition is appropriate for the experimental errors.

The Huynh–Feldt condition of equal variances for all possible differences between repeated measures is the least restrictive assumption for the experimental errors. It will be assumed the Huynh–Feldt condition is satisfied for the data.

The Analysis of Variance for Crossover Designs

The Analysis Without Carryover Effects

If the crossover design is a balanced row–column design the analysis of variance described in Chapter 8 can be used *in the absence of carryover effects*. The subjects and time periods are the rows and columns of the design, and the direct treatment effects are orthogonal to the rows and columns. The sums of squares for rows, columns, treatments, and experimental error can be computed with the analysis of variance shown in Table 8.10.

The Analysis with Carryover Effects

The analysis of variance for the model with treatment carryover effects in Equation (16.1) is outlined in Table 16.2. The separation of the sums of squares

Table 16.2 Analysis of variance table for a crossover design with n sequences, p periods, t treatments, and r_i subjects in the ith sequence; $N = \Sigma_i^n r_i$

Source Variation	Degrees of Freedom	Sum of Squares	Mean Square
Total	$Np - 1$	SS Total	
Between subjects:			
Sequence	$n - 1$	SSS	MSS
Subjects within sequence	$(N - n)$	SSW	MSW
Within subjects:			
Period	$p - 1$	SSP	MSP
Treatments (direct)	$t - 1$	SST	MST
Treatments (carryover)	$t - 1$	SSC	MSC
Error	$(N - 1)(p - 1) - 2(t - 1)$	SSE	MSE

partitions into between- and within-subjects groupings indicates the correspondence to a repeated measures split-plot univariate analysis. The subjects are the whole plots and the repeated measures over the p periods are the subplots.

Treatment and Carryover Effects Are Nonorthogonal

When carryover effects are present the direct treatment effects and carryover effects are not orthogonal, nor are the carryover effects orthogonal to the subject blocks in the balanced row–column design. The relationship between the effects can be observed in Display 16.2. The periods and sequences are each complete block designs for the direct treatment effects. The sequences form a balanced incomplete block design for the carryover effects. Each pair of carryover effects occur together with two of the sequences. The direct and carryover treatment effects are not orthogonal because they do not occur in all possible combinations. The direct and carryover effects of the same treatment never occur together in the same observation.

Display 16.2   Model Effects, τ_i and λ_i, for the Roughage Diet Study

			Sequence			
Period	1	2	3	4	5	6
I	τ_1	τ_2	τ_3	τ_1	τ_2	τ_3
II	$\tau_2 + \lambda_1$	$\tau_3 + \lambda_2$	$\tau_1 + \lambda_3$	$\tau_3 + \lambda_1$	$\tau_1 + \lambda_2$	$\tau_2 + \lambda_3$
III	$\tau_3 + \lambda_2$	$\tau_1 + \lambda_3$	$\tau_2 + \lambda_1$	$\tau_2 + \lambda_3$	$\tau_3 + \lambda_1$	$\tau_1 + \lambda_2$

The significance of carryover effects must be determined before inferences can be made about differences among the direct effects of treatments. Estimates of differences between treatment means, $\mu_i - \mu_j$, need to be adjusted for the carryover effects if they are present in the study.

Adjusted Sums of Squares for Nonorthogonal Effects

The sums of squares for direct and carryover treatment effects in the analysis of variance each must be adjusted for the other.

The adjusted sums of squares are computed from the differences between sums of squares for experimental error from full and reduced models. The sum of squares for carryover effects adjusted for direct effects requires the error sums of squares from the models:

$$\text{Reduced model} \qquad y_{ijk} = \mu + \alpha_i + b_{ij} + \gamma_k + \tau_d + e_{ijk}$$
$$\text{Full model} \qquad y_{ijk} = \mu + \alpha_i + b_{ij} + \gamma_k + \tau_d + \lambda_c + e_{ijk}$$

$$(16.2)$$

The sum of squares for carryover effects adjusted for direct effects is $SSC = SSE_{r(\lambda)} - SSE_f$, where $SSE_{r(\lambda)}$ is the error sum of squares computed from the reduced model without λ_c and SSE_f is computed from the full model.

The sum of squares for direct effects adjusted for carryover effects requires the error sums of squares from the models:

$$\text{Reduced model} \qquad y_{ijk} = \mu + \alpha_i + b_{ij} + \gamma_k + \lambda_c + e_{ijk}$$
$$\text{Full model} \qquad y_{ijk} = \mu + \alpha_i + b_{ij} + \gamma_k + \lambda_c + \tau_d + e_{ijk}$$

$$(16.3)$$

The sum of squares for direct effects adjusted for carryover effects is $SST = SSE_{r(\tau)} - SSE_f$, where $SSE_{r(\tau)}$ is the error sum of squares computed from the reduced model without τ_d and SSE_f is computed from the full model.

Data Coding for Computing Programs Requires Special Attention

Many statistical computer programs compute the required sums of squares partitions for balanced or unbalanced designs. Special attention must be given to coding the data file to include the carryover effects in the model. Alternative coding strategies for the carryover effects can be found in Ratkowsky, Alldredge, and Cotton (1990) and Milliken and Johnson (1984). The detailed coding for the data file of Example 16.1 is shown in Appendix 16A.1. Formulae for manual calculations are available for selected balanced designs in Cochran and Cox (1957), Petersen (1985), and Gill (1978). A brief outline of critical formulae for adjusted treatment sums of squares for balanced designs appears in Appendix 16A.2.

The Analysis of Variance for Roughage Diets

The analysis of variance for the roughage diet study is shown in Table 16.3. The two analyses illustrate the alternative fitting of full and reduced models to obtain the adjusted sums of squares for carryover and direct treatment effects.

Table 16.3 Analysis of variance for NDF digestion from the roughage diet study in a crossover design

Source of Variation	Degrees of Freedom	Sum of Squares	Mean Square	F	Pr > F
Sequence	5	331.67	66.33	3.48	.080
Steers in sequence	6	114.33	19.06		
Period	2	288.17	144.08	16.01	.000
Diet	2	559.50	279.75*	31.09	.000
Carryover	2	18.37	9.19†	1.02	.380
Error	18	161.96	9.00		

*Mean Square for direct treatment effects unadjusted for carryover effects, MST(unadj.)
†Mean Square for carryover treatment effects adjusted for direct effects, MSC(adj.)

Source of Variation	Degrees of Freedom	Sum of Squares	Mean Square	F	Pr > F
Sequence	5	331.67	66.33	3.48	.080
Steers in sequence	6	114.33	19.06		
Period	2	288.17	144.08	16.01	.000
Carryover	2	129.60	64.80*	7.20	.005
Diet	2	448.28	224.14†	24.91	.000
Error	18	161.96	9.00		

*Mean Square for carryover treatment effects unadjusted for direct effects, MSC(unadj.)
†Mean Square for direct treatment effects adjusted for carryover effects, MST(adj.)

The first analysis in Table 16.3 provides the mean square for carryover effects adjusted for direct effects, MS(Carryover, adj.) $= 9.19$, and the second analysis provides the mean square for direct treatment effects adjusted for carryover effects, MS(Diet, adj.) $= 224.14$. The differences between the adjusted and unadjusted mean squares reflect the nonorthogonal relationship between the carryover effects and direct effects of treatments. The unadjusted mean square for direct treatment effects in the first analysis is MS(Diet, unadj.) $= 279.75$, and the unadjusted mean square for carryover effects in the second analysis is MS(Carryover, unadj.) $= 64.80$.

Tests of Hypotheses for Direct and Carryover Effects of Treatments

The null hypothesis for carryover effects is H_0: $\lambda_1 = \lambda_2 = \lambda_3 = 0$. It is the hypothesis of initial interest for the study because inferences about the direct treatment effects are dependent on the presence of carryover effects. The test statistic for the hypothesis of no carryover effects is

$$F_0 = \frac{MS(\text{Carryover, adj.})}{MSE} = \frac{9.19}{9.00} = 1.02$$

with 2 and 18 degrees of freedom. The level of significance for the test is $Pr > F = .380$, and the null hypothesis is not rejected. The carryover effects are not significant in the roughage diet study.

The statistic $F_0 = MS(\text{Diet, adj.})/MSE = 224.14/9.00 = 24.90$ with 2 and 18 degrees of freedom tests the null hypothesis for direct treatment effects. The null hypothesis is rejected with significance level $Pr > F = .000$. There are significant differences in NDF digestion among the three roughage diets.

The significance of carryover effects unadjusted for direct treatment effects in the second analysis of variance in Table 16.3 again reflects the nonorthogonal relationship between direct and carryover effects. The correlation between their estimates leads to difficulties in distinguishing between their separate contributions. The sum of squares for carryover effects unadjusted for direct effects,

$$SS(\text{Carryover, unadj.}) = 129.60$$

is considerably larger than the adjusted sum of squares,

$$SS(\text{Carryover, adj.}) = 18.37$$

because the direct effects are intermingled with the carryover effects in the unadjusted sum of squares. Consequently, the significance of the unadjusted mean square is due to a combination of variation in direct and carryover effects, and it is not possible to determine which set of effects is contributing to the significance.

If the carryover effects are not significant it is common practice to base all inferences on the treatment means unadjusted for carryover effects. The analysis for Latin square designs in Chapter 8 would be used in this case. However, a detailed study by Abeyasekera and Curnow (1984) on the consequences of ignoring any adjustments for carryover effects led them to recommend always adjusting the treatment means for carryover effects regardless of the significance of the test for carryover effects.

Interpretation of Treatments with Multiple Contrasts

According to the research hypothesis the rumen digestion efficiency is dependent on the source of roughage. The standard or control diet A included 35% chopped alfalfa hay. The altered diets had half that amount of alfalfa and 17.5% wheat straw (diet B) or 17.5% cottonseed hulls (diet C). The null hypothesis of no change in NDF digestion by the altered diets can be tested with comparisons of the control diet with the altered diets using the Dunnett method. The estimates of the contrasts between the adjusted means of altered diets and diet A, $\hat{\mu}_A - \hat{\mu}_B$ and $\hat{\mu}_A - \hat{\mu}_C$, are shown in Table 16.4, along with the 95% simultaneous confidence intervals for comparisons of control with other treatments using the Dunnett method.

Table 16.4   Estimates of contrasts between the control and altered diets for the roughage diet study in a crossover design

Contrast	*Estimate*	*Standard Error*	*95% SCI (L, U)*
Diet A–Diet B	4.06	1.37	(0.77, 7.35)
Diet A–Diet C	9.63	1.37	(6.34, 12.92)

The estimated standard error of the difference between two adjusted means is 1.37 in the third column of Table 16.4. The two-sided Dunnett statistic for two comparisons is $d_{.05,2,18} = 2.40$. The 95% SCI indicate that the replacement of half the alfalfa hay by cottonseed hulls (diet C) in the diet reduced the NDF digestion to a greater extent than replacement by wheat straw (diet B) since the lower limit of the interval for diet C is further removed from 0 than is that for diet B.

The critical value for a confident inequalities test with the Dunnett method is $|\hat{\mu}_A - \hat{\mu}_i| > D(2, .05) = 1.37(2.40) = 3.28$. The difference for diet A versus diet B, 4.06, and the difference for diet A versus diet C, 9.63, are both significant because they exceed the critical value. The replacement of half the alfalfa hay by cottonseed hulls (diet C) in the diet reduces the NDF digestion by an estimated 9.63%, and the replacement by wheat straw (diet B) reduces the NDF digestion by an estimated 4.06%.

16.3   Balanced Designs for Crossover Studies

Practical considerations dictate whether a crossover study is appropriate for the research problem. The designs are most effective when treatments manifest an effect on the subject in a reasonably short period of time to provide a study time of manageable length. Extensive discussions of crossover studies can be found in Jones and Kenward (1989) and Ratkowsky, Evans, and Alldredge (1993).

When Crossover Designs Can Be Successful

Studies that measure responses to changes in animal diets have been successfully completed with crossover designs. Crossover designs reduce the costs of maintaining the larger numbers of animals required to provide equivalent treatment replication with other designs. The cost savings are especially pertinent in large animal studies. Human factors studies can use crossover designs to remove subject- to-subject variability in stress or anxiety responses to treatments. Differential physiological or psychological stresses are measured readily on each subject in experiments designed to compare different systems of component assembly or instrument operation with each subject.

When Crossover Designs May Not Be Successful

The designs should not be used in medical clinical trials for acute conditions such as postoperative pain. If a treatment in the sequence cures the acute condition nothing remains to treat in the succeeding periods. The designs can be used effectively to study the treatment of persistent conditions such as arthritis when treatments alleviate the symptoms rather than cure when administered to the individual.

Practical Considerations Influence Design Choice

Other practical considerations influence the choice of a design if the crossover study is feasible. The likelihood of subject loss during the course of the study increases with the number of periods. The more desirable designs, those balanced for carryover effects, increase in complexity as more treatments are added to the study. Consequently, the correct order of treatment administration to subjects becomes more difficult to manage in large studies. The ability to interpret results can diminish with the complexity of the design.

Many different designs have been developed for crossover studies to meet specific requirements for the practical research problem or to have good statistical properties. Three general categories of designs introduced in this section are useful for a variety of research studies that can appropriately use crossover studies. They are restricted to balanced designs for three or more treatments and three or more periods. Two-period designs are not efficient for estimation of treatment means. Designs for two treatments require special orchestration of treatment sequences to estimate all of the effects of interest in the study. Designs for two treatments are discussed in a separate section.

The categories of designs discussed in the remainder of this section are based on the relationship between the number of treatments t and the number of periods p. They are designs with $p = t, p = t + 1$, and $p < t$.

Compare Efficiency of Crossover Designs with Latin Squares

The efficiency of a design for statistical estimation of direct and carryover effects is an important consideration in the choice of design. The variance of the difference between two treatment effects, σ_d^2, forms the basis for efficiency measures. The Latin square design, in the absence of carryover effects, provides the minimum variance, $2\sigma^2/r$, and is the standard for comparison. The ratio of the variance for the Latin square to the crossover design measures the relative efficiency of the crossover design as

$$RE = \frac{2\sigma^2/r}{\sigma_d^2} \times 100 \qquad \textbf{(16.4)}$$

The relative efficiency measure is applied to either direct or crossover effects. The variance of the difference, σ_d^2, pertains to either treatment effect adjusted for the other.

The most desirable designs have variance balance where the variance of the difference between two treatment effects, direct or carryover, is the same for all treatment pairs. The balance defined by the relationship between direct and carryover effects is the other desirable feature of a design. The design is balanced if the direct effect of each treatment is associated equally frequently with the first-order carryover effect of each other treatment. Patterson and Lucas (1962) presented some general formulae to compute efficiencies for balanced designs.

Designs for Equal Numbers of Periods and Treatments, $p = t$

The Williams Designs Constructed from Latin Squares

Williams (1949) presented methods to construct balanced crossover designs from Latin square arrangements. If the number of treatments is even, $t = 4, 6, \ldots$, a balanced design can be constructed from one particular Latin square. Two particular squares are required if the number of treatments is an odd number, $t = 3, 5, 7, \ldots$.

Two 3×3 Latin squares that produced six treatment sequences were used for the design of the roughage diet study of Example 16.1. The design is reproduced in Display 16.3 showing the two 3×3 Latin squares. One Latin square was used for the first three sequences of treatments in the experiment, while the second square was used for the last three sequences.

Display 16.3   A Balanced Crossover Design from Two 3×3 Latin Squares

	Square 1			Square 2		
Period	*1*	*2*	*3*	*4*	*5*	*6*
I	A	B	C	A	B	C
	↓	↓	↓	↓	↓	↓
II	B	C	A	C	A	B
	↓	↓	↓	↓	↓	↓
III	C	A	B	B	C	A
		$E_d = 80.0$			$E_c = 44.44$	

Williams designs for $t = 4$, 5, and 6 treatments and balanced for carryover effects are shown in Display 16.4. Each treatment follows each of the other treatments one time in the designs with one Latin square for an even number of treatments, and the direct effects of each treatment associate once with the first-order carryover effects of all other treatments. For the designs with an odd number of treatments each treatment follows the others twice, and the direct effects of each treatment associate twice with the first-order carryover effect of all other treatments. The relative efficiency of the Williams designs for direct effects E_d and the relative efficiency E_c for carryover effects are shown below each of the designs in Displays 16.3 and 16.4.

Display 16.4   Balanced Latin Square Crossover Designs for Four, Five, or Six Treatments

Four Treatments

	Sequence Group			
Period	1	2	3	4
I	A	B	C	D
II	D	A	B	C
III	B	C	D	A
IV	C	D	A	B

$E_d = 90.91$ $E_c = 62.50$

Six Treatments

	Sequence Group					
Period	1	2	3	4	5	6
I	A	B	C	D	E	F
II	C	D	E	F	A	B
III	B	C	D	E	F	A
IV	E	F	A	B	C	D
V	F	A	B	C	D	E
VI	D	E	F	A	B	C

$E_d = 96.55$ $E_c = 77.78$

Five Treatments

	Sequence Group 1					Sequence Group 2				
Period	1	2	3	4	5	6	7	8	9	10
I	A	B	C	D	E	A	B	C	D	E
II	B	C	D	E	A	C	D	E	A	B
III	D	E	A	B	C	B	C	D	E	A
IV	E	A	B	C	D	E	A	B	C	D
V	C	D	E	A	B	D	E	A	B	C

$E_d = 94.74$ $E_c = 72.00$

Designs from Orthogonal Sets of Latin Squares

A complete set of orthogonal Latin squares provides a crossover design balanced for all orders of carryover effects. An orthogonal set of Latin squares for t treatments requires $t - 1$ squares. Two Latin squares are orthogonal when the two squares are superimposed and each treatment of one square occurs once with each treatment of the other square.

Two orthogonal 3×3 Latin squares were used for the roughage diet study. They make up the Williams design shown in Display 16.3.

Note when square 1 is imposed over square 2 that the treatment label A in square 1 will occur one time each with treatment labels A, B, and C in square 2.

The same occurrence pattern is true of the labels B and C from square 1 when superimposed on square 2.

The complete set of orthogonal squares requires $t(t-1)$ subjects for any number of treatments. For example, the orthogonal set of three Latin squares for four treatments requires 12 subject groups.

The efficiency of the design for estimating direct and carryover effects from the complete set of orthogonal squares is the same as that for the Williams designs. The Williams designs require only $t-1$ subjects for t even and $2(t-1)$ subjects for t odd, thus the only real advantage of a complete orthogonal set of Latin square designs over the Williams designs is that the complete set is balanced for all orders of carryover effects up to order $(p-1)$. For example, the design with four treatments is balanced for first-, second-, and third-order carryover effects.

When only first-order carryover effects are present in the study the Williams designs have a clear savings in number of treatment sequences to maintain with more than three treatments.

Extra-Period Designs for Orthogonal Direct and Carryover Effects, $p = t + 1$

The direct and carryover effects are not orthogonal in designs where the number of periods equals the number of treatments. Lucas (1957) pointed out that the non-orthogonality of the balanced designs resulted in much lower efficiency for estimating carryover effects than for estimating direct effects of treatments. The differences in efficiencies for estimating the two effects can be seen in Display 16.4. The addition of an extra period to the balanced design can remove the non-orthogonality and provide designs with independent estimates of direct and carryover effects that have more equal precision for estimates of direct and carryover effects. Lucas (1957) provided the first formal description of extra-period crossover designs.

The simplest extra-period design is derived from a balanced crossover design by repeating in period $(p+1)$ the treatment administered to the subject in period p. An extra-period design for $t = 4$ treatments in $p = 5$ periods derived from the Williams design for four treatments is shown in Display 16.5.

Each treatment administered to a subject in period IV is repeated on the same subject in period V. Each treatment is preceded by each of the other treatments in the design including itself. For example, treatment A is preceded by itself in sequence 3, by B in sequence 2, by C in sequence 4, and by D in sequence 3.

The extra-period Williams design is completely balanced for carryover effects since each carryover effect occurs an equal number of times in each sequence and with each direct effect. Thus, the estimates of the direct effects are the same whether or not the carryover effects are in the model, and vice versa. The addition of an extra period makes the design unbalanced with respect to sequences and direct effects because treatments appear an unequal number of times within each sequence. Note in Display 16.5 that one of the treatments appears two times in a sequence while each of the other treatments appears only once in the same sequence.

The efficiencies for orthogonal sets of Latin squares with one extra period are the same as those for the Williams designs with one extra period. The advantage of

```
┌─────────────────────────────────────────────────────────┐
│                                                         │
│   Display 16.5    Extra-Period Crossover Design         │
│                   for Four Treatments                   │
│                                                         │
│                        Sequence Group                   │
│            Period     1     2     3     4               │
│            ─────────────────────────────────           │
│              I        A     B     C     D               │
│                       ↓     ↓     ↓     ↓               │
│              II       D     A     B     C               │
│                       ↓     ↓     ↓     ↓               │
│              III      B     C     D     A               │
│                       ↓     ↓     ↓     ↓               │
│              IV       C     D     A     B               │
│                       ↓     ↓     ↓     ↓               │
│              V        C     D     A     B               │
│            E_d = 96.00          E_c = 80.00             │
│                                                         │
└─────────────────────────────────────────────────────────┘
```

increasing the precision for estimating carryover effects must be weighed against the disadvantage of extending the length and cost of the experiment with the extra period. The designs should only be used if there is a strong indication that the carryover effects are likely to occur in the study. Some bias in the responses can occur if either the subjects or the investigators are aware that the last two measurement periods involve the same treatment.

Designs for Less than a Full Treatment Cycle, $p < t$

Each subject received all of the treatments in the balanced designs discussed in the previous sections. Practical or ethical considerations can prevent adminstration of the full cycle of treatments to each of the subjects. It may be more practical to administer only part of the treatments to each subject to reduce the chances of subjects dropping out of the study. Ethical considerations may prevent the investigator from administering too many treatments to patients in a clinical trial.

Patterson (1951, 1952) described methods to construct balanced designs for studies that required the number of periods to be less than the number of treatments, $p < t$. The subjects are incomplete blocks in the design because they receive less than the full complement of treatments. The incomplete designs may be derived by deleting one or more periods from a complete set of orthogonal Latin squares. The incomplete designs also may be obtained from the incomplete Latin squares or Youden squares discussed in Chapter 9 for row–column designs. Finally, they can be constructed from regular balanced incomplete block designs. For illustration, a design for seven treatments and four periods is shown in Display 16.6.

Display 16.6   Incomplete Design for Seven Treatments and Four Periods

	Sequence						
Period	1	2	3	4	5	6	7
1	A	B	C	D	E	F	G
2	B	C	D	E	F	G	A
3	D	E	F	G	A	B	C
4	G	A	B	C	D	E	F

	Sequence						
Period	8	9	10	11	12	13	14
1	A	B	C	D	E	F	G
2	G	A	B	C	D	E	F
3	E	F	G	A	B	C	D
4	B	C	D	E	F	G	A

Tables of incomplete Latin square designs for crossover studies can be found in Petersen (1985) and Patterson and Lucas (1962). Jones and Kenward (1989) listed the most efficient incomplete designs for $t = 3, 4, 5, 6$, and 7 treatments.

16.4   Crossover Designs for Two Treatments

Perhaps the most widely used design for crossover studies has been the two-treatment, two-sequence, two-period design. One group receives the treatment sequence A→B and another group receives the treatment sequence B→A, as shown in Display 16.7. Each of the treatments appears in each sequence and in each period of the 2×2 Latin square arrangement. Thus, the treatment effects are not confounded with the effects of sequences or periods, the row and column equivalents of the Latin square design.

Display 16.7   The 2 × 2 Crossover Design

	Period	
Sequence	I	II
1	A $\longrightarrow$ B	
2	B $\longrightarrow$ A	

The Linear Model for Two Treatments

The linear model for the design with two treatments is

$$y_{ijk} = \mu + \alpha_i + b_{ij} + \gamma_k + \tau_d + \lambda_c + e_{ijk} \qquad \textbf{(16.5)}$$

$$i, k, d, c = 1, 2 \quad j = 1, 2, \ldots, r_i$$

where μ is the general mean, α_i is the sequence effect, b_{ij} is the random subject effect with mean 0 and variance σ_b^2, γ_k is the period effect, τ_d is the direct treatment effect, λ_c is the carryover effect, and e_{ijk} is the independent, random error with mean 0 and variance σ^2.

Carryover Effects Confounded in the 2 × 2 Design

The 2 × 2 crossover design has one major deficiency. The carryover effects are confounded with other effects in the study, and the significance of the carryover effects cannot be tested from an analysis of the model in Equation (16.5). The observed means for the subjects in each period of each sequence of the design, $\overline{y}_{11}$, $\overline{y}_{12}, \overline{y}_{21}, \overline{y}_{22}$, contain all of the information available in the study to estimate the design factor effects for sequences and periods and the direct and carryover treatment effects. The expected values of the cell means, $E(\overline{y}_{ik}) = \mu_{ik} = \mu + \alpha_i + \gamma_k + \tau_d + \lambda_c$, are shown in Display 16.8.

Display 16.8   Expected Cell Means for the 2 × 2 Crossover Design

	Expected Cell Mean, μ_{ij}	
Sequence	*Period I*	*Period II*
A→B	$\mu + \alpha_1 + \gamma_1 + \tau_1$	$\mu + \alpha_1 + \gamma_2 + \tau_2 + \lambda_1$
B→A	$\mu + \alpha_2 + \gamma_1 + \tau_2$	$\mu + \alpha_2 + \gamma_2 + \tau_1 + \lambda_2$

Simplified expectations follow upon application of the relationship $\tau_1 + \tau_2 = 0$ based on our definition of treatment effects. The sum of the two treatment effects is 0 so that $\tau_2 = -\tau_1$ and $-\tau_1$ can be substituted for τ_2 in the expectations. Similar substitutions can be made for the other factor effects. Given the substitutions $\alpha_2 = -\alpha_1, \gamma_2 = -\gamma_1$, and $\lambda_2 = -\lambda_1$, the expectations are

$$E(\overline{y}_{11}) = \mu_{11} = \mu + \alpha_1 + \gamma_1 + \tau_1$$

$$E(\overline{y}_{12}) = \mu_{12} = \mu + \alpha_1 - \gamma_1 - \tau_1 + \lambda_1$$

$$E(\overline{y}_{21}) = \mu_{21} = \mu - \alpha_1 + \gamma_1 - \tau_1$$

$$E(\overline{y}_{22}) = \mu_{22} = \mu - \alpha_1 - \gamma_1 + \tau_1 - \lambda_1$$

The five parameters to be estimated are μ, α_1, γ_1, τ_1, and λ_1, and only four observed means are available for estimation of the parameters. It is not possible to obtain estimates of parameters for some of the effects without involving the parameters of other effects. In particular, it is not possible to disentangle the sequence effect α_1 or the direct treatment effect τ_1 from the carryover effect λ_1 in the analysis. If one of the parameters can be eliminated from the model, the remaining parameters can be estimated from the four cell means.

Eliminate Some Model Parameters to Estimate Effects

A number of alternative models and methods of analysis have been proposed for the 2×2 crossover design (Grizzle, 1965, 1974; Hills & Armitage, 1979; Milliken & Johnson, 1984). Regardless of the model one parameter has to be eliminated to estimate the remaining parameters and test hypotheses about them. Under certain circumstances it may be possible to provide arguments for the elimination of sequence or period effects from the model. Eliminating either of those parameters puts the analysis at risk because inferences are made under the generous assumption of either no period or no sequence effect. Effects are left in the model because it is necessary to test their significance in the study and not because it is believed they are actually present in the study.

The test of significance for carryover effects from the 2×2 crossover design is possible only under the questionable assumption that some other effect is nonexistent. Under these circumstances the design should not be used. Jones and Kenward (1989) consider the design to be appropriate if the sequence, period, and carryover effects are negligible in size relative to the direct treatment effect. On the other hand, a panel for the U.S. Food and Drug Administration has recommended the design not be used in drug evaluation.

A Little Help from Baseline and Washout Observations

Adjuncts to the 2×2 crossover design reviewed in detail by Jones and Kenward (1989) include the addition of baseline observations and washout period observations. The baseline observations taken prior to any treatment administration enable the estimation of carryover effects and the test for their significance. However, the direct and carryover effects are still nonorthogonal in the design and the estimates of the direct and carryover effects are highly correlated, $\rho = 0.87(\sigma^2/n)$. Although the two effects are no longer confounded, the high correlation between their estimates leads to difficulties in distinguishing between their separate contributions. When estimates of effects are highly correlated, contradictory tests of significance can occur in the analysis. One effect may be nonsignificant when a second effect is in the model but significant when the second effect is removed from the model.

The addition of observations on the washout period between the two-treatment period reduces the correlation between estimates of the two effects to $\rho = 0.50(\sigma^2/n)$. Although the correlation is reduced, substantial nonorthogonality remains between the direct and carryover effects. The estimates are uncorrelated if observations are included from a second washout period after the second treatment is terminated. Thus, a completely orthogonal design for two treatments is only possible with added baseline and washout data that require a total of five measurement periods.

Add an Extra Period to Estimate Carryover Effects

The other adjuncts for the 2×2 crossover design reviewed by Jones and Kenward (1989) included designs for the two treatments administered in various combinations of sequences and periods. The addition of extra sequences or extra periods of treatment increases the ability to estimate the important effects in the crossover study. Two of the designs for two sequences in three periods will be discussed in this section.

The *switchback* design for two treatments, shown in Display 16.9, has been used frequently as an adjunct to the 2×2 design to obtain estimates of the direct and carryover effects. However, the *extra- period* design for two treatments, also shown in Display 16.9, is a better design for the estimation of the treatment effects.

Display 16.9   Switchback and Extra Period Designs for Two Treatments

	Switchback Period				Extra Period		
Sequence	I	II	III	Sequence	I	II	III
1	A	B	A	1	A	B	B
2	B	A	B	2	B	A	A

The first two periods of both designs provide the basic 2×2 crossover design, but the designs differ in the treatment administered during the third period. The third period treatment reverts to the first period treatment in the switchback design, whereas the second period treatment is repeated in the third period of the extra-period design.

Both designs have six cell means for the estimation of design and treatment factor effects, and the analysis of variance in Table 16.2 based on the model in Equation (16.5) can be used for the sum of squares partitions. An analysis of either design provides valid tests for all of the effects in the model.

Extra-Period Design Superior to the Switchback Design

Recall from the previous section on balanced designs that direct and carryover treatment effects are orthogonal in the extra-period design so that the estimates of direct and carryover treatment effects are uncorrelated. The effects are not orthogonal in the switchback design, and the correlation between the direct and carryover effects is $\rho = 0.87(\sigma^2/n)$.

The additional superiority of the extra-period design resides in its greater efficiency for the estimation of direct and carryover treatment effects. The variances of the estimates for direct and carryover effects in the extra-period design are

$$\sigma_{\hat{\tau}}^2 = 0.19\left(\frac{\sigma^2}{n}\right) \quad \text{and} \quad \sigma_{\hat{\lambda}}^2 = 0.25\left(\frac{\sigma^2}{n}\right)$$

whereas the variances for the effects in the switchback design are

$$\sigma_{\hat{\tau}}^2 = 0.75\left(\frac{\sigma^2}{n}\right) \quad \text{and} \quad \sigma_{\hat{\lambda}}^2 = 1.0\left(\frac{\sigma^2}{n}\right)$$

The statistical properties of the extra-period design are superior to those of the switchback design since the estimates of the direct and carryover treatment effects are uncorrelated and the variances of the estimates are smaller.

The choice of design to this point has focused on estimation and significance tests for the primary design and treatment factors in the crossover design: the sequence, period, direct treatment, and carryover treatment effects. Other designs can be constructed to facilitate the estimation of additional effects that are potentially important in some studies. These effects include interactions between the direct and carryover treatment effects and interactions between the treatment effects and the design factors such as periods and sequences. Discussions on designs for the estimation and tests for these additional model effects can be found in Jones and Kenward (1989).

EXERCISES FOR CHAPTER 16

1. A pharmaceutical manufacturer conducted a crossover study with an anticonvulsant drug (DPH) used in the management of grand mal and psychomotor seizures. A single dose of DPH was given to a subject, and the plasma level of the drug was measured 12 hours after the drug was administered. The four treatments were (A) 100 mg generic DPH product in solution, (B) 100 mg manufacturer DPH in capsule, (C) 100 mg generic DPH product in capsule, and (D) 300 mg manufacturer DPH in capsule.

    A 4 × 4 Latin square crossover design was used for the study. Two subjects were assigned to each of the four treatment sequences. A single dose of each treatment was administered to each subject with an interval of two weeks between treatments. The subject's plasma levels of DPH are shown in the table at the end of the exercise.
    a. Is the design balanced for carryover effects? Explain.
    b. Write a linear model for the study, explain the terms, and state the assumptions necessary for the analysis.
    c. Compute the analysis of variance for the data, and test the significance of the direct and carryover effects.
    d. The manufacturer wanted to know if the bioavailability of DPH in the capsule form was as high as that for the liquid form, whether the bioavailability of the generic product was

equivalent to their product, and whether there was a dosage effect (300 mg versus 100 mg) on the bioavailability. Devise a meaningful set of contrasts among the four treatments, test their significance, and interpret the results.

e. Obtain the residual plots for the analysis, and interpret them.

Sequence	Subject	Period			
		I	II	III	IV
ABDC	1	2.3	1.7	1.8	4.5
	2	1.0	1.2	1.1	3.3
BCAD	3	1.9	4.4	1.4	1.5
	4	0.9	2.1	1.0	0.8
CDBA	5	3.2	0.8	0.8	0.9
	6	2.2	0.8	0.9	0.8
DACB	7	1.4	1.2	3.6	1.1
	8	0.9	0.8	2.9	0.8

Source: K. S. Albert et al. (1974), Bioavailability of diphenylhydantoin. *Clinical Pharmacology and Therapeutics,* 16, 727–735.

2. A crossover design was used to compare drugs for the control of hypertension. Two drugs, A and B, were used alone and in combination. The combination of the two drugs was labeled as drug C in the experiment. Subjects were randomly assigned to one of six sequences of the drug treatments. Each treatment period lasted four weeks with no washout period between treatments. The systolic blood pressure of the subjects measured at the end of each period is shown in the table at the top of the next page.

a. Is the design balanced for carryover effects? Explain.

b. Write a linear model for the study, explain the terms, and state the assumptions necessary for the analysis.

c. Compute the analysis of variance for the data, and test the significance of the direct and carryover effects.

d. Compute the contrasts between drug C (the combination of A and B) and drugs A and B separately along with the standard error. Is the systolic blood pressure for the combination of the two drugs significantly different from that of the two drugs alone? Do drugs A and B differ significantly with respect to systolic blood pressure?

e. Obtain the residual plots for the analysis, and interpret them.

3. A digestion trial with beef steers was conducted in an extra period Latin square crossover design to evaluate the effects of low-quality roughage on feed digestion. The low-quality roughages used in the trial were (A) cottonseed hull, (B) bermuda straw, and (C) wheat straw, and the high-quality roughage used as a control was (D) alfalfa hay. One steer was randomly assigned to each sequence of four diets. The steer remained on each diet for 30 days and measurements on dry matter digestion were made during the last week of the trial allowing a 21-day adjustment to each diet. The

Sequence	Subject	Period		
		I	*II*	*III*
ABC	1	174	146	164
	2	145	125	130
	3	230	174	200
	4	240	130	195
ACB	5	192	150	160
	6	194	208	160
	7	175	152	175
	8	202	160	180
BAC	9	184	192	176
	10	140	150	150
	11	155	230	226
	12	180	185	190
BCA	13	136	132	138
	14	145	154	166
	15	194	210	190
	16	180	180	190
CAB	17	206	220	210
	18	160	180	145
	19	188	200	190
	20	185	197	182
CBA	21	180	180	208
	22	210	160	226
	23	185	180	200
	24	190	145	160

Source: B. J. Jones and M. G. Kenward (1989), *Design and Analysis of Cross-Over Trials*. London: Chapman Hall.

roughage diet fed in the fourth period was repeated during the fifth period. The data on dry matter digestion for each steer in each sequence are shown in the table.

Steer	Period				
	I	*II*	*III*	*IV*	*V*
1	75(A)	76(B)	79(C)	81(D)	79(D)
2	79(C)	73(A)	79(D)	75(B)	77(B)
3	81(D)	79(C)	75(B)	72(A)	73(A)
4	76(B)	79(D)	72(A)	76(C)	73(C)

a. Is the design balanced for carryover effects? Explain.
b. Is there an advantage to this design over a four-period design? Explain.
c. The linear model for this extra-period crossover design is

$$y_{ij} = \mu + \alpha_i + \gamma_j + \tau_d + \lambda_c + e_{ij}$$

$$i = 1, 2, \ldots, 4 \quad j = 1, 2, \ldots, 5 \quad d, c = 1, 2, \ldots, 4$$

where μ is the general mean, α_i is the steer sequence effect, γ_j is the period effect, τ_d is the direct treatment effect, λ_c is the carryover effect, and e_{ij} is the normally distributed independent random error with variance σ^2.

d. Compute the analysis of variance for the data. Make the computations with your program using a sequential fit of carryover and direct effects. Notice the sums of squares for direct and carryover effects are the same regardless of their order with respect to one another in the sequential fit of terms. Why is this so?
e. Test the significance of the direct and carryover effects.
f. Use the Dunnett method to compare the control diet, alfalfa hay, with each of the low-roughage diets, and interpret the results.
g. Obtain the residual plots for the analysis and interpret them.

4. An extra-period crossover design was used to compare two drugs for the control of hypertension. Subjects were randomly assigned to one of two sequences of the drug treatments. Each treatment period lasted six weeks with a one-week washout period between treatments. The diastolic blood pressures of the subjects measured at the end of each period are shown in the table.

| Sequence | Subject | Period | | |
		I	II	III
ABB	1	73	92	75
	2	90	90	80
	3	95	75	75
	4	80	80	90
	5	90	90	70
	6	45	60	45
	7	70	60	80
	8	122	101	90
BAA	9	60	70	100
	10	100	85	80
	11	50	45	70
	12	65	70	70
	13	88	88	88
	14	70	70	80
	15	60	90	70

a. Is the design balanced for carryover effects? Explain.

b.   Is there an advantage to this design over a simple two-period design without using the third period as shown in the data table? Explain.

c.   Write a linear model for the study, explain the terms, and state the assumptions necessary for the analysis.

d.   Compute the analysis of variance for the data, and test the significance of the direct and carry-over effects.

e.   Compute the contrast between drugs A and B along with the standard error. Do drugs A and B differ significantly with respect to diastolic blood pressure?

f.   Obtain the residual plots for the analysis, and interpret them.

5.   A crossover study was conducted to evaluate four keyboard layouts. Twelve volunteers experienced in a common keyboard configuration were used in the study. Each subject used the four test layouts in sequence. Each subject was randomly assigned to a sequence of layouts. Each layout was used for four days in their ordinary data and text entry activities. On the fifth day they were all given a common task to perform with their assigned layout, and the number of errors on the task were recorded. None of the subjects knew they were being tested on the final day. The number of errors recorded for each subject on each layout are shown in the table.

	Period			
Subject	I	II	III	IV
1	7(D)	2(B)	1(A)	5(C)
2	1(A)	4(C)	6(D)	3(B)
3	6(C)	1(A)	3(B)	7(D)
4	3(B)	6(D)	3(C)	1(A)
5	4(C)	5(D)	1(A)	2(B)
6	6(D)	4(C)	2(B)	0(A)
7	1(A)	3(B)	4(C)	5(D)
8	2(B)	2(A)	7(D)	4(C)
9	5(D)	0(A)	3(C)	3(B)
10	0(A)	4(D)	2(B)	3(C)
11	3(C)	2(B)	7(D)	0(A)
12	2(B)	4(C)	0(A)	6(D)

a.   Is the design balanced for carryover effects? Explain.

b.   Write a linear model for the study, explain the terms, and state the assumptions necessary for the analysis. Do you think an assumption of normal distribution for the observations is valid for this study? Explain.

c.   Compute the analysis of variance for the data, and test the significance of the direct and carry-over effects.

d.   Obtain the residual plots for the analysis, and interpret them. Do you think a transformation of the data will improve the analysis? If so, which transformation might be appropriate?

16A.1   Appendix: Coding Data Files for Crossover Studies

Carryover Effects Coded as Factor Levels

This coding scheme was suggested for the GLM program in SAS by Ratkowsky, Alldredge, and Cotton (1990). The correct sum of squares partitions are produced also by the MANOVA program in SPSS with this coding scheme. The estimates of the adjusted treatment means may differ by some constant amount between programs because of their differing computing algorithms. However, estimates of contrasts among the treatment means and standard errors for the contrasts will be consistent among the programs. Some programs will not complete the analysis with this coding scheme.

The basic codes for levels of a factor in a data file ordinarily are a sequence of integers, $1, 2, 3, \ldots$. A code for the level of each factor is required for each observation in the data file. However, the carryover effects do not occur with observations in the first period of the crossover study, and there is no "level" to code for carryover effects. Some programs will allow a code for the factor even though the factor does not occur for the observation. For the examples in this chapter the levels of the carryover factor were identified with the levels $0, 1, 2, \ldots, t$, where the "0" code was used for the first-period observations because there were no carryover effects present in the first period. The observation was given a code of "1" if treatment 1 occurred in the previous period, a code of "2" if treatment 2 occurred in the previous period, and so forth.

The complete coding for the data file used in Example 16.1 included columns of codes for the period (P), sequence (S), animal within sequence (A), diet (D), and carryover (C) factors as well as the column of data for the response variable NDF.

P	S	A	D	C	NDF	P	S	A	D	C	NDF	P	S	A	D	C	NDF
1	1	1	1	0	50	2	1	1	2	1	61	3	1	1	3	2	53
1	1	2	1	0	55	2	1	2	2	1	63	3	1	2	3	2	57
1	2	1	2	0	44	2	2	1	3	2	42	3	2	1	1	3	57
1	2	2	2	0	51	2	2	2	3	2	45	3	2	2	1	3	59
1	3	1	3	0	35	2	3	1	1	3	55	3	3	1	2	1	47
1	3	2	3	0	41	2	3	2	1	3	56	3	3	2	2	1	50
1	4	1	1	0	54	2	4	1	3	1	48	3	4	1	2	3	51
1	4	2	1	0	58	2	4	2	3	1	51	3	4	2	2	3	54
1	5	1	2	0	50	2	5	1	1	2	57	3	5	1	3	1	51
1	5	2	2	0	55	2	5	2	1	2	59	3	5	2	3	1	55
1	6	1	3	0	41	2	6	1	2	3	56	3	6	1	1	2	58
1	6	2	3	0	46	2	6	2	2	3	58	3	6	2	1	2	61

Note the coding for the carryover factor C is "0" in the first period ($P = 1$). The carryover factor code in the second period ($P = 2$) is identical to the diet (D)

code for the same steer in the previous period ($P = 1$). Likewise, the diet codes for the second period are those used for the carryover code in the third period.

Carryover Effects Coded as Covariates

Milliken and Johnson (1984) introduced the carryover effects into the model as covariates. The method of coding is sometimes referred to as *indicator* or *dummy* variable codes. The coding method is based on the knowledge that a restriction must be placed on the solutions for factor effects in the normal equations. One restriction on the relationship among the solutions that can be used for the carryover effects is $\widehat{\lambda}_1 + \widehat{\lambda}_2 + \cdots + \widehat{\lambda}_t = 0$. The restriction indicates that knowledge of $t - 1$ effects automatically provides the value for the remaining effect. The relationship is $\widehat{\lambda}_1 + \widehat{\lambda}_2 + \widehat{\lambda}_3 = 0$ with three-carryover treatment effects. Knowledge of the values for two-carryover effects automatically provides the value for the third. Therefore, $\widehat{\lambda}_3 = -\widehat{\lambda}_1 - \widehat{\lambda}_2$ and carryover effects have 2 degrees of freedom.

Only $t - 1$ covariates are required for the linear model to include parameters for carryover effects because of the restrictions on the carryover effects. The model for the roughage diet study is used as an example. The model expressed with two covariates for carryover effects is

$$y_{ijk} = \mu + \alpha_i + b_{ij} + \gamma_k + \tau_d + \lambda_1 X_1 + \lambda_2 X_2 + e_{ijk} \tag{16A.1}$$

$$i = 1, 2, \ldots, 6 \quad j = 1, 2 \quad k = 1, 2, 3 \quad d = 1, 2, 3$$

One method of coding the covariates for the carryover effects in period k is

$$X_1 = \begin{cases} 1 & \text{if } d = 1 \text{ in period } (k - 1) \\ -1 & \text{if } d = 3 \text{ in period } (k - 1) \\ 0 & \text{otherwise} \end{cases}$$

Thus, $X_1 = 1$ when diet A occurs in the previous period and $X_1 = -1$ when diet C occurs in the previous period.

Also,

$$X_2 = \begin{cases} 1 & \text{if } d = 2 \text{ in period } (k - 1) \\ -1 & \text{if } d = 3 \text{ in period } (k - 1) \\ 0 & \text{otherwise} \end{cases}$$

Thus, $X_2 = 1$ when diet B occurs in the previous period and $X_2 = -1$ when diet C occurs in the previous period.

The complete coding for the data file for Example 16.1 includes columns of codes for the period (P), sequence (S), animal within sequence (A), diet (D), and covariates for the carryover effects X_1 and X_2, as well as the column of data for the response variable NDF.

There is no carryover effect in the first period ($P = 1$), and the covariates are coded $X_1 = X_2 = 0$. The observation on diet B in the second period ($P = 2$) of the

P	S	A	D	X_1	X_2	NDF	P	S	A	D	X_1	X_2	NDF	P	S	A	D	X_1	X_2	NDF
1	1	1	1	0	0	50	2	1	1	2	1	0	61	3	1	1	3	0	1	53
1	1	2	1	0	0	55	2	1	2	2	1	0	63	3	1	2	3	0	1	57
1	2	1	2	0	0	44	2	2	1	3	0	1	42	3	2	1	1	−1	−1	57
1	2	2	2	0	0	51	2	2	2	3	0	1	45	3	2	2	1	−1	−1	59
1	3	1	3	0	0	35	2	3	1	1	−1	−1	55	3	3	1	2	1	0	47
1	3	2	3	0	0	41	2	3	2	1	−1	−1	56	3	3	2	2	1	0	50
1	4	1	1	0	0	54	2	4	1	3	1	0	48	3	4	1	2	−1	−1	51
1	4	2	1	0	0	58	2	4	2	3	1	0	51	3	4	2	2	−1	−1	54
1	5	1	2	0	0	50	2	5	1	1	0	1	57	3	5	1	3	1	0	51
1	5	2	2	0	0	55	2	5	2	1	0	1	59	3	5	2	3	1	0	55
1	6	1	3	0	0	41	2	6	1	2	−1	−1	56	3	6	1	1	0	1	58
1	6	2	3	0	0	46	2	6	2	2	−1	−1	58	3	6	2	1	0	1	61

first sequence ($S = 1$) follows diet A from the first period; thus, a carryover effect for diet A is required and the covariate codes are $X_1 = 1$ and $X_2 = 0$. The observation on diet A in the third period ($P = 3$) of the second sequence ($S = 2$) follows diet C from the second period; thus, a carryover effect for diet C is required with covariate codes $X_1 = -1$ and $X_2 = -1$.

The sum of squares for the two covariates, X_1 and X_2, will be the sum of squares for carryover effects adjusted for direct effects. This method for the analysis can be used with any statistical program that allows the inclusion of covariates in the model. Each program will have unique syntax for the representation of added covariates in the model.

16A.2 Appendix: Treatment Sum of Squares for Balanced Designs

The quantities shown next for manual calculation of the adjusted sums of squares for direct and carryover effects in the balanced crossover designs and balanced extra-period designs are valid for equal numbers of subjects per treatment sequence. Following the notation of Cochran and Cox (1957) the required formulae are

T_i = treatment totals, $i = 1, 2, \ldots, t$
R_i = total of observations in the periods immediately following treatment i
F_i = total of sequences in which treatment i is the final treatment
P_1 = total of period 1
G = grand total of all observations

Also, let t = the number of treatments, r = the number of subjects per sequence, and m = the number of Latin squares used for the design.

Williams Designs: For the Williams designs compute

$$\widehat{T}_i = (t^2 - t - 1)T_i + tR_i + F_i + P_1 - tG$$

for the adjusted sum of squares for direct treatment effects

$$SST\text{(adjusted)} = \frac{\Sigma \widehat{T}_i^2}{rmt(t^2 - t - 1)(t + 1)(t - 2)}$$

Compute

$$\widehat{R}_i = tT_i + t^2 R_i + tF_i + tP_1 - (t + 2)G$$

for the adjusted sum of squares for carryover treatment effects

$$SSC\text{(adjusted)} = \frac{\Sigma \widehat{R}_i^2}{rmt^3(t + 1)(t - 2)}$$

Compute the usual unadjusted sums of squares for periods, sequences, subjects within sequences, and treatments according to the Latin rectangle design in Chapter 8. The sum of squares for experimental error to complete the sum of squares partition for the analysis of variance as shown in Table 16.2 is

$$SSE = \text{Total } SS - SSS - SSW - SSP - SST\text{(unadjusted)} - SSC$$

The estimates of differences between the adjusted treatment means can be found as

$$\widehat{\mu}_i - \widehat{\mu}_j = \frac{\widehat{T}_i - \widehat{T}_j}{rmt(t + 1)(t - 2)}$$

The standard error estimate for the contrast with adjusted means from a balanced crossover design is

$$s_{(\widehat{\mu}_i - \widehat{\mu}_j)} = \sqrt{\frac{2MSE}{N} \cdot \frac{(t^2 - t - 1)}{(t + 1)(t - 2)}}$$

where N is the number of replications per treatment.

Extra-Period Williams Design: Adjustments to the calculations required for the extra-period Williams designs are

$$\widehat{T}_i = (t + 1)T_i - F_i - G$$

with the adjusted sum of squares for treatments as

$$SST\text{(adjusted)} = \frac{\Sigma \widehat{T}_i^2}{mt(t + 1)(t + 2)}$$

Compute

$$\widehat{R}_i = tR_i + P_1 - G$$

for the adjusted sum of squares for carryover effects as

$$SSC\text{(adjusted)} = \frac{\Sigma \widehat{R}_i^2}{mt^3}$$

The estimated difference between two adjusted treatment means is

$$\widehat{\mu}_i - \widehat{\mu}_j = \frac{\widehat{T}_i - \widehat{T}_j}{mt(t+2)}$$

with standard error estimate

$$s_{(\widehat{\mu}_i - \widehat{\mu}_j)} = \sqrt{\frac{2MSE}{mt} \cdot \frac{(t+1)}{(t+2)}}$$

17   Analysis of Covariance

The use of additional information in the experimental units as a local control practice to reduce the estimate of experimental error is the primary subject of this chapter. The values for treatment means in a research study may depend on covariates that vary among the experimental units and have a significant relationship with the primary response variable. The analysis of covariance is used in this chapter to remove the influence of the covariates on treatment comparisons in completely randomized and complete block designs.

17.1   Local Control with a Measured Covariate

A number of local control techniques are used in experiments to control experimental error variance. Local control practices reduce experimental error variance and increase the precision for estimates of treatment means and tests of hypotheses. Concomitant variables are often used to select and group units to control experimental error variation.

Many concomitant variables, or *covariates,* can be measured at any time during the course of the experiment, and their influence on the response variable can be assessed when analyzing the results. The analysis of covariance, combining regression methodology with the analysis of variance, evaluates the influence of the covariate on the response variable and enables the comparison of treatments on a common basis relative to the values of the covariates.

Often, many factors external to the treatment factors influence the response variable. Blocking on the basis of these influential factors is one of the primary means used by researchers to control experimental error. When blocks of units are constructed with similar values for the factors, the experimental treatments can be compared with one another in a more homogeneous environment.

Frequently, the experimental setting may prohibit blocking of like units for a variety of reasons. There may be incomplete knowledge about the experimental material or the effects of external factors may not appear until after the experiment has begun. Too few units of like value may exist for adequate blocking. Even though the investigator may have knowledge of the influential factors, the number of additional factors may prohibit the use of all of them as blocking criteria.

Typical studies with additional variables that influence treatment comparisons include:

- Clinical trials where age, weight, sex, previous medical history, or occupation of patients can influence their response to the treatments. Although the investigators can block on two or three of these factors, it is not possible to ignore the influence of the remaining variables.

- Trials with fruit trees, which are blocked on the basis of soil or irrigation gradients, but the influence of previous treatments and historical yield records cannot be ignored.

- Feeding trials where blocks of animals can be constructed on the basis of litter and initial weights, but the amount of feed consumed during the course of the study will also have influence on the measured weight gains.

The study described in Example 17.1 illustrates an experiment in which the response variable is affected not only by the treatment applied to the subjects of the study but also by a covariate that was measured on each subject prior to the study.

Example 17.1   Effects of Exercise on Oxygen Ventilation

A common clinical method to evaluate an individual's cardiovascular capacity is through treadmill exercise testing. One of the measures obtained during treadmill testing, maximal oxygen uptake, is considered the best index of work capacity and maximal cardiovascular function.

The measured maximal oxygen uptake by an individual depends on a number of factors, including the mode of testing, test protocol, and the subjects' physical condition and age. A common test protocol on the treadmill is the inclined protocol where grade and speed incrementally increase until exhaustion occurs.

Research Hypothesis: Researchers in exercise physiology were of the opinion that the conditions for treadmill testing should simulate as closely as possible the mode of subject cardiovascular training to attain their maximal oxygen uptake during the test. It was hypothesized that step aerobic training was better simulated by the inclined protocol than a flat terrain running regimen.

Treatment Design: Two treatments selected for the study were a 12-week step aerobic training program and a 12-week outdoor running regimen on flat terrain. The subjects were to be tested on a treadmill before and after the 12-week training period with the inclined protocol. If the hypothesis were true, the subjects from the step aerobic training would show greater increases in maximal oxygen uptake than the subjects trained only on flat terrain. Those trained on flat terrain might be limited by localized muscle fatigue when tested on inclined protocols and thus not attain maximal oxygen uptake before exhaustion.

Experiment Design: The subjects were 12 healthy males who did not participate in a regular exercise program. Six men were randomly assigned to each group in a completely randomized design. Various respiratory measurements were made on the subjects while on the treadmill before the 12-week period. There were no differences in the respiratory measurements of the two groups of subjects prior to treatment.

The measurement of interest for this example is the change in maximal ventilation (liters/minute) of oxygen for the 12-week period. The observations on the 12 subjects and their ages are shown in Table 17.1, along with the group means and standard errors.

Table 17.1 Maximal ventilation change (liters/minute) following a 12-week exercise program

Group	Age	Change	Group	Age	Change
Aerobic	31	17.05	Running	23	− 0.87
	23	4.96		22	− 10.74
	27	10.40		22	− 3.27
	28	11.05		25	− 1.97
	22	0.26		27	7.50
	24	2.51		20	− 7.25
Mean	25.83	7.71		23.17	− 2.77
Std. Err.	1.40	2.55		1.01	2.54

Source: D. Allen, Exercise Physiology, University of Arizona.

Is Treadmill Performance Related to Age?

A graph of the change in maximal ventilation (y) for each of the subjects with their age (x) on the horizontal axis is shown in Figure 17.1. The plot suggests a strong positive relationship between the age of the subjects and their change in maximal ventilation on the treadmill regardless of the treatment group to which they belong. Thus, there appears to be considerable experimental error variation within each group associated with age differences.

The study protocol required the eligible subjects to be healthy males between the ages of 20 and 35 with a sedentary lifestyle. Although the two groups of subjects did not differ in their average pretest maximal ventilation measures, a one-way analysis of variance reveals a significant difference between the two groups in the maximal ventilation change after training. The aerobic exercise group had a larger change in ventilation rate than the running group, but the aerobic group consists of

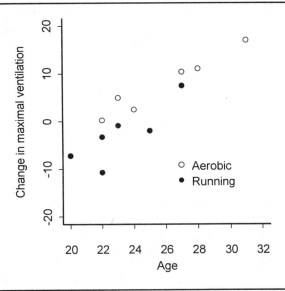

Figure 17.1 The relationship between maximal ventilation and age in the exercise physiology study

an older group of males. It must be determined whether the difference is a result of the exercise or the age differences in the groups. That is, suppose the mean changes in maximal ventilation were compared at the same age for both groups. Would the aerobic exercise group mean still be significantly greater than the running group mean? The analysis of covariance can be used to help answer that question and to determine whether the relationship between maximal ventilation change and age contributes significantly to experimental error variation.

17.2 Analysis of Covariance for Completely Randomized Designs

The Linear Model and Analysis of Covariance

The experiment on exercise physiology was conducted in a completely randomized design with two treatment groups. Assuming a linear relationship between the response variable y and a covariate x the linear model for the completely randomized design is

$$y_{ij} = \mu_i + \beta(x_{ij} - \overline{x}_{..}) + e_{ij} \qquad (17.1)$$

$$i = 1, 2, \ldots, t \quad j = 1, 2, \ldots, r$$

where μ_i is the treatment mean, β is the coefficient for the linear regression of y_{ij} on x_{ij}, and the e_{ij} are independent, normally distributed random experimental errors with mean 0 and variance σ^2. Two additional key assumptions for this model are that the regression coefficient β is the same for all treatment groups, and the treatments do not influence the covariate x.

The first objective of the covariance analysis is to determine whether the addition of the covariate has reduced the estimate of experimental error variance. If the reduction is significant, then we obtain estimates of the treatment group means $\bar{y}_{ia}$ adjusted to the same value of the covariate x for each of the treatment groups and determine the significance of treatment differences on the basis of the adjusted treatment means.

Alternative Models to Evaluate the Covariate Contribution

The analysis will require least squares estimates of the parameters for the alternative full and reduced models, which are

- the full model $y_{ij} = \mu_i + \beta(x_{ij} - \bar{x}_{..}) + e_{ij}$
- a reduced model without the covariate $y_{ij} = \mu_i + e_{ij}$
- a reduced model without treatment effects $y_{ij} = \mu + \beta(x_{ij} - \bar{x}_{..}) + e_{ij}$

The reduced model without the covariate is required to assess the influence of the covariate, and the reduced model without the treatment effects is required to assess the significance of treatment effects in the presence of the covariate.

Least squares estimates of the parameters are derived for the full model to obtain

$$SSE_f = \Sigma[y_{ij} - \hat{\mu}_i - \hat{\beta}(x_{ij} - \bar{x}_{..})]^2$$

with $N - t - 1$ degrees of freedom; the reduced model without the covariate for

$$SSE_r = \Sigma\,(y_{ij} - \hat{\mu}_i)^2$$

with $N - t$ degrees of freedom; and the reduced model without treatment effects for

$$SSE_r^* = \Sigma\,[y_{ij} - \hat{\mu} - \hat{\beta}(x_{ij} - \bar{x}_{..})]^2$$

with $N - 2$ degrees of freedom.

The sum of squares reduction due to the addition of the covariate x to the model is obtained as the difference

$$SS(\text{Covariate}) = SSE_r - SSE_f$$

with 1 degree of freedom. The adjusted treatment sum of squares after fitting the covariate is

$$SST(\text{adjusted}) = SSE_r^* - SSE_f$$

with $t - 1$ degrees of freedom.

The SSE for each of the three models fit to the exercise physiology data (Table 17.1) are $SSE_f = 70.39$, $SSE_r = 389.30$, and $SSE_r^* = 142.18$.

Sum of Squares Partitions for the Analysis of Covariance

These sum of squares reductions can be computed by most computer programs for the analysis of covariance, and they will produce the required information found in the following discussion. The sum of squares partitions for the change in maximal ventilation rate with the age covariate in the exercise physiology study is shown in the analysis of variance in Table 17.2. Notice without the age covariate the estimated experimental error variance is $MSE_r = SSE_r/(N - t) = 389.3/10 = 38.93$. The addition of the covariate has reduced the estimate to $MSE = 7.82$ in Table 17.2. Thus, use of the age covariate as local control to reduce σ^2 through covariates appears to be effective. The gain in efficiency due to the covariate is illustrated later in this section.

Table 17.2 Analysis of covariance for maximal ventilation change with age covariate in an exercise physiology study

Source of Variation	Degrees of Freedom	Sum of Squares	Mean Square	F	Pr > F
Regression	1	318.91	318.91*	40.78	.000
Group	1	71.79	71.79†	9.18	.014
Error	9	70.39	7.82‡		

*MS(Covariate) †MST(adjusted) ‡MSE_f

Tests of Hypotheses About Covariates and Treatments

Determine the significance of the reduction due to the covariate with a test of the null hypothesis H_0: $\beta = 0$. The test statistic is

$$F_0 = \frac{MS(\text{Covariate})}{MSE} = \frac{318.91}{7.82} = 40.78 \qquad (17.2)$$

with critical value $F_{.05,1,9} = 5.12$. The null hypothesis is rejected with $Pr > F = .000$ in Table 17.2; the addition of the covariate has significantly reduced experimental error variance.

The significant relationship between change in maximal ventilation rate and age of the subjects indicates the necessity to assess the significance of the treatment effects after the covariance adjustment. The null hypothesis for the adjusted treatment means is H_0: $\mu_1 = \mu_2$, and the test statistic is

$$F_0 = \frac{MST(\text{adjusted})}{MSE} = \frac{71.79}{7.82} = 9.18 \qquad (17.3)$$

with critical value $F_{.05,1,9} = 5.12$. The null hypothesis is rejected with $Pr > F = 0.014$ in Table 17.2, and we conclude that treatment means adjusted for age are different.

Treatment Means Adjusted to a Common Covariate Value

The estimates of the treatment means are adjusted to a common value for the covariate if inclusion of the covariate in the model significantly reduced experimental error variance. The treatment means can be adjusted to any value of the covariate, but ordinarily they are adjusted to the overall mean $\bar{x}_{..}$ as

$$\bar{y}_{ia} = \bar{y}_{i.} - \hat{\beta}(\bar{x}_{i.} - \bar{x}_{..}) \qquad (17.4)$$

The estimate of the regression coefficient, which most programs compute automatically, is

$$\hat{\beta} = \frac{\sum_{i=1}^{t}\sum_{j=1}^{r}(x_{ij} - \bar{x}_{i.})(y_{ij} - \bar{y}_{i.})}{\sum_{i=1}^{t}\sum_{j=1}^{r}(x_{ij} - \bar{x}_{i.})^2} = \frac{169.10}{89.67} = 1.886 \qquad (17.5)$$

and the estimated regression equation for the ith treatment group will be

$$\hat{y}_i = \bar{y}_{i.} + \hat{\beta}(x_i - \bar{x}_{i.}) \qquad (17.6)$$

and are

$$\hat{y}_{1j} = 7.70 + 1.886(x_{1j} - 25.83)$$

for the aerobic group and

$$\hat{y}_{2j} = -2.77 + 1.886(x_{2j} - 23.17)$$

for the running group. The regression line for each of the treatment groups is shown in Figure 17.2.

The treatment means adjusted to the mean age $\bar{x}_{..} = 24.5$ are

$$\bar{y}_{1a} = 7.70 - 1.886(25.83 - 24.5) = 5.19$$

$$\bar{y}_{2a} = -2.77 - 1.886(23.17 - 14.5) = -0.26$$

The unadjusted and adjusted treatment means are shown with the computed regression line for each of the treatment groups in Figure 17.2.

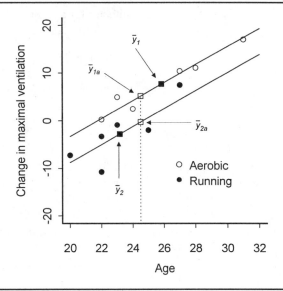

Figure 17.2 Regression between maximal ventilation and age in the exercise physiology study with adjusted treatment means

The difference between the unadjusted treatment means $\bar{y}_{1.} - \bar{y}_{2.} = 7.71 - (-2.77) = 10.48$ was much greater than the difference between the treatment means adjusted to a common age, $\bar{y}_{1A} - \bar{y}_{2A} = 5.19 - (-0.26) = 5.45$. Part of the difference between the unadjusted means was due to the difference of over two years in the average ages of the subjects in the two treatment groups, $\bar{x}_{1.} = 25.83$ and $\bar{x}_{2.} = 23.17$. The adjusted means are estimates of the mean maximal ventilation change at a common age. Thus, the difference between the adjusted means reflects only the effect of exercise on the change in maximal ventilation rate.

Standard Errors for the Adjusted Treatment Means

Two quantities useful for calculation of standard errors and relative efficiency in the analysis of covariance are the sums of squares for treatments, T_{xx}, and experimental error, E_{xx}, from an analysis of variance for the age covariate, x. These sums of squares and their values for this study are

$$T_{xx} = r \sum_{i=1}^{t} (\bar{x}_{i.} - \bar{x}_{..})^2 = 21.31$$

and

$$E_{xx} = \sum_{i=1}^{t} \sum_{j=1}^{r} (x_{ij} - \bar{x}_{i.})^2 = 89.67$$

The standard error estimator for an adjusted treatment mean is

$$s_{\bar{y}_{ia}} = \sqrt{MSE\left[\frac{1}{r_i} + \frac{(\bar{x}_{i.} - \bar{x}_{..})^2}{E_{xx}}\right]} \qquad (17.7)$$

The estimated standard error for the adjusted aerobic exercise group mean is

$$s_{\bar{y}_{1a}} = \sqrt{7.82\left[\frac{1}{6} + \frac{(25.83 - 24.50)^2}{89.67}\right]} = 1.21$$

The standard error estimate for the adjusted running group mean will be the same as that for the exercise group mean since the quantity $(\bar{x}_{i.} - \bar{x}_{..})^2$ in Equation (17.7) is the same for both treatment groups.

The standard error estimator for the difference between two adjusted treatment means is not always available from computer programs without some clever programming on the part of the user. It is calculated as

$$s_{(\bar{y}_{ia} - \bar{y}_{ja})} = \sqrt{MSE\left[\frac{1}{r_i} + \frac{1}{r_j} + \frac{(\bar{x}_{i.} - \bar{x}_{j.})^2}{E_{xx}}\right]} \qquad (17.8)$$

The difference between the adjusted treatment means for the aerobic exercise and running groups is $\bar{y}_{1a} - \bar{y}_{2a} = 5.19 - (-0.26) = 5.45$ with standard error estimate

$$s_{(\bar{y}_{1a} - \bar{y}_{2a})} = \sqrt{7.82\left[\frac{1}{6} + \frac{1}{6} + \frac{(25.83 - 23.17)^2}{89.67}\right]} = 1.80$$

Even when all r_i are equal, the standard error of the difference will vary among the pairs of treatments with more than two treatment groups because of the term $(\bar{x}_{i.} - \bar{x}_{j.})$ in Equation (17.8). In practice the variation in the estimate is slight. A single average standard error of the difference suggested by Finney (1946) to simplify the analysis with equal replication numbers is

$$s_{(\bar{y}_{ia} - \bar{y}_{ja})} = \sqrt{\frac{2MSE}{r}\left[1 + \frac{T_{xx}}{(t - 1)E_{xx}}\right]} \qquad (17.9)$$

The substitution of T_{xx} for $(\bar{x}_{i.} - \bar{x}_{..})^2$ in Equation (17.7) provides the average standard error for the adjusted treatment means.

Was There Increased Efficiency with a Covariate?

Whether the covariance adjustment has been worth the required effort depends on the gain in efficiency of the estimated means. The efficiency of the covariance adjustment relative to the analysis without the adjustment is based on the ratio of respective variances for the estimates of treatment means. The estimate of experimental error variance from the reduced model without the covariate is $MSE_r = SSE_r/(N - t) = 389.3/10 = 38.93$. The average variance suggested by

Finney (1946) can be used for the estimate with the covariance adjustment. The efficiency is calculated as

$$E = \frac{MSE_r}{MSE\left[1 + \dfrac{T_{xx}}{(t-1)E_{xx}}\right]} \tag{17.10}$$

Given $MSE_r = 38.93$, $T_{xx} = 21.33$, $E_{xx} = 89.67$, $MSE = 7.82$, and $t = 2$ the efficiency of the covariance adjustment is $E = 38.93/9.68 = 4.0$. Thus, without covariance adjustment for age, four times as many subjects would be required for the exercise study to achieve the same precision on the estimated treatment means.

Critical Assumptions for a Valid Covariance Analysis

The validity of inferences from the analysis of variance requires an assumption of independent and homogeneous normally distributed errors. An evaluation of the assumptions regarding homogeneous and normally distributed experimental errors can be achieved with an estimate of the residuals for each observation as $\hat{e}_{ij} = y_{ij} - \bar{y}_{i.} - \hat{\beta}(x_{ij} - \bar{x}_{i.})$. Many computer programs will supply estimates of the residuals that can be used for the normal plots and plots of residuals versus estimated values to evaluate the assumptions as outlined in Chapter 4.

Additional assumptions critical to the validity of inferences from the analysis of covariance are (1) the covariate x is unaffected by the treatments, (2) a linear relationship exists between the response variable y and the covariate x, and (3) the regression coefficient β is the same for all treatment groups.

Do the Treatments Affect the Covariate?

If the covariate x as well as the primary response variable y is affected by the treatments the resultant response is multivariate, and the covariance adjustment for treatment means is inappropriate. In these cases, an analysis of the bivariate response (x, y) utilizing multivariate analysis methods is in order. Adjustment for the covariate is appropriate if it is measured prior to treatment administration since the treatments have not yet had the opportunity to affect its value. If the covariate is measured concurrently with the response variable, then it must be decided whether it could be affected by the treatments before the covariance adjustments are considered.

Is the Regression Coefficient the Same for All Treatments?

Comparisons among adjusted treatment means are independent of the covariate value if the regression lines for the treatment groups are parallel. If the relationship between y and x differs among the treatment groups as shown in Figure 17.3, then differences between adjusted treatment means depend crucially on the level of x chosen for the adjustment.

This heterogeneity of regression coefficients among treatment groups resembles interaction between factors in the standard factorial treatment design. In

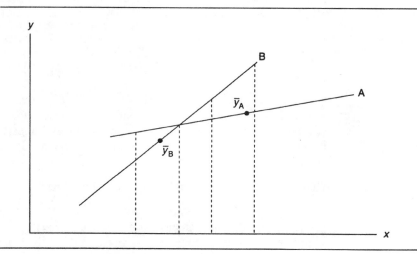

Figure 17.3 Differing regression relationships between the response variable y and the covariate x

that case comparisons are made with simple effects of one factor at each level of other factors. A similar strategy must be used in the case of continuous covariates with different regression coefficients for the treatment groups.

Similar difficulties occur with nonlinear relationships. Under these circumstances the inferences regarding the responses must include a complete description involving the effects of the treatments and the covariate.

Evaluation of the Separate Regressions Model

The Linear Model with Separate Regressions for Each Treatment

The linear model with different regression coefficients for each of the treatment groups is

$$y_{ij} = \mu_i + \beta_i(x_{ij} - \overline{x}_{i.}) + e_{ij} \tag{17.11}$$

$$i = 1, 2, \ldots, t \quad j = 1, 2, \ldots, r$$

where β_i is the regression coefficient for the ith treatment group. A test for the equality of the regression coefficients, H_0: $\beta_1 = \beta_2 = \cdots = \beta_t$, can be derived from an analysis of the two alternative models, the model with a common regression for all treatments and the model with separate regressions for the treatments. Different regressions imply the presence of an interaction between the treatments and the covariate; thus, the model alternatively can be written to include a term for interaction between the covariate and treatments.

Example 17.2   Auditory Discrimination and Cultural Environment

Hendrix, Carter, and Scott (1982) described a study conducted to determine the difference in the ability to discriminate aurally between environmental sounds with respect to several factors. The study was designed to test the effects of a treatment on auditory discrimination. Subjects belonging to two different cultural groups were given pre-treatment and post-treatment tests on auditory discrimination.

Only a portion of the data is used for this example to illustrate the effects of heterogeneous regression coefficients for the covariate. The analysis will be conducted to determine whether the gain in auditory score between the pre- and post-treatment administration differed between the subjects with pre-test scores as a covariate. The data for gain in scores and pre-test scores from female subjects in the auditory treatment group are shown in Table 17.3.

Table 17.3   Pre-test and gain scores for auditory discrimination in two cultural groups

Culture	Pre-Test	Gain	Culture	Pre-Test	Gain
1	64	7	2	52	2
	39	32		50	12
	69	2		59	6
	56	20		42	10
	67	4		62	1
	39	26		35	9
	32	34		41	6
	62	8		36	8
	64	4		37	8
	66	2		64	6
Mean	55.8	13.6		47.8	6.8

A Sum of Squares Partition for Homogeneity of Regressions

Let the model $y_{ij} = \mu_i + \beta(x_{ij} - \bar{x}_{..}) + e_{ij}$ in Equation (17.1), assuming a common slope, be model 1 with the usual $t + 1$ parameters, μ_i and β. Let the model $y_{ij} = \mu_i + \beta_i(x_{ij} - \bar{x}_{i.}) + e_{ij}$ in Equation (17.11), assuming different slopes for each treatment, be model 2 with $2t$ parameters, μ_i and β_i. The sum of squares for experimental error from model 1, say SSE_1, will have $(N - t - 1)$ degrees of freedom, and the sum of squares for experimental error from model 2, say SSE_2, will have $(N - 2t)$ degrees of freedom.

The analysis of variance of the auditory discrimination data for model 1 in Table 17.4 contains $SSE_1 = 519.38$ with 17 degrees of freedom. The analysis of variance for model 2 is shown in Table 17.5 with 2 degrees of freedom for the sum of squares reduction for separate regressions calculated within each culture group.

The error sum of squares has been reduced to $SSE_2 = 147.22$ with 16 degrees of freedom.

Table 17.4   Analysis of covariance for score gain with pre-test score covariate in the auditory discrimination study

Source of Variation	Degrees of Freedom	Sum of Squares	Mean Square	F	Pr > F
Regression	1	1061.12	1061.12	34.73	.000
Group	1	647.47	647.47	21.19	.000
Error	17	519.38*	30.55		

*SSE_1

Table 17.5   Analysis of covariance score gain assuming separate regressions of gain on pre-test scores for each cultural group

Source of Variation	Degrees of Freedom	Sum of Squares	Mean Square	F	Pr > F *
Group	1	610.24	610.24	66.33	.000
Regression within groups	2	1433.28	716.64	77.89	.000
Error	16	147.22*	9.20		

*SSE_2

The sum of squares to test homogeneity of regression coefficients for the treatment groups is the difference in the experimental error sum of squares for the two models or

$$SS(\text{Homogeneity}) = SSE_1 - SSE_2 \qquad (17.12)$$

$$= 519.38 - 147.22$$

$$= 372.16$$

with $(N - t - 1) - (N - 2t) = (t - 1)$ or $17 - 16 = 1$ degrees of freedom. The F_0 statistic to test the null hypothesis of equal regression coefficients, H_0: $\beta_1 = \beta_2 = \cdots = \beta_t$ is

$$F_0 = \frac{MS(\text{Homogeneity})}{MSE_2} \qquad (17.13)$$

with critical value $F_{\alpha,(t-1),(N-2t)}$. The test for the exercise study is $F_0 = (372.16/1)/9.20 = 40.45$ exceeds $F_{.05,1,16} = 4.49$, and the null hypothesis of equal regression coefficients for the two cultural groups is rejected. The regression of gain in score on pre-test scores is different for the two cultural groups, and adjustment of the cultural group means to the same pre-test score with a common regression is not appropriate.

Many statistical programs are capable of directly computing the required sums of squares for homogeneity of regression coefficients in Equation (17.12) when specified as a *treatment by covariate* interaction effect in the program. The analysis of variance for the covariance model with a treatment by covariate interaction term is shown in Table 17.6. Note the sum of squares for the Groups × Regression interaction is equivalent to the sum of squares for homogeneity, Equation (17.12), derived from the alternative models with common and different regressions for the treatment groups.

Table 17.6 Analysis of covariance for score gain, including the interaction between the pre-test score covariate and cultural groups

Source of Variation	Degrees of Freedom	Sum of Squares	Mean Square	F	Pr > F
Group	1	610.24	610.24	66.32	.000
Regression	1	770.63	770.63	83.75	.000
Groups × Regression	1	372.16	372.16	40.45	.000
Error	16	147.22	9.20		

A Regression Coefficient Estimate for Each Treatment

If the regressions are significantly different among the treatment groups the least squares estimate of the regression coefficient for the ith treatment group is computed from the x_{ij} and y_{ij} values within the treatment group as

$$\widehat{\beta}_i = \frac{\sum_{j=1}^{r} (x_{ij} - \overline{x}_{i.})(y_{ij} - \overline{y}_{i.})}{\sum_{j=1}^{r} (x_{ij} - \overline{x}_{i.})^2} \tag{17.14}$$

The estimated regression equation for the ith treatment group is $\widehat{y}_i = \overline{y}_{i.} + \widehat{\beta}_i(x_{ij} - \overline{x}_{i.})$. The estimates of the regression coefficients for the two cultural groups and their standard errors (in parentheses) are $\widehat{\beta}_1 = -0.91(0.07)$ and $\widehat{\beta}_2 = -0.16(0.09)$. Both of these estimates are available from computer programs for linear models. The standard error estimate for the estimated regression coefficient, $\widehat{\beta}_i$, is the square root of $MSE/\sum_{j=1}^{r} (x_{ij} - \overline{x}_{i.})^2$.

The estimated regression equations for the two cultural groups are

$$\text{Group 1}: \quad \widehat{y}_1 = 55.8 - 0.91(x - 13.6)$$

$$\text{Group 2}: \quad \widehat{y}_2 = 47.8 - 0.16(x - 6.8)$$

The regression equations are plotted in Figure 17.4 along with the observations for each cultural group. Clearly, the regression lines are different from one another and a comparison of cultural group means adjusted to the overall pre-test mean of $\bar{x} = 51.8$ would be relatively meaningless.

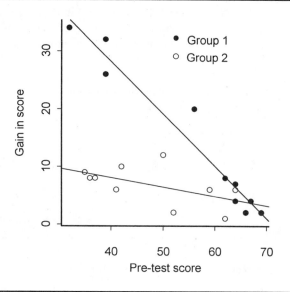

Figure 17.4 Regression between gain in score and pre-test scores with separate regressions for cultural groups

Rather, consider the regression lines as estimates of the simple effect of pre-test scores at each level of cultural group and make inferences on that basis. In both cultural groups the gain in score decreased with higher pre-test scores, but the rate of decrease was much greater with Group 1 than with Group 2. In fact a simple Student t test for $H_0 : \beta_2 = 0$ is $t_0 = -0.16/0.09 = -1.78$, which would lead to nonrejection of the null hypothesis; so no convincing evidence exists to indicate gain changes with pre-test scores in Group 2.

It is evident that the gain in scores from treatment in Group 1 are much greater when the subjects had low initial auditory discrimination and appear to benefit much more from training than do those in Group 2 with similar low initial auditory discrimination. When subjects from either group had high initial auditory discrimination on the pre-test the gain from training was relatively low or negligible.

Confidence intervals or tests of hypotheses between the estimated group means for any given value of pretest score $x = x_0$, say , $\widehat{y}_{1|x_0} - \widehat{y}_{2|x_0}$, can be constructed with the variance of the contrast as

$$s^2_{\widehat{y}_{1|x_0} - \widehat{y}_{2|x_0}} = MSE\left[\frac{1}{r_1} + \frac{1}{r_2} + \frac{(\overline{x}_0 - \overline{x}_{1.})^2}{\Sigma\,(x_{1j} - \overline{x}_{1.})^2} + \frac{(\overline{x}_0 - \overline{x}_{2.})^2}{\Sigma\,(x_{2j} - \overline{x}_{2.})^2}\right] \quad \textbf{(17.15)}$$

17.3 The Analysis of Covariance for Blocked Experiment Designs

The analysis of covariance can be applied to any experimental design with a straightforward extension of the principles applied to the completely randomized design in the previous section. However, a test for equality of regression coefficients among treatment groups is not possible with blocked designs unless there is more than one experimental unit for each treatment within each block. The analysis is illustrated with a randomized complete block design.

Example 17.3   Nutrient Availability Tests with Barley in the Greenhouse

Management methods on forest and range watersheds affect the nutrient status and availability of nutrients in any vegetation and soil-type complex. Knowledge of the soil, plant, and nutrient relationships is essential to properly manage watershed vegetation and soils.

The availability of certain soil nutrients in these watershed soils is evaluated by a pot culture technique in a greenhouse with barley plants. In principle, the method is based on the law of limiting factors. The test plants are grown in the soil fertilized to an optimum level and in the soil fertilized in identical fashion but without the nutrient in question. If the nutrient is deficient in the soil, the plants cultured in the complete nutrient soil will exhibit more plant growth than those cultured in the soil with the nutrient omitted from the fertilizer.

Research Objective: In one such study an investigator wanted to determine the availability of nitrogen and phosphorus in a watershed dominated by chaparral vegetation. He had collected soil samples from under the canopies of the dominant vegetation in the watershed, mountain mahogany, and composited the samples for a pot culture evaluation of nitrogen and phosphorus availability.

Treatment Design: Four nutrient treatments used for the study were (1) check, no fertilizer added; (2) full, a complete fertilizer; (3) N_0, nitrogen omitted from full, and (4) P_0, phosphorus omitted from full. The nutrient treatments were added as solutions to the soil, mixed, and placed in plastic pots in the greenhouse.

Experimental Design: The treatment pots were arranged on a greenhouse bench in a randomized complete block design to control experimental error variation caused by gradients in light and temperature in the greenhouse.

The barley plants were grown in the pots for seven weeks when plants were harvested, dried, and weighed. A leaf blight infected the plants part way through the experiment. It was assumed the blight would affect the growth of the plants and at the end of the experiment the percentage of the leaf area affected by the blight was measured in each container before the barley plants were harvested. The total dry weight of the barley plants and the percent leaf area affected with the blight is shown in Table 17.7 for each container in the experiment.

Table 17.7 Total dry matter y in grams and percent blighted leaf area x of barley plants

	Treatment							
	Check		*Full*		N_0		P_0	
Block	y	x	y	x	y	x	y	x
1	23.1	13	30.1	7	26.4	10	26.2	8
2	20.9	12	31.8	5	27.2	9	25.3	9
3	28.3	7	32.4	6	28.6	6	29.7	7
4	25.0	9	30.6	7	28.5	6	26.0	7
5	25.1	8	27.5	9	30.8	5	24.9	9
Mean	24.48	9.8	30.48	6.8	28.3	7.2	26.42	8.0

Source: Dr. J. Klemmedson, Renewable Natural Resources, University of Arizona.

The Linear Model for a Randomized Complete Block Design

The effects model for the randomized complete block design can be expressed as

$$y_{ij} = \mu + \tau_i + \rho_j + \beta(x_{ij} - \overline{x}_{..}) + e_{ij} \qquad (17.15)$$

$$i = 1, 2, \ldots, t \quad j = 1, 2, \ldots, r$$

where μ is the general mean, τ_i is the treatment effect, ρ_j is the block effect, β is the regression of y on x, and the e_{ij} are independent and normally distributed random errors with mean 0 and variance σ^2. It is further assumed that the covariate is unaffected by the treatments or blocks and the regression is the same for all treatments.

Alternative Models to Evaluate the Covariate Influence

Alternative full and reduced models are used to evaluate the influence of the covariate and also the significance of the treatment effects after adjustment for the covariate if necessary. The required models and their experimental error sums of squares are

- the full model, $y_{ij} = \mu + \tau_i + \rho_j + \beta(x_{ij} - \overline{x}_{..}) + e_{ij}$, and SSE_f with $(r-1)(t-1) - 1$ degrees of freedom

- the usual randomized complete block model with no covariate, $y_{ij} = \mu + \tau_i + \rho_j + e_{ij}$, and SSE_r with $(r-1)(t-1)$ degrees of freedom

- no treatment effects, $y_{ij} = \mu + \rho_j + \beta(x_{ij} - \overline{x}_{..}) + e_{ij}$, and SSE_r^* with $(N - r - 1)$ degrees of freedom

Finally, although not entirely necessary for the analysis, the adjusted sum of squares for blocks can be computed using the model with

- no block effects, $y_{ij} = \mu + \tau_i + \beta(x_{ij} - \overline{x}_{..}) + e_{ij}$, and SSE_r^{**} with $(N - t - 1)$ degrees of freedom

The SSE for each of the models fit to the barley data, Table 17.7, are $SSE_f = 12.577$, $SSE_r = 39.167$, $SSE_r^* = 37.437$, and $SSE_r^{**} = 20.363$.

Sum of Squares Partitions for the Analysis of Covariance

The sum of squares reduction upon adding the covariate x to the usual randomized complete block model is obtained as the difference

$$SS(\text{Covariate}) = SSE_r - SSE_f$$

$$= 39.167 - 12.577 = 26.590$$

with 1 degree of freedom. The adjusted treatment sum of squares after fitting the covariate and block effects is

$$SST(\text{adjusted}) = SSE_r^* - SSE_f$$

$$= 37.437 - 12.577 = 24.860$$

with $(t - 1) = 3$ degrees of freedom. The adjusted block sum of squares after fitting the covariate and treatment effects is

$$SSB(\text{adjusted}) = SSE_r^{**} - SSE_f$$

$$= 20.363 - 12.577 = 7.786$$

The analysis of covariance for the nutrient availability test with barley is shown in Table 17.8.

The significance of the covariate requires a test of the null hypothesis H_0: $\beta = 0$ with the statistic

Table 17.8 Analysis of covariance for dry matter production of barley plants with percent blight damaged leaf area as a covariate

Source of Variation	Degrees of Freedom	Sum of Squares	Mean Square	F	$Pr > F$
Regression	1	26.590	26.590	23.263	0.001
Block	4	7.786	1.947	1.703	0.219
Treatment	3	24.860	8.287	7.250	0.006
Error	11	12.577	1.143		

$$F_0 = \frac{MS(\text{Covariate})}{MSE} = \frac{26.590}{1.143} = 23.263$$

which is significant with $Pr > F = .001$ in Table 17.8. Thus, the relationship between the percent blight damaged leaf area and dry matter production of the barley plants is significant.

The estimate of the regression coefficient, which will be provided by most computer programs, is $\widehat{\beta} = -0.863$. The negative coefficient indicates the dry matter production decreases with an increase in the incidence of the disease on the plant.

The null hypothesis of no differences among the treatment means is tested with the statistic

$$F_0 = \frac{MST(\text{adjusted})}{MSE} = \frac{8.287}{1.143} = 7.25$$

and the null hypothesis is rejected with $Pr > F = .006$ in Table 17.8.

Adjusted Treatment Means and Their Standard Errors

The adjusted treatment means are calculated the same as for the completely randomized design with Equation (17.4) and are shown in Table 17.9 along with their standard errors. The standard errors shown in Equations (17.7) and (17.9) can be used for the adjusted treatment means with the sums of squares for treatments, $T_{xx} = 26.55$, and error, $E_{xx} = 35.70$, from the analysis of variance for the covariate, blight infection, using the data in Table 17.7. The estimated average standard error of the difference between two adjusted treatment means is

$$s_{(\bar{y}_{ia} - \bar{y}_{ja})} = \sqrt{\frac{2(1.143)}{5}\left[1 + \frac{26.55}{(3)35.7}\right]} = 0.76 \qquad \textbf{(17.16)}$$

Interpretations with Multiple Contrasts

The one-sided Dunnett 95% simultaneous confidence with the "Full" treatment as a control can be used to determine whether the soil was deficient in nutrients. The

Table 17.9 Adjusted means and their standard errors from the analysis of covariance for dry matter production of barley plants with percent blight damaged leaf area as a covariate

Treatment	Adjusted Mean	Standard Error
Check	26.08	0.58
Full	29.49	0.52
N_0	27.65	0.50
P_0	26.46	0.48

soil is deficient if the plant grown in the control or Full treatment exceeds that for any treatments deficient in one or more of the nutrients; therefore, if the upper bound of the interval for treatment *minus* control is negative the treatment is deficient in nutrients.

From Appendix Table VI the critical value of the Dunnett statistic for a one-sided interval is $d_{.05,3,11} = 2.31$. The Dunnett criterion is $D(3, .05) = 2.31(0.76) = 1.76$. The 95% SCI upper bounds for differences between the treatments and the control, Full, treatment are shown in Table 17.10.

Table 17.10   Upper bounds 95% simultaneous confidence intervals using the Dunnett method to compare the control treatment, Full, to other treatments

Treatment	Adjusted Mean $(\bar{y}_i)$	$(\bar{y}_i - \bar{y}_c)$	95% SCI Upper Bounds
Check	26.08	-3.41	-1.65
Full	$\bar{y}_c = 29.49$	$-$	$-$
N_0	27.65	-1.84	-0.08
P_0	26.46	-3.03	-1.27

The difference between the Check treatment with no added nutrients and the control treatment is $26.08 - 29.29 = -3.41$ with an upper bound, -1.65, on the one-sided confidence interval. Thus, the soil is deficient in some unspecified nutrients. Specific tests for deficiencies in nitrogen and phosphorus require the comparisons between the full treatment and the N_0 and P_0 treatments.   The difference between the nitrogen deficient treatment N_0 and the control treatment is $27.65 - 29.49 = -1.84$ with an upper bound of -0.08. The difference between the phosphorus deficient treatment P_0 and the control is $26.46 - 29.49 = -3.03$ with an upper bound of -1.27. Although the soil was deficient in both nitrogen and phosphorous, the upper bound for the phosphorous comparison was much farther removed from 0 than that for the nitrogen comparison, indicating a greater deficiency in phosphorous than in nitrogen.

17.4 Practical Consequences of Covariance Analysis

Practical application of the analysis of covariance has been demonstrated only with completely randomized and randomized complete block designs. However, the use of covariates can be extended to any treatment and experiment design as well as to comparative observational studies of complex structure and studies requiring the use of multiple covariates for adjustment. The objective in this chapter was to introduce the basic ideas underlying the use of additional information on basic units of the study. The specific manual formulae for the analysis will depend on the specific treatment and experiment design employed for the study. However, in all cases the use of full and reduced models, as illustrated with the two designs in this chapter, will enable an assessment of the covariates influence on the reduction of experimental error and the significance of adjusted treatment means.

Extensive discussions on the uses and misuses of covariates in research studies were provided in two special issues of *Biometrics* (1957), Volume 13, No. 3 and (1982), Volume 38, No. 3. Of particular interest are articles by Cochran (1957), Smith (1957), and Cox and McCullagh (1982). A number of issues arise relevant to the use of covariates. Among those concerns are the applicability in certain situations and the relationship between blocking and covariates.

Analysis of Covariance Combines the Features of Two Models

The analysis of covariance combines the features of models for the analysis of variance and regression to partition the total variation into components ascribable to (1) the treatment effects, (2) the effects attributable to any covariates, and (3) random experimental error as well as the variation associated with any design blocking factors. The basic intention is to compare treatments at a common value for the covariate.

When Covariates Are Superior to Blocks for Error Control

On the surface, covariance analysis appears to offer an alternative to blocking for reducing experimental error. Blocking designs restrict the number of criteria that reasonably can be used for local control. On the other hand, covariance permits the use of any number of factors thought necessary. Covariance also makes better use of exact values for quantitative factors, whereas blocking groups the same factors into classes of values. The advantage of covariates seems great when there are more than a few factors available as potential candidates for blocking variables.

When Blocks Are Superior to Covariates for Error Control

Covariance may be at a distinct disadvantage without blocking because random allocation of treatments to experimental units can result in an uneven distribution of treatments among the covariate classes. Any association between the covariate and the treatments confounds the effects of both on the response variable. Blocking on the values of the covariates distributes the treatments evenly among the covariate classes and avoids confounding the effects of the treatments and covariates.

Blocking is most effective with qualitative factors that produce recognizable variation among the experimental units. These factors include study management practices when tasks have to be performed by several technicians or on different days. Batches of raw materials provide natural effective qualitative blocking criteria.

Whenever the systematic differences are highly recognizable blocking is a most effective means of reducing experimental error variation that maintains the orthogonality required to avoid confounding the effects of the covariates with the treatments. In summary, blocking is recommended as a first course to reduce experimental error variance with adjustments on additional information if necessary with the analysis of covariance to further improve precision.

Comparative Observational Studies at a Disadvantage

The analysis of comparative observational studies can benefit equally from an analysis that includes covariates to reduce error variance and adjust group means for differences in their covariate values. Observational studies suffer from the disadvantage that units cannot be randomized to the defined treatment groups. The possibility exists for an influence on the response by additional unobserved covariates that are associated with the treatment groups, thus introducing an unknown bias into the group comparisons. Experimental studies have the advantage that the effects of these variables are distributed among the units by randomization and their influence is much less likely to be confounded with the effects of the treatments.

The Dangers of Extrapolation Beyond the Data

Finally, caution must be used when treatment means are adjusted to a common value for the covariate. Even though the regressions are parallel and there is no possibility that the treatments affect the covariate, the values of the covariate could be quite different for the treatment groups. If the covariate values are widely separated for the treatment groups, then adjustments would have to be extrapolated to a value of the covariate that is not common to either of the groups. An extreme example for illustration is a situation where income is used as a covariate for adjustment in comparing a group of corporate executives to entry-level clerical workers. Clearly, there would be no overlap of the income levels for the two groups. The adjustment would apply to the extrapolated region of an average income not included in either group and a comparison would be made between two groups in a nonsense setting. Even if the extrapolated adjustments were valid the standard errors of extrapolated values would be quite large.

EXERCISES FOR CHAPTER 17

1. An experiment was conducted on the shear strength of spot welds for three types of steel alloy. Six welds were made on each of the alloys and the force required to shear the weld was measured. The diameter of the weld was measured because it was believed that the strength of the weld was affected by its diameter. The data are shown in the table where y = weld strength and x = weld diameter.

Alloy	y	x	Alloy	y	x	Alloy	y	x
1	37.5	12.5	2	57.5	16.5	3	38.0	15.5
1	40.5	14.0	2	69.5	17.5	3	44.5	16.0
1	49.0	16.0	2	87.0	19.0	3	53.0	19.0
1	51.0	15.0	2	92.0	19.5	3	55.0	18.0
1	61.5	18.0	2	107.0	24.0	3	58.5	19.0
1	63.0	19.5	2	119.5	22.5	3	60.0	20.5

   a. Use weld diameter as a covariate for weld strength, and write a linear model for the experiment, identify each of the terms in the model, and state the assumptions for the model.

   b. Conduct the analysis of covariance, and test the significance of the covariate and adjusted treatment means.

   c. Compute the adjusted treatment means, their standard errors, and an average standard error of the difference between two adjusted means.

   d. Plot the regression line for each alloy showing the observed means and adjusted means for each alloy.

   e. Compute the efficiency of the covariance adjustment.

   f. Test the hypothesis of homogeneous regressions for each of the alloys.

   g. Discuss the results of the experiment and the effectiveness of the covariance adjustment.

   h. The significance level of the test for homogeneous regressions was $Pr > F = .068$. When the source of variation for Group × Regression is not significant we end up with the sum of squares for error in the regular analysis of covariance assuming homogeneous regression coefficient. Effectively we are pooling the Group × Regression sum of squares partition with the sum of squares for error from the analysis of covariance for the model with separate regression coefficients for each treatment group. Hendrix et al. (1982) suggested in that case we should use a significance level of $\alpha = .20$ or $.25$ since it was similar to the problem of testing incompletely specified models as discussed by Bozivich, Bancroft, and Hartley (1956). What do you think about this strategy?

   i. Suppose you subscribe to the philosophy of a significance level of $\alpha = .20$ or $.25$ when testing the null hypothesis of equal regression coefficients for all treatment groups. The null hypothesis in part (f) is then rejected. Conduct the analysis of covariance with separate regression estimates for each treatment group, and compute the estimated regression coefficients and their standard errors for each alloy. What is your inference from the study at this point?

2. A nutrition scientist conducted an experiment to evaluate the effects of four vitamin supplements on the weight gain of laboratory animals. The experiment was conducted in a completely

randomized design with five separately caged animals for each treatment. The caloric intake will differ among animals and influence weight gain so the investigator measured the caloric intake of each animal. The data on weight gain (y = grams) and calorie intake (x = calories/10) are shown in the table.

Diet	y	x	Diet	y	x	Diet	y	x	Diet	y	x
1	48	35	2	65	40	3	79	51	4	59	53
1	67	44	2	49	45	3	52	41	4	50	52
1	78	44	2	37	37	3	63	47	4	59	52
1	69	51	2	73	53	3	65	47	4	42	51
1	53	47	2	63	42	3	67	48	4	34	43

a. Determine whether diet influenced calorie intake to the extent that the latter would be invalidated as a covariate.

b. Use calorie intake as a covariate for weight gain, and write a linear model for the experiment. Identify each of the terms in the model, and state the assumptions for the model.

c. Conduct the analysis of covariance, and test the significance of the covariate and adjusted treatment means.

d. Compute the adjusted treatment means, their standard errors, and an average standard error of the difference between two adjusted means.

e. Plot the regression line for each diet showing the observed means and adjusted means for each diet.

f. Compute the efficiency of the covariance adjustment.

g. Test the hypothesis of homogeneous regressions for each of the diets.

h. Discuss the results of the experiment and the effectiveness of the covariance adjustment.

3. A plant scientist conducted an experiment to study the effects of drip irrigation water level on sweet corn growth, yield, and quality. Three levels of irrigation were used in the experiment (15.8, 24.0, and 28.5 inches of water applied), and the experiment was arranged in a randomized complete block design to control for soil variability in the field. One of the response variables measured was the weight of culls per plot or the amount of sweet corn in the plot that was unsuitable for market. The number of plants per plot varied and this would affect the crop yield on the plot. Since the soil moisture was optimized to establish the crop stand the variation in plants per plot was not affected by the irrigation levels imposed after the crop was established. The observed yield of culls (y = metric tons/hectare) and x = plants in a 40-foot section of row are shown in the table.

	Irrigation Level					
	1		2		3	
Block	y	x	y	x	y	x
1	1.5	45	1.9	54	1.1	43
2	3.1	58	1.8	57	1.8	60
3	3.8	61	2.9	55	3.7	71
4	3.3	59	2.3	56	1.8	48

a. Use plants-per-40-feet of row as a covariate for yield of culls, and write a linear model for the experiment, identify each of the terms in the model, and state the assumptions for the model.
b. Conduct the analysis of covariance, and test the significance of the covariate and adjusted treatment means.
c. Show the value of the regression coefficient. Compute the adjusted treatment means, their standard errors, and an average standard error of the difference between two adjusted means.
d. Compute the efficiency of the covariance adjustment.
e. Discuss the results of the experiment and the effectiveness of the covariance adjustment.

4. An experiment was conducted in a randomized complete block design to study the effect of natural control, *Bacillus*, and a standard chemical insecticide for control of hornworm infestations on a crop plant. The treatments included four sources of *Bacillus* (Treatments 1–4), a standard chemical treatment (Treatment 5), and a control of no treatment (Treatment 6). The treatments were applied to plants grown in field plots in the field. The number of hornworms (*count*) on each plant were counted prior to treatment. The number of live hornworms (*live*) were counted 20 hours after application of the treatment. The data for each plot are shown in the table.

	Block							
	1		2		3		4	
Treatment	Count	Live	Count	Live	Count	Live	Count	Live
Bacillus 1	15	17	25	26	18	21	23	26
Bacillus 2	19	18	21	22	20	19	19	20
Bacillus 3	19	19	19	21	21	23	25	22
Bacillus 4	22	14	31	26	17	17	19	19
Chemical	17	5	22	6	26	13	18	10
Control	22	25	14	19	22	26	23	27

a. Use the count of hornworms before treatment as a covariate for the number of live hornworms 20 hours after treatment, and write a linear model for the experiment, identify each of the terms in the model, and state the assumptions for the model.
b. Conduct the analysis of covariance, and test the significance of the covariate and adjusted treatment means.
c. Show the value of the regression coefficient. Compute the adjusted treatment means, their standard errors, and an average standard error of the difference between two adjusted means.
d. Compute the efficiency of the covariance adjustment.
e. Discuss the results of the experiment and the effectiveness of the covariance adjustment.
f. The response variable is a count measure and likely does not have a normal distribution. Did you check the assumptions of homogeneous variance and normal distributions of experimental errors for the model? If not, do so now, and if necessary take corrective actions (according to discussions in Chapter 4) and repeat parts (b) through (e).

5. The analysis of covariance can be used to estimate a missing value in a blocked design and provide an unbiased estimate of the treatment sum of squares to test the hypothesis of no differences among the treatment means with a missing value (Coons, 1957).

A covariate x is introduced for the missing value. The covariate is assigned the value $x = -1$ for the missing y value and $x = 0$ for all other values of y that are not missing. A value of $y = 0$ is assigned to the missing value. Compute a regular analysis of covariance with the values assigned to y and the covariate x. The estimate of the missing value is the estimate of the regression coefficient $\hat{\beta} = E_{xy}/E_{xx}$, and the adjusted treatment sum of squares in the analysis of covariance is the correct sum of squares to test the null hypothesis of no differences among the treatment means.

As a demonstration of the technique use the data for the randomized complete block design in Exercise 17.3 and ignore the covariate, $x =$ number of plants, for this exercise. Assume the observation $y = 2.9$ on irrigation level 2 in block 3 is missing and assign it the value $y = 0$. Construct the new covariate with $x = -1$ for the missing value and $x = 0$ for all other values of y, and conduct the analysis of covariance. Estimate the missing value, and use the adjusted mean square for treatments to test the hypothesis of no difference among the treatment means.

6. Describe a study in your own field in which a covariate is used in addition to or in place of a blocking criterion to reduce the error variance and adjust treatment means. Provide justification for the use of the covariate values directly in the statistical model in place of blocking on the covariate, and justify its use on the basis that it is not affected by the treatments. Use an example from a journal article or your own research experiences if possible.

References

Abeyasekera, S., and Curnow, R. N. (1984). "The desirability of adjusting for residual effects in a crossover design." *Biometrics 40*, 1071–1078.

Addelman, S. (1970). "Variability of treatments and experimental units in the design and analysis of experiments." *Journal of the American Statistical Association 65*, 1095–1108.

Anderson, R. L. (1960). "Uses of variance component analysis in the interpretation of biological experiments." *Bulletin of the International Statistical Institute 37*, 71–90.

Anderson, V. L., and McLean, R. A. (1974). *Design of experiments: A realistic approach.* New York: Marcel Dekker.

Anscombe, F. J. (1948). "The transformation of Poisson, binomial, and negative-binomial data." *Biometrika 35*, 246–254.

Bailey, R. A. (1986). "Randomization, constrained." In *Encyclopedia of Statistical Sciences,* vol. 7, 524–530. S. Kotz and N. L. Johnson (eds.). New York: Wiley.

Bailey, R. A. (1987). "Restricted randomization. A practical example." *Journal of the American Statistical Association 82*, 712–719.

Bainbridge, T. R. (1963). "Staggered, nested designs for estimating variance components." *ASQC Annual Convention Transactions,* 93–103.

Bartlett, M. S. (1947). "The use of transformations." *Biometrics 3*, 39–52.

Bennett, C. A., and Franklin, N. L. (1954). *Statistical analysis in chemistry and the chemical industry.* New York: Wiley.

Berk, K. N., and Picard, R. R. (1991). "Significance tests for saturated orthogonal arrays." *Journal of Quality Technology 23*, 79–89.

Bicking, C. A. (1954). "Some uses of statistics in the planning of experiments." *Industrial Quality Control 10*, 20–24.

Biometrics (1957). "Special Issue on the Analysis of Covariance." *Biometrics 13*, 261–405.

Biometrics (1982). "Special Issue: Analysis of Covariance." *Biometrics 38*, 540–753.

Boik, R. J. (1981). "A priori tests in repeated measures designs: Effects of non-sphericity." *Psychometrika* 46, 241–255.

Bose, R. C., Clatworthy, W. H., and Shrikhande, S. S. (1954). "Tables of partially balanced designs with two associate classes." North Carolina Agricultural Experiment Station. *Technical Bulletin*, no. 107.

Bose, R. C., and Nair, K. R. (1939). "Partially balanced incomplete block designs." *Sankhya* 4, 337–372.

Box, G. E. P. (1854a). "Some theorems on quadratic forms applied in the study of analysis of variance problems, I. Effect of inequality of variance in the one-way classification." *Annals of Mathematical Statistics* 25, 290–302.

Box, G. E. P. (1954b). "Some theorems on quadratic forms applied in the study of analysis of variance problems, II. Effect of inequality of variances and correlation between errors in the two-way classification." *Annals of Mathematical Statistics* 25, 484–498.

Box, G. E. P., and Behnken, D. W. (1960). "Some new three level designs for the study of quantitative variables." *Technometrics* 2, 455–476.

Box, G. E. P., and Cox, D. R. (1964). "An analysis of transformations." *Journal of the Royal Statistical Society, B* 26, 211–243.

Box, G. E. P., and Draper, N. R. (1987). *Empirical model-building and response surfaces.* New York: Wiley.

Box, G. E. P., and Hunter, J. S. (1957). "Multifactor experimental designs for exploring response surfaces." *Annals of Mathematical Statistics* 28, 195–241.

Box, G. E. P., Hunter, W. G, and Hunter, J. S. (1978). *Statistics for experimenters. An introduction to design, data analysis, and model building.* New York: Wiley.

Box, G. E. P., and Meyer, R. D. (1986). "An analysis for unreplicated fractional factorials." *Technometrics* 28, 11–18.

Box, G. E. P., and Wilson, K. G. (1951). "On the experimental attainment of optimum conditions." *Journal of the Royal Statistical Society, B* 13, 1–45.

Box, J. F. (1978). *R. A. Fisher, the life of a scientist.* New York: Wiley.

Box, M. J., and Draper, N. R. (1974). "On minimum-point second-order designs." *Technometrics* 16, 613–616.

Bozivich, H., Bancroft, T. A., and Hartley, H. O. (1956). Power of analysis of variance test procedures for incompletely specified models. *Annals of Mathematical Statistics* 27, 1017–1043.

Bradu, D., and Gabriel, K. R. (1978). "The biplot as a diagnostic tool for models of two-way tables." *Technometrics* 20, 47–68.

Brown, M. B., and Forsythe, A. B. (1974). "Robust tests for the equality of variances." *Journal of the American Statistical Association* 69, 364–367.

Carmer, S. G., Nyquist, W. E., and Walker, W. M. (1989). "Least significant differences for combined analyses of experiments with two- or three-factor treatment designs." *Agronomy Journal* 81, 665–672.

Carmer, S. G., and Swanson, M. R. (1973). "An evaluation of ten pairwise multiple comparison procedures by Monte Carlo methods." *Journal of the American Statistical Association* 68, 66–74.

Carmer, S. G., and Walker, W. M. (1985). "Pairwise multiple comparisons of treatment means in agronomic research." *Journal of Agronomic Education* 14, 19–26.

Clatworthy, W. H. (1973). "Tables of two-associate-class partially balanced designs." National Bureau of Standards, *Applied Mathematics Series*, no. 63. Washington, D.C.

Cleveland, W. S. (1993). *Visualizing data.* Summit, N.J.: Hobart Press.

Cochran, W. G. (1947). "Some consequences when the assumptions for the analysis of variance are not satisfied." *Biometrics* 3, 22–38.

Cochran, W. G. (1957). "Analysis of covariance: Its nature and uses." *Biometrics* 13, 261–281.

Cochran, W. G. (1965). *Sampling techniques,* 2d ed. New York: Wiley.

Cochran, W. G. (1983). *Planning and analysis of observational studies,* L. E. Moses and F. Mosteller (eds.). New York: Wiley.

Cochran, W. G., and Cox, G. M. (1957). *Experimental designs,* 2d ed. New York: Wiley.

Collett, D. (1991). *Modelling binary data.* London: Chapman and Hall.

Conner, W. S., and Zelen, M. (1959). "Fractional factorial experimental designs for factors at three levels." National Bureau of Standards, *Applied Mathematics Series*, no. 54. Washington, D.C.

Conover, W. J., Johnson, M. E., and Johnson, M. M. (1981). "A comparative study of tests for homogeneity of variances, with applications to the outer continental shelf bidding data." *Technometrics* 23, 351–361.

Cook, R. D., and Weisberg, S. (1982). *Residuals and influence in regression.* New York: Chapman and Hall.

Coons, I. (1957). "The analysis of covariance as a missing plot technique." *Biometrics* 13, 387–405.

Coons, J. M., Kuehl, R. O., Oebker, N. F., and Simons, N. R. (1989). "Germination of eleven tomato phenotypes at constant or alternating high temperatures." *HortScience* 24(6), 927–930.

Cornell, J. A. (1990). *Experiments with mixtures: Designs, models, and the analysis of mixture data,* 2d ed. New York: Wiley.

Cox, D. R. (1958). *Planning of experiments.* New York: Wiley.

Cox, D. R. (1970). *Analysis of binary data.* London: Chapman and Hall.

Cox, D. R., and McCullagh, P. (1982). "Some aspects of analysis of covariance." *Biometrics* 38, 541–561.

Daniel, C. (1959). "Use of half-normal plots in interpreting factorial two-level experiments." *Technometrics* 1, 311–342.

Daniel, C. (1976). *Applications of statistics to industrial experimentation.* New York: Wiley.

Daniel, C. (1978). "Patterns in residuals in the two-way layout." *Technometrics* 20, 385–395.

Diamond, W. J. (1989). *Practical experiment designs for engineers and scientists,* 2d ed. New York: Van Nostrand Reinhold.

Diggle, P. J., Liang, K., and Zeger, S. L. (1994). *Analysis of longitudinal data.* Oxford: Oxford University Press.

Dobson, A. J. (1990). *An introduction to generalized linear models.* London: Chapman and Hall.

Draper, N. R. (1985). "Small composite designs." *Technometrics* 27, 173–180.

Draper, N. R., and Lin, D. K. J. (1990). "Small response-surface designs." *Technometrics* 32, 187–194.

Dunnett, C. W. (1955). "A multiple comparisons procedure for comparing several treatments with a control." *Journal of the American Statistical Association* 509, 1096–1121.

Dunnett, C. W. (1964). "New tables for multiple comparisons with a control." *Biometrics* 20, 482–491.

Edgington, E. S. (1987). *Randomization tests,* 2d ed. New York: Marcel Dekker.

Einot, I., and Gabriel, K. R. (1975). "A study of the powers of several methods of multiple comparisons." *Journal of the American Statistical Association* 70, 574–583.

Eisenhart, C. (1947). "The assumptions underlying the analysis of variance." *Biometrics* 3, 1–21.

Federer, W. T. (1955). *Experimental design: Theory and application.* New York: Macmillan.

Federer, W. T. (1975). "The misunderstood split-plot." In *Applied Statistics*, R. P. Gupta (ed.). Amsterdam: North Holland.

Finner, H. (1990). "On the modified S-method and directional errors." *Communications in Statistics—Theory and Methods*, A19, 41–53.

Finney, D. J. (1946). "Standard errors of yields adjusted for regression on an independent measurement." *Biometrics* 2, 53–55.

Finney, D. J. (1978). *Statistical methods in biological assay,* 3d ed. London: Griffin.

Fisher, R. A. (1925). *Statistical methods for research workers.* Edinburgh: Oliver and Boyd.

Fisher, R. A. (1926). "The arrangement of field experiments." *Journal of the Ministry of Agriculture of Great Britain* 33, 503–513.

Fisher, R. A. (1935). *The design of experiments.* Edinburgh: Oliver and Boyd.

Fisher, R. A. (1960). *The design of experiments,* 7th ed. New York: Hafner.

Fisher, R. A., and Yates, F. (1963). *Statistical tables for biological, agricultural and medical research,* 6th ed. Edinburgh: Oliver and Boyd.

Fleiss, J. L. (1981). *Statistical methods for rates and proportions,* 2d ed. New York: Wiley.

Fleiss, J. L. (1986). *The design and analysis of clinical experiments.* New York: Wiley.

Gates, C. E. (1995). "What really is experimental error in block designs?" *The American Statistician* 49, 362–363.

Gates, C. E., and Shiue, C. (1962). "The analysis of variance of the S-stage hierarchical classification." *Biometrics* 18, 529–536.

Gaylor, D. W., and Hopper, F. N. (1969). "Estimating the degrees of freedom for linear combinations of mean squares by Satterthwaite's formula." *Technometrics* 11, 691–706.

Gill, J. L. (1978). *Design and analysis of experiments in the animal and medical sciences.* Vol. 1. Ames, Iowa: Iowa State University Press.

Goldsmith, C. H., and Gaylor, D. W. (1970). "Three stage nested designs for estimating variance components." *Technometrics* 12, 487–498.

Grandage, A. (1958). "Orthogonal coefficients for unequal intervals. Query 130." *Biometrics* 14, 287–289.

Graybill, F. A. (1961). *An introduction to linear statistical models.* Vol. 1. New York: McGraw-Hill.

Graybill, F. A. (1983). *Matrices with applications in statistics,* 2d ed. Belmont, CA: Wadsworth.

Greenberg, B. G. (1951). "Why randomize?" *Biometrics* 7, 309–322.

Greenhouse, S. W., and Geisser, S. (1959). "On methods in the analysis of profile data." *Psychometrika,* 24, 95–112.

Grizzle, J. E. (1965). "The two-period change-over design and its use in clinical trials." *Biometrics* 21, 467–480.

Grizzle, J. E. (1974). "Corrigenda to Grizzle (1965)." *Biometrics* 30, 727.

Haaland, P. D. (1989). *Experimental design in biotechnology.* New York: Marcel Dekker.

Hader, R. J. (1973). "An improper method of randomization in experimental design." *The American Statistician* 27, 82–84.

Harshbarger, B. (1949). "Triple rectangular lattices." *Biometrics* 5, 1–13.

Harshbarger, B., and Davis, L. L. (1952). "Latinized rectangular lattices." *Biometrics* 8, 73–84.

Hartley, H. O. (1950). "The maximum F-ratio as a shortcut test for heterogeneity of variance." *Biometrika* 37, 308–312.

Hartley, H. O., and Searle, S. R. (1969). "A discontinuity in mixed-model analysis." *Biometrics* 25, 573–576.

Hartley, H. O., and Rao, J. N. K. (1967). "Maximum likelihood estimation for the mixed analysis of variance model." *Biometrika* 54, 93–108.

Hayter, A. J. (1984). "A proof of the conjecture that the Tukey–Kramer multiple comparisons procedure is conservative." *Annals of Statistics* 12, 61–75.

Hayter, A. J. (1986). "The maximum familywise error rate of Fisher's least significant difference test." *Journal of the American Statistical Association* 81, 1000–1004.

Hayter, A. J. (1990). "A one-sided Studentized range test for testing against a simple order alternative." *Journal of the American Statistical Association* 85, 778–785.

Hayter, A. J., and Liu, W. (1996). "Exact calculations for the one-sided Studentized range test for testing against a simple ordered alternative." *Computational Statistics & Data Analysis* 22, 17–25.

Henderson, C. R. (1953). "Estimation of variance and covariance components." *Biometrics* 9, 226–252.

Hendrix, L. J., Carter, M. W., and Scott, D. T. (1982). Covariance analysis with heterogeneity of slopes in fixed models. *Biometrics* 38, 641–650.

Hering, F. and Mejza, S. (1997). "Incomplete split-block designs." *Biometrical Journal* 39, 227–238.

Hicks, C. R. (1973). *Fundamental concepts in the design of experiments,* 2d ed. New York: Holt, Rinehart & Winston.

Hills, M., and Armitage, P. (1979). "The two-period cross-over clinical trial." *British Journal of Clinical Pharmacology* 8, 7–20.

Hinkelmann, K., and Kempthorne, O. (1994). *Design and Analysis of Experiments.* New York: Wiley.

Hochberg, Y., and Tamhane, A. C. (1987). *Multiple comparison procedures.* New York: Wiley.

Hocking, R. R. (1973). "A discussion of the two-way mixed model." *American Statistician* 27, 148–152.

Hocking, R. R. (1985). *The analysis of linear models.* Monterey, Calif.: Brooks/Cole.

Hocking, R. R., and Speed, F. M. (1975). "A full rank analysis of some linear model problems." *Journal of the American Statistical Association* 70, 706–712.

Hsu, J. C. (1984). "Constrained simultaneous confidence intervals for multiple comparisons with the best." *Annals of Statistics* 12, 1136–1144.

Hsu, J. C. (1996). *Multiple comparisons. Theory and methods.* London: Chapman and Hall.

Hung, H. M., O'Neill, R. T., Bauer, P., and Köhne, K. (1977). "The behavior of the *P*-value when the alternative hypothesis is true." *Biometrics* 53, 11–22.

Hunter, J. S. (1985). "Statistical design applied to product design." *Journal of Quality Technology* 17, 210–221.

Hunter, J. S. (1989). "Let's all beware the Latin Square." *Quality Engineering* 1(4), 453–465.

Hurlbert, S. H. (1984). "Pseudoreplication and the design of ecological field experiments." *Ecological Monographs* 54, 187–211.

Huynh, H. (1978). "Some approximate tests for repeated measurement designs." *Psychometrika* 43, 161–175.

Huynh, H., and Feldt, L. S. (1970). "Conditions under which mean square ratios in repeated measurements designs have exact F–distributions." *Journal of the American Statistical Association* 65, 1582–1589.

Huynh, H., and Feldt, L. S. (1976). "Estimation of the Box correction for degrees of freedom from sample data in the randomized block and split-plot designs." *Journal of Educational Statistics* 1, 69–82.

Ito, P. K. (1980). "Robustness of ANOVA and MANOVA test procedures." In *Handbook of Statistics*, Vol. 1. P. R. Krishnaiah (ed.), 199–236. Amsterdam: North-Holland.

Jarrett, R. G. (1978). "The analysis of designed experiments with missing observations." *Applied Statistics* 27, 38–46.

John, J. A. (1966). "Cyclic incomplete block designs." *Journal of the Royal Statistical Society, B* 28, 345–360.

John, J. A. (1987). *Cyclic designs.* London: Chapman and Hall.

John, J. A., and Williams, E. R. (1995). *Cyclic and computer-generated designs,* 2d ed., London: Chapman and Hall.

John, P. W. M. (1971). *Statistical design and analysis of experiments.* New York: Macmillan.

Johnson, D., and Graybill, F. (1972). "An analysis of a two-way model with interaction and no replication." *Journal of the American Statistical Association* 67, 862–868.

Johnson, N. R., and Leone, F. C. (1977). *Statistics and experimental design in engineering and the physical sciences,* Vol. II. New York: Wiley.

Jones, B., and Kenward, M. G. (1989). *Design and analysis of cross-over trials.* London: Chapman and Hall.

Jones, D. (1984). "Use, misuse, and role of multiple-comparison procedures in ecological and agricultural entomology." *Environmental Entomology* 13, 635–649.

Kempthorne, O. (1952). *The design and analysis of experiments.* New York: Wiley.

Kempthorne, O. (1966). "Some aspects of experimental inference." *Journal of the American Statistical Association* 61, 11–34.

Kempthorne, O. (1975). "Inference from Experiments and Randomization." In *A Survey of Statistical Design and Linear Models*, 303–331. J. N. Srivastava (ed.). Amsterdam: North Holland.

Keuls, M. (1952). "The use of the Studentized range in connection with an analysis of variance." *Euphytica* 1, 112–122.

Kish, L. (1987). *Statistical design for research.* New York: Wiley.

Koch, G. G. (1983). "Intraclass Correlation Coefficient." In *Encyclopedia of Statistical Sciences,* Vol. 4, 212–217. S. Kotz and N. L. Johnson (eds.). New York: Wiley.

Kramer, C. Y. (1956). "Extension of multiple range tests to group means with unequal numbers of replications." *Biometrics* 12, 309–310.

Lenth, R. V. (1989). "Quick and easy analysis of unreplicated factorials." *Technometrics* 31, 469–473.

Lentner, M., Arnold, J. C., and Hinkelmann, K. (1989). "The efficiency of blocking: How to use MS(Blocks)/MS(Error) correctly." *The American Statistician* 43, 106–108.

Leone, F. C., Nelson, L. S., Johnson, N. L., and Eisenstat, S. (1968). "Sampling distributions of variance components. II. Empirical studies of unbalanced nested designs." *Technometrics* 10, 719–737.

Levene, H. (1960). "Robust tests for equality of variances." In *Contributions to Probability and Statistics*, 278–292. I. Olkin (ed.). Stanford, Calif.: Stanford University Press.

Little, R. J. A., and Rubin, D. B. (1987). *Statistical analysis with missing data.* New York: Wiley.

Lucas, H. L. (1957). "Extra-period Latin-square change-over designs." *Journal of Dairy Science* 40, 225–239.

Mandel, J. (1971). "A new analysis of variance model for non-additive data." *Technometrics* 13, 1–18.

Manly, B. F. J. (1991). *Randomization and Monte Carlo methods in biology.* London: Chapman and Hall.

Marvel, J. T., Berry, J. W., Kuehl, R. O., and Deutschmann, A. J. (1969). "Vinylation of Methyl a–D–Glucopyranoside." *Carbohydrate Research* 9, 295–303.

Mason, R. L., Gunst, R. F., and Hess, J. L. (1989). *Statistical design and analysis of experiments with applications to engineering and science.* New York: Wiley.

Mauchly, J. W. (1940). "Significance test for sphericity of a normal n–variate distribution." *Annals of Mathematical Statistics* 11, 204–209.

McCullagh, P., and Nelder, J. A. (1989). *Generalized linear models,* 2d ed. London: Chapman and Hall.

McIntosh, M. S. (1983). "Analysis of combined experiments." *Agronomy Journal* 75, 153–155.

Mead, R. (1988). *The design of experiments. Statistical principles for practical application.* Cambridge, UK: Cambridge University Press.

Mejza, I., and Mejza, S. (1996). "Incomplete split-plot designs generated by GDPBIBD(2)." *Calcutta Statistical Association Bulletin* 46, 117–127.

Milliken, G. A., and Johnson, D. E. (1984). *Analysis of messy data. Vol. I: Designed experiments.* Belmont, Calif.: Lifetime Learning Publications.

Montgomery, D. C. (1997). *Design and analysis of experiments*, 4th ed. New York: Wiley.

Moon, T. E., Levine, N., Cartmel, B., Bangert, J., Rodney, S., Schreiber, M., Peng, Y., Ritenbaugh, C., Meyskens, F., Alberts, D., and the Southwest Skin Cancer Prevention Study Group. (1995). "Design and recruitment for retinoid skin cancer prevention (SKICAP) trials." *Cancer Epidemiology, Biomarkers, and Prevention* 4, 661–669.

Mosteller, F., and Tukey, J. W. (1977). *Data analysis and regression. A second course in statistics.* Reading, Mass.: Addison-Wesley.

Myers, R. H., and Montgomery, D. C. (1995). *Response surface methodology. Process and product optimization using designed experiments.* New York: Wiley.

Myers, R. H., Vining, G. G., Giovannitti-Jensen, A., and Myers, S. L. (1992). "Variance dispersion properties of second-order response surface designs." *Journal of Quality Technology* 24, 1–11.

Nelder, J. A., and Wedderburn, R. W. M. (1972). "Generalized linear models." *Journal of the Royal Statistical Society* A 135, 370–384.

Nelson, L. A., and Rawlings, J .O. (1983). "Ten common misuses of statistics in agronomic research and reporting." *Journal of Agronomic Education* 12, 100–105.

Newman, D. (1939). "The distribution of the range in samples from a normal population, expressed in terms of an independent estimate of standard deviation." *Biometrika* 31, 20–30.

Notz, W. (1982). "Minimal-point second-order designs." *Journal of Statistical Planning and Inference* 6, 47–58.

Ostle, B., and Mensing, R. W. (1975). *Statistics in research,* 3d ed. Ames, Iowa: The Iowa State University Press.

Patterson, H. D. (1951). "Change-over trials." *Journal of the Royal Statistical Society, B* 13, 256–271.

Patterson, H. D. (1952). "The construction of balanced designs for experiments involving sequences of treatments." *Biometrika* 39, 32–48.

Patterson, H. D., and Lucas, H. L. (1962). "Change–over designs." North Carolina Agricultural Experiment Station *Technical Bulletin*, no. 147.

Patterson, H. D., and Thompson, R. (1971). Recovery of interblock information when block sizes are unequal. *Biometrika* 58, 545–554.

Patterson, H. D., and Williams, E. R. (1976). "A new class of resolvable incomplete block designs." *Biometrika* 63, 83–92.

Patterson H. D., Williams, E. R., and Hunter, E. A. (1978). "Block designs for variety trials." *Journal of Agricultural Science* 90, 395–400.

Petersen, R. G. (1985). *Design and analysis of experiments*. New York: Marcel Dekker.

Pignatiello, J. J., and Ramberg, J. S. (1985). *Journal of Quality Technology* 17, 198–206. Discussion of article by R. N. Kackar (1985). "Off-line quality control, parameter design, and the Taguchi method." *Journal of Quality Technology* 17, 176–188.

Pignatiello, J. J., and Ramberg, J. S. (1991). "Top ten triumphs and tragedies of Genichi Taguchi." *Quality Engineering* 4(2), 211–225.

Plackett, R. L., and Burman, J. P. (1946). "The design of optimum multifactorial experiments." *Biometrika* 33, 305–325.

Preece, D. A. (1983). "Latin squares, Latin cubes, Latin rectangles, etc." In *Encyclopedia of Statistical Sciences*, Vol. 4, 504–510. S. Kotz and N. L. Johnson (eds.). New York: Wiley.

Ratkowsky, D. A., Alldredge, J. R., and Cotton, J. W. (1990). "Analyzing balanced or unbalanced Latin squares and other repeated measures designs for carryover effects using the GLM procedure." *SAS Users Group International, 15th Annual Conference Proceedings*, 1353–1358.

Ratkowsky, D. A., Evans, M. A., and Alldredge, J. R. (1993). *Cross-over experiments: Design, analysis, and application*. New York: Marcel Dekker.

Rawlings, J. O. (1988). *Applied regression analysis: A research tool*. Pacific Grove, Calif.: Brooks/Cole.

Robinson, J. (1967). "Incomplete split-plot designs." *Biometrics* 23, 793–802.

Roquemore, K. G. (1976). "Hybrid designs for quadratic response surfaces." *Technometrics* 18, 419–423.

Satterthwaite, F. E. (1946). "An approximate distribution of estimates of variance components." *Biometrics* 2, 110–114.

Saville, D. J. (1990). "Multiple comparison procedures: The practical solution." *The American Statistician* 44, 174–180.

Scheffé, H. (1953). "A method for judging all contrasts in the analysis of variance." *Biometrika* 40, 87–104.

Scheffé, H. (1959). *The analysis of variance*. New York: Wiley.

Searle, S. R. (1971). *Linear models*. New York: Wiley.

Searle, S. R., Casella, G., and McCulloch, C. E. (1992). *Variance components*. New York: Wiley.

Searle, S. R. (1987). *Linear models for unbalanced data*. New York: Wiley.

Searle, S. R., Speed, F. M., and Henderson, H. V. (1981). "Some computational and model equivalencies in analysis of variance of unequal-subclass-numbers data." *The American Statistician* 35, 16–33.

Smith, H. F. (1957). "Interpretation of adjusted treatment means and regressions in analysis of covariance." *Biometrics* 13, 282–308.

Smith, J. R., and Beverly, J. M. (1981). "The use and analysis of staggered nested factorial designs." *Journal of Quality Technology* 13, 166–173.

Speed, F. M., and Hocking, R. R. (1976). "The use of the R( )-notation with unbalanced data." *The American Statistician* 30, 30–33.

Speed, F. M., Hocking, R. R., and Hackney, O. P. (1978). "Methods of analysis of linear models with unbalanced data." *Journal of the American Statistical Association* 73, 105–112.

Steel, R. G. D., and Torrie, J. H. (1980). *Principles and procedures statistics: A biometrical approach*, 2d ed. New York: McGraw Hill.

Taguchi, G. (1986). *Introduction to quality engineering.* Tokyo: Asian Productivity Organization.

Tanur, J. F., Mosteller, F., Kruskal, W., Link, R., Pietero, R., Rising, G., and Lehman, E. (1978). *Statistics: A guide to the unknown.* San Francisco: Holden-Day.

Thompson, W. A., Jr. (1962). The problem with negative estimates of variance components. *Annals of Mathematical Statistics* 33, 273–289.

Tukey, J. W. (1949a). "Comparing individual means in the analysis of variance." *Biometrics* 5, 99–114.

Tukey, J. W. (1949b). "One degree of freedom for non-additivity." *Biometrics* 5, 232–242.

Tukey, J. W. (1955). "Answer to a query on non-additivity in Latin squares." *Biometrics* 11, 111–113.

Tukey, J. W. (1977). *Exploratory data analysis.* Reading, Mass.: Addison-Wesley.

Tukey, J. W. (1994). "The problem of multiple comparisons." In H. L. Braun (ed.), *The collected works of John W. Tukey.* Vol. VIII, Chapter 1, pp. 1–300. New York: Chapman and Hall.

Urquhart, N. S., and Weeks, D. L. (1978). "Linear models in messy data: Some problems and alternatives." *Biometrics* 34, 696–705.

Urquhart, N. S., Weeks, D. L., and Henderson, C. R. (1973). "Estimation associated with linear models." *Communications in Statistics* 1, 303–330.

Velleman, P. F., and Hoaglin, D. C. (1981). *Applications, basics, and computing of exploratory data analysis.* Boston, Mass.: Duxbury Press.

Vining, G. G. (1998). *Statistical methods for engineers.* Pacific Grove, Calif.: Duxbury Press.

Voss, D. T. (1988). "Generalized modulus-ratio tests for analysis of factorial designs with zero degrees of freedom for error." *Communications in Statistics—Theory and Methods* 17, 3345–3359.

Whitaker, E. R., Williams, E. R., and John, J. A. (1998). *CycDesigN,* CSIRO Forestry and Forest Products, Canberra, Australia, and The University of Waikato, Hamilton, New Zealand.

Williams, E. J. (1949). "Experimental designs balanced for the estimation of residual effects of treatments." *Australian Journal of Scientific Research* 2, 149–168.

Williams, E. R. (1986). Row and column designs with contiguous replicates. *Australian Journal of Statistics* 28, 154–163.

Williams, E. R., and Talbot, M. (1993). ALPHA+: *Experimental designs for variety trials. Design user manual*. CSIRO, Canberra and SASS, Edinburgh.

Williams, J. S. (1962). "A confidence interval for variance components." *Biometrika* 49, 278–281.

Yates, F. (1934). "The analysis of multiple classifications with unequal numbers in the different classes." *Journal of the American Statistical Association* 29, 51–66.

Yates, F. (1936a). "Incomplete randomized blocks." *Annals of Eugenics* 7, 121–140.

Yates, F. (1936b). "A new method of arranging variety trials involving a large number of varieties." *Journal of Agricultural Science* 26, 424–455.

Yates, F. (1940a). "The recovery of interblock information in balanced incomplete block designs." *Annals of Eugenics* 10, 317–325.

Yates, F. (1940b). "Lattice squares." *Journal of Agricultural Science* 30, 672–687.

Yates, F. (1948). "Contribution to the discussion of 'The Validity of Comparative Experiments' by F. J. Anscombe." *Journal of the Royal Statistical Society, A* 111, 204–205.

Youden, W. J. (1937). "Use of incomplete block replications in estimating tobacco-mosaic virus." *Contributions from Boyce Thompson Institute* 9, 41–48.

Youden, W. J. (1940). "Experimental designs to increase accuracy of greenhouse studies." *Contributions from Boyce Thompson Institute* 11, 219–228.

Youden, W. J. (1956). Special invited paper delivered to the Institute of Mathematical Statistics, December 8, 1956.

Youden, W. J. (1972). "Randomization and experimentation." *Technometrics* 14, 13–22.

Zahn, D. A. (1975a). "Modifications of and revised critical values for the half-normal plot." *Technometrics* 17, 189–200.

Zahn, D. A. (1975b). "Empirical study of the half-normal plot." *Technometrics* 17, 201–211.

Appendix

TABLES

Table I Standard normal distribution; entries are $P(Z \geq z) = \alpha$

z	.00	.01	.02	.03	.04	.05	.06	.07	.08	.09
0.00	.5000	.4960	.4920	.4880	.4840	.4801	.4761	.4721	.4681	.4641
0.10	.4602	.4562	.4522	.4483	.4443	.4404	.4364	.4325	.4286	.4247
0.20	.4207	.4168	.4129	.4090	.4052	.4013	.3974	.3936	.3897	.3859
0.30	.3821	.3783	.3745	.3707	.3669	.3632	.3594	.3557	.3520	.3483
0.40	.3446	.3409	.3372	.3336	.3300	.3264	.3228	.3192	.3156	.3121
0.50	.3085	.3050	.3015	.2981	.2946	.2912	.2877	.2843	.2810	.2776
0.60	.2743	.2709	.2676	.2643	.2611	.2578	.2546	.2514	.2483	.2451
0.70	.2420	.2389	.2358	.2327	.2296	.2266	.2236	.2206	.2177	.2148
0.80	.2119	.2090	.2061	.2033	.2005	.1977	.1949	.1922	.1894	.1867
0.90	.1841	.1814	.1788	.1762	.1736	.1711	.1685	.1660	.1635	.1611
1.00	.1587	.1562	.1539	.1515	.1492	.1469	.1446	.1423	.1401	.1379
1.10	.1357	.1335	.1314	.1292	.1271	.1251	.1230	.1210	.1190	.1170
1.20	.1151	.1131	.1112	.1093	.1075	.1056	.1038	.1020	.1003	.0985
1.30	.0968	.0951	.0934	.0918	.0901	.0885	.0869	.0853	.0838	.0823
1.40	.0808	.0793	.0778	.0764	.0749	.0735	.0721	.0708	.0694	.0681
1.50	.0668	.0655	.0643	.0630	.0618	.0606	.0594	.0582	.0571	.0559
1.60	.0548	.0537	.0526	.0516	.0505	.0495	.0485	.0475	.0465	.0455
1.70	.0446	.0436	.0427	.0418	.0409	.0401	.0392	.0384	.0375	.0367
1.80	.0359	.0351	.0344	.0336	.0329	.0322	.0314	.0307	.0301	.0294
1.90	.0287	.0281	.0274	.0268	.0262	.0256	.0250	.0244	.0239	.0233
2.00	.0228	.0222	.0217	.0212	.0207	.0202	.0197	.0192	.0188	.0183
2.10	.0179	.0174	.0170	.0166	.0162	.0158	.0154	.0150	.0146	.0143
2.20	.0139	.0136	.0132	.0129	.0125	.0122	.0119	.0116	.0113	.0110
2.30	.0107	.0104	.0102	.0099	.0096	.0094	.0091	.0089	.0087	.0084
2.40	.0082	.0080	.0078	.0075	.0073	.0071	.0069	.0068	.0066	.0064
2.50	.0062	.0060	.0059	.0057	.0055	.0054	.0052	.0051	.0049	.0048
2.60	.0047	.0045	.0044	.0043	.0041	.0040	.0039	.0038	.0037	.0036
2.70	.0035	.0034	.0033	.0032	.0031	.0030	.0029	.0028	.0027	.0026
2.80	.0026	.0025	.0024	.0023	.0023	.0022	.0021	.0021	.0020	.0019
2.90	.0019	.0018	.0018	.0017	.0016	.0016	.0015	.0015	.0014	.0014
3.00	.0013	.0013	.0013	.0012	.0012	.0011	.0011	.0011	.0010	.0010

Adapted with permission from *CRC Standard Probability and Statistics Tables and Formulae* (1991), edited by W. H. Beyer, CRC Press, Boca Raton, Florida.

Table II   Student's t distribution; $t_{\alpha,\nu}$ for $P(t \geq t_{\alpha,\nu}) = \alpha$

$\nu \backslash^{\alpha}$	0.40	0.25	0.10	0.05	0.025	0.01	0.005	0.0005
1	.325	1.000	3.078	6.314	12.706	31.821	63.657	636.309
2	.289	.816	1.886	2.920	4.303	6.965	9.925	31.598
3	.277	.765	1.638	2.353	3.182	4.541	5.841	12.924
4	.271	.741	1.533	2.132	2.776	3.747	4.604	8.610
5	.267	.727	1.476	2.015	2.571	3.365	4.032	6.869
6	.265	.718	1.440	1.943	2.447	3.143	3.707	5.959
7	.263	.711	1.415	1.895	2.365	2.998	3.499	5.408
8	.262	.706	1.397	1.860	2.306	2.896	3.355	5.041
9	.261	.703	1.383	1.833	2.262	2.821	3.250	4.781
10	.260	.700	1.372	1.812	2.228	2.764	3.169	4.587
11	.260	.697	1.363	1.796	2.201	2.718	3.106	4.437
12	.259	.695	1.356	1.782	2.179	2.681	3.055	4.318
13	.259	.694	1.350	1.771	2.160	2.650	3.012	4.221
14	.258	.692	1.345	1.761	2.145	2.624	2.977	4.140
15	.258	.691	1.341	1.753	2.131	2.602	2.947	4.073
16	.258	.690	1.337	1.746	2.120	2.583	2.921	4.015
17	.257	.689	1.333	1.740	2.110	2.567	2.898	3.965
18	.257	.688	1.330	1.734	2.101	2.552	2.878	3.922
19	.257	.688	1.328	1.729	2.093	2.539	2.861	3.883
20	.257	.687	1.325	1.725	2.086	2.528	2.845	3.850
21	.257	.686	1.323	1.721	2.080	2.518	2.831	3.819
22	.256	.686	1.321	1.717	2.074	2.508	2.819	3.792
23	.256	.685	1.319	1.714	2.069	2.500	2.807	3.768
24	.256	.685	1.318	1.711	2.064	2.492	2.797	3.745
25	.256	.684	1.316	1.708	2.060	2.485	2.787	3.725
26	.256	.684	1.315	1.706	2.056	2.479	2.779	3.707
27	.256	.684	1.314	1.703	2.052	2.473	2.771	3.690
28	.256	.683	1.313	1.701	2.048	2.467	2.763	3.674
29	.256	.683	1.311	1.699	2.045	2.462	2.756	3.659
30	.256	.683	1.310	1.697	2.042	2.457	2.750	3.646
40	.255	.681	1.303	1.684	2.021	2.423	2.704	3.551
60	.254	.679	1.296	1.671	2.000	2.390	2.660	3.460
120	.254	.677	1.289	1.658	1.980	2.358	2.617	3.373
∞	.253	.674	1.282	1.645	1.960	2.326	2.576	3.291

Computed with MINITAB by R. O. Kuehl.

Table III Chi-square distribution; $\chi^2_{\alpha,\nu}$ for $P(\chi^2 \geq \chi^2_{\alpha,\nu}) = \alpha$

ν \ α	0.995	0.99	0.975	0.95	0.90	0.75	0.50	0.25	0.10	0.05	0.025	0.01	0.005
1	.0000393	.000157	.000982	.00393	.0158	.102	.455	1.32	2.71	3.84	5.02	6.63	7.88
2	.0100	.0201	.0506	.103	.211	.575	1.39	2.77	4.61	5.99	7.38	9.21	10.6
3	.0717	.115	.216	.352	.584	1.21	2.37	4.11	6.25	7.81	9.35	11.3	12.8
4	.207	.297	.484	.711	1.06	1.92	3.36	5.39	7.78	9.49	11.1	13.3	14.9
5	.412	.554	.831	1.15	1.61	2.67	4.35	6.63	9.24	11.1	12.8	15.1	16.7
6	.676	.872	1.24	1.64	2.20	3.45	5.35	7.84	10.6	12.6	14.4	16.8	18.5
7	.989	1.24	1.69	2.17	2.83	4.25	6.35	9.04	12.0	14.1	16.0	18.5	20.3
8	1.34	1.65	2.18	2.73	3.49	5.07	7.34	10.2	13.4	15.5	17.5	20.1	22.0
9	1.73	2.09	2.70	3.33	4.17	5.90	8.34	11.4	14.7	16.9	19.0	21.7	23.6
10	2.16	2.56	3.25	3.94	4.87	6.74	9.34	12.5	16.0	18.3	20.5	23.2	25.2
11	2.60	3.05	3.82	4.57	5.58	7.58	10.3	13.7	17.3	19.7	21.9	24.7	26.8
12	3.07	3.57	4.40	5.23	6.30	8.44	11.3	14.8	18.5	21.0	23.3	26.2	28.3
13	3.57	4.11	5.01	5.89	7.04	9.30	12.3	16.0	19.8	22.4	24.7	27.7	29.8
14	4.07	4.66	5.63	6.57	7.79	10.2	13.3	17.1	21.1	23.7	26.1	29.1	31.3
15	4.60	5.23	6.26	7.26	8.55	11.0	14.3	18.2	22.3	25.0	27.5	30.6	32.8
16	5.14	5.81	6.91	7.96	9.31	11.9	15.3	19.4	23.5	26.3	28.8	32.0	34.3
17	5.70	6.41	7.56	8.67	10.1	12.8	16.3	20.5	24.8	27.6	30.2	33.4	35.7
18	6.26	7.01	8.23	9.39	10.9	13.7	17.3	21.6	26.0	28.9	31.5	34.8	37.2
19	6.84	7.63	8.91	10.1	11.7	14.6	18.3	22.7	27.2	30.1	32.9	36.2	38.6
20	7.43	8.26	9.59	10.9	12.4	15.5	19.3	23.8	28.4	31.4	34.2	37.6	40.0
21	8.03	8.90	10.3	11.6	13.2	16.3	20.3	24.9	29.6	32.7	35.5	38.9	41.4
22	8.64	9.54	11.0	12.3	14.0	17.2	21.3	26.0	30.8	33.9	36.8	40.3	42.8
23	9.26	10.2	11.7	13.1	14.8	18.1	22.3	27.1	32.0	35.2	38.1	41.6	44.2
24	9.89	10.9	12.4	13.8	15.7	19.0	23.3	28.2	33.2	36.4	39.4	43.0	45.6
25	10.5	11.5	13.1	14.6	16.5	19.9	24.3	29.3	34.4	37.7	40.6	44.3	46.9
26	11.2	12.2	13.8	15.4	17.3	20.8	25.3	30.4	35.6	38.9	41.9	45.6	48.3
27	11.8	12.9	14.6	16.2	18.1	21.7	26.3	31.5	36.7	40.1	43.2	47.0	49.6
28	12.5	13.6	15.3	16.9	18.9	22.7	27.3	32.6	37.9	41.3	44.5	48.3	51.0
29	13.1	14.3	16.0	17.7	19.8	23.6	28.3	33.7	39.1	42.6	45.7	49.6	52.3
30	13.8	15.0	16.8	18.5	20.6	24.5	29.3	34.8	40.3	43.8	47.0	50.9	53.7

Adapted from *Biometrika Tables for Statisticians*, Vol. I, 1966, edited by E. S. Pearson and H. O. Hartley, with permission of the Biometrica Trustees.

Table IV F distribution; F_{α,ν_1,ν_2} for $P(F \geq F_{\alpha,\nu_1,\nu_2}) = \alpha$

$\alpha = 0.10$

$\nu_2 \backslash \nu_1$	1	2	3	4	5	6	7	8	9	10	12	15	20	24	30	40	60	120	∞
1	39.86	49.50	53.59	55.83	57.24	58.20	58.91	59.44	59.86	60.19	60.71	61.22	61.74	62.00	62.26	62.53	62.79	63.06	63.33
2	8.53	9.00	9.16	9.24	9.29	9.33	9.35	9.37	9.38	9.39	9.41	9.42	9.44	9.45	9.46	9.47	9.47	9.48	9.49
3	5.54	5.46	5.39	5.34	5.31	5.28	5.27	5.25	5.24	5.23	5.22	5.20	5.18	5.18	5.17	5.16	5.15	5.14	5.13
4	4.54	4.32	4.19	4.11	4.05	4.01	3.98	3.95	3.94	3.92	3.90	3.87	3.84	3.83	3.82	3.80	3.79	3.78	3.76
5	4.06	3.78	3.62	3.52	3.45	3.40	3.37	3.34	3.32	3.30	3.27	3.24	3.21	3.19	3.17	3.16	3.14	3.12	3.10
6	3.78	3.46	3.29	3.18	3.11	3.05	3.01	2.98	2.96	2.94	2.90	2.87	2.84	2.82	2.80	2.78	2.76	2.74	2.72
7	3.59	3.26	3.07	2.96	2.88	2.83	2.78	2.75	2.72	2.70	2.67	2.63	2.59	2.58	2.56	2.54	2.51	2.49	2.47
8	3.46	3.11	2.92	2.81	2.73	2.67	2.62	2.59	2.56	2.54	2.50	2.46	2.42	2.40	2.38	2.36	2.34	2.32	2.29
9	3.36	3.01	2.81	2.69	2.61	2.55	2.51	2.47	2.44	2.42	2.38	2.34	2.30	2.28	2.25	2.23	2.21	2.18	2.16
10	3.29	2.92	2.73	2.61	2.52	2.46	2.41	2.38	2.35	2.32	2.28	2.24	2.20	2.18	2.16	2.13	2.11	2.08	2.06
11	3.23	2.86	2.66	2.54	2.45	2.39	2.34	2.30	2.27	2.25	2.21	2.17	2.12	2.10	2.08	2.05	2.03	2.00	1.97
12	3.18	2.81	2.61	2.48	2.39	2.33	2.28	2.24	2.21	2.19	2.15	2.10	2.06	2.04	2.01	1.99	1.96	1.93	1.90
13	3.14	2.76	2.56	2.43	2.35	2.28	2.23	2.20	2.16	2.14	2.10	2.05	2.01	1.98	1.96	1.93	1.90	1.88	1.85
14	3.10	2.73	2.52	2.39	2.31	2.24	2.19	2.15	2.12	2.10	2.05	2.01	1.96	1.94	1.91	1.89	1.86	1.83	1.80
15	3.07	2.70	2.49	2.36	2.27	2.21	2.16	2.12	2.09	2.06	2.02	1.97	1.92	1.90	1.87	1.85	1.82	1.79	1.76
16	3.05	2.67	2.46	2.33	2.24	2.18	2.13	2.09	2.06	2.03	1.99	1.94	1.89	1.87	1.84	1.81	1.78	1.75	1.72
17	3.03	2.64	2.44	2.31	2.22	2.15	2.10	2.06	2.03	2.00	1.96	1.91	1.86	1.84	1.81	1.78	1.75	1.72	1.69
18	3.01	2.62	2.42	2.29	2.20	2.13	2.08	2.04	2.00	1.98	1.93	1.89	1.84	1.81	1.78	1.75	1.72	1.69	1.66
19	2.99	2.61	2.40	2.27	2.18	2.11	2.06	2.02	1.98	1.96	1.91	1.86	1.81	1.79	1.76	1.73	1.70	1.67	1.63
20	2.97	2.59	2.38	2.25	2.16	2.09	2.04	2.00	1.96	1.94	1.89	1.84	1.79	1.77	1.74	1.71	1.68	1.64	1.61
21	2.96	2.57	2.36	2.23	2.14	2.08	2.02	1.98	1.95	1.92	1.87	1.83	1.78	1.75	1.72	1.69	1.66	1.62	1.59
22	2.95	2.56	2.35	2.22	2.13	2.06	2.01	1.97	1.93	1.90	1.86	1.81	1.76	1.73	1.70	1.67	1.64	1.60	1.57
23	2.94	2.55	2.34	2.21	2.11	2.05	1.99	1.95	1.92	1.89	1.84	1.80	1.74	1.72	1.69	1.66	1.62	1.59	1.55
24	2.93	2.54	2.33	2.19	2.10	2.04	1.98	1.94	1.91	1.88	1.83	1.78	1.73	1.70	1.67	1.64	1.61	1.57	1.53
25	2.92	2.53	2.32	2.18	2.09	2.02	1.97	1.93	1.89	1.87	1.82	1.77	1.72	1.69	1.66	1.63	1.59	1.56	1.52
26	2.91	2.52	2.31	2.17	2.08	2.01	1.96	1.92	1.88	1.86	1.81	1.76	1.71	1.68	1.65	1.61	1.58	1.54	1.50
27	2.90	2.51	2.30	2.17	2.07	2.00	1.95	1.91	1.87	1.85	1.80	1.75	1.70	1.67	1.64	1.60	1.57	1.53	1.49
28	2.89	2.50	2.29	2.16	2.06	2.00	1.94	1.90	1.87	1.84	1.79	1.74	1.69	1.66	1.63	1.59	1.56	1.52	1.48
29	2.89	2.50	2.28	2.15	2.06	1.99	1.93	1.89	1.86	1.83	1.78	1.73	1.68	1.65	1.62	1.58	1.55	1.51	1.47
30	2.88	2.49	2.28	2.14	2.05	1.98	1.93	1.88	1.85	1.82	1.77	1.72	1.67	1.64	1.61	1.57	1.54	1.50	1.46
40	2.84	2.44	2.23	2.09	2.00	1.93	1.87	1.83	1.79	1.76	1.71	1.66	1.61	1.57	1.54	1.51	1.47	1.42	1.38
60	2.79	2.39	2.18	2.04	1.95	1.87	1.82	1.77	1.74	1.71	1.66	1.60	1.54	1.51	1.48	1.44	1.40	1.35	1.29
120	2.75	2.35	2.13	1.99	1.90	1.82	1.77	1.72	1.68	1.65	1.60	1.55	1.48	1.45	1.41	1.37	1.32	1.26	1.19
∞	2.71	2.30	2.08	1.94	1.85	1.77	1.72	1.67	1.63	1.60	1.55	1.49	1.42	1.38	1.34	1.30	1.24	1.17	1.00

Table IV F distribution; F_{α,ν_1,ν_2} for $P(F \geq F_{\alpha,\nu_1,\nu_2}) = \alpha$

$\alpha = 0.05$

$\nu_2 \backslash \nu_1$	1	2	3	4	5	6	7	8	9	10	12	15	20	24	30	40	60	120	∞
1	161.4	199.5	215.7	224.6	230.2	234.0	236.8	238.9	240.5	241.9	243.9	245.9	248.0	249.1	250.1	251.1	252.2	253.3	254.3
2	18.51	19.00	19.16	19.25	19.30	19.33	19.35	19.37	19.38	19.40	19.41	19.43	19.45	19.45	19.46	19.47	19.48	19.49	19.50
3	10.13	9.55	9.28	9.12	9.01	8.94	8.89	8.85	8.81	8.79	8.74	8.70	8.66	8.64	8.62	8.59	8.57	8.55	8.53
4	7.71	6.94	6.59	6.39	6.26	6.16	6.09	6.04	6.00	5.96	5.91	5.86	5.80	5.77	5.75	5.72	5.69	5.66	5.63
5	6.61	5.79	5.41	5.19	5.05	4.95	4.88	4.82	4.77	4.74	4.68	4.62	4.56	4.53	4.50	4.46	4.43	4.40	4.36
6	5.99	5.14	4.76	4.53	4.39	4.28	4.21	4.15	4.10	4.06	4.00	3.94	3.87	3.84	3.81	3.77	3.74	3.70	3.67
7	5.59	4.74	4.35	4.12	3.97	3.87	3.79	3.73	3.68	3.64	3.57	3.51	3.44	3.41	3.38	3.34	3.30	3.27	3.23
8	5.32	4.46	4.07	3.84	3.69	3.58	3.50	3.44	3.39	3.35	3.28	3.22	3.15	3.12	3.08	3.04	3.01	2.97	2.93
9	5.12	4.26	3.86	3.63	3.48	3.37	3.29	3.23	3.18	3.14	3.07	3.01	2.94	2.90	2.86	2.83	2.79	2.75	2.71
10	4.96	4.10	3.71	3.48	3.33	3.22	3.14	3.07	3.02	2.98	2.91	2.85	2.77	2.74	2.70	2.66	2.62	2.58	2.54
11	4.84	3.98	3.59	3.36	3.20	3.09	3.01	2.95	2.90	2.85	2.79	2.72	2.65	2.61	2.57	2.53	2.49	2.45	2.40
12	4.75	3.89	3.49	3.26	3.11	3.00	2.91	2.85	2.80	2.75	2.69	2.62	2.54	2.51	2.47	2.43	2.38	2.34	2.30
13	4.67	3.81	3.41	3.18	3.03	2.92	2.83	2.77	2.71	2.67	2.60	2.53	2.46	2.42	2.38	2.34	2.30	2.25	2.21
14	4.60	3.74	3.34	3.11	2.96	2.85	2.76	2.70	2.65	2.60	2.53	2.46	2.39	2.35	2.31	2.27	2.22	2.18	2.13
15	4.54	3.68	3.29	3.06	2.90	2.79	2.71	2.64	2.59	2.54	2.48	2.40	2.33	2.29	2.25	2.20	2.16	2.11	2.07
16	4.49	3.63	3.24	3.01	2.85	2.74	2.66	2.59	2.54	2.49	2.42	2.35	2.28	2.24	2.19	2.15	2.11	2.06	2.01
17	4.45	3.59	3.20	2.96	2.81	2.70	2.61	2.55	2.49	2.45	2.38	2.31	2.23	2.19	2.15	2.10	2.06	2.01	1.96
18	4.41	3.55	3.16	2.93	2.77	2.66	2.58	2.51	2.46	2.41	2.34	2.27	2.19	2.15	2.11	2.06	2.02	1.97	1.92
19	4.38	3.52	3.13	2.90	2.74	2.63	2.54	2.48	2.42	2.38	2.31	2.23	2.16	2.11	2.07	2.03	1.98	1.93	1.88
20	4.35	3.49	3.10	2.87	2.71	2.60	2.51	2.45	2.39	2.35	2.28	2.20	2.12	2.08	2.04	1.99	1.95	1.90	1.84
21	4.32	3.47	3.07	2.84	2.68	2.57	2.49	2.42	2.37	2.32	2.25	2.18	2.10	2.05	2.01	1.96	1.92	1.87	1.81
22	4.30	3.44	3.05	2.82	2.66	2.55	2.46	2.40	2.34	2.30	2.23	2.15	2.07	2.03	1.98	1.94	1.89	1.84	1.78
23	4.28	3.42	3.03	2.80	2.64	2.53	2.44	2.37	2.32	2.27	2.20	2.13	2.05	2.01	1.96	1.91	1.86	1.81	1.76
24	4.26	3.40	3.01	2.78	2.62	2.51	2.42	2.36	2.30	2.25	2.18	2.11	2.03	1.98	1.94	1.89	1.84	1.79	1.73
25	4.24	3.39	2.99	2.76	2.60	2.49	2.40	2.34	2.28	2.24	2.16	2.09	2.01	1.96	1.92	1.87	1.82	1.77	1.71
26	4.23	3.37	2.98	2.74	2.59	2.47	2.39	2.32	2.27	2.22	2.15	2.07	1.99	1.95	1.90	1.85	1.80	1.75	1.69
27	4.21	3.35	2.96	2.73	2.57	2.46	2.37	2.31	2.25	2.20	2.13	2.06	1.97	1.93	1.88	1.84	1.79	1.73	1.67
28	4.20	3.34	2.95	2.71	2.56	2.45	2.36	2.29	2.24	2.19	2.12	2.04	1.96	1.91	1.87	1.82	1.77	1.71	1.65
29	4.18	3.33	2.93	2.70	2.55	2.43	2.35	2.28	2.22	2.18	2.10	2.03	1.94	1.90	1.85	1.81	1.75	1.70	1.64
30	4.17	3.32	2.92	2.69	2.53	2.42	2.33	2.27	2.21	2.16	2.09	2.01	1.93	1.89	1.84	1.79	1.74	1.68	1.62
40	4.08	3.23	2.84	2.61	2.45	2.34	2.25	2.18	2.12	2.08	2.00	1.92	1.84	1.79	1.74	1.69	1.64	1.58	1.51
60	4.00	3.15	2.76	2.53	2.37	2.25	2.17	2.10	2.04	1.99	1.92	1.84	1.75	1.70	1.65	1.59	1.53	1.47	1.39
120	3.92	3.07	2.68	2.45	2.29	2.17	2.09	2.02	1.96	1.91	1.83	1.75	1.66	1.61	1.55	1.50	1.43	1.35	1.25
∞	3.84	3.00	2.60	2.37	2.21	2.10	2.01	1.94	1.88	1.83	1.75	1.67	1.57	1.52	1.46	1.39	1.32	1.22	1.00

Table IV F distribution; F_{α,ν_1,ν_2} for $P(F \geq F_{\alpha,\nu_1,\nu_2}) = \alpha$

$\alpha = 0.05$

ν_2＼ν_1	1	2	3	4	5	6	7	8	9	10	12	15	20	24	30	40	60	120	∞
1	161.4	199.5	215.7	224.6	230.2	234.0	236.8	238.9	240.5	241.9	243.9	245.9	248.0	249.1	250.1	251.1	252.2	253.3	254.3
2	18.51	19.00	19.16	19.25	19.30	19.33	19.35	19.37	19.38	19.40	19.41	19.43	19.45	19.45	19.46	19.47	19.48	19.49	19.50
3	10.13	9.55	9.28	9.12	9.01	8.94	8.89	8.85	8.81	8.79	8.74	8.70	8.66	8.64	8.62	8.59	8.57	8.55	8.53
4	7.71	6.94	6.59	6.39	6.26	6.16	6.09	6.04	6.00	5.96	5.91	5.86	5.80	5.77	5.75	5.72	5.69	5.66	5.63
5	6.61	5.79	5.41	5.19	5.05	4.95	4.88	4.82	4.77	4.74	4.68	4.62	4.56	4.53	4.50	4.46	4.43	4.40	4.36
6	5.99	5.14	4.76	4.53	4.39	4.28	4.21	4.15	4.10	4.06	4.00	3.94	3.87	3.84	3.81	3.77	3.74	3.70	3.67
7	5.59	4.74	4.35	4.12	3.97	3.87	3.79	3.73	3.68	3.64	3.57	3.51	3.44	3.41	3.38	3.34	3.30	3.27	3.23
8	5.32	4.46	4.07	3.84	3.69	3.58	3.50	3.44	3.39	3.35	3.28	3.22	3.15	3.12	3.08	3.04	3.01	2.97	2.93
9	5.12	4.26	3.86	3.63	3.48	3.37	3.29	3.23	3.18	3.14	3.07	3.01	2.94	2.90	2.86	2.83	2.79	2.75	2.71
10	4.96	4.10	3.71	3.48	3.33	3.22	3.14	3.07	3.02	2.98	2.91	2.85	2.77	2.74	2.70	2.66	2.62	2.58	2.54
11	4.84	3.98	3.59	3.36	3.20	3.09	3.01	2.95	2.90	2.85	2.79	2.72	2.65	2.61	2.57	2.53	2.49	2.45	2.40
12	4.75	3.89	3.49	3.26	3.11	3.00	2.91	2.85	2.80	2.75	2.69	2.62	2.54	2.51	2.47	2.43	2.38	2.34	2.30
13	4.67	3.81	3.41	3.18	3.03	2.92	2.83	2.77	2.71	2.67	2.60	2.53	2.46	2.42	2.38	2.34	2.30	2.25	2.21
14	4.60	3.74	3.34	3.11	2.96	2.85	2.76	2.70	2.65	2.60	2.53	2.46	2.39	2.35	2.31	2.27	2.22	2.18	2.13
15	4.54	3.68	3.29	3.06	2.90	2.79	2.71	2.64	2.59	2.54	2.48	2.40	2.33	2.29	2.25	2.20	2.16	2.11	2.07
16	4.49	3.63	3.24	3.01	2.85	2.74	2.66	2.59	2.54	2.49	2.42	2.35	2.28	2.24	2.19	2.15	2.11	2.06	2.01
17	4.45	3.59	3.20	2.96	2.81	2.70	2.61	2.55	2.49	2.45	2.38	2.31	2.23	2.19	2.15	2.10	2.06	2.01	1.96
18	4.41	3.55	3.16	2.93	2.77	2.66	2.58	2.51	2.46	2.41	2.34	2.27	2.19	2.15	2.11	2.06	2.02	1.97	1.92
19	4.38	3.52	3.13	2.90	2.74	2.63	2.54	2.48	2.42	2.38	2.31	2.23	2.16	2.11	2.07	2.03	1.98	1.93	1.88
20	4.35	3.49	3.10	2.87	2.71	2.60	2.51	2.45	2.39	2.35	2.28	2.20	2.12	2.08	2.04	1.99	1.95	1.90	1.84
21	4.32	3.47	3.07	2.84	2.68	2.57	2.49	2.42	2.37	2.32	2.25	2.18	2.10	2.05	2.01	1.96	1.92	1.87	1.81
22	4.30	3.44	3.05	2.82	2.66	2.55	2.46	2.40	2.34	2.30	2.23	2.15	2.07	2.03	1.98	1.94	1.89	1.84	1.78
23	4.28	3.42	3.03	2.80	2.64	2.53	2.44	2.37	2.32	2.27	2.20	2.13	2.05	2.01	1.96	1.91	1.86	1.81	1.76
24	4.26	3.40	3.01	2.78	2.62	2.51	2.42	2.36	2.30	2.25	2.18	2.11	2.03	1.98	1.94	1.89	1.84	1.79	1.73
25	4.24	3.39	2.99	2.76	2.60	2.49	2.40	2.34	2.28	2.24	2.16	2.09	2.01	1.96	1.92	1.87	1.82	1.77	1.71
26	4.23	3.37	2.98	2.74	2.59	2.47	2.39	2.32	2.27	2.22	2.15	2.07	1.99	1.95	1.90	1.85	1.80	1.75	1.69
27	4.21	3.35	2.96	2.73	2.57	2.46	2.37	2.31	2.25	2.20	2.13	2.06	1.97	1.93	1.88	1.84	1.79	1.73	1.67
28	4.20	3.34	2.95	2.71	2.56	2.45	2.36	2.29	2.24	2.19	2.12	2.04	1.96	1.91	1.87	1.82	1.77	1.71	1.65
29	4.18	3.33	2.93	2.70	2.55	2.43	2.35	2.28	2.22	2.18	2.10	2.03	1.94	1.90	1.85	1.81	1.75	1.70	1.64
30	4.17	3.32	2.92	2.69	2.53	2.42	2.33	2.27	2.21	2.16	2.09	2.01	1.93	1.89	1.84	1.79	1.74	1.68	1.62
40	4.08	3.23	2.84	2.61	2.45	2.34	2.25	2.18	2.12	2.08	2.00	1.92	1.84	1.79	1.74	1.69	1.64	1.58	1.51
60	4.00	3.15	2.76	2.53	2.37	2.25	2.17	2.10	2.04	1.99	1.92	1.84	1.75	1.70	1.65	1.59	1.53	1.47	1.39
120	3.92	3.07	2.68	2.45	2.29	2.17	2.09	2.02	1.96	1.91	1.83	1.75	1.66	1.61	1.55	1.50	1.43	1.35	1.25
∞	3.84	3.00	2.60	2.37	2.21	2.10	2.01	1.94	1.88	1.83	1.75	1.67	1.57	1.52	1.46	1.39	1.32	1.22	1.00

Table IV F distribution; F_{α,ν_1,ν_2} for $P(F \geq F_{\alpha,\nu_1,\nu_2}) = \alpha$

$\alpha = 0.01$

$\nu_2 \backslash \nu_1$	1	2	3	4	5	6	7	8	9	10	12	15	20	24	30	40	60	120	∞
1	4052	4999.5	5403	5625	5764	5859	5928	5982	6022	6056	6106	6157	6209	6235	6261	6287	6313	6339	6366
2	98.50	99.00	99.17	99.25	99.30	99.33	99.36	99.37	99.39	99.40	99.42	99.43	99.45	99.46	99.47	99.47	99.48	99.49	99.50
3	34.12	30.82	29.46	28.71	28.24	27.91	27.67	27.49	27.35	27.23	27.05	26.87	26.69	26.60	26.50	26.41	26.32	26.22	26.13
4	21.20	18.00	16.69	15.98	15.52	15.21	14.98	14.80	14.66	14.55	14.37	14.20	14.02	13.93	13.84	13.75	13.65	13.56	13.46
5	16.26	13.27	12.06	11.39	10.97	10.67	10.46	10.29	10.16	10.05	9.89	9.72	9.55	9.47	9.38	9.29	9.20	9.11	9.02
6	13.75	10.92	9.78	9.15	8.75	8.47	8.26	8.10	7.98	7.87	7.72	7.56	7.40	7.31	7.23	7.14	7.06	6.97	6.88
7	12.25	9.55	8.45	7.85	7.46	7.19	6.99	6.84	6.72	6.62	6.47	6.31	6.16	6.07	5.99	5.91	5.82	5.74	5.65
8	11.26	8.65	7.59	7.01	6.63	6.37	6.18	6.03	5.91	5.81	5.67	5.52	5.36	5.28	5.20	5.12	5.03	4.95	4.86
9	10.56	8.02	6.99	6.42	6.06	5.80	5.61	5.47	5.35	5.26	5.11	4.96	4.81	4.73	4.65	4.57	4.48	4.40	4.31
10	10.04	7.56	6.55	5.99	5.64	5.39	5.20	5.06	4.94	4.85	4.71	4.56	4.41	4.33	4.25	4.17	4.08	4.00	3.91
11	9.65	7.21	6.22	5.67	5.32	5.07	4.89	4.74	4.63	4.54	4.40	4.25	4.10	4.02	3.94	3.86	3.78	3.69	3.60
12	9.33	6.93	5.95	5.41	5.06	4.82	4.64	4.50	4.39	4.30	4.16	4.01	3.86	3.78	3.70	3.62	3.54	3.45	3.36
13	9.07	6.70	5.74	5.21	4.86	4.62	4.44	4.30	4.19	4.10	3.96	3.82	3.66	3.59	3.51	3.43	3.34	3.25	3.17
14	8.86	6.51	5.56	5.04	4.69	4.46	4.28	4.14	4.03	3.94	3.80	3.66	3.51	3.43	3.35	3.27	3.18	3.09	3.00
15	8.68	6.36	5.42	4.89	4.56	4.32	4.14	4.00	3.89	3.80	3.67	3.52	3.37	3.29	3.21	3.13	3.05	2.96	2.87
16	8.53	6.23	5.29	4.77	4.44	4.20	4.03	3.89	3.78	3.69	3.55	3.41	3.26	3.18	3.10	3.02	2.93	2.84	2.75
17	8.40	6.11	5.18	4.67	4.34	4.10	3.93	3.79	3.68	3.59	3.46	3.31	3.16	3.08	3.00	2.92	2.83	2.75	2.65
18	8.29	6.01	5.09	4.58	4.25	4.01	3.84	3.71	3.60	3.51	3.37	3.23	3.08	3.00	2.92	2.84	2.75	2.66	2.57
19	8.18	5.93	5.01	4.50	4.17	3.94	3.77	3.63	3.52	3.43	3.30	3.15	3.00	2.92	2.84	2.76	2.67	2.58	2.49
20	8.10	5.85	4.94	4.43	4.10	3.87	3.70	3.56	3.46	3.37	3.23	3.09	2.94	2.86	2.78	2.69	2.61	2.52	2.42
21	8.02	5.78	4.87	4.37	4.04	3.81	3.64	3.51	3.40	3.31	3.17	3.03	2.88	2.80	2.72	2.64	2.55	2.46	2.36
22	7.95	5.72	4.82	4.31	3.99	3.76	3.59	3.45	3.35	3.26	3.12	2.98	2.83	2.75	2.67	2.58	2.50	2.40	2.31
23	7.88	5.66	4.76	4.26	3.94	3.71	3.54	3.41	3.30	3.21	3.07	2.93	2.78	2.70	2.62	2.54	2.45	2.35	2.26
24	7.82	5.61	4.72	4.22	3.90	3.67	3.50	3.36	3.26	3.17	3.03	2.89	2.74	2.66	2.58	2.49	2.40	2.31	2.21
25	7.77	5.57	4.68	4.18	3.85	3.63	3.46	3.32	3.22	3.13	2.99	2.85	2.70	2.62	2.54	2.45	2.36	2.27	2.17
26	7.72	5.53	4.64	4.14	3.82	3.59	3.42	3.29	3.18	3.09	2.96	2.81	2.66	2.58	2.50	2.42	2.33	2.23	2.13
27	7.68	5.49	4.60	4.11	3.78	3.56	3.39	3.26	3.15	3.06	2.93	2.78	2.63	2.55	2.47	2.38	2.29	2.20	2.10
28	7.64	5.45	4.57	4.07	3.75	3.53	3.36	3.23	3.12	3.03	2.90	2.75	2.60	2.52	2.44	2.35	2.26	2.17	2.06
29	7.60	5.42	4.54	4.04	3.73	3.50	3.33	3.20	3.09	3.00	2.87	2.73	2.57	2.49	2.41	2.33	2.23	2.14	2.03
30	7.56	5.39	4.51	4.02	3.70	3.47	3.30	3.17	3.07	2.98	2.84	2.70	2.55	2.47	2.39	2.30	2.21	2.11	2.01
40	7.31	5.18	4.31	3.83	3.51	3.29	3.12	2.99	2.89	2.80	2.66	2.52	2.37	2.29	2.20	2.11	2.02	1.92	1.80
60	7.08	4.98	4.13	3.65	3.34	3.12	2.95	2.82	2.72	2.63	2.50	2.35	2.20	2.12	2.03	1.94	1.84	1.73	1.60
120	6.85	4.79	3.95	3.48	3.17	2.96	2.79	2.66	2.56	2.47	2.34	2.19	2.03	1.95	1.86	1.76	1.66	1.53	1.38
∞	6.63	4.61	3.78	3.32	3.02	2.80	2.64	2.51	2.41	2.32	2.18	2.04	1.88	1.79	1.70	1.59	1.47	1.32	1.00

Table V Bonferroni t statistic; $t_{\alpha/2,k,\nu}$ for $P\left(|t| \geq t_{\alpha/2,k,\nu}\right) = \alpha$

$\alpha = 0.01$

$\nu \backslash k$	2	3	4	5	6	7	8	9	10	11	12	13	14	15	20
5	4.77	5.25	5.60	5.89	6.14	6.35	6.54	6.71	6.87	7.01	7.15	7.27	7.39	7.50	7.98
6	4.32	4.70	4.98	5.21	5.40	5.56	5.71	5.84	5.96	6.07	6.17	6.26	6.35	6.43	6.79
7	4.03	4.36	4.59	4.79	4.94	5.08	5.20	5.31	5.41	5.50	5.58	5.66	5.73	5.80	6.08
8	3.83	4.12	4.33	4.50	4.64	4.76	4.86	4.96	5.04	5.12	5.19	5.25	5.32	5.37	5.62
9	3.69	3.95	4.15	4.30	4.42	4.53	4.62	4.71	4.78	4.85	4.91	4.97	5.02	5.08	5.29
10	3.58	3.83	4.00	4.14	4.26	4.36	4.44	4.52	4.59	4.65	4.71	4.76	4.81	4.85	5.05
11	3.50	3.73	3.89	4.02	4.13	4.22	4.30	4.37	4.44	4.49	4.55	4.60	4.64	4.68	4.86
12	3.43	3.65	3.81	3.93	4.03	4.12	4.19	4.26	4.32	4.37	4.42	4.47	4.51	4.55	4.72
13	3.37	3.58	3.73	3.85	3.95	4.03	4.10	4.16	4.22	4.27	4.32	4.36	4.40	4.44	4.60
14	3.33	3.53	3.67	3.79	3.88	3.96	4.03	4.09	4.14	4.19	4.23	4.28	4.31	4.35	4.50
15	3.29	3.48	3.62	3.73	3.82	3.90	3.96	4.02	4.07	4.12	4.16	4.20	4.24	4.27	4.42
16	3.25	3.44	3.58	3.69	3.77	3.85	3.91	3.96	4.02	4.06	4.10	4.14	4.18	4.21	4.35
17	3.22	3.41	3.54	3.65	3.73	3.80	3.86	3.92	3.97	4.01	4.05	4.09	4.12	4.15	4.29
18	3.20	3.38	3.51	3.61	3.69	3.76	3.82	3.87	3.92	3.96	4.00	4.04	4.07	4.10	4.23
19	3.17	3.35	3.48	3.58	3.66	3.73	3.79	3.84	3.88	3.93	3.96	4.00	4.03	4.06	4.19
20	3.15	3.33	3.46	3.55	3.63	3.70	3.75	3.80	3.85	3.89	3.93	3.96	3.99	4.02	4.15
25	3.08	3.24	3.36	3.45	3.52	3.58	3.64	3.68	3.73	3.76	3.80	3.83	3.86	3.88	4.00
30	3.03	3.19	3.30	3.39	3.45	3.51	3.56	3.61	3.65	3.68	3.71	3.74	3.77	3.80	3.90
40	2.97	3.12	3.23	3.31	3.37	3.43	3.47	3.51	3.55	3.58	3.61	3.64	3.67	3.69	3.79
60	2.91	3.06	3.16	3.23	3.29	3.34	3.39	3.43	3.46	3.49	3.52	3.54	3.57	3.59	3.68
120	2.86	3.00	3.09	3.16	3.22	3.26	3.31	3.34	3.37	3.40	3.43	3.45	3.47	3.49	3.58
∞	2.81	2.94	3.02	3.09	3.14	3.19	3.23	3.26	3.29	3.32	3.34	3.36	3.38	3.40	3.48

k = number of comparisons

From B. J. R. Bailey (1977) "Tables of the Bonferroni t Statistic." *Journal of the American Statistical Association* 72, 469–478. Reprinted with permission from *Journal of the American Statistical Association*. Copyright 1977 by the American Statistical Association. All rights reserved.

Table V Bonferroni t statistic; $t_{\alpha/2,k,\nu}$ for $P(|t| \geq t_{\alpha/2,k,\nu}) = \alpha$

$\alpha = 0.05$

$\nu \backslash k$	2	3	4	5	6	7	8	9	10	11	12	13	14	15	20
5	3.16	3.53	3.81	4.03	4.22	4.38	4.53	4.66	4.77	4.88	4.98	5.08	5.16	5.25	5.60
6	2.97	3.29	3.52	3.71	3.86	4.00	4.12	4.22	4.32	4.40	4.49	4.56	4.63	4.70	4.98
7	2.84	3.13	3.34	3.50	3.64	3.75	3.86	3.95	4.03	4.10	4.17	4.24	4.30	4.36	4.59
8	2.75	3.02	3.21	3.36	3.48	3.58	3.68	3.76	3.83	3.90	3.96	4.02	4.07	4.12	4.33
9	2.69	2.93	3.11	3.25	3.36	3.46	3.55	3.62	3.69	3.75	3.81	3.86	3.91	3.95	4.15
10	2.63	2.87	3.04	3.17	3.28	3.37	3.45	3.52	3.58	3.64	3.69	3.74	3.79	3.83	4.00
11	2.59	2.82	2.98	3.11	3.21	3.29	3.37	3.44	3.50	3.55	3.60	3.65	3.69	3.73	3.89
12	2.56	2.78	2.93	3.05	3.15	3.24	3.31	3.37	3.43	3.48	3.53	3.57	3.61	3.65	3.81
13	2.53	2.75	2.90	3.01	3.11	3.19	3.26	3.32	3.37	3.42	3.47	3.51	3.55	3.58	3.73
14	2.51	2.72	2.86	2.98	3.07	3.15	3.21	3.27	3.33	3.37	3.42	3.46	3.50	3.53	3.67
15	2.49	2.69	2.84	2.95	3.04	3.11	3.18	3.23	3.29	3.33	3.37	3.41	3.45	3.48	3.62
16	2.47	2.67	2.81	2.92	3.01	3.08	3.15	3.20	3.25	3.30	3.34	3.38	3.41	3.44	3.58
17	2.46	2.66	2.79	2.90	2.98	3.06	3.12	3.17	3.22	3.27	3.31	3.34	3.38	3.41	3.54
18	2.45	2.64	2.77	2.88	2.96	3.03	3.09	3.15	3.20	3.24	3.28	3.32	3.35	3.38	3.51
19	2.43	2.63	2.76	2.86	2.94	3.01	3.07	3.13	3.17	3.22	3.26	3.29	3.32	3.35	3.48
20	2.42	2.61	2.74	2.85	2.93	3.00	3.06	3.11	3.15	3.20	3.23	3.27	3.30	3.33	3.46
25	2.38	2.57	2.69	2.79	2.86	2.93	2.99	3.03	3.08	3.12	3.15	3.19	3.22	3.24	3.36
30	2.36	2.54	2.66	2.75	2.82	2.89	2.94	2.99	3.03	3.07	3.10	3.13	3.16	3.19	3.30
40	2.33	2.50	2.62	2.70	2.78	2.84	2.89	2.93	2.97	3.01	3.04	3.07	3.10	3.12	3.23
60	2.30	2.46	2.58	2.66	2.73	2.79	2.83	2.88	2.91	2.95	2.98	3.01	3.03	3.06	3.16
120	2.27	2.43	2.54	2.62	2.68	2.74	2.78	2.82	2.86	2.89	2.92	2.95	2.97	3.00	3.09
∞	2.24	2.39	2.50	2.58	2.64	2.69	2.73	2.77	2.81	2.84	2.87	2.89	2.91	2.94	3.02

k = number of comparisons

Table VI   Multiple comparisons with the best and Dunnett tests;
$d_{\alpha,k,\nu}$ for $P(d \geq d_{\alpha,k,\nu}) = \alpha$

$\nu \backslash k$	2	3	4	5	6	7	8	9	10	11	12	15	20
5	2.44	2.68	2.85	2.98	3.08	3.16	3.24	3.30	3.36	3.41	3.45	3.57	3.72
6	2.34	2.56	2.71	2.83	2.92	3.00	3.07	3.12	3.17	3.22	3.26	3.37	3.50
7	2.27	2.48	2.62	2.73	2.82	2.89	2.95	3.01	3.05	3.10	3.13	3.23	3.36
8	2.22	2.42	2.55	2.66	2.74	2.81	2.87	2.92	2.96	3.01	3.04	3.14	3.25
9	2.18	2.37	2.50	2.60	2.68	2.75	2.81	2.86	2.90	2.94	2.97	3.06	3.18
10	2.15	2.34	2.47	2.56	2.64	2.70	2.76	2.81	2.85	2.89	2.92	3.01	3.12
11	2.13	2.31	2.44	2.53	2.60	2.67	2.72	2.77	2.81	2.85	2.88	2.96	3.07
12	2.11	2.29	2.41	2.50	2.58	2.64	2.69	2.74	2.78	2.81	2.84	2.93	3.03
13	2.09	2.27	2.39	2.48	2.55	2.61	2.66	2.71	2.75	2.78	2.82	2.90	3.00
14	2.08	2.25	2.37	2.46	2.53	2.59	2.64	2.69	2.72	2.76	2.79	2.87	2.97
15	2.07	2.24	2.36	2.44	2.51	2.57	2.62	2.67	2.70	2.74	2.77	2.85	2.95
16	2.06	2.23	2.34	2.43	2.50	2.56	2.61	2.65	2.69	2.72	2.75	2.83	2.93
17	2.05	2.22	2.33	2.42	2.49	2.54	2.59	2.64	2.67	2.71	2.74	2.81	2.91
18	2.04	2.21	2.32	2.41	2.48	2.53	2.58	2.62	2.66	2.69	2.72	2.80	2.89
19	2.03	2.20	2.31	2.40	2.47	2.52	2.57	2.61	2.65	2.68	2.71	2.79	2.88
20	2.03	2.19	2.30	2.39	2.46	2.51	2.56	2.60	2.64	2.67	2.70	2.77	2.87
24	2.01	2.17	2.28	2.36	2.43	2.48	2.53	2.57	2.60	2.64	2.66	2.74	2.83
30	1.99	2.15	2.25	2.33	2.40	2.45	2.50	2.54	2.57	2.60	2.63	2.70	2.79
40	1.97	2.13	2.23	2.31	2.37	2.42	2.47	2.51	2.54	2.57	2.60	2.67	2.75
60	1.95	2.10	2.21	2.28	2.35	2.39	2.44	2.48	2.51	2.54	2.56	2.63	2.72
120	1.93	2.08	2.18	2.26	2.32	2.37	2.41	2.45	2.48	2.51	2.53	2.60	2.68
∞	1.92	2.06	2.16	2.23	2.29	2.34	2.38	2.42	2.45	2.48	2.50	2.56	2.64

$\alpha = 0.05$ (one-sided)

From C. W. Dunnett (1955) "A Multiple Comparison Procedure for Comparing Several Treatments with a Control." *Journal of the American Statistical Association* 50, 1112–1118. Reprinted with permission from *Journal of the American Statistical Association*. Copyright 1955 by the American Statistical Association. All rights reserved. C. W. Dunnett (1964) "New Tables for Multiple Comparisons with a Control." *Biometrics* 20, 482–491; and additional tables produced by C. W. Dunnett in 1980.

Table VI Multiple comparisons with the best and Dunnett tests;
$d_{\alpha,k,\nu}$ for $P(|d| \geq d_{\alpha,k,\nu}) = \alpha$

					$\alpha = 0.05$ (two-sided)								
$\nu \backslash k$	2	3	4	5	6	7	8	9	10	11	12	15	20
5	3.03	3.29	3.48	3.62	3.73	3.82	3.90	3.97	4.03	4.09	4.14	4.26	4.42
6	2.86	3.10	3.26	3.39	3.49	3.57	3.64	3.71	3.76	3.81	3.86	3.97	4.11
7	2.75	2.97	3.12	3.24	3.33	3.41	3.47	3.53	3.58	3.63	3.67	3.78	3.91
8	2.67	2.88	3.02	3.13	3.22	3.29	3.35	3.41	3.46	3.50	3.54	3.64	3.76
9	2.61	2.81	2.95	3.05	3.14	3.20	3.26	3.32	3.36	3.40	3.44	3.53	3.65
10	2.57	2.76	2.89	2.99	3.07	3.14	3.19	3.24	3.29	3.33	3.36	3.45	3.57
11	2.53	2.72	2.84	2.94	3.02	3.08	3.14	3.19	3.23	3.27	3.30	3.39	3.50
12	2.50	2.68	2.81	2.90	2.98	3.04	3.09	3.14	3.18	3.22	3.25	3.34	3.45
13	2.48	2.65	2.78	2.87	2.94	3.00	3.06	3.10	3.14	3.18	3.21	3.29	3.40
14	2.46	2.63	2.75	2.84	2.91	2.97	3.02	3.07	3.11	3.14	3.18	3.26	3.36
15	2.44	2.61	2.73	2.82	2.89	2.95	3.00	3.04	3.08	3.12	3.15	3.23	3.33
16	2.42	2.59	2.71	2.80	2.87	2.92	2.97	3.02	3.06	3.09	3.12	3.20	3.30
17	2.41	2.58	2.69	2.78	2.85	2.90	2.95	3.00	3.03	3.07	3.10	3.18	3.27
18	2.40	2.56	2.68	2.76	2.83	2.89	2.94	2.98	3.01	3.05	3.08	3.16	3.25
19	2.39	2.55	2.66	2.75	2.81	2.87	2.92	2.96	3.00	3.03	3.06	3.14	3.23
20	2.38	2.54	2.65	2.73	2.80	2.86	2.90	2.95	2.98	3.02	3.05	3.12	3.22
24	2.35	2.51	2.61	2.70	2.76	2.81	2.86	2.90	2.94	2.97	3.00	3.07	3.16
30	2.32	2.47	2.58	2.66	2.72	2.77	2.82	2.86	2.89	2.92	2.95	3.02	3.11
40	2.29	2.44	2.54	2.62	2.68	2.73	2.77	2.81	2.85	2.87	2.90	2.97	3.06
60	2.27	2.41	2.51	2.58	2.64	2.69	2.73	2.77	2.80	2.83	2.86	2.92	3.00
120	2.24	2.38	2.47	2.55	2.60	2.65	2.69	2.73	2.76	2.79	2.81	2.87	2.95
∞	2.21	2.35	2.44	2.51	2.57	2.61	2.65	2.69	2.72	2.74	2.77	2.83	2.91

Table VI   Multiple comparisons with the best and Dunnett tests; $d_{\alpha,k,\nu}$ for $P(d \geq d_{\alpha,k,\nu}) = \alpha$

						$\alpha = 0.01$ (one-sided)							
$\nu \backslash k$	2	3	4	5	6	7	8	9	10	11	12	15	20
5	3.90	4.21	4.43	4.60	4.73	4.85	4.94	5.03	5.11	5.17	5.24	5.39	5.59
6	3.61	3.88	4.07	4.21	4.33	4.43	4.51	4.59	4.64	4.70	4.76	4.89	5.06
7	3.42	3.66	3.83	3.96	4.07	4.15	4.23	4.30	4.35	4.40	4.45	4.57	4.72
8	3.29	3.51	3.67	3.79	3.88	3.96	4.03	4.09	4.14	4.19	4.23	4.34	4.48
9	3.19	3.40	3.55	3.66	3.75	3.82	3.89	3.94	3.99	4.04	4.08	4.18	4.31
10	3.11	3.31	3.45	3.56	3.64	3.71	3.78	3.83	3.88	3.92	3.96	4.06	4.18
11	3.06	3.25	3.38	3.48	3.56	3.63	3.69	3.74	3.79	3.83	3.86	3.96	4.08
12	3.01	3.19	3.32	3.42	3.50	3.56	3.62	3.67	3.71	3.75	3.79	3.88	3.99
13	2.97	3.15	3.27	3.37	3.44	3.51	3.56	3.61	3.65	3.69	3.73	3.81	3.92
14	2.94	3.11	3.23	3.32	3.40	3.46	3.51	3.56	3.60	3.64	3.67	3.76	3.87
15	2.91	3.08	3.20	3.29	3.36	3.42	3.47	3.52	3.56	3.60	3.63	3.71	3.82
16	2.88	3.05	3.17	3.26	3.33	3.39	3.44	3.48	3.52	3.56	3.59	3.67	3.78
17	2.86	3.03	3.14	3.23	3.30	3.36	3.41	3.45	3.49	3.53	3.56	3.64	3.74
18	2.84	3.01	3.12	3.21	3.27	3.33	3.38	3.42	3.46	3.50	3.53	3.61	3.71
19	2.83	2.99	3.10	3.18	3.25	3.31	3.36	3.40	3.44	3.47	3.50	3.58	3.68
20	2.81	2.97	3.08	3.17	3.23	3.29	3.34	3.38	3.42	3.45	3.48	3.56	3.65
24	2.77	2.92	3.03	3.11	3.17	3.22	3.27	3.31	3.35	3.38	3.41	3.48	3.57
30	2.72	2.87	2.97	3.05	3.11	3.16	3.21	3.24	3.28	3.31	3.34	3.41	3.50
40	2.68	2.82	2.92	2.99	3.05	3.10	3.14	3.18	3.21	3.24	3.27	3.34	3.42
60	2.64	2.78	2.87	2.94	3.00	3.04	3.08	3.12	3.15	3.18	3.20	3.27	3.35
120	2.60	2.73	2.82	2.89	2.94	2.99	3.03	3.06	3.09	3.12	3.14	3.20	3.28
∞	2.56	2.68	2.77	2.84	2.89	2.93	2.97	3.00	3.03	3.06	3.08	3.14	3.21

Table VI Multiple comparisons with the best and Dunnett tests;
$d_{\alpha,k,\nu}$ for $P(|d| \geq d_{\alpha,k,\nu}) = \alpha$

						$\alpha = 0.01$ (two-sided)							
$\nu \backslash k$	2	3	4	5	6	7	8	9	10	11	12	15	20
5	4.63	4.98	5.22	5.41	5.56	5.69	5.80	5.89	5.98	6.05	6.12	6.30	6.52
6	4.21	4.51	4.71	4.87	5.00	5.10	5.20	5.28	5.35	5.41	5.47	5.62	5.81
7	3.95	4.21	4.39	4.53	4.64	4.74	4.82	4.89	4.95	5.01	5.06	5.19	5.36
8	3.77	4.00	4.17	4.29	4.40	4.48	4.56	4.62	4.68	4.73	4.78	4.90	5.05
9	3.63	3.85	4.01	4.12	4.22	4.30	4.37	4.43	4.48	4.53	4.57	4.68	4.82
10	3.53	3.74	3.88	3.99	4.08	4.16	4.22	4.28	4.33	4.37	4.42	4.52	4.65
11	3.45	3.65	3.79	3.89	3.98	4.05	4.11	4.16	4.21	4.25	4.29	4.39	4.52
12	3.39	3.58	3.71	3.81	3.89	3.96	4.02	4.07	4.12	4.16	4.19	4.29	4.41
13	3.33	3.52	3.65	3.74	3.82	3.89	3.94	3.99	4.04	4.08	4.11	4.20	4.32
14	3.29	3.47	3.59	3.69	3.76	3.83	3.88	3.93	3.97	4.01	4.05	4.13	4.24
15	3.25	3.43	3.55	3.64	3.71	3.78	3.83	3.88	3.92	3.95	3.99	4.07	4.18
16	3.22	3.39	3.51	3.60	3.67	3.73	3.78	3.83	3.87	3.91	3.94	4.02	4.13
17	3.19	3.36	3.47	3.56	3.63	3.69	3.74	3.79	3.83	3.86	3.90	3.98	4.08
18	3.17	3.33	3.44	3.53	3.60	3.66	3.71	3.75	3.79	3.83	3.86	3.94	4.04
19	3.15	3.31	3.42	3.50	3.57	3.63	3.68	3.72	3.76	3.79	3.83	3.90	4.00
20	3.13	3.29	3.40	3.48	3.55	3.60	3.65	3.69	3.73	3.77	3.80	3.87	3.97
24	3.07	3.22	3.32	3.40	3.47	3.52	3.57	3.61	3.64	3.68	3.70	3.78	3.87
30	3.01	3.15	3.25	3.33	3.39	3.44	3.49	3.52	3.56	3.59	3.62	3.69	3.78
40	2.95	3.09	3.19	3.26	3.32	3.37	3.41	3.44	3.48	3.51	3.53	3.60	3.68
60	2.90	3.03	3.12	3.19	3.25	3.29	3.33	3.37	3.40	3.42	3.45	3.51	3.59
120	2.85	2.97	3.06	3.12	3.18	3.22	3.26	3.29	3.32	3.35	3.37	3.43	3.51
∞	2.79	2.92	3.00	3.06	3.11	3.15	3.19	3.22	3.25	3.27	3.29	3.35	3.42

Table VII Studentized range; $q_{\alpha,k,\nu}$ for $P(q \geq q_{\alpha,k,\nu}) = \alpha$

$$\alpha = 0.05$$

$\nu \backslash k$	2	3	4	5	6	7	8	9	10	11	12	13	14	15	16	17	18	19	20
1	18.1	26.7	32.8	37.2	40.5	43.1	45.4	47.3	49.1	50.6	51.9	53.2	54.3	55.4	56.3	57.2	58.0	58.8	59.6
2	6.09	8.28	9.80	10.89	11.73	12.43	13.03	13.54	13.99	14.39	14.75	15.08	15.38	15.65	15.91	16.14	16.36	16.57	16.77
3	4.50	5.88	6.83	7.51	8.04	8.47	8.85	9.18	9.46	9.72	9.95	10.16	10.35	10.52	10.69	10.84	10.98	11.12	11.24
4	3.93	5.00	5.76	6.31	6.73	7.06	7.35	7.60	7.83	8.03	8.21	8.37	8.52	8.67	8.80	8.92	9.03	9.14	9.24
5	3.64	4.60	5.22	5.67	6.03	6.33	6.58	6.80	6.99	7.17	7.32	7.47	7.60	7.72	7.83	7.93	8.03	8.12	8.21
6	3.46	4.34	4.90	5.31	5.63	5.89	6.12	6.32	6.49	6.65	6.79	6.92	7.04	7.14	7.24	7.34	7.43	7.51	7.59
7	3.34	4.16	4.68	5.06	5.35	5.59	5.80	5.99	6.15	6.29	6.42	6.54	6.65	6.75	6.84	6.93	7.01	7.08	7.16
8	3.26	4.04	4.53	4.89	5.17	5.40	5.60	5.77	5.92	6.05	6.18	6.29	6.39	6.48	6.57	6.65	6.73	6.80	6.87
9	3.20	3.95	4.42	4.76	5.02	5.24	5.43	5.60	5.74	5.87	5.98	6.09	6.19	6.28	6.36	6.44	6.51	6.58	6.65
10	3.15	3.88	4.33	4.66	4.91	5.12	5.30	5.46	5.60	5.72	5.83	5.93	6.03	6.12	6.20	6.27	6.34	6.41	6.47
11	3.11	3.82	4.26	4.58	4.82	5.03	5.20	5.35	5.49	5.61	5.71	5.81	5.90	5.98	6.06	6.14	6.20	6.27	6.33
12	3.08	3.77	4.20	4.51	4.75	4.95	5.12	5.27	5.40	5.51	5.61	5.71	5.80	5.88	5.95	6.02	6.09	6.15	6.21
13	3.06	3.73	4.15	4.46	4.69	4.88	5.05	5.19	5.32	5.43	5.53	5.63	5.71	5.79	5.86	5.93	6.00	6.06	6.11
14	3.03	3.70	4.11	4.41	4.64	4.83	4.99	5.13	5.25	5.36	5.46	5.56	5.64	5.72	5.79	5.86	5.92	5.98	6.03
15	3.01	3.67	4.08	4.37	4.59	4.78	4.94	5.08	5.20	5.31	5.40	5.49	5.57	5.65	5.72	5.79	5.85	5.91	5.96
16	3.00	3.65	4.05	4.34	4.56	4.74	4.90	5.03	5.15	5.26	5.35	5.44	5.52	5.59	5.66	5.73	5.79	5.84	5.90
17	2.98	3.62	4.02	4.31	4.52	4.70	4.86	4.99	5.11	5.21	5.31	5.39	5.47	5.55	5.61	5.68	5.74	5.79	5.84
18	2.97	3.61	4.00	4.28	4.49	4.67	4.83	4.96	5.07	5.17	5.27	5.35	5.43	5.50	5.57	5.63	5.69	5.74	5.79
19	2.96	3.59	3.98	4.26	4.47	4.64	4.79	4.92	5.04	5.14	5.23	5.32	5.39	5.46	5.53	5.59	5.65	5.70	5.75
20	2.95	3.58	3.96	4.24	4.45	4.62	4.77	4.90	5.01	5.11	5.20	5.28	5.36	5.43	5.50	5.56	5.61	5.66	5.71
24	2.92	3.53	3.90	4.17	4.37	4.54	4.68	4.81	4.92	5.01	5.10	5.18	5.25	5.32	5.38	5.44	5.50	5.55	5.59
30	2.89	3.48	3.84	4.11	4.30	4.46	4.60	4.72	4.83	4.92	5.00	5.08	5.15	5.21	5.27	5.33	5.38	5.43	5.48
40	2.86	3.44	3.79	4.04	4.23	4.39	4.52	4.63	4.74	4.82	4.90	4.98	5.05	5.11	5.17	5.22	5.27	5.32	5.36
60	2.83	3.40	3.74	3.98	4.16	4.31	4.44	4.55	4.65	4.73	4.81	4.88	4.94	5.00	5.06	5.11	5.15	5.20	5.24
120	2.80	3.36	3.69	3.92	4.10	4.24	4.36	4.47	4.56	4.64	4.71	4.78	4.84	4.90	4.95	5.00	5.04	5.09	5.13
∞	2.77	3.32	3.63	3.86	4.03	4.17	4.29	4.39	4.47	4.55	4.62	4.68	4.74	4.80	4.84	4.89	4.93	4.97	5.01

Adapted from *Biometrika Tables for Statisticians*, Vol. 1, 1966, edited by E. S. Pearson and H. O. Hartley, with permission of the Biometrica Trustees.

Table VII Studentized range; $q_{\alpha,k,\nu}$ for $P(q \geq q_{\alpha,k,\nu}) = \alpha$

$\alpha = 0.01$

ν\k	2	3	4	5	6	7	8	9	10	11	12	13	14	15	16	17	18	19	20
1	90.0	135	164	186	202	216	227	237	246	253	260	266	272	272	282	286	290	294	298
2	14.0	19.0	22.3	24.7	26.6	28.2	29.5	30.7	31.7	32.6	33.4	34.1	34.8	35.4	36.0	36.5	37.0	37.5	37.9
3	8.26	10.6	12.2	13.3	14.2	15.0	15.6	16.2	16.7	17.1	17.5	17.9	18.2	18.5	18.8	19.1	19.3	19.5	19.8
4	6.51	8.12	9.17	9.96	10.6	11.1	11.5	11.9	12.3	12.6	12.8	13.1	13.3	13.5	13.7	13.9	14.1	14.2	14.4
5	5.70	6.97	7.80	8.42	8.91	9.32	9.67	9.97	10.24	10.48	10.70	10.89	11.08	11.24	11.40	11.55	11.68	11.81	11.93
6	5.24	6.33	7.03	7.56	7.97	8.32	8.61	8.87	9.10	9.30	9.49	9.65	9.81	9.95	10.08	10.21	10.32	10.43	10.54
7	4.95	5.92	6.54	7.01	7.37	7.68	7.94	8.17	8.37	8.55	8.71	8.86	9.00	9.12	9.24	9.35	9.46	9.55	9.65
8	4.74	5.63	6.20	6.63	6.96	7.24	7.47	7.68	7.87	8.03	8.18	8.31	8.44	8.55	8.66	8.76	8.85	8.94	9.03
9	4.60	5.43	5.96	6.35	6.66	6.91	7.13	7.32	7.49	7.65	7.78	7.91	8.03	8.13	8.23	8.32	8.41	8.49	8.57
10	4.48	5.27	5.77	6.14	6.43	6.67	6.87	7.05	7.21	7.36	7.48	7.60	7.71	7.81	7.91	7.99	8.07	8.15	8.22
11	4.39	5.14	5.62	5.97	6.25	6.48	6.67	6.84	6.99	7.13	7.25	7.36	7.46	7.56	7.65	7.73	7.81	7.88	7.95
12	4.32	5.04	5.50	5.84	6.10	6.32	6.51	6.67	6.81	6.94	7.06	7.17	7.26	7.36	7.44	7.52	7.59	7.66	7.73
13	4.26	4.96	5.40	5.73	5.98	6.19	6.37	6.53	6.67	6.79	6.90	7.01	7.10	7.19	7.27	7.34	7.42	7.48	7.55
14	4.21	4.89	5.32	5.63	5.88	6.08	6.26	6.41	6.54	6.66	6.77	6.87	6.96	7.05	7.12	7.20	7.27	7.33	7.39
15	4.17	4.83	5.25	5.56	5.80	5.99	6.16	6.31	6.44	6.55	6.66	6.76	6.84	6.93	7.00	7.07	7.14	7.20	7.26
16	4.13	4.78	5.19	5.49	5.72	5.92	6.08	6.22	6.35	6.46	6.56	6.66	6.74	6.82	6.90	6.97	7.03	7.09	7.15
17	4.10	4.74	5.14	5.43	5.66	5.85	6.01	6.15	6.27	6.38	6.48	6.57	6.66	6.73	6.80	6.87	6.94	7.00	7.05
18	4.07	4.70	5.09	5.38	5.60	5.79	5.94	6.08	6.20	6.31	6.41	6.50	6.58	6.65	6.72	6.79	6.85	6.91	6.96
19	4.05	4.67	5.05	5.33	5.55	5.73	5.89	6.02	6.14	6.25	6.34	6.43	6.51	6.58	6.65	6.72	6.78	6.84	6.89
20	4.02	4.64	5.02	5.29	5.51	5.69	5.84	5.97	6.09	6.19	6.29	6.37	6.45	6.52	6.59	6.65	6.71	6.76	6.82
24	3.96	4.54	4.91	5.17	5.37	5.54	5.69	5.81	5.92	6.02	6.11	6.19	6.26	6.33	6.39	6.45	6.51	6.56	6.61
30	3.89	4.45	4.80	5.05	5.24	5.40	5.54	5.65	5.76	5.85	5.93	6.01	6.08	6.14	6.20	6.26	6.31	6.36	6.41
40	3.82	4.37	4.70	4.93	5.11	5.27	5.39	5.50	5.60	5.69	5.77	5.84	5.90	5.96	6.02	6.07	6.12	6.17	6.21
60	3.76	4.28	4.60	4.82	4.99	5.13	5.25	5.36	5.45	5.53	5.60	5.67	5.73	5.79	5.84	5.89	5.93	5.98	6.02
120	3.70	4.20	4.50	4.71	4.87	5.01	5.12	5.21	5.30	5.38	5.44	5.51	5.56	5.61	5.66	5.71	5.75	5.79	5.83
∞	3.64	4.12	4.40	4.60	4.76	4.88	4.99	5.08	5.16	5.23	5.29	5.35	5.40	5.45	5.49	5.54	5.57	5.61	5.65

Table VIII $FMax$ test; $F_\alpha Max$ for $P(FMax \geq F_\alpha Max) = \alpha$ for t treatment groups and ν degrees of freedom

$\alpha = 0.05$

$\nu \backslash t$	2	3	4	5	6	7	8	9	10	11	12
2	39.0	87.5	142	202	266	333	403	475	550	626	704
3	15.4	27.8	39.2	50.7	62.0	72.9	83.5	93.9	104	114	124
4	9.60	15.5	20.6	25.2	29.5	33.6	37.5	41.1	44.6	48.0	51.4
5	7.15	10.8	13.7	16.3	18.7	20.8	22.9	24.7	26.5	28.2	29.9
6	5.82	8.38	10.4	12.1	13.7	15.0	16.3	17.5	18.6	19.7	20.7
7	4.99	6.94	8.44	9.70	10.8	11.8	12.7	13.5	14.3	15.1	15.8
8	4.43	6.00	7.18	8.12	9.03	9.78	10.5	11.1	11.7	12.2	12.7
9	4.03	5.34	6.31	7.11	7.80	8.41	8.95	9.45	9.91	10.3	10.7
10	3.72	4.85	5.67	6.34	6.92	7.42	7.87	8.28	8.66	9.01	9.34
12	3.28	4.16	4.79	5.30	5.72	6.09	6.42	6.72	7.00	7.25	7.48
15	2.86	3.54	4.01	4.37	4.68	4.95	5.19	5.40	5.59	5.77	5.93
20	2.46	2.95	3.29	3.54	3.76	3.94	4.10	4.24	4.37	4.49	4.59
30	2.07	2.40	2.61	2.78	2.91	3.02	3.12	3.21	3.29	3.36	3.39
60	1.67	1.85	1.96	2.04	2.11	2.17	2.22	2.26	2.30	2.33	2.36
∞	1.00	1.00	1.00	1.00	1.00	1.00	1.00	1.00	1.00	1.00	1.00

$\alpha = 0.01$

$\nu \backslash t$	2	3	4	5	6	7	8	9	10	11	12
2	199	448	729	1036	1362	1705	2063	2342	2813	3204	3605
3	47.5	85	120	151	184	21(6)	24(9)	28(1)	31(0)	33(7)	36(1)
4	23.2	37	49	59	69	79	89	97	106	113	120
5	14.9	22	28	33	38	42	46	50	54	57	60
6	11.1	15.5	19.1	22	25	27	30	32	34	36	37
7	8.89	12.1	14.5	16.5	18.4	20	22	23	24	26	27
8	7.50	9.9	11.7	13.2	14.5	15.8	16.6	17.9	18.9	19.8	21
9	6.54	8.5	9.9	11.1	12.1	13.1	13.9	14.7	15.3	16.0	16.6
10	5.85	7.4	8.6	9.6	10.4	11.1	11.8	12.4	12.9	13.4	13.9
12	4.91	6.1	6.9	7.6	8.2	8.7	9.1	9.5	9.9	10.2	10.6
15	4.07	4.9	5.5	6.0	6.4	6.7	7.1	7.3	7.5	7.8	8.0
20	3.32	3.8	4.3	4.6	4.9	5.1	5.3	5.5	5.6	5.8	5.9
30	2.63	3.0	3.3	3.4	3.6	3.7	3.8	3.9	4.0	4.1	4.2
60	1.96	2.2	2.3	2.4	2.4	2.5	2.5	2.6	2.6	2.7	2.7
∞	1.00	1.0	1.0	1.0	1.0	1.0	1.0	1.0	1.0	1.0	1.0

Table IX *F* test power curves for fixed effects model analysis of variance

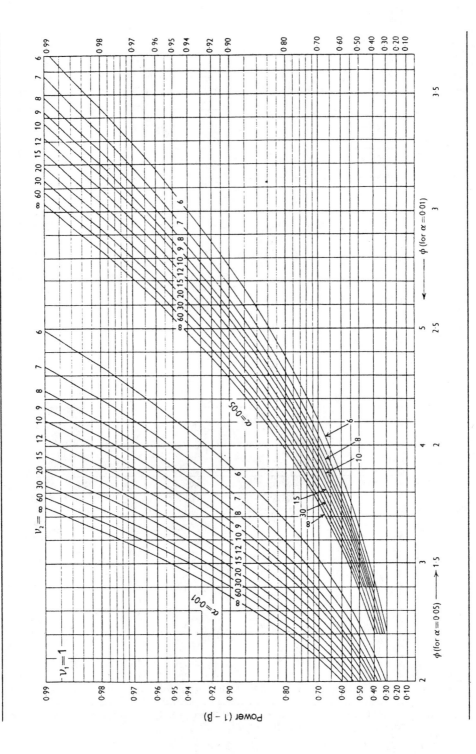

Adapted with permission from *Biometrika Tables for Statisticians*, Vol. 1, 1966, edited by E. S. Pearson and H. O. Hartley.

Table IX *F* test power curves for fixed effects model analysis of variance

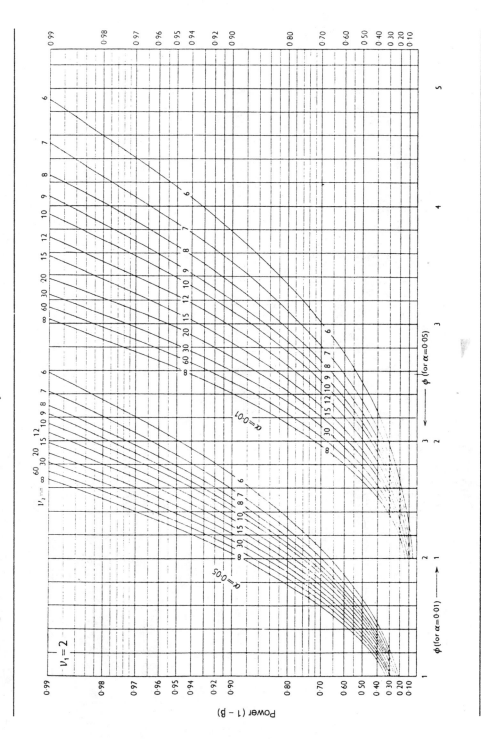

Table IX *F* test power curves for fixed effects model analysis of variance

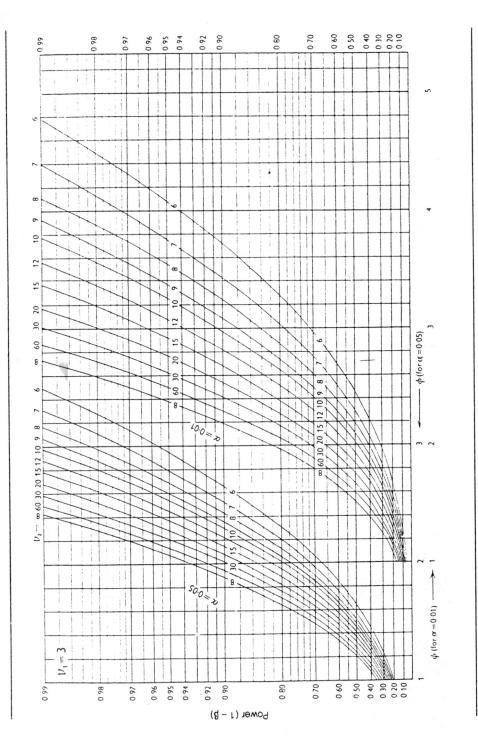

606

Table IX *F* test power curves for fixed effects model analysis of variance

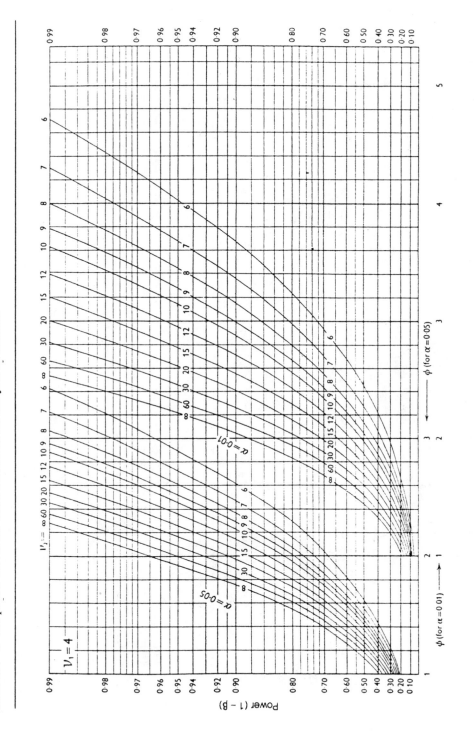

Table IX *F* test power curves for fixed effects model analysis of variance

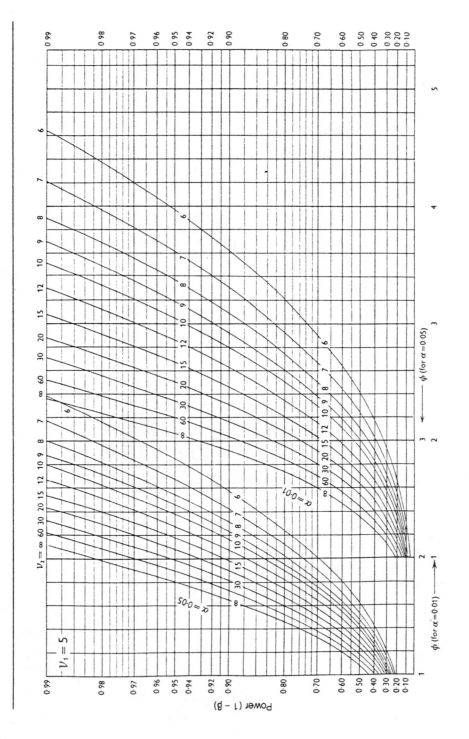

Table IX *F* test power curves for fixed effects model analysis of variance

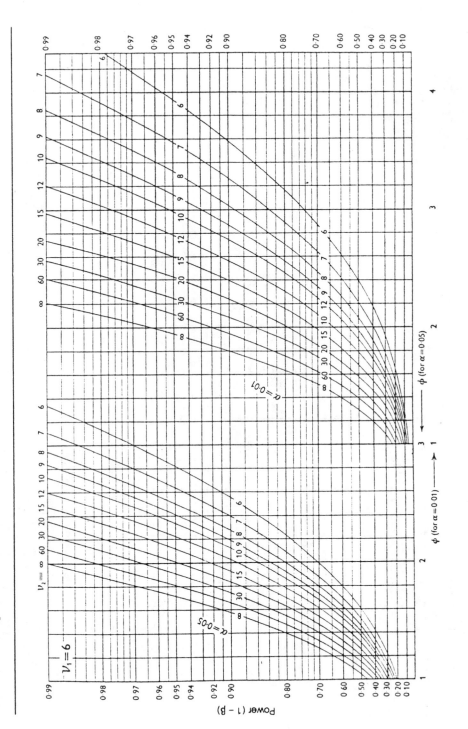

Table IX *F* test power curves for fixed effects model analysis of variance

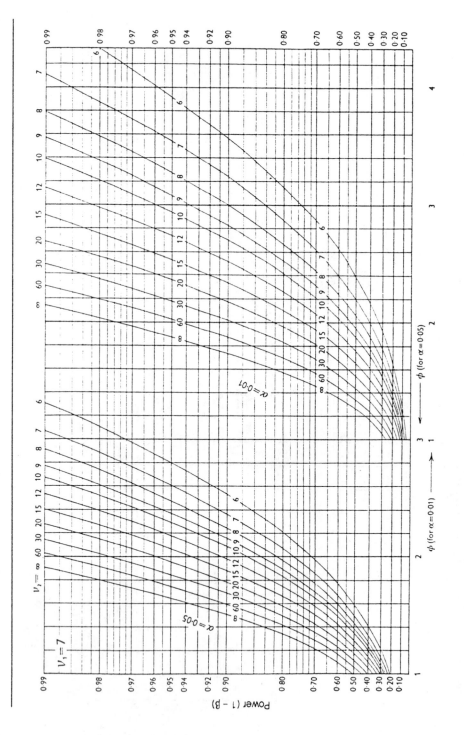

Table IX F test power curves for fixed effects model analysis of variance

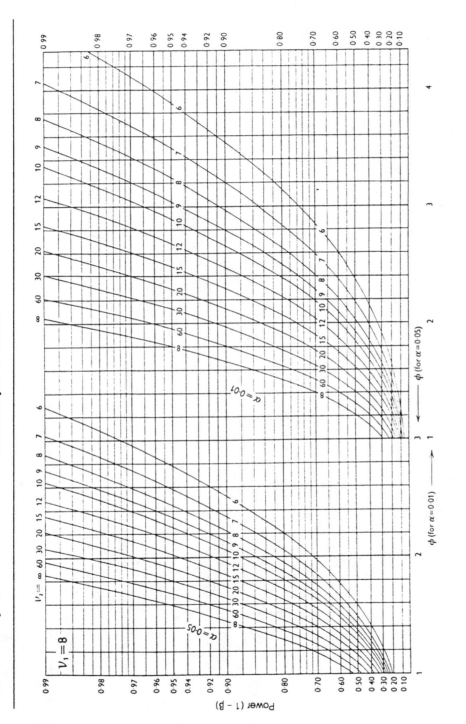

Table IX *F* test power curves for fixed effects model analysis of variance

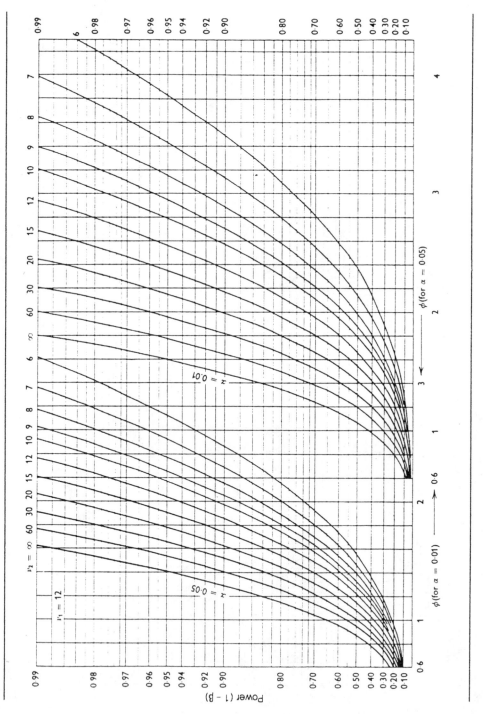

Table IX *F* test power curves for fixed effects model analysis of variance

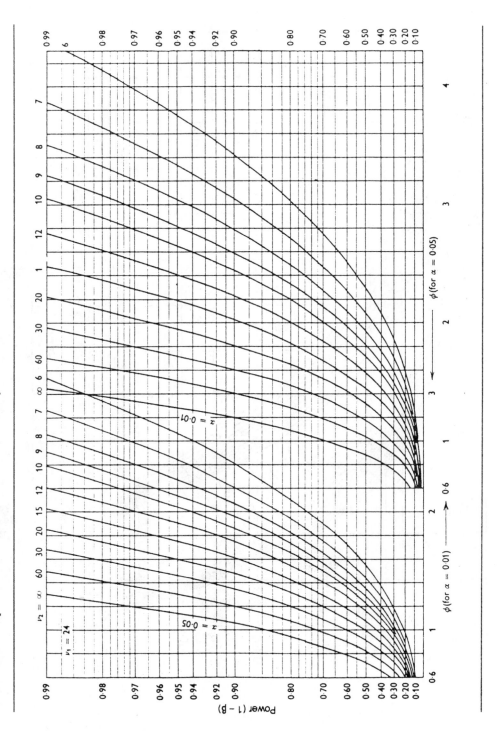

Table X *F* test power curves for random effects model analysis of variance

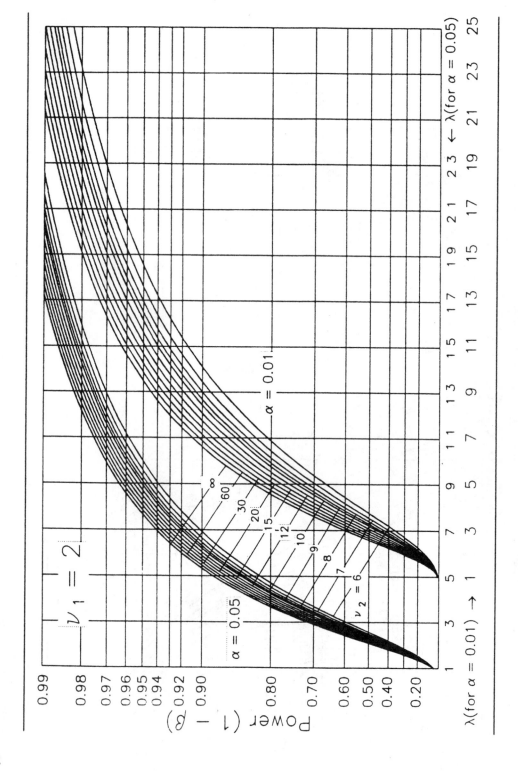

Table X *F* test power curves for random effects model analysis of variance

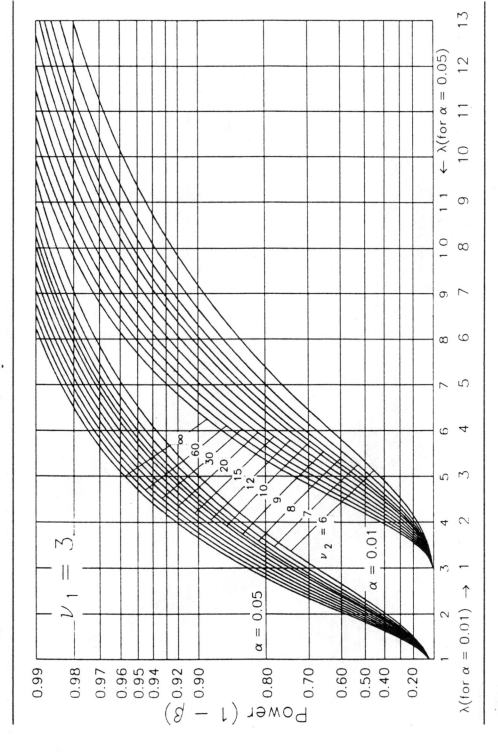

Table X *F* test power curves for random effects model analysis of variance

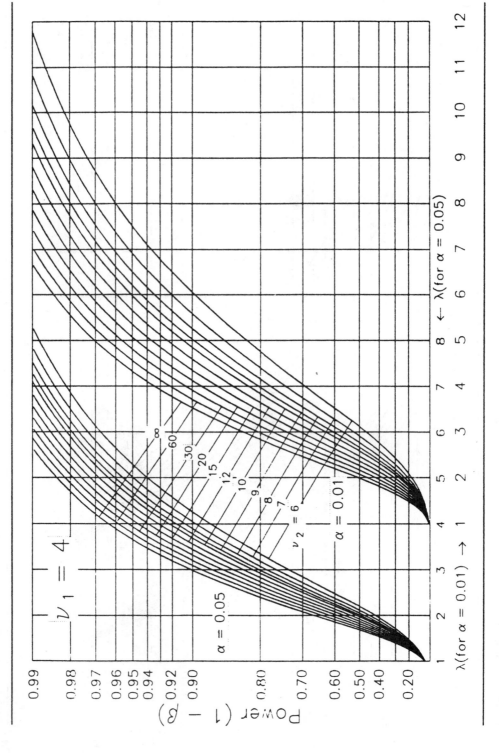

Table X *F* test power curves for random effects model analysis of variance

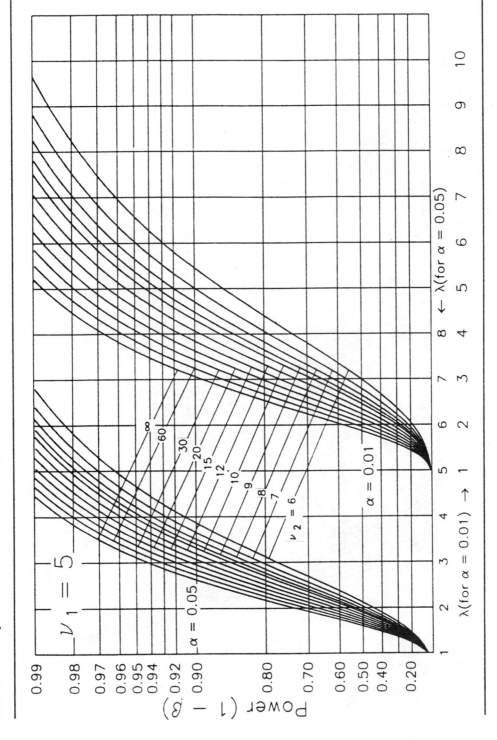

Table X F test power curves for random effects model analysis of variance

Table X *F* test power curves for random effects model analysis of variance

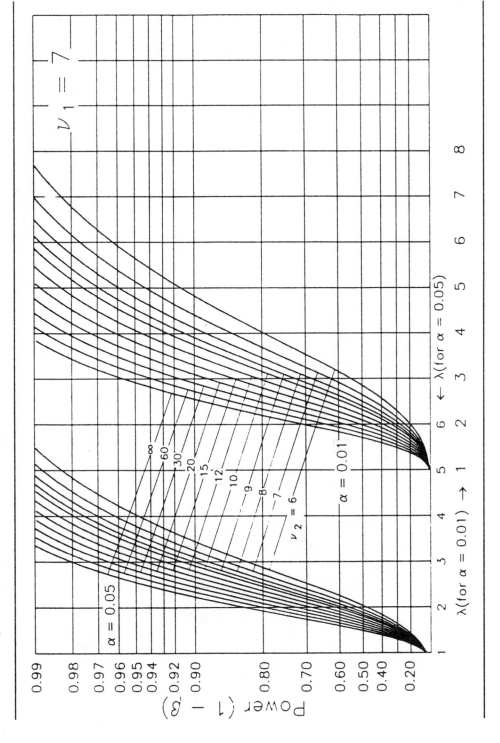

Table X *F* test power curves for random effects model analysis of variance

Table X *F* test power curves for random effects model analysis of variance

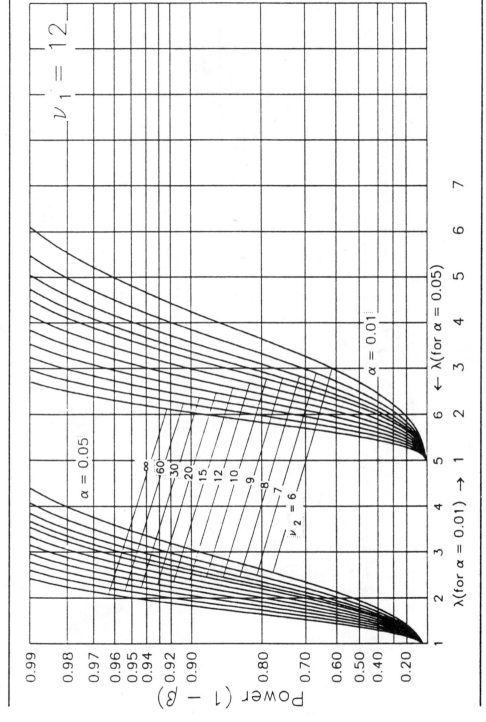

Table X *F* test power curves for random effects model analysis of variance

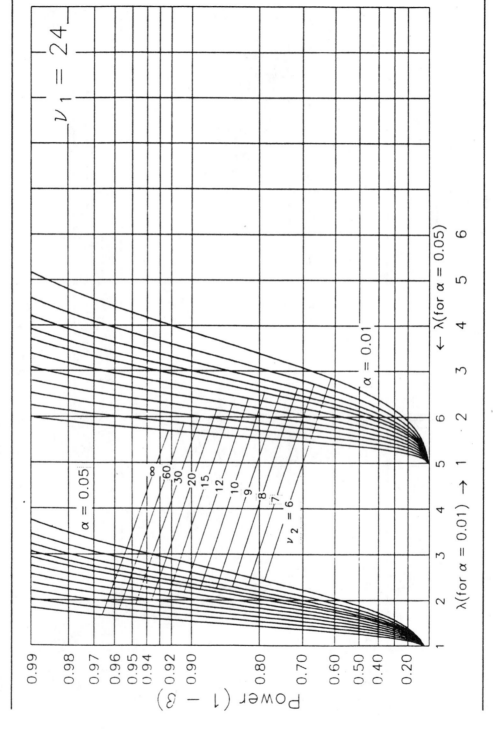

Table XI Orthogonal polynomials

X_j	$t=3$		$t=4$			$t=5$				$t=6$					$t=7$					
	P_1	P_2	P_1	P_2	P_3	P_1	P_2	P_3	P_4	P_1	P_2	P_3	P_4	P_5	P_1	P_2	P_3	P_4	P_5	P_6
1	-1	1	-3	1	-1	-2	2	-1	1	-5	5	-5	1	-1	-3	5	-1	3	-1	1
2	0	-2	-1	-1	3	-1	-1	2	-4	-3	-1	7	-3	5	-2	0	1	-7	4	-6
3	1	1	1	-1	-3	0	-2	0	6	-1	-4	4	2	-10	-1	-3	1	1	-5	15
4			3	1	1	1	-1	-2	-4	1	-4	-4	2	10	0	-4	0	6	0	-20
5						2	2	1	1	3	-1	-7	-3	-5	1	-3	-1	1	5	15
6										5	5	5	1	1	2	0	-1	-7	-4	-6
7															3	5	1	3	1	1
$\sum_{j=1}^{t}\{P_i(X_j)\}^2$	2	6	20	4	20	10	14	10	70	70	84	180	28	252	28	84	6	154	84	924
λ	1	3	2	1	$\tfrac{10}{3}$	1	1	$\tfrac{5}{6}$	$\tfrac{35}{12}$	2	$\tfrac{3}{2}$	$\tfrac{5}{3}$	$\tfrac{7}{12}$	$\tfrac{21}{10}$	1	1	$\tfrac{1}{6}$	$\tfrac{7}{12}$	$\tfrac{7}{20}$	$\tfrac{77}{60}$

X_j	$t=8$						$t=9$						$t=10$					
	P_1	P_2	P_3	P_4	P_5	P_6	P_1	P_2	P_3	P_4	P_5	P_6	P_1	P_2	P_3	P_4	P_5	P_6
1	-7	7	-7	7	-7	1	-4	28	-14	14	-4	4	-9	6	-42	18	-6	3
2	-5	1	5	-13	23	-5	-3	7	7	-21	11	-17	-7	2	14	-22	14	-11
3	-3	-3	7	-3	-17	9	-2	-8	13	-11	-4	22	-5	-1	35	-17	-1	10
4	-1	-5	3	9	-15	-5	-1	-17	9	9	-9	1	-3	-3	31	3	-11	6
5	1	-5	-3	9	15	-5	0	-20	0	18	0	-20	-1	-4	12	18	-6	-8
6	3	-3	-7	-3	17	9	1	-17	-9	9	9	1	1	-4	-12	18	6	-8
7	5	1	-5	-13	-23	-5	2	-8	-13	-11	4	22	3	-3	-31	3	11	6
8	7	7	7	7	7	1	3	7	-7	-21	-11	-17	5	-1	-35	-17	1	10
9							4	28	14	14	4	4	7	2	-14	-22	-14	-11
10													9	6	42	18	6	3
$\sum_{j=1}^{t}\{P_i(X_j)\}^2$	168	168	264	616	2184	264	60	2772	990	2002	468	1980	330	132	8580	2860	780	660
λ	2	1	$\tfrac{2}{3}$	$\tfrac{7}{12}$	$\tfrac{7}{10}$	$\tfrac{11}{60}$	1	3	$\tfrac{5}{6}$	$\tfrac{7}{12}$	$\tfrac{3}{20}$	$\tfrac{11}{60}$	2	$\tfrac{1}{2}$	$\tfrac{5}{3}$	$\tfrac{5}{12}$	$\tfrac{1}{10}$	$\tfrac{11}{240}$

Table XII 10,000 random digits

40092	81747	62038	32614	91062	95535	85635	46183	52525	86502
04040	77141	68126	39738	00241	90805	90476	31866	43819	72869
27744	59344	15764	27705	86115	10818	33029	47185	78385	47791
03012	35878	31415	02956	73710	73199	71014	68069	42556	25033
03756	14835	94850	14777	02637	98934	18678	52385	53701	97816
32815	29692	41490	78596	23967	53733	32292	33595	89471	51662
13702	22046	10820	74519	09094	42165	96372	43048	82936	34352
90988	43962	72505	00500	42042	73894	50219	71653	90329	61575
62355	05504	69481	95689	06051	27611	72996	91049	80893	71328
49076	28508	71392	13453	51243	18476	54021	05696	11779	15863
36273	09222	95599	32931	21871	98668	99815	94235	42693	51724
68993	03491	80984	48526	90843	29395	51200	97425	39987	36822
39920	31267	60848	05034	03234	64340	09434	53143	05462	34473
54411	06333	36913	08937	96649	36916	29084	26470	41677	97141
36589	59006	78749	47682	81371	87395	46397	88578	53407	45513
50992	10021	61655	46972	77136	72126	75735	54573	21754	34005
89336	50439	97392	55854	86657	02737	25120	67498	40731	63664
54553	95650	25878	12743	41984	83063	34507	74733	54853	04773
87225	55691	25090	31227	86467	65431	19301	21752	57547	99668
89701	46487	48959	69554	59786	04082	30302	18492	82721	90393
04307	47780	07824	26546	52978	22552	95034	57905	33641	63382
73569	68313	08308	32795	27608	20956	92853	60375	10364	56966
14979	67281	49874	26421	25443	48250	15627	85412	67329	71950
63914	98384	78123	83289	46311	38100	89622	74531	91274	67461
86390	62009	13480	13275	85547	67966	55299	24147	21527	38419
32007	57426	35585	85429	18445	49946	93822	81152	08889	94365
01836	33548	68505	64965	21008	99708	75050	86951	00673	73955
32452	64986	81436	50212	35425	13971	11767	48191	08463	83295
91424	82977	32735	80372	33184	56791	72463	94831	11268	26694
38244	80884	35719	07613	73294	39929	89473	92680	85910	06938
26010	21397	37765	63680	04611	37636	92158	23594	14873	09599
67054	42605	18159	00922	60210	80042	76776	48738	64415	99442
83513	91260	18937	89770	87417	66732	54208	99378	77020	84131
65512	49102	25594	51608	64913	61490	73883	29741	66253	68670
77810	94342	98901	37379	05225	97931	51528	10666	37760	89043
07968	43404	73783	69894	01357	34726	89661	89342	30408	82356
85370	05580	25323	86903	31950	82648	57589	58573	58332	55619
60923	13031	43571	62495	96389	46218	61748	42218	95879	97246
85495	42421	04882	67652	77291	12909	73833	45871	71388	82924
06328	30206	35409	78693	67795	64053	64309	51064	02516	29597
76447	35060	98444	14312	04058	85851	61351	22310	36618	90115
78866	29368	84660	13189	86648	88606	30739	67032	70300	45249
02448	59524	35486	74847	64272	26363	89711	29296	66690	35704
90016	85192	20040	97348	38337	39697	74895	21585	09245	63855
30806	23228	45362	65937	86967	98883	50230	70750	34638	60422
61224	84197	05618	34422	46881	35190	65108	83574	91763	74682
69058	80489	83177	79446	28867	06358	21001	13830	51949	33314
29731	88669	90396	86789	06896	21048	21903	65161	69464	18954
55831	36845	20465	09528	42037	57535	15875	83455	15612	99552
54654	11577	44676	07215	37288	90974	48352	07127	62359	36357

Computed by R. D. Axelson and R. O. Kuehl.

Table XII   10,000 random digits

26862	23753	16576	48185	26188	37728	25811	97801	49522	37980
88683	98306	07900	23707	99948	33525	74273	47346	28068	67382
11649	19469	97566	73268	84821	17286	78963	69433	27656	20853
11539	08341	75699	97873	28867	21957	62406	20655	43287	20888
56151	48566	02341	00975	07214	50747	88664	98610	27313	64162
69530	17535	27838	40610	50380	78140	07497	36527	50167	46282
15818	76764	92511	03502	03169	06702	59346	35361	77128	18245
50336	94817	84848	65960	76977	95100	63306	19380	46687	31515
28406	21331	85488	88274	48787	44075	86769	09610	02927	33135
72796	02583	10970	77788	83695	69263	20800	32501	91294	98923
19569	04841	41380	89265	01799	11059	93494	66138	13304	12325
75557	66470	34334	82621	74506	43534	26514	11624	42412	76386
11018	38364	01933	32767	56338	85843	84565	62142	29687	17325
91226	45783	50106	60482	03949	91614	06628	65204	65570	32278
14651	00799	86285	40491	42792	51441	91259	33280	21469	14226
45076	54205	08493	36125	46555	08566	12084	62827	00321	56568
26297	93649	18706	64045	25840	79460	46865	92165	93820	34790
92147	37854	71738	19132	03564	68174	93762	37616	05231	55992
78686	88798	72430	69712	59483	93455	12339	77474	55033	80771
39337	46482	68067	62770	22840	32935	56812	14135	19428	19452
22252	15585	55870	52572	14044	26189	89418	52296	38387	75269
57240	75441	73297	63779	59276	17017	85257	30527	12722	35094
04322	54208	11657	11593	70223	60145	99744	91446	70523	31993
27666	19664	78042	51445	47023	30784	69771	82770	55057	52078
53574	41181	17109	61114	83876	95597	41259	37891	84897	89452
57403	53175	09644	44756	43387	57407	48145	42504	63603	37123
47663	74298	42107	67427	31764	81212	60736	57428	14755	42171
23443	29258	09203	79058	55323	12976	32399	17237	85784	02217
51643	09345	84572	28560	87805	02876	45418	73275	39984	96349
80051	75581	23727	58813	75587	23116	83055	86758	00086	80536
43926	60090	60696	72918	51179	49157	07915	65483	78143	10451
10254	05670	92734	24229	38248	84834	02502	09713	34985	31628
59972	80565	94063	83430	54499	83544	62526	58198	51422	43211
60281	89538	02140	93302	13432	15221	69413	13619	12440	49242
16811	18472	29709	39684	72700	32277	03202	66143	91079	90161
01902	59532	16742	87759	74717	38255	98729	42024	47121	20304
41733	54639	64241	70303	57220	78894	00192	03386	38487	27908
43355	69142	53361	09024	11162	46052	93213	17886	91801	98031
70729	22517	60942	41250	37148	24255	03358	38681	54169	01229
99082	83541	65511	79195	35411	30115	34826	92891	16193	49832
06263	94134	73079	02018	93129	25999	62186	62716	39844	71598
25600	31706	45184	04039	12757	17543	57061	35372	14213	75793
38681	10679	56604	69704	00663	87405	30603	90223	22174	06601
80481	16829	91125	21262	03530	61672	46950	15458	10962	44808
47010	50624	57003	15278	31486	44583	73363	00197	45102	14852
77358	22444	92701	91049	78909	10623	57688	61120	83266	83874
38510	77101	96660	96889	78941	07529	22497	96254	04580	48105
81400	38106	00865	27113	91974	34324	39319	71900	27482	66416
99793	34087	52787	02700	29651	15165	64316	69420	08167	94154
55128	29344	80171	25092	80960	14721	59328	06345	91581	82399

Table XII 10,000 random digits

27716	39430	74892	55062	77137	37327	49419	30022	98867	91330
88339	55349	44443	05790	77996	60410	41369	22078	31121	13179
37968	25059	37540	04687	53545	26673	40810	48176	06334	98994
15917	55975	42519	36245	01377	03940	29064	34000	65999	20130
70360	58659	09025	65477	56751	44808	61307	88623	89037	68398
00645	21621	36761	53781	73718	58716	81236	17964	70404	57782
89194	42646	63316	59445	18655	46159	16175	17398	72326	11634
55980	75632	55491	14356	56454	00347	62540	86968	68038	58485
13171	81146	28515	59825	88041	74881	65046	53098	59673	49321
86112	27709	22965	16439	21368	60597	39390	78060	05766	87653
43520	50388	33109	80118	52631	28818	77139	37368	90715	76626
38819	71158	07453	94947	52652	20698	05598	87513	45302	94056
33517	96172	08551	91814	11724	65092	65214	17879	79958	06629
83251	07374	39275	31071	41215	42198	15736	20401	62220	15274
41449	51749	75858	33478	82088	93677	32579	56769	13978	17131
84803	30272	85826	18935	24913	79617	72859	79433	94633	40520
84633	75447	06239	52081	94605	29414	90088	86603	01489	17407
01638	69018	93805	73036	07877	41875	73920	87707	71362	30956
28674	59616	07748	09703	49503	74360	14030	35940	10022	42009
83293	38075	26568	69884	37289	14617	18927	93178	30591	23286
68589	63091	88890	38333	71939	37194	21497	13995	35084	15634
45432	44958	68132	99054	63817	30182	73247	57265	53261	25571
11054	89885	77436	86004	68590	73162	92891	11720	71241	89156
42514	56335	12990	36130	46050	93096	63546	78406	62320	51346
07047	90954	35462	86360	24284	09102	25426	15567	25015	88667
17561	05233	52549	90437	92488	57833	83542	05540	91798	98934
05911	53885	56556	40850	14137	97683	84795	51261	66590	77263
17612	20461	62032	47587	17035	83371	94995	50315	41561	21500
06111	59316	99244	53028	21979	10783	44488	22348	98002	18472
74570	09252	44243	51742	74563	07363	56710	15735	73850	23924
44776	03402	35319	10026	27044	19091	51864	70008	56725	34529
83366	42305	34398	66549	99205	74699	57967	96962	17378	43224
01073	59291	56232	91521	98063	96809	25210	96350	46086	22857
42767	58685	31976	44953	26620	49569	85394	15279	74548	74456
13285	58416	71593	12007	65202	61687	48650	27987	23498	72343
76273	17196	84114	42240	14314	49976	68714	33538	06533	87191
24056	80981	20168	29047	20856	65202	16533	28520	11420	96331
49676	14882	65467	73855	03075	61191	64955	48555	91374	80971
64991	57009	77897	83153	95013	86277	70494	18046	56932	20434
28089	98413	31829	62562	48896	20935	94369	48925	97773	37099
96534	97921	13604	70506	29624	44005	92916	41939	35505	90644
50836	96073	16019	48998	21487	50844	90073	76007	87108	54137
28746	47219	30895	23263	60178	50722	15734	60946	89808	82700
80584	53948	44644	76912	98636	95840	78725	77378	30973	15414
32514	27970	73626	19182	31168	63792	07136	97354	59750	85256
15164	71201	39626	58961	33178	38641	85647	45973	19310	84457
39324	98893	22000	96405	15322	77189	92508	53154	47615	14436
83484	60576	62727	96233	01802	98190	76160	59963	84478	45070
42484	28641	47335	80096	84105	19111	74979	42318	84907	52870
65911	27235	53577	04477	95852	70219	82456	19107	99280	08322

Table XII   10,000 random digits

94221	07150	80463	03658	07013	55333	56464	04052	26053	81893
84431	13952	08149	80802	09136	71238	55048	09919	40389	61875
05455	88879	13449	32774	53964	18829	21872	74294	85282	92231
70079	21122	27004	58731	21160	80238	68780	82384	20602	92898
11446	25521	48438	72828	21550	35582	32768	69773	38451	79113
45708	85141	04684	34025	71208	19992	51615	71231	88276	58194
79659	33506	87595	03537	59265	92924	83027	41286	01315	14180
29884	97311	35812	45807	77277	94148	67863	24496	87510	42172
43279	67351	27095	17012	94794	30437	68917	13450	03254	05915
76413	06486	82936	33337	78926	25558	33856	14237	17551	59814
22492	13665	36925	48848	50459	49600	52743	20066	76869	79832
58291	55383	67639	66799	62393	04016	17872	55180	62652	10594
27126	62422	62990	56929	84904	19431	98746	55471	68157	65921
00842	24512	32501	35772	28232	10170	72564	21429	20642	27757
73267	31540	46725	32423	36226	89466	48202	25074	75094	60558
00964	87753	51908	44409	93930	08466	98989	17080	34531	77828
19944	78118	79260	23252	34690	76692	19405	11712	51394	48362
70288	68077	60880	23921	11896	20005	08466	86560	31270	69837
79468	67331	29400	11097	23247	71989	00401	05865	67119	80106
46373	06322	61880	71922	24235	47410	66986	26995	12239	86031
61516	21667	78846	71110	41710	64814	57890	26749	11543	88532
41099	45697	50877	86382	62166	07811	91211	05631	26154	32851
11024	75195	61800	59089	42760	38788	66395	47948	35330	74353
70868	80366	38791	71706	26600	01965	70249	43459	66862	15575
87057	86347	29685	79945	79152	46353	73810	50543	60423	98381
22388	64789	68609	63841	44728	45553	38269	97970	56182	01730
46942	66348	10183	50169	19373	54011	74267	03039	40453	44193
29806	89043	20745	66576	16824	75842	76690	65355	04887	68928
14617	79295	99744	62723	68156	47838	17085	81836	43124	47199
25809	85025	49691	68786	43859	50193	72474	26608	21241	66151
34780	60969	42518	07858	17204	98628	46176	04768	96909	23402
94677	73718	64271	59301	31606	66479	97437	87625	69066	53509
12250	10164	28403	22306	32881	58027	08882	01821	67718	74001
15253	78715	01384	02794	91983	00765	88148	56337	31053	17782
74787	90110	69156	62945	69569	13906	55002	99567	49338	95426
57379	47097	40942	61028	56704	29036	32246	57688	76317	67112
25663	48264	05635	98825	24427	77318	19674	44417	51022	69841
88330	25095	40796	63642	62300	23016	32759	43912	14635	06138
27651	22887	84669	92353	24804	40367	78290	45034	17498	93457
44555	85772	01754	36006	91187	72110	23796	63114	12759	12726
43299	87460	52469	24806	79248	53238	39640	49745	48627	56065
08740	06533	80838	38622	15149	08813	83839	96731	26326	13098
01029	77011	70206	54340	62941	29887	87785	78345	90186	65065
77885	85449	53717	52475	89906	40846	13143	50758	29288	77939
09852	08312	18396	34775	60762	52712	97343	75834	22023	87492
00298	02548	91069	14424	10383	88365	27797	58311	75041	45545
22831	43108	29166	08517	07100	78997	90983	13085	19481	86602
43584	61723	29354	96769	77087	33578	93149	23121	12969	04072
69116	44955	37-85	63158	44994	41991	43607	82729	72299	06758
49035	19029	73264	77376	96988	72124	70367	68128	89698	06584

Table XIII 400 random permutations of first 10 integers

9	5	6	1	6	5	3	1	6	6	1	3	1	7	2	7	6	10	10	3
7	8	9	5	1	1	8	8	5	5	9	10	2	4	5	5	4	2	4	2
10	7	1	8	5	6	1	3	3	9	3	1	3	6	9	9	9	4	3	5
6	6	7	2	3	10	4	10	9	8	8	2	5	1	7	1	2	1	9	1
3	4	4	4	9	3	10	5	1	7	7	8	7	5	6	8	1	3	8	10
2	10	2	9	7	9	5	6	7	2	5	5	4	9	1	6	8	8	2	8
8	9	3	3	2	4	2	7	4	4	4	6	10	2	4	10	5	9	1	7
1	2	8	7	10	8	7	9	8	3	6	9	9	3	3	4	10	7	6	9
4	3	10	6	8	2	6	4	10	1	2	4	6	10	10	3	3	5	7	6
5	1	5	10	4	7	9	2	2	10	10	7	8	8	8	2	7	6	5	4
5	5	10	3	9	3	3	2	9	9	3	10	2	6	9	2	2	7	9	5
4	9	8	4	7	5	10	5	6	3	10	7	7	3	7	10	4	3	2	2
7	8	7	2	2	6	8	1	10	10	9	4	8	1	3	6	7	6	6	1
3	7	1	9	10	9	9	6	4	4	4	1	3	5	4	3	8	5	1	9
10	3	3	10	1	8	1	10	1	8	5	5	6	4	8	5	1	1	3	8
2	1	2	6	5	4	4	9	2	5	6	9	10	10	2	1	9	10	8	10
8	10	5	5	3	7	2	3	3	7	7	8	5	7	1	9	6	4	7	7
1	2	6	7	6	10	5	4	5	6	8	6	9	9	6	8	5	2	5	3
6	6	4	1	4	2	7	8	7	1	2	3	4	8	10	7	10	9	4	4
9	4	9	8	8	1	6	7	8	2	1	2	1	2	5	4	3	8	10	6
1	9	4	9	2	3	10	10	7	6	9	5	3	7	8	3	1	6	1	8
4	6	10	10	5	6	7	5	4	3	7	7	5	9	7	4	7	1	8	10
9	8	9	4	8	1	2	9	8	2	10	4	10	6	2	5	5	5	3	4
2	3	7	3	7	10	5	1	1	4	1	6	1	1	5	9	2	7	2	3
10	4	8	5	1	7	9	4	5	7	6	9	4	10	1	8	8	10	9	1
8	10	6	1	3	5	1	8	3	1	8	10	6	2	9	10	4	2	7	5
5	7	3	2	10	2	4	2	10	10	2	3	2	3	10	6	6	4	6	6
7	5	2	7	6	4	3	6	6	9	5	1	7	5	3	7	10	9	4	9
3	2	1	6	4	9	6	3	9	8	3	8	8	8	4	1	9	8	10	2
6	1	5	8	9	8	8	7	2	5	4	2	9	4	6	2	3	3	5	7
3	8	4	4	8	7	8	10	8	2	1	2	10	4	7	7	2	2	7	9
9	7	8	9	6	6	6	7	1	4	8	4	7	6	4	8	5	1	4	4
2	1	6	1	3	3	10	6	10	5	2	5	3	2	9	4	3	10	1	3
1	10	2	3	2	4	7	2	2	8	9	7	4	10	6	6	8	7	6	2
5	6	9	7	9	1	2	9	4	10	3	9	9	8	3	2	4	8	9	8
8	2	10	8	10	2	1	3	9	1	7	3	1	3	8	9	1	3	10	7
10	9	3	6	1	9	4	8	3	6	6	1	8	5	2	1	9	5	5	6
6	5	7	2	4	8	5	1	6	9	4	10	6	1	1	3	6	6	3	1
4	4	1	5	7	10	9	4	5	3	5	6	2	9	5	10	7	4	8	5
7	3	5	10	5	5	3	5	7	7	10	8	5	7	10	5	10	9	2	10
1	5	7	5	4	8	8	2	3	4	3	2	6	5	1	7	8	5	1	1
5	3	4	9	9	7	3	6	3	2	1	1	8	4	3	9	4	7	8	9
4	1	9	8	2	9	7	8	7	7	6	9	9	3	8	2	9	1	9	5
10	7	1	1	6	1	9	2	1	1	10	8	4	6	7	6	5	9	4	10
2	2	3	3	10	2	4	9	8	9	8	3	7	10	6	3	7	6	7	6
3	10	10	10	5	4	2	7	3	3	4	4	5	1	5	5	3	4	10	2
8	6	5	6	8	5	5	10	5	5	2	5	2	7	2	8	1	10	2	8
6	9	8	2	3	10	6	5	6	8	7	10	10	2	10	4	10	8	5	4
7	4	2	7	1	3	1	1	10	10	5	7	1	8	4	1	2	2	6	7
9	8	6	4	7	6	10	4	4	6	9	6	3	9	9	10	6	3	3	3

Computed by R. D. Axelson and R. O. Kuehl.

Table XIII   400 random permutations of first 10 integers

5	2	4	1	5	4	9	3	5	3	9	2	1	8	9	4	4	7	5	4
8	7	1	8	8	3	7	10	9	9	10	3	2	10	7	3	7	9	10	6
7	1	7	9	7	7	10	1	8	5	7	1	7	3	3	8	2	4	7	10
1	3	2	2	9	8	1	6	4	1	3	9	9	8	6	2	6	5	4	2
9	4	5	6	1	1	6	8	6	2	5	6	6	4	8	10	10	2	8	5
2	9	6	4	2	9	4	2	7	7	1	4	5	9	4	9	9	6	6	7
10	8	9	10	4	10	2	9	3	6	4	7	8	5	1	1	8	8	3	8
4	5	8	5	10	2	8	5	1	8	2	8	3	2	2	7	3	1	2	9
6	10	3	3	3	5	3	4	10	4	8	5	10	1	5	5	5	3	1	3
3	6	10	7	6	6	5	7	2	10	6	10	4	7	10	6	1	10	9	1

8	5	2	4	5	8	2	7	10	7	5	1	10	5	2	4	6	3	2	3
3	7	8	3	8	5	9	2	1	2	7	10	3	9	6	9	8	2	7	10
4	8	10	9	4	3	8	3	6	3	8	5	4	7	7	7	9	6	4	2
6	9	1	10	3	10	4	1	4	10	3	8	1	3	3	2	2	1	9	1
5	4	7	7	10	4	6	6	3	8	6	9	6	6	1	10	7	8	1	5
7	1	4	2	7	2	7	5	8	9	10	6	5	8	4	6	4	4	5	9
1	6	5	1	9	9	10	10	5	1	4	2	9	10	5	8	3	5	10	6
2	2	3	8	6	6	1	8	9	6	2	7	7	2	8	5	10	7	8	4
10	3	9	5	2	7	3	9	7	4	9	3	2	4	9	3	5	9	3	7
9	10	6	6	1	1	5	4	2	5	1	4	8	1	10	1	1	10	6	8

9	5	3	4	2	9	9	4	5	9	3	2	8	10	3	6	7	7	10	4
6	2	1	9	4	6	3	7	9	3	9	4	3	2	4	7	6	9	4	1
1	8	7	1	1	5	8	1	1	6	5	5	2	6	9	2	8	3	6	5
7	4	5	2	10	3	7	10	4	7	10	10	5	1	2	8	10	10	2	8
8	3	8	10	5	10	1	3	7	4	7	9	7	4	1	4	9	5	5	3
4	1	10	6	3	7	4	6	3	1	6	8	4	8	5	3	5	2	7	2
10	10	6	7	8	4	5	5	6	5	8	1	10	9	7	9	3	8	8	9
2	7	9	3	9	2	2	9	2	2	1	6	6	3	10	5	1	4	3	7
5	9	2	5	6	8	10	2	10	8	4	3	9	5	8	1	4	6	1	6
3	6	4	8	7	1	6	8	8	10	2	7	1	7	6	10	2	1	9	10

1	9	1	7	3	3	2	9	9	3	4	7	5	8	7	6	6	2	1	4
4	6	9	2	2	5	1	5	7	8	10	9	3	4	5	1	10	1	6	2
10	4	6	4	5	1	6	7	8	9	2	10	8	2	2	5	9	3	7	9
2	2	5	6	8	4	8	4	5	6	5	3	9	10	4	3	4	7	9	1
5	7	7	3	6	10	4	1	6	1	1	4	4	7	9	9	2	6	10	8
9	3	3	5	1	2	5	6	1	5	9	8	1	6	10	10	3	8	4	10
8	1	4	1	10	8	10	10	10	2	7	2	2	1	1	4	7	4	2	6
7	5	10	9	9	6	9	3	3	7	8	5	6	3	6	2	8	5	8	5
3	10	8	10	4	7	3	2	4	10	6	1	10	5	3	7	1	10	3	7
6	8	2	8	7	9	7	8	2	4	3	6	7	9	8	8	5	9	5	3

6	8	9	9	6	7	9	2	5	3	8	6	8	9	6	10	3	2	7	1
8	6	7	10	4	8	7	8	8	1	10	9	9	2	4	2	1	10	6	3
9	10	10	8	5	2	5	1	6	9	7	7	1	5	2	8	9	1	10	5
10	2	2	2	2	1	3	5	10	7	5	8	4	4	10	1	4	5	9	10
1	5	5	6	7	9	8	9	3	2	4	3	6	3	8	6	5	9	3	9
7	7	4	4	1	6	1	4	7	8	3	2	7	7	1	4	10	6	4	6
3	9	1	3	9	3	2	6	2	6	1	10	10	10	7	9	6	8	5	4
4	1	3	7	8	4	6	7	4	5	9	4	2	8	3	7	7	3	2	2
5	3	6	5	10	5	4	3	9	10	2	1	5	1	5	3	2	4	8	7
2	4	8	1	3	10	10	10	1	4	6	5	3	6	9	5	8	7	1	8

Table XIII 400 random permutations of first 10 integers

10	8	8	2	8	1	7	6	7	6	1	7	2	5	4	4	9	1	3	7
4	1	4	8	6	6	8	4	1	8	4	8	1	8	2	8	1	3	10	4
7	4	3	3	10	4	5	1	8	2	2	6	7	2	9	1	5	10	4	10
1	10	10	4	4	7	3	9	6	10	3	5	5	8	5	7	8	4	1	5
6	9	5	10	2	9	2	5	9	9	9	1	8	3	8	9	4	2	8	8
2	2	6	6	1	3	9	2	3	5	10	4	4	9	7	6	7	6	9	6
3	6	7	5	5	10	1	10	5	3	7	9	10	7	1	5	3	8	5	9
8	5	9	1	9	8	4	7	10	1	8	10	3	10	3	2	2	9	6	3
9	3	2	9	3	2	10	3	4	4	5	3	9	6	10	3	6	5	7	2
5	7	1	7	7	5	6	8	2	7	6	2	6	1	6	10	10	7	2	1
10	1	10	9	5	5	1	6	5	8	9	4	3	3	7	10	8	8	1	10
7	5	3	10	4	10	3	9	3	10	4	3	5	4	10	6	9	4	7	2
4	3	7	4	2	8	6	8	1	4	2	6	8	9	2	5	4	9	6	7
5	2	9	1	3	2	7	10	2	2	7	8	1	10	9	4	10	2	4	6
9	10	2	3	6	9	9	4	7	5	10	10	6	5	6	2	7	1	10	1
6	9	8	7	9	7	10	7	8	9	5	2	9	1	8	9	1	10	2	3
3	6	6	6	10	1	2	2	4	6	6	7	10	6	3	7	6	6	8	8
2	4	1	2	7	3	8	1	6	1	8	5	4	7	4	1	5	7	9	5
1	8	5	8	1	6	5	5	9	7	3	9	7	8	5	3	3	3	5	9
8	7	4	5	8	4	4	3	10	3	1	1	2	2	1	8	2	5	3	4
3	3	2	8	9	1	10	7	10	9	1	4	8	2	5	9	9	7	1	1
2	8	10	5	3	7	3	1	2	3	7	3	9	1	3	4	3	2	9	4
9	7	5	9	1	2	4	3	9	7	6	9	7	7	2	1	2	10	8	9
6	2	6	2	2	10	1	5	6	5	9	10	4	3	10	2	10	8	7	10
7	9	4	7	10	4	5	6	3	6	3	5	1	5	8	7	6	3	4	8
5	1	3	1	7	5	7	8	8	1	2	1	5	9	4	6	8	1	2	6
8	4	7	6	8	8	8	9	4	10	10	6	2	4	9	8	1	4	5	2
4	6	1	4	5	6	2	10	5	2	5	8	10	6	7	3	4	6	10	7
1	5	8	10	6	3	9	2	7	4	4	2	3	10	6	10	7	5	3	3
10	10	9	3	4	9	6	4	1	8	8	7	6	8	1	5	5	9	6	5
7	3	5	2	9	2	2	1	6	7	1	7	9	7	2	1	1	7	2	2
4	5	10	9	10	9	1	9	1	6	3	5	8	6	4	7	5	5	7	10
3	7	6	10	5	8	4	6	7	3	5	4	10	5	10	10	2	9	5	1
5	1	9	4	4	7	10	7	10	1	10	6	5	3	8	3	7	4	1	3
2	2	3	8	1	6	8	2	5	9	8	1	3	1	6	8	8	2	3	4
10	4	2	7	2	3	7	4	2	5	4	2	1	9	9	9	3	1	9	8
1	9	1	5	7	1	6	5	3	8	2	8	4	8	7	6	10	10	8	6
8	10	4	6	6	4	5	3	4	2	6	9	7	2	1	2	4	8	10	7
6	8	8	3	8	5	3	10	8	10	7	10	2	10	5	4	9	6	6	5
9	6	7	1	3	10	9	8	9	4	9	3	6	4	3	5	6	3	4	9
2	3	4	6	2	5	4	4	1	4	1	3	6	1	10	6	5	2	3	9
4	4	1	8	5	6	5	10	10	5	10	7	4	9	9	1	6	4	6	6
8	10	3	2	8	4	2	6	2	1	3	5	2	5	3	9	1	6	7	8
6	5	10	4	3	2	3	8	7	8	9	1	9	4	6	8	7	3	10	5
9	1	2	1	10	3	1	3	9	6	4	8	10	10	4	5	8	5	4	4
5	8	7	3	6	9	6	9	4	9	8	2	3	2	7	3	9	7	8	10
7	7	9	9	4	8	8	5	6	2	2	10	5	7	8	10	2	1	2	3
1	6	5	10	7	7	9	1	5	10	7	4	1	3	1	7	4	10	5	7
3	2	6	7	9	1	10	2	3	7	6	6	7	6	5	2	3	8	9	1
10	9	8	5	1	10	7	7	8	3	5	9	8	8	2	4	10	9	1	2

Table XIII   400 random permutations of first 10 integers

```
 3   9   7   9   1      2  10   5   3   5      5   5   3   5   4      2   5   3   7   4
 5   7   9   5   5      7   5   4   6  10     10   2   5  10   2      3   3   7   6   7
 9   4   6  10   6      4   8   9   5   8      3  10  10   2   7     10   4   1   3   6
10   3   3   2  10      8   2   6   1   9      4   4   9   8   5      5   8   4   8   8
 7   2   5   6   9      1   9   1  10   3      9   3   1   4   6      9  10   5   4  10
 2   8   4   3   8      6   6   3   4   6      6   7   4   1   1      7   6   6   9   5
 4  10   2   4   3      9   7   7   8   2      1   6   2   6   3      6   1  10   2   1
 1   5  10   7   4      5   3   8   7   7      7   8   7   9   9      8   7   8  10   2
 8   1   8   1   7     10   1  10   9   4      8   9   6   3  10      1   9   9   1   3
 6   6   1   8   2      3   4   2   2   1      2   1   8   7   8      4   2   2   5   9

 8   4   7   3  10      1   8   4   9   4      4   8   7   9   3      1  10  10   9   3
 4   6   9   1   6     10   4   1   6   7      8   5  10   6   6      7   5   7  10   9
 6   2   8   5   7      2   2   6   4   8      9  10   4   4  10     10   7   8   5   5
 2   1   5   7   8      8   1   5   5   5      6   7   6   3   8      8   1   9   1   1
10   9   6   9   9      3   5   9   2   2      3   6   8  10   7      5   4   3   2   2
 3   7   1   6   5      9   6   3   1   6      7   9   3   2   9      3   2   4   6  10
 7   3   4   8   3      6   3   7   8   3      5   4   1   5   2      2   8   5   3   4
 9   8  10   2   2      7  10  10  10   9      2   1   2   8   4      6   3   6   7   7
 1  10   3  10   4      4   7   2   3  10      1   2   5   7   1      9   6   2   4   6
 5   5   2   4   1      5   9   8   7   1     10   3   9   1   5      4   9   1   8   8

 6   5   5   1   1      6   9   8   7   2      2   2   1   6   5      7   3   6   7   8
 2   7   2  10   8      5   5   7   3   9      6   3   9   5   8      1   6   2  10   9
 4   6  10   3  10      9   3   2   8   7      7   8   3   7  10      3   5   8   6   5
 3   9   7   5   7     10   7   6   9   3      3   5   6   4   7     10   8   1   2   7
10   8   1   6   4      7   2   4   4   6     10   1  10   1   4      9  10   3   1   2
 9   1   4   9   2      8   6   3   5   5      5   9   2  10   6      5   1   5   4   1
 5   2   8   8   5      3   8   5   1  10      1   6   5   9   1      8   9   7   5  10
 1  10   6   4   3      4   4  10   2   1      8   7   8   2   3      2   4   9   8   3
 8   3   3   7   6      1   1   1   6   4      4   4   4   3   9      6   2   4   9   6
 7   4   9   2   9      2  10   9  10   8      9  10   7   8   2      4   7  10   3   4

 4   1   5   1   4      5   2   3  10   4      2   4   4   5   8      5   6   2  10   2
10   9   1   7   2     10   4   6   1  10      4   8   1   9   5      4   1   6   6   4
 7   4   9   8   1      9   6   1   9   2      6   2   7   7  10      1   3   4   7   7
 6   2  10   3   7      7   1   4   3   7     10  10   5   8   7     10  10   8   1   9
 3   7   4   2  10      3   5   8   5   1      7   6   8   2   4      6   5   5   9   3
 9  10   6  10   9      6   3   7   7   6      5   9   9   1   3      8   8   9   8   6
 5   8   2   6   5      2   7   5   2   9      1   7  10  10   6      3   4   3   2   8
 8   6   3   4   8      1   9   9   8   8      8   5   3   3   1      9   9  10   4   1
 1   5   8   5   6      4  10   2   4   3      3   3   2   6   2      7   7   7   5   5
 2   3   7   9   3      8   8  10   6   5      9   1   6   4   9      2   2   1   3  10

 9   2   2   2   3      1   5   7   2   2     10   1   4   1   6      2   2   9  10   5
 7   5  10   6   4      5   7   6   6   1      8   5   2   5   3      3   8  10   9   1
 4   3   1   7  10      9   1   2   8   3      3  10   9   7  10      8   7   7   8   2
 6  10   9   9   6      3   8   1   1   6      6   2   7   9   4     10   6   5   6   6
10   1   7  10   2      2   6   3   9  10      4   9  10   4   9      9   4   6   5   8
 2   7   8   3   1      6   4   9  10   5      2   6   3   2   6      6   3   8   2   7
 3   6   6   5   8      8   9   8   7   7      7   8   6   3   8      4   1   1   7   9
 8   9   5   4   9     10  10  10   4   9      1   7   8  10   2      5  10   2   1   3
 5   8   3   1   5      7   3   5   3   8      5   3   1   8   5      1   5   4   3   4
 1   4   4   8   7      4   2   4   5   4      9   4   5   6   7      7   9   3   4  10
```

Answers to Selected Exercises

Chapter 2

1. **a.** $y_{ij} = \mu_i + e_{ij}$ $i = 1, 2, 3$ $j = 1, 2, \ldots, 5$
   $\mu_i =$ signal type mean, $e_{ij} =$ experimental error
   b. e_{ij} random, independent with mean 0 and variance σ^2 the same for all signals
   c.

Source	df	SS	MS
Signal	2	1,202.63	601.31
Error	12	137.83	11.49
Total	14	1,340.46	

   d. 35.5, 20.9, 14.0   standard error $= 1.52$
   e. (32.2, 38.8), (17.6, 24.2), (10.7, 17.3)
   f. $F_0 = 52.4$, $F(.05, 2, 12) = 3.89$, Reject
   g. $5\mu_1 = 177.4$, $5\mu_2 = 104.5$, $5\mu_3 = 70.0$

3. **a.** $y_{ij} = \mu_i + e_{ij}$ $i = 1, 2, \ldots, 5$ $j = 1, 2, \ldots, 5$
   $\mu_i =$ treatment mean, $e_{ij} =$ experimental error
   b. e_{ij} random, independent with mean 0 and variance σ^2 for all treatments
   c.

Source	df	SS	MS
Treatment	4	48,569	12,142
Error	20	3,110	156
Total	24	51,679	

   d. 86.37, 112.85, 206.16, 110.69, 87.97   standard error $= 5.59$
   e. (74.71, 98.03), (101.19, 124.51), (194.50, 217.82), (99.03, 122.35), (76.31, 99.63)
   f. $F_0 = 78.08$, $F(.05, 4, 20) = 2.87$, Reject
   g. $5\mu_1 = 431.87$, $5\mu_2 = 564.26$, $5\mu_3 = 1030.78$, $5\mu_4 = 553.47$, $5\mu_5 = 439.86$

5. **a.** $y_{ij} = \mu_i + e_{ij}$ $i = 1, 2, \ldots, 5$ $j = 1, 2, \ldots, r_i$ $r_1 = 4, r_2 = r_3 = r_4 = r_5 = 2$
   $\mu_i =$ treatment mean, $e_{ij} =$ experimental error
   b. e_{ij} random, independent with mean 0 and variance σ^2 for all treatments
   c.

Source	df	SS	MS
Treatment	4	0.175969	0.043992
Error	7	0.000094	0.000013
Total	11	0.176062	

d. mean(std err), 0.049(0.0018), 0.086(0.0026), 0.116(0.0026), 0.187(0.0026), 0.397(0.0026)
e. (0.045, 0.053), (0.080, 0.092), (0.110, 0.122), (0.181, 0.193), (0.391, 0.403)
f. $F_0 = 3284.75$, $F(.05, 4, 7) = 4.12$, Reject
g. $4\mu_1 = 0.197$, $2\mu_2 = 0.171$, $2\mu_3 = 0.231$, $2\mu_4 = 0.374$, $2\mu_5 = 0.794$

7. Use $D = 30, t = 5, \sigma^2 = 156$ in Equation (2.26). If $r = 7$, $\Phi = 2$ and power is approximately 0.93 in Appendix IX for $\nu_1 = 4$ and $\nu_2 = 30$

Chapter 3

1. **a.** estimate (std err), (i) 18.03(1.86), (ii) 6.90(2.14)
b. (i), 1083.603, (ii) 119.025, SST = 1202.628
c. $t(.025, 12) = 2.179$, (i) $t_0 = 9.71$, Reject, (ii) $t_0 = 3.22$, Reject
d. $MSE = 11.49$, $F(.05, 1, 12) = 4.75$, (i) $F_0 = 94.34$, Reject, (ii) $F_0 = 10.36$, Reject
e. $t_0^2 = F_0$

3.

Source	df	SS	MS	F_0
Nitrogen	4	4,994.8	1,248.7	5.61
Linear	1	2,958.4	2,958.4	13.29
Quadratic	1	1,783.1	1,783.1	8.01
Cubic	1	67.6	67.6	0.30
Quartic	1	185.7	185.7	0.83
Error	15	3,338.0	222.5	

F_0 for linear and quadratic exceeds $F(.05, 1, 1.5) = 4.54$; thus quadratic is the best-fitting polynomial.
Estimated coefficient (std err): $\bar{y}_{..} = 142.60(3.34)$, $a_1 = 8.60(2.36)$, $a_2 = -5.64(1.99)$, $a_3 = 1.30(2.36)$, $a_4 = -0.81(0.89)$
Quadratic regression equation is $\hat{y} = 114.11 + 0.6234N - 0.0023N^2$

Nitrogen	Mean	Predicted	Std Err	95% CI
0	112.0	114.1	7.0	(99.2,129.0)
50	145.5	139.6	4.5	(130.0,149.2)
100	149.0	153.9	5.2	(142.8,165.0)
150	157.5	156.8	4.5	(147.2,166.4)
200	149.0	148.5	7.0	(133.6,163.4)

5. **a.** $d(.05, 7, 8) = 2.81$ $M = 2.81\sqrt{2(6.02)/2} = 6.89$

Patient	$\bar{y}_i$	$\max\limits_{j \neq i} \bar{y}_j$	D_i	$D_i - M$	$D_i + M$	95% SCI (L, U)
1	167.0	244.6	−77.6	−84.5	−70.7	(−84.5, 0)
2	185.5	244.6	−59.1	−66.0	−52.2	(−66.0, 0)
3	104.0	244.6	−140.6	−147.5	−133.7	(−147.5, 0)
4	214.9	244.6	−29.7	−36.6	−22.8	(−36.6, 0)
5	149.0	244.6	−95.6	−102.5	−88.7	(−102.5, 0)
6	169.4	244.6	−75.2	−82.1	−68.3	(−82.1, 0)
7	160.5	244.6	−84.1	−91.0	−77.2	(−91.0, 0)
8	244.6	214.9	29.7	22.8	36.6	(0, 36.6)

b. Patient 8, with $D_i + M > 0$, is selected as the subject with the highest cholesterol.

7. $d(.05, 3, 16) = 2.59$, $D(.05, 3) = 2.59\sqrt{2(6.875)/5} = 4.3$

| | | | 95% SCI |
Treatment	Mean	$\bar{y}_i - \bar{y}_c$	(L, U)
Control	42.2	—	—
3% Glucose	29.0	-13.2	$(-17.5, -8.9)$
3% Fructose	27.6	-14.6	$(-18.9, -10.3)$
3% Sucrose	34.0	-8.2	$(-12.5, -3.9)$

All treatments differ significantly from the control since none of the intervals include 0. The 3% fructose is most removed from the control, and the 3% sucrose is least removed from the control based on the upper limits.

9. $F(.05, 2, 12) = 3.89$

Contrast	Estimate	Std Err	S(.05)
Pretimed vs. others	18.03	1.86	5.19
Semi vs. Fully	6.90	2.14	5.97

Both contrasts are significant since the estimates exceed $S(.05)$.

11. $s^2 = 155.51$ 20 df
LSD(.05) = 16.45, HSD(5, .05) = 23.60, SNK(2, .05) = 16.45,
SNK(3, .05) = 19.95, SNK(4, .05) = 22.08, SNK(5, .05) = 23.60

| | | Significant Pairs by Each Criterion* | | |
Treatment	Mean	LSD(.05)	HSD(5, .05)	SNK(k, .05)
1. Premolt	86.37	2,3,4	2,3,4	2,3,4
2. Fasting	112.85	1,3,5	1,3,5	1,3,5
3. 60 gm bran	206.16	1,2,4,5	1,2,4,5	1,2,4,5
4. 80 gm bran	110.69	1,3,5	1,3	1,3,5
5. Laying mash	87.97	2,3,4	2,3	2,3,4

*Each row of numbers identifies those treatments declared significantly different from the treatment shown in the first column of that row.

e. The HSD provides greater strength of inference with confident inequalities than does LSD or SNK; they provide tests of homogeneity only with the weak sense experimentwise error rates.

f.

| | | Tukey 95% SCI |
Contrast	Estimate	(L, U)
Bran60 − Bran80	95.5	(71.9, 119.0)
Bran60 − Fasting	93.3	(69.7, 117.0)
Bran60 − Mash	118.0	(94.6, 142.0)
Bran60 − Premolt	120.0	(96.2, 143.0)
Bran80 − Fasting	−2.2	(−25.8, 21.4)
Bran80 − Mash	22.7	(−0.9, 46.3)
Bran80 − Premolt	24.3	(0.7, 47.9)
Fasting − Mash	24.9	(1.3, 48.5)
Fasting − Premolt	26.5	(2.9, 50.1)
Mash − Premolt	1.6	(−22.0, 25.2)

g. Width of interval estimate as well as direction and magnitude of separation between two means. Clearly, the separation between the Bran60 stage and each of the other stages is much greater than the separation between any other pair of stages.

Chapter 4

1. **a.**

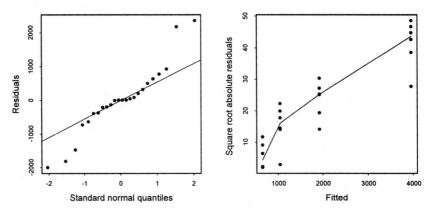

Levene(Med) test for homogeneous variances, $F_0 = 23.42$, $F(.05, 3, 20) = 3.10$, Reject

b. $\widehat{\beta} = 1.73$, $\widehat{p} = 1 - 1.73 = -0.73$, use reciprocal transformation from ladder of powers, $x = 1/y$

c. Residual plots after reciprocal transformation

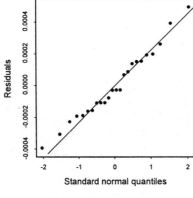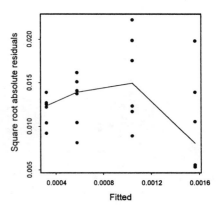

Levene(Med) test for homogeneous variances, $F_0 = 1.56$, $F(.05, 3, 20) = 3.10$, Do not reject

d. Conduct analysis on coded value $x = 1000 \times (1/y)$

Source	df	SS	MS	F_0
Linear	1	4.73	4.73	87.02
Quadratic	1	0.63	0.63	11.55
Cubic	1	0.04	0.04	0.69
Error	20	1.09	0.05	

Quadratic polynomial is significant, $F(.05, 1, 20) = 4.35$

3. **a.** Binomial

b. $\sin^{-1}\sqrt{\widehat{\pi}}$, where $\widehat{\pi} = (y + 3/8)/(n + 3/4)$

c.

Source	df	SS	MS	F_0
Clones	8	0.05	0.01	1.00
Error	9	0.05	0.01	

Means: 0.47, 0.45, 0.40, 0.32, 0.49, 0.44, 0.46, 0.40, 0.48

d. $d(.05, 8, 9) = 2.81$ $M = 2.81\sqrt{2(0.01)/2} = 0.281$
All clones are selected, indicating there are no differences between the highest rooting clones and the others.

5.   a. f-values are 0.0333, 0.1000, 0.1667, 0.233, 0.3000, 0.3667, 0.4333, 0.5000, 0.5667, 0.6333, 0.7000, 0.7667, 0.8333, 0.9000, 0.9667
$q(f_i)$ are $-1.83,\ -1.28,\ -0.97,\ -0.73,\ -0.52,\ -0.34,\ -0.17,\ 0.00,\ 0.17,\ 0.34,\ 0.52,$ 0.73, 0.97, 1.28, 1.83

b.

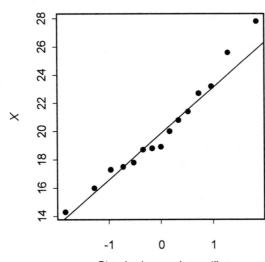

Standard normal quantiles

Chapter 5

1.   a. $y_{ij} = \mu + a_i + e_{ij}$ $i = 1, 2, ..., 5$ $j = 1, 2, ..., 8$
a_i = random effects of sires with mean 0 and variance σ_a^2
e_{ij} = random effects of progeny with mean 0 and variance σ_e^2

Source	df	SS	MS	E(MS)
Sires	4	5,591.1	1,397.8	$\sigma_e^2 + 8\sigma_a^2$
Progeny	35	16,232.7	463.8	σ_e^2

b. $\hat{\sigma}_e^2 = 463.8,\ \hat{\sigma}_a^2 = 116.7$
$\hat{\sigma}_e^2$ 90% CI: (326, 722), $\hat{\sigma}_a^2$ 90% CI: $(-4, 1387)$
c. $F_0 = 3.01,\ F(.05, 4, 35) = 2.65$, Reject
d. $\hat{\rho}_I = 0.20$, 90% CI: (.02, 0.67)

4.   a. $y_{ij} = \mu + a_i + e_{ij}$ $i = 1, 2, ..., 8$ $j = 1, 2, 3, 4$
a_i = random effects of cottonseed lots with mean 0 and variance σ_a^2
e_{ij} = random effects of seed samples within lots with mean 0 and variance σ_e^2

Source	df	SS	MS	E(MS)
Lots	7	13,696.5	1,956.6	$\sigma_e^2 + 4\sigma_a^2$
Samples	24	5,548.2	231.2	σ_e^2

b. $\hat{\sigma}_e^2 = 231.2$, $\hat{\sigma}_a^2 = 431.4$

c. 662.6

d. 34.9% to samples within lots and 65.1% to lots

e. 25.7

f. See Section 5.8 or Section 5.10 for applications. Either determines optimum allocations of samples based on cost or number of samples per lot to achieve a test of hypothesis power.

6. **a.** $y_{ij} = \mu + \tau_i + e_{ij} + d_{ijk}$ $i = 1, 2$ $j = 1, 2, 3$ $k = 1, 2, 3$
τ_i = fixed effect of manufacturers
e_{ij} = random effects of filters with mean 0 and variance σ_e^2
d_{ijk} = random effects of filter tests with mean 0 and variance σ_d^2

Source	df	SS	MS	E(MS)
Manufacturer	1	1.8689	1.8689	$\sigma_d^2 + 3\sigma_e^2 + 9\theta_\tau^2$
Filters	4	3.7413	0.9353	$\sigma_d^2 + 3\sigma_e^2$
Tests	12	0.3047	0.0254	σ_d^2

b. $F_0 = 2.0$, $F(.05, 1, 4) = 7.71$,   Do not reject

c. Mean (std err): 0.47(0.32), 1.12(0.32)
95% CI: $(-0.42, 1.36)$, $(0.23, 2.01)$

d. $c_1 = 200, c_2 = 1$, $\hat{\sigma}_d^2 = 0.0254$, $\hat{\sigma}_e^2 = 0.3033$
$n = 5$ tests, $r = 8$ filters

8. **a.** $y_{ij} = \mu + a_i + e_{ij} + d_{ijk}$ $i = 1, 2, ..., 15$ $j = 1, 2, ..., r_i$ $k = 1, 2, ..., n_{ij}$
a_i = random field effects with mean 0 and variance σ_a^2
e_{ij} = random section effects with mean 0 and variance σ_e^2
d_{ijk} = random soil sample effects with mean 0 and variance σ_d^2

Source	df	SS	MS	E(MS)
Fields	14	14.43	1.03	$\sigma_d^2 + 1.19\sigma_e^2 + 2.38\sigma_a^2$
Sections	15	11.53	0.77	$\sigma_d^2 + 1.2\sigma_e^2$
Samples	6	8.80	1.47	σ_d^2

b. $\hat{\sigma}_d^2 = 1.47$, $\hat{\sigma}_e^2 = -0.58$, $\hat{\sigma}_a^2 = 0.11$

c. $M = 0.78$, $\nu = 15$, $F_0 = 0.78/0.77 = 1.01$, $F(.05, 14, 15) = 2.42$, Do not reject

d. $F = 0.77/1.47 = 0.52$, $F(.05, 15, 6) = 3.94$, Do not reject

Chapter 6

1. **a.** $y_{ijk} = \mu + \alpha_i + \beta_j + (\alpha\beta)_{ij} + e_{ijk}$ $i = 1, 2, 3$ $j = 1, 2$ $k = 1, 2, 3, 4$
α_i = fixed effect of alcohol, β_i = fixed effect of base
$(\alpha\beta)_{ij}$ = interaction effect between alcohol and base
e_{ijk} = experimental error, mean 0, variance σ^2

Source	df	SS	MS
Alcohol	2	5.40	2.70
Base	1	6.51	6.51
Interaction	2	22.57	11.28
Error	18	36.76	2.04

b. Alcohol marginal means: 89.98, 91.04, 90.10; standard error $= 0.50$
Base marginal means: 89.85, 90.89; standard error $= 0.41$

Base	Alcohol 1	2	13
1	90.83	89.80	88.93
2	89.13	92.28	91.28

Cell means standard error $= 0.71$

c. $F_0 = 5.53$, $F(.05, 2, 18) = 3.55$. Reject

d. Contrast: (Base 1 − Base 2) for each alcohol, $s_c = 1.01$,
   Bonferroni $t(.025, 3, 18) = 2.64$

Alcohol	c	t_0	
1	1.70	1.68	do not reject
2	− 2.48	− 2.46	do not reject
3	− 2.35	− 2.33	do not reject

e. Levene(Med) test, $F_0 = 0.51$, Do not reject equal treatment variances hypothesis

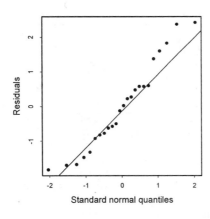
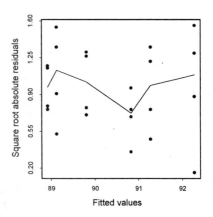

3. **a.** $y_{ijk} = \mu + \alpha_i + \beta_j + (\alpha\beta)_{ij} + e_{ijk}$ $i = 1, 2, 3, 4$ $j = 1, 2, 3, 4$ $k = 1, 2$
α_i = fixed effect of fabric, β_i = fixed effect of temperature
$(\alpha\beta)_{ij}$ = interaction effect between fabric and temperature
e_{ijk} = experimental error, mean 0, variance σ^2

Source	df	SS	MS
Fabric	3	41.88	13.96
Temperature	3	283.94	94.65
T linear	1	262.14	262.14
T quadratic	1	21.78	21.78
Fabric × Temperature	9	15.86	1.76
F × *T* linear	3	11.31	3.77
F × *T* quadratic	3	2.69	0.90
Error	16	0.80	0.05

b. $F_0 = 35.24$, $F(.05, 9, 16) = 2.54$, Reject

c. Temperature linear $F_0 = 5242.9$, Temperature quadratic $F_0 = 435.6$,
$F(.05, 1, 16) = 4.49$, Reject both null hypotheses

d. Fabric × Temperature linear $F_0 = 75.4$
Fabric × Temperature quadratic $F_0 = 18.0$
$F(.05, 3, 16) = 3.24$, Reject for Fabric × Temperature linear and reject for Fabric × Temperature quadratic

e.

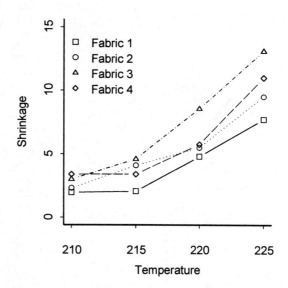

5.   a.

Source	df	MS	F_0
Insecticide (I)	4	199.84	
Herbicide (H)	4	171.70	
I linear	1	4.96	0.03
I quadratic	1	553.01	3.18
I cubic	1	162.90	0.94
H linear	1	572.91	3.29
H quadratic	1	108.75	0.63
H cubic	1	4.65	0.03
I lin × H lin	1	3,102.49	17.83
I lin × H quad	1	236.60	1.36
I lin × H cub	1	6.76	0.04
I quad × H lin	1	1,180.30	6.78
I quad × H quad	1	195.00	1.12
I quad × H cub	1	37.03	0.21
I cub × H lin	1	379.80	2.18
I cub × H quad	1	23.41	0.13
I cub × H cub	1	18.49	0.11
Error	75	174.00	

b.   Reject null hypothesis if $F_0 \geq F(.05, 1, 7.5) = 3.97$.

c.   wgt $= 105 - 1.17I - 40.7H + 0.00749I^2 + 1.72(I)(H) - 0.0145(I^2)(H)$

7.   a. $y_{ijkl} = \mu + \alpha_i + \beta_j + (\alpha\beta)_{ij} + \gamma_k + (\alpha\gamma)_{ik} + (\beta\gamma)_{jk} + (\alpha\beta\gamma)_{ijk} + e_{ijkl}$
$i, j, k, = 1, 2, 3 \quad l = 1, 2$
α_i = fixed soil texture effect, β_j = fixed salinity effect, γ_k = fixed water effect, $(\alpha\beta)_{ij}$ = interaction between texture and salinity, $(\alpha\gamma)_{ik}$ = interaction between texture and water, $(\beta\gamma)_{jk}$ = interaction between salinity and water, $(\alpha\beta\gamma)_{ijk}$ = three factor interaction, e_{ijkl} = experimental error with mean 0 and variance σ^2

Source	df	SS	MS	F_0
Texture (T)	2	12.46	6.23	110.98
Salinity (S)	2	58.13	29.07	517.77
Water (W)	2	26.28	13.14	234.09
$T \times S$	4	1.28	0.32	5.72
$T \times W$	4	3.04	0.76	13.52
$S \times W$	4	20.62	5.15	91.82
$T \times S \times W$	8	1.65	0.21	3.68
Error	27	1.52	0.06	

b. Soil × Salinity × Water means

Salinity	2			8			10		
Water	0	5	15	0	5	15	0	5	15
L. Sand	0.54	1.85	3.39	0.09	0.10	0.13	0.07	0.11	0.20
Loam	0.96	2.35	5.40	0.21	0.29	0.58	0.14	0.29	0.39
Clay	1.44	3.08	5.56	0.62	0.95	1.64	0.49	0.80	2.41

Cell means standard error = 0.17

Texture × Salinity Means

	Salinity		
Texture	2	8	16
L. Sand	1.93	0.10	0.12
Loam	2.90	0.36	0.27
Clay	3.36	1.07	1.23

Cell means standard error = 0.10

Texture × Water Means

Water		
0	5	15
0.23	0.69	1.24
0.43	0.98	2.12
0.85	1.61	3.20

Cell means standard error = 0.10

Salinity × Water Means

	Water		
Salinity	0	5	15
2	0.98	2.42	4.78
8	0.30	0.45	0.78
16	0.23	0.40	1.00

Cell means standard error = 0.10

Texture means: 0.72, 1.18, 1.89, standard error = 0.06
Salinity means: 2.73, 0.51, 0.54, standard error = 0.06
Water means: 0.50, 1.09, 2.19, standard error = 0.06

c. Reject for Texture, Salinity, and Water, $F(.05, 2, 27) = 3.35$
Reject for all two factor interactions, $F(.05, 4, 27) = 2.73$
Reject for three factor interaction, $F(.05, 8, 27) = 2.31$

d.

Source	df	SS	MS	F_0
T	2	12.46	6.23	110.98*
S (lin)	1	38.55	38.55	686.70*
S (quad)	1	19.58	19.58	348.84*
W (lin)	1	26.28	26.28	468.04*
W (quad)	1	0.01	0.01	0.13
$T \times S(1)$	2	0.96	0.48	8.55*
$T \times S(q)$	2	0.32	0.16	2.89
$T \times W(1)$	2	2.98	1.49	26.57*
$T \times W(q)$	2	0.05	0.03	0.48
$S(1) \times W(1)$	1	12.18	12.18	216.90*
$S(1) \times W(q)$	1	0.13	0.13	2.29
$S(q) \times W(1)$	1	8.30	8.30	147.79*
$S(q) \times W(q)$	1	0.02	0.02	0.30
$T \times S(1) \times W(1)$	2	1.27	0.63	11.27*
$T \times S(1) \times W(q)$	2	0.19	0.09	1.67
$T \times S(q) \times W(1)$	2	0.18	0.09	1.58
$T \times S(q) \times W(q)$	2	0.02	0.01	0.18

*Reject the null hypothesis if $F_0 \geq F(.05, 1, 27) = 4.21$
or $F_0 \geq F(.05, 2, 27) = 3.35$.

11. a. The **SS Yates** and **MS Yates** columns are the Weighted Squares of Means calculations, which can be produced by programs (MINITAB GLM with "Adj. SS," SAS GLM with "Type III SS," or Splus with the "summary.aov" function). Notice the **SS sequential** column. The Interaction and the Error sums of squares are identical to those for the **Yates** columns. The interaction and error sums of squares partitions are obtained by a *sequential* fit of terms to the model in the order presented in the **Source** column and may be used equivalently to test interaction. Most programs also produce the sequential fit (MINITAB GLM with "Seq SS," SAS GLM with "Type I SS," and the default for Splus "summary.aov" function).

Source	df	SS sequential	SS Yates	MS Yates
Base	1	8.897	14.242	14.242
Alcohol	2	4.105	7.675	3.838
Interaction	2	29.240	29.240	14.620
Error	10	19.873	19.873	1.987

$F_0 = MS(\text{Interaction})/MSE = 14.620/1.987 = 7.36$, Reject the null hypothesis of no interaction since $F_0 > F(.05, 2, 27) = 3.35$. Not necessary to test main effects.

b. Mean(std. error)
Base: 89.8(0.49), 91.8(0.61). Alcohol: 90.0(0.64), 92.0(0.81),90.4(0.54)
Interaction: $\hat{\mu}_{ij}$ = observed cell mean and $s_{\hat{\mu}_{ij}} = \sqrt{MSE/r_{ij}}$
91.1(1.00), 89.3(0.81), 88.9(0.70), 89.0(0.81), 94.7(1.41), 91.8(0.81)

c. Alcohol level, difference (std. error), t_0
Alcohol 1, 2.02 (1.29), 1.57:Alcohol 2, $-5.43(1.63)$, -3.33:Alcohol 3, $-2.84(1.08)$, -2.64
Use the Bonferroni t for 3 comparisons and 10 degrees of freedom and Reject H_0 if $|t_0| > 2.87$

d. Cell means remain the same. Observed marginal means for Base: 89.5, 91.0 and for Alcohol: 89.8, 90.6, 91.8.

Chapter 7

1. a. $y_{ijk} = \mu + a_i + b_j + (ab)_{ij} + e_{ijk}$ $i = 1, 2, 3, 4$ $j = 1, 2, 3, 4, 5$ $k = 1, 2$
a_i = random run effect, mean 0, variance σ_a^2

b_j = random patient effect, mean 0, variance σ_b^2
$(ab)_{ij}$ = random run × patient interaction effect, mean 0, variance σ_{ab}^2
e_{ijk} = experimental error, mean 0, variance σ^2

Source	df	MS	E(MS)
Runs (**R**)	3	364.46	$\sigma^2 + 2\sigma_{ab}^2 + 10\sigma_a^2$
Patients (**P**)	4	13,152.03	$\sigma^2 + 2\sigma_{ab}^2 + 8\sigma_b^2$
R × **P**	12	9.12	$\sigma^2 + 2\sigma_{ab}^2$
Error	20	6.61	σ^2

c. 35.53; 1,642.86; 1.26
d. Interaction: H_0: $\sigma_{ab}^2 = 0$, H_a:$\sigma_{ab}^2 \neq 0$, $F_0 = 1.38$, $F(.05, 12, 20) = 2.28$, Do not reject
Patients: H_0: $\sigma_b^2 = 0$, H_a: $\sigma_b^2 \neq 0$, $F_0 = 1,442.11$, $F(.05, 4, 12) = 3.26$, Reject
Runs: H_0: $\sigma_a^2 = 0$, H_a: $\sigma_a^2 \neq 0$, $F_0 = 39.96$, $F(.05, 3, 12) = 3.49$, Reject

3. a. $y_{ijk} = \mu + \alpha_i + b_{j(i)} + c_{k(ij)}$ $i = 1, 2, 3$ $j = 1, 2, 3, 4$ $k = 1, 2$
α_i = fixed effect of alloys
$b_{j(i)}$ = random effect of castings within alloys, mean 0, variance $\sigma_{b(a)}^2$
$c_{k(ij)}$ = random effect of bars within castings, mean 0, variance $\sigma_{c(b)}^2$

b.

Source	df	MS	E(MS)
Alloys	2	14.69	$\sigma_{c(b)}^2 + 2\sigma_{b(a)}^2 + 8\theta_a^2$
Castings (Alloys)	9	0.52	$\sigma_{c(b)}^2 + 2\sigma_{b(a)}^2$
Bars (Castings)	12	0.36	$\sigma_{c(b)}^2$

c. $F_0 = 28.25$, $F(.05, 2, 9) = 4.26$, Reject
d. Means (std err): 14.7(0.25), 16.7(0.25), 14.1(0.25)
95% CI: (14.1, 15.3), (16.1, 17.3), (13.5, 14.7)
e. Bars: 0.36, Castings: 0.08

5. a. $y_{ijkl} = \mu + \alpha_i + b_j + (ab)_{ij} + c_{k(j)} + (ac)_{ik(j)} + e_{ijkl}$
α_i = fixed machine effect
b_j = random plot effect, mean 0, variance σ_b^2
$(ab)_{ij}$ = random effect for machine by plot interaction, mean 0, variance σ_{ab}^2
$c_{k(j)}$ = random sample within plot effect, mean 0, variance $\sigma_{c(b)}^2$
$(ac)_{ik(j)}$ = random effect for machine by sample interaction nested within plot, mean 0, variance $\sigma_{ac(b)}^2$
e_{ijkl} = random error, mean 0, variance σ^2

Source	df	MS*	E(MS)
Machine (**M**)	1	25.29	$\sigma^2 + 2\sigma_{ac(b)}^2 + 4\sigma_{ab}^2 + 12\theta_a^2$
Plot (**P**)	2	24.82	$\sigma^2 + 2\sigma_{ac(b)}^2 + 4\sigma_{c(b)}^2 + 4\sigma_{ab}^2 + 8\sigma_b^2$
M × **P**	2	0.10	$\sigma^2 + 2\sigma_{ac(b)}^2 + 4\sigma_{ab}^2$
Sample (**S**) within **P**	3	8.86	$\sigma^2 + 2\sigma_{ac(b)}^2 + 4\sigma_{c(b)}^2$
M × **S** within **P**	3	0.54	$\sigma^2 + 2\sigma_{ac(b)}^2$
Error	12	0.117	σ^2

*Mean square value × 10^3

c. $F_0 = MSM/MS(MP) = 252.9$, $F(.05, 1, 2) = 18.51$, Reject

Chapter 8

1. **a.** $y_{ij} = \mu + \tau_i + \rho_j + e_{ij}$ $i = 1, 2, \ldots, 6$ $j = 1, 2, \ldots, 8$
τ_i = fixed irrigation method effect, ρ_j = fixed block effect
e_{ij} = experimental error, mean 0, variance σ^2

Source	df	SS	MS
Blocks	7	457,507	65,358
Irrigation	5	47,842	9,568
Error	35	16,888	4,197

b. Experimental errors are random and independent with mean 0 and variance σ^2. Treatment and block effects are additive. Randomization justifies assumption of random, independent errors. The relationship between treatment effects is consistent among the blocks.

c. 22.9, 32.4

d. $d(.05, 5, 35) = 2.64$ $D(5, .05) = 85.5$

Irrigation	Mean	$\bar{y}_i - \bar{y}_c$	95% SCI (L, U)
Flood	229.6	control	
Trickle	299.6	70.0	$(-15.5, 155.5)$
Basin	290.4	60.8	$(-24.7, 146.3)$
Spray	223.8	-5.8	$(-91.3, 79.7)$
Sprinkler	292.0	62.4	$(-23.1, 147.9)$
Sprinkler+spray	291.0	61.4	$(-24.1, 146.9)$

No treatment is significantly different from flood, since all intervals include 0.

e. $RE = 0.99(3.17) = 3.14$. More than three times as many replications required by the completely randomized design to have same variance of treatment mean.

2. **a.** $y_{ij} = \mu + \tau_i + \rho_j + e_{ij}$ $i = 1, 2, \ldots, 5$ $j = 1, 2, \ldots, 5$
τ_i = fixed fertilizer effect, ρ_j = fixed block effect
e_{ij} = experimental error, mean 0, variance σ^2

Source	df	SS	MS
Blocks	4	4.95	1.24
Fertilizer	4	72.12	18.03
Error	16	12.76	0.80

b.

Contrast	SS	F_0
None vs. others $(1, -\frac{1}{4}, -\frac{1}{4}, -\frac{1}{4}, -\frac{1}{4})$	3.13	3.93
Nitrogen $(0, -1, +1, -1, +1)$	4.32	5.42*
P_2O_5 $(0, -1, -1, +1, +1)$	64.44	80.79*
Interaction $(0, +1, -1, -1, +1)$	0.22	0.28

*Reject null hypothesis if $F_0 \geq F(.05, 1, 16) = 4.49$

c. 0.45, 0.80, 0.80, 0.80

d. $RE = 0.98(1.09) = 1.07$

5. **a.** $y_{ij} = \mu + \rho_i + \gamma_j + \tau_k + e_{ij}$ $i, j, k = 1, 2, 3, 4$
ρ_i = fixed time period (row) effect, γ_j = fixed technician (column) effect
τ_k = fixed construction method effect
e_{ij} = experimental error, mean 0, variance σ^2

Source	df	SS	MS
Time	3	467.19	155.73
Technician	3	17.19	5.73
Method	3	145.69	48.56
Error	6	22.88	3.81

b. 0.98, 1.38

c. $q(.05, 4, 6) = 4.90$, $HSD(4, .05) = 4.80$, 95% SCI are $\bar{y}_i - \bar{y}_j \pm 4.80$
95% SCI for differences that do not include 0 are (A, D), (A, C)

d. $RE = .93(11) = 10.2$

7. a. $y_{ijk} = \mu + \tau_i + \rho_j + (\tau\rho)_{ij} + e_{ijk}$ $i = 1, 2, \ldots, 5$ $j, k = 1, 2$
τ_i = fixed nitrogen rate effect, ρ_j = fixed block effect
$(\tau\rho)_{ij}$ = fixed rate × block interaction effect
e_{ijk} = experimental error, mean 0, variance σ^2
The interaction effect can be included in the model and its sum of squares partition, computed by virtue of the fact that two plots for each treatment were included in each block. Thus, experimental error variance can be estimated separately as the variance among plots treated alike within each block.

Source	df	SS	MS
Blocks	1	1022.45	1022.45
Rates	4	4813.00	1203.25
Rate linear	1	2856.10	2856.10
Rate quadratic	1	1716.07	1716.07
Rate cubic	1	60.03	60.03
Block × Rate	4	287.80	71.95
Error	10	422.50	42.25

b. $F_0 = 1.70$, $F(.05, 4, 10) = 3.48$, Do not reject

c. See *SS* partitions in analysis of variance table.
Linear: $F_0 = 67.6$, Quadratic: $F_0 = 40.62$, $F(.05, 1, 10) = 4.96$
Reject null hypothesis for linear and quadratic effects

d. $F_0 = 1.42$, Do not reject null hypothesis; cubic not significant

9. a.

Source	df	SS	MS
Total	45	614,708	
Blocks (unadj.)	7	432,384	61,769
Irrigation (adj.)	5	51,923	10,385
Error	33	130,402	3,952

b. $F_0 = 2.63$, $F(.05, 5, 33) = 2.50$, Reject

c. Mean(std err): 290.1(24.1), 290.4(22.2), 223.8(22.2), 292.0(22.2), 291.0(22.2), 213.7(24.1)

d. Larger than those treatments without lost plots

e. (i) 34.1, (ii) 32.8, (iii) 32.8, (iv) 31.4

f. Standard errors increased if either or both treatments had missing plots; decreased power

10. a.

Source	df	Year 1 Spring	Year 1 Winter	Year 2 Spring	Year 2 Winter
Treatment	3	32.55	41.05	193.15	48.65
Error	4	12.99	18.11	12.19	45.31

*Mean square value × 10^3

b. $F\,max = 45.31/12.19 = 3.72$, $F_{.05}\,max = 20.6$, $t = 4, \nu = 4$
Do not reject hypothesis of equal variances

c.

Source	df	MS	E(MS)
Year (Y)	1	13.61	$\sigma^2 + 2\sigma_{tys}^2 + 4\sigma_{ty}^2 + 8\sigma_{ys}^2 + 16\sigma_y^2$
Season (S)	1	1,911.01	$\sigma^2 + 2\sigma_{tys}^2 + 8\sigma_{ys}^2 + 16\theta_s^2$
Treatment (T)	3	257.70	$\sigma^2 + 2\sigma_{tys}^2 + 4\sigma_{ty}^2 + 8\theta_t^2$
$Y \times S$	1	94.61	$\sigma^2 + 2\sigma_{tys}^2 + 8\sigma_{ys}^2$
$T \times S$	3	15.94	$\sigma^2 + 2\sigma_{tys}^2 + 4\theta_{ts}^2$
$T \times Y$	3	28.47	$\sigma^2 + 2\sigma_{tys}^2 + 4\sigma_{ty}^2$
$T \times Y \times S$	3	13.27	$\sigma^2 + 2\sigma_{tys}^2$
Error	16	22.15	σ^2

*Mean square value $\times 10^3$

d. Season: $F_0 = 1911.01/94.61 = 20.2$, $F(.05, 1, 1) = 161.4$, Do not reject
Treatments: $F_0 = 257.7/28.47 = 9.05$, $F(.05, 3, 3) = 9.28$, Do not reject
$Y \times S$: $F_0 = 94.61/13.27 = 7.13$, $F(.05, 1, 3) = 10.13$, Do not reject
$T \times S$: $F_0 = 15.94/13.27 = 1.20$, $F(.05, 3, 3) = 9.28$, Do not reject
$T \times Y$: $F_0 = 28.47/13.27 = 2.15$, $F(.05, 3, 3) = 9.28$, Do not reject
$T \times Y \times S$: $F_0 = 13.27/22.15 = 0.60$, $F(.05, 3, 16) = 3.24$, Do not reject

Chapter 9

1. **a.** $\lambda = 1$

b. $E = \frac{2}{3}$

c. $y_{ij} = \mu + \tau_i + \rho_j + e_{ij}$ $i = 1, 2, 3, 4$ $j = 1, 2, 3, 4, 5, 6$
$\tau_i =$ fixed temperature effect, $\rho_j =$ fixed block effect
$e_{ij} =$ experimental error, mean 0, variance σ^2

Source	df	SS	MS
Blocks (unadj.)	5	613.66	122.73
Temperature (adj.)	3	718.29	239.43
Temp linear	1	677.08	677.08
Temp quadratic	1	7.33	7.33
Temp cubic	1	33.87	33.87
Error	3	41.00	13.67

Test for temperature: $F_0 = 17.51$, $F(.05, 3, 3) = 9.28$. Reject

d. means: 26.46, 23.82, 10.07, 3.61; standard error $= 2.50$

e. 3.70

f. See analysis of variance table for SS partitions
Linear: $F_0 = 49.53$, $F(.05, 1, 3) = 10.13$, Reject
Quadratic: $F_0 = 0.54$, $F(.05, 1, 3) = 10.13$, Do not reject

g. No. F_0 for cubic $= 2.48$, $F(.05, 1, 3) = 10.13$, Do not reject

3. **a.** $y_{ijm} = \mu + \tau_i + \rho_j + \gamma_m + e_{ijm}$ $i, m = 1, 2, \ldots, 7$ $j = 1, 2, 3, 4$
$\tau_i =$ fixed intersection type effect, $\rho_j =$ fixed observer effect
$\gamma_m =$ fixed city effect
$e_{ijm} =$ experimental error, mean 0, variance σ^2

Source	df	SS	MS
Observer	3	146.51	48.84
City (unadj.)	6	1,342.07	233.68
Intersection (adj.)	6	9,972.46	1,662.08
Error	12	368.85	30.74

b. 4.19
c. Plan 9B.5, $E = 0.88$
d. $d(.05, 6, 12) = 2.58$, $M = 2.58(4.19) = 10.8$

Intersection	$\bar{y}_i$	max $\bar{y}_j$ $j \neq i$	D_i	$D_i - M$	$D_i + M$	95% SCI (L, U)
1	68.4	28.6	39.88	29.0	50.6	$(0, 50.6)$
2	28.6	28.7	-0.1	-10.9	10.77	$(-10.9, 10.7)$
3	28.7	28.6	0.1	-10.7	10.9	$(-10.7, 10.9)$
4	40.4	28.6	11.8	1.0	22.6	$(0, 22.6)$
5	44.8	28.6	16.2	5.4	27.0	$(0, 27.0)$
6	85.2	28.6	56.6	45.8	67.4	$(0, 67.4)$
7	32.6	28.6	4.0	-6.8	14.8	$(-6.8, 14.8)$

Select intersection types 2, 3, and 7 with $P(\text{CS}) = 0.95$.

5. a. $t = 5, r = 4, k = 4, b = 5$
   b $\lambda = 3$

Chapter 10

1. a. $y_{ijk} = \mu + \tau_i + \gamma_j + \rho_{m(j)} + e_{ijm}$ $i = 1, 2, \ldots, 8$ $j = 1, 2, \ldots, 7$ $m = 1, 2$
   τ_i = variety effect, γ_j = replication effect, $\rho_{m(j)}$ = block within replication effect
   e_{ijm} = experimental error, mean 0, variance σ^2

Source	df	SS	MS
Replication	6	67.35	11.22
Block (unadj.)	7	336.90	48.13
Variety (adj.)	7	1,245.99	178.00
Error	35	358.19	10.23

Null hypothesis test for variety: $F_0 = 17.4$, $F(.05, 7, 35) = 2.29$, Reject
Variety means $(\hat{\mu})$: 43.3, 54.4, 38.4, 51.8, 47.5, 49.4, 42.5, 51.0
Standard error $= 1.29$
   b. 1.85
   c. 0.86

3. a. $y_{ijlm} = \mu + \tau_i + \beta_m + \rho_{i(m)} + \gamma_{l(m)} + e_{ijlm}$
   τ_i = variety effect, β_m = replication effect, $\rho_{i(m)}$ = row within replication effect
   $\gamma_{l(m)}$ = column within replication effect
   e_{ijm} = experimental error, mean 0, variance σ^2

Source	df	SS	MS
Replication	3	5.12	1.71
Column (unadj.)	8	29.30	3.66
Row (unadj.)	8	37.50	4.69
Variety (adj.)	8	130.87	16.36
Error	8	6.04	0.75

Null hypothesis test for variety: $F_0 = 21.81$, $F(.05, 8, 8) = 3.44$, Reject
Results below obtained from GLM in SAS:
Variety means $(\widehat{\mu}_i)$: 59.1, 53.4, 55.1, 53.0, 55.2, 54.2, 54.4, 48.7, 56.9
Standard error = 0.60

b. 0.87 from GLM in SAS

c. 0.50

Chapter 11

1. Block I, Units 1–8: $(1), ab, ac, ad, bc, bd, cd, abcd$
Block II, Units 9–16: $a, b, c, d, abc, abd, acd, bcd$

3. Confound ABC in Rep I, ABD in Rep II, ACD in Rep III, and BCD in Rep IV.
Rep I, Block I: $(1), d, ab, ac, bc, abd, acd, bcd$
Rep I, Block II: $a, b, c, ad, bd, cd, abc, abcd$
Rep II, Block I: $(1), c, ab, ad, bd, abc, acd, bcd$
Rep II, Block II: $a, b, d, ac, bc, cd, abd, abcd$
Rep III, Block I: $(1), b, ac, ad, cd, abc, abd, bcd$
Rep III, Block II: $a, c, d, ab, bc, bd, acd, abcd$
Rep IV, Block I: $(1), a, bc, bd, cd, abc, abd, acd$
Rep IV, Block II: $b, c, d, ab, ac, ad, bcd, abcd$

5. **a.** B, AC, ABC, CDE
b. Yes, because another choice for defining contrasts will not leave main effects or two-factor interactions confounded.
c. Use the plan in Table 11A.1 with $ACE, ABEF$, and $ABCD$ as defining contrasts. No main effects or two-factor interactions are confounded.

7. **a.**

Source	df	SS	MS	F_0
Replication	1	11.28	11.28	
Blocks	2	56.31	28.16	
A (speed)	1	639.03	639.03	80.28*
B (angle)	1	11.28	11.28	1.42
C (lubrication)	1	52.53	52.53	6.60*
D (alloy)	1	124.03	124.03	15.58*
AB	1	132.03	132.03	16.59*
AC	1	3.78	3.78	0.48
AD	1	34.03	34.03	4.28
BC	1	2.53	2.53	0.32
BD	1	0.03	0.03	0.00
CD	1	26.28	26.28	3.30
ABC	1	205.03	205.03	25.76*
ABD	1	175.78	175.78	22.08*
ACD	1	7.03	7.03	0.88
BCD	1	7.03	7.03	0.88
Error	14	111.44	7.96	

*Reject null hypothesis if $F_0 \geq F(.05, 1, 14) = 4.60$

b. Effect (estimate):
$A(-8.94), B(-1.19), C(-2.56), D(3.94), AB(4.06), AC(0.69), AD(-2.06),$
$BC(-0.56), BD(-0.06), CD(1.81), ABC(-5.06), ABD(4.69), ACD(-0.94),$
$BCD(-0.94)$, standard error $= 1.00$

c. Prepare a plot of speed × angle cell means for each level of lubrication and for each level of alloy to interpret the speed × angle interaction as levels of lubrication or levels of alloy change. Although the main effects of speed, lubrication, and alloy are significant, as is the speed × angle interaction, the modification of their effects by the significant three factor interactions must be examined.

9. a. generalized interactions are AC and B^2

b. generalized interactions are AB^2C and BC
Choose (b) to avoid confounding a main effect.

Chapter 12

1. 1. 1/2, 1, 0, 1, 16
3. 1/4, 2, 1, 3, 16
5. 1/4, 2, 1, 3, 32
7. 1/16, 4, 11, 15, 8
9. 1/32, 5, 26, 31, 16
11. 1/64, 6, 57, 63, 32

3. a. $ABEF$
b. IV
c. $A = BCD, ACDEF, BEF; B = ACD, BCDEF, AEF; C = ABD, DEF, ABCEF$
$D = ABC, CEF, ABDEF; E = ABF, CDF, ABCDE; F = ABE, CDE, ABCDF$
$AB = CD, EF, ABCDEF; AC = BD, ADEF, BCEF; AD = BC, ACEF, BDEF$
$AE = BF, ACDF, BCDE; AF = BE, ACDE, BCDF; BC = AD, BDEF, ACEF$
$BD = AC, BCEF, ADEF; BE = AF, ACED, BCDF; BF = AE, ACDF, BCDE$
$CD = AB, EF, ABCDEF; CE = DF, ABCF, ABED; CF = DE, ABDF, ABCE$
$DE = CF, ABCE, ABDF; DF = CE, ABDE, ABCF; EF = AB, CD, ABCDEF$

5. Yes. Suppose the 2_{III}^{5-2} design with factors $A, B, C, D,$ and E using 8 units with levels of D and E generated as $D = AB,$ and $E = AC.$ Fold over the design by adding a second 2_{III}^{5-2} design with all signs reversed from the first design, including the I column so that $I = F,$ a sixth factor. Then $D = ABF$ and $E = ACF.$ Design generators have four letters for a Resolution IV design using 16 units for a 2_{IV}^{6-2} design.

7. a. $D, E,$ and BE since they are no longer confounded with other main effects or two-factor interactions
b. Use $I = ABC = CDE.$ All main effects are no longer confounded with significant two-factor interactions or other main effects. However, $AD = BE$ and $AE = BD,$ and potential two-factor interactions are grouped as pairs.

9. 2_{IV}^{4-1}

11. a. V
b. $E = -ABCD$
c. See Table 12.3 and change signs on alias
d. No three-factor or higher interaction
e. Effect (estimate):
$A(0.145), B(0.087.5), C(0.0375), C(0.0375), D(-0.0375), E(-0.47), AB(0.015), AC(0.095),$

$AD(0.03)$, $AE(-0.1525)$, $BC(-0.0675)$, $BD(0.1625)$, $BE(-0.405)$, $CD(0.0725)$, $CE(0.135)$, $DE(-0.315)$

Choose E, BE, and DE as significant effects from normal scores plot. $MSE = 0.03978$ from analysis of variance. Effect standard error $= 0.0997$.

Source	df	SS	MS
E	1	0.88360	0.88360
BE	1	0.65610	0.65610
DE	1	0.39690	0.39690
Error	12	0.47738	0.03978

The normal probability plot to evaluate this model reveals no unusual residuals from assumed normal distribution.

Chapter 13

1. **a.** $s^2 = 6.3$, 5 d.f.

  b. $\widehat{\beta}_0 = 50.25$, $\widehat{\beta}_1 = 7.5$, $\widehat{\beta}_2 = 10.75$, standard error $= 0.89$
  H_0: $\beta_1 = 0$, $t_0 = 8.42$; H_0: $\beta_2 = 0$, $t_0 = 12.08$; $t(.025, 5) = 2.571$, Reject both; linear effects significant

  c. $\widehat{\beta}_{12} = 1.5$, standard error $= 0.89$, $t_0 = 1.69$, $t(.025, 5) = 2.571$, Do not reject; No significant interaction

  d. $\bar{y}_f - \bar{y}_c = -0.25$, standard error $= 1.98$, $t_0 = -0.13$, $t(.025, 5) = 2.571$, Do not reject; No significant departure from linear surface

  e. (A, B), $(25, 35)$, $(40, 56.5)$, $(55, 78)$, $(70, 99.5)$, $(85, 121)$, $(100, 142.5)$

3.

	2^n Design				Axial		
x_1	x_2	x_3	x_4	x_1	x_2	x_3	x_4
-1	-1	-1	-1	-2	0	0	0
$+1$	-1	-1	-1	$+2$	0	0	0
-1	$+1$	-1	-1	0	-2	0	0
$+1$	$+1$	-1	-1	0	$+2$	0	0
-1	-1	$+1$	-1	0	0	-2	0
$+1$	-1	$+1$	-1	0	0	$+2$	0
-1	$+1$	$+1$	-1	0	0	0	-2
$+1$	$+1$	$+1$	-1	0	0	0	$+2$
-1	-1	-1	$+1$				
$+1$	-1	-1	$+1$				
-1	$+1$	-1	$+1$		Center*		
$+1$	$+1$	-1	$+1$	x_1	x_2	x_3	x_4
-1	-1	$+1$	$+1$	0	0	0	0
$+1$	-1	$+1$	$+1$		*7 replications		
-1	$+1$	$+1$	$+1$				
$+1$	$+1$	$+1$	$+1$				

5. **a.** $\widehat{y} = 431 + 4.04x_1 + 64.75x_2 - 8.80x_1^2 - 44.55x_2^2 + 1.75x_1x_2$

Source	df	SS	MS
Regression	5	47,552	9,510
Linear	2	33,669	16,835
Quadratic	3	13,883	4,628
Error	7	2,795	399
Lack of fit	3	1,964	655
Pure error	4	831	208

b. Complete model: $F_0 = 45.7, F(.05, 5, 4) = 6.266,$ Reject
   Quadratic components: $F_0 = 22.3, F(.05, 3, 4) = 6.59,$ Reject
   Linear components: $F_0 = 80.9, F(.05, 2, 4) = 6.94,$ Reject
   Lack of fit: $F_0 = 3.2, F(.05, 3, 4) = 6.59,$ Do not reject
   Quadratic model is significant with no significant lack of fit to quadratic

c. Methionine: $x_1 = 0.302, M = 0.986$
   Carotene: $x_2 = 0.733, C = 75.91$

d. $\hat{y} = 455 - 44.572_1^2 - 8.742_2^2$
   $Z_1 = -0.725 - 0.024x_1 + 1.000x_2$
   $Z_2 = -0.320 + 1.000x_1 + 0.024x_2$
   From design specified in Table 13.4 for near rotatable and orthogonal blocking designs

Block I			**Block II**		
-1	-1	-1	1.682	0	0
$+1$	-1	-1	-1.682	0	0
-1	$+1$	-1	0	1.682	0
$+1$	$+1$	-1	0	-1.682	0
-1	-1	$+1$	0	0	1.682
$+1$	-1	$+1$	0	0	-1.682
-1	$+1$	$+1$	0	0	0
$+1$	$+1$	$+1$	0	0	0
0	0	0			
0	0	0			

$\alpha = 1.789$ for Equation (13.7)

10. a.

Sulphur	Asphalt	Sand
0.3	0.2	0.5
0.1	0.4	0.5
0.1	0.2	0.7
0.2	0.3	0.5
0.2	0.2	0.6
0.1	0.3	0.6

b. 2.02, 6 df

c. Psuedocomponent equation:
   $\hat{y} = 12.85\tilde{x}_1 + 3.00\tilde{x}_2 + 3.45\tilde{x}_3 + 39.70\tilde{x}_1\tilde{x}_2 + 40.40\tilde{x}_1\tilde{x}_3 + 10.90\tilde{x}_2\tilde{x}_3$
   Actual component equation:
   $\hat{y} = -559.70x_1 - 140.95x_2 - 58.70x_3 + 992.50x_1x_2 + 1010.00x_1x_3 + 272.50x_2x_3$

Source	df	SS	MS
Regression	5	491.62	98.32
Linear	2	262.42	131.21
Quadratic	3	229.20	76.40
Error	6	12.10	2.02

Test for linear: $F_0 = 64.96$, $F(.05, 2, 6) = 5.14$, Reject
Test for quadratic: $F_0 = 37.82$, $F(.05, 3, 6) = 4.76$, Reject

Chapter 14

1. a. $y_{ijk} = \mu + \alpha_i + \rho_k + d_{ik} + \beta_j + (\alpha\beta)_{ij} + e_{ijk}$ $i, k = 1, 2, 3, 4$ $j = 1, 2, 3$
α_i = fixed plant population density effect, β_j = fixed hybrid effect, $(\alpha\beta)_{ij}$ = fixed population × hybrid interaction effect, ρ_k = fixed block effect, d_{ik} = whole plot random error with mean 0 and σ_d^2, e_{ijk} = subplot random error with mean 0 and variance σ_e^2

Source	df	SS	MS
Blocks	3	408.98	136.33
Population (*P*)	3	6,429.39	2,143.13
P linear	1	5,658.23	5,658.23
P quadratic	1	767.64	767.64
P cubic	1	3.51	3.51
Error(1)	9	466.54	51.84
Hybrid (*H*)	2	881.41	440.70
$H \times P$	6	207.51	34.58
$H \times P$ linear	2	72.74	36.37
$H \times P$ quadratic	2	92.30	46.15
$H \times P$ cubic	2	42.47	21.24
Error(2)	24	595.70	24.82

b.

Mean Head Seed Weight (g)

Hybrid	Plants per Meter of Row				Hybrid Means
	10	15	25	40	
TAM 680	38.6	32.6	18.4	13.5	25.8
RS 671	45.2	35.5	24.0	17.1	30.4
Tx 399 × Tx 2536	57.3	40.4	25.9	21.6	36.3
Population means	47.0	36.1	22.8	17.4	

Factor(standard error of mean): Population(2.1), Hybrid(1.2), Cells(2.5)

c (i)) 1.8, (ii) 2.9, (iii) 4.11
d. Interaction: $F_0 = 1.39$, $F(.05, 6, 24) = 2.51$, Do not reject
Population: $F_0 = 41.34$, $F(.05, 3, 9) = 3.86$, Reject
Hybrid: $F_0 = 17.76$, $F(.05, 2, 24) = 3.40$, Reject
e. Subplot: $RE = 0.98(1.3) = 1.27$
Whole plot: $RE = 0.88(0.62) = 0.55$
f. See analysis of variance table for polynomial sums of squares partitions. No interaction partitions are significant with $F(.05, 2, 24) = 3.40$ as a critical value.
Population linear: $F_0 = 109.15$, Population quadratic: $F_0 = 14.81$, Population cubic: $F_0 = 0.07$, $F(.05, 1, 9) = 5.12$, Reject for linear and quadratic

3. a. $y_{ijklm} = \mu + \alpha_i + \beta_j + (\alpha\beta)_{ij} + \rho_k + \gamma_l + d_{kl} + \theta_m + (\alpha\theta)_{im} + (\beta\theta)_{jm}$
$+ (\alpha\beta\theta)_{ijm} + e_{klm}$ $i, j = 1, 2$ $k, l = 1, 2, 3, 4$ $m = 1, 2, 3$
α_i = fixed A effect, β_j = fixed B effect, $(\alpha\beta)_{ij}$ = fixed $A \times B$ interaction effect, ρ_k = fixed row effect, γ_l = fixed column effect, d_{kl} = random whole plot error with mean 0 and variance σ_d^2, θ_m = fixed C effect, $(\alpha\theta)_{im}$ = fixed $A \times C$ interaction effect, $(\beta\theta)_{jm}$ = fixed $B \times C$ interaction effect, $(\alpha\beta\theta)_{ijm}$ = fixed $A \times B \times C$ interaction effect, e_{klm} = random subplot error with mean 0 and variance σ_e^2

b.

Source	df	Expected Mean Square
Rows	3	
Columns	3	
A	1	$\sigma_e^2 + 3\sigma_d^2 + 24\theta_a^2$
B	1	$\sigma_e^2 + 3\sigma_d^2 + 24\theta_b^2$
AB	1	$\sigma_e^2 + 3\sigma_d^2 + 12\theta_{ab}^2$
Error(1)	6	$\sigma_e^2 + 3\sigma_d^2$
C	2	$\sigma_e^2 + 16\theta_c^2$
AC	2	$\sigma_e^2 + 8\theta_{ac}^2$
BC	2	$\sigma_e^2 + 8\theta_{bc}^2$
ABC	2	$\sigma_e^2 + 4\theta_{abc}^2$
Error(2)	24	σ_e^2

7. a. $y_{ijkl} = \mu + \alpha_i + \rho_j + d_{ij} + \beta_k + \gamma_l + (\beta\gamma)_{kl} + (\alpha\beta)_{ik} + (\alpha\gamma)_{il} + (\alpha\beta\gamma)_{ikl} + e_{ijkl}$
$i, j = 1, 2$ $k, l = 1, 2, 3$
α_i = fixed phosphorus effect, ρ_j = fixed block effect, d_{ij} = random whole-plot error with mean 0 and variance σ_d^2, β_k = fixed nitrogen effect, γ_l = fixed water effect, $(\beta\gamma)_{kl}$ = fixed nitrogen × water interaction effect, $(\alpha\beta)_{ik}$ = fixed phosphorus × nitrogen interaction effect, $(\alpha\gamma)_{il}$ = fixed phosphorus × water interaction effect, $(\alpha\beta\gamma)_{ikl}$ = fixed phosphorus × nitrogen × water interaction effect, e_{ijkl} = random subplot error with mean 0 and variance σ_e^2

Source	df	SS	MS
Blocks	1	0.67	0.67
Phosphorus (P)	1	1.25	1.25
Error(1)	1	27.56	27.56
Nitrogen (N)	2	2,768.65	1,384.32
Water (W)	2	751.84	375.92
NW	4	242.08	60.52
PN	2	12.71	6.35
PW	2	0.81	0.40
PNW	4	13.87	3.47
Error(2)	16	101.13	6.32

b.

Water	Nitrogen	Phosphorus 0	Phosphorus 245	$N \times W$ Means
16	0	8.4	12.6	10.5
	130	35.3	33.7	34.55
	260	37.7	36.7	37.2
22	0	7.6	8.6	8.1
	130	20.7	20.8	20.8
	260	28.6	27.3	27.9
28	0	7.6	8.4	8.0
	130	20.6	19.8	20.2
	260	21.3	23.1	22.2

Water	Phosphorus 0	Phosphorus 245
16	27.1	27.6
22	18.9	18.9
28	16.5	17.1

Nitrogen	Phosphorus 0	Phosphorus 245
0	7.8	9.8
130	25.5	24.8
260	29.22	29.00

Means(std err): $PNW(1.8)$, $NW(1.3)$, $PW(1.0)$, $PN(1.0)$
P means: 20.8, 21.2, standard error $= 1.2$
N means: 8.8, 25.1, 29.1, standard error $= 0.7$
W means: 27.4, 18.9, 16.8, standard error $= 0.7$

c. (i) 1.7, (ii) 1.0, (iii) 1.0, (iv) 1.8

d. PNW: $F_0 = 0.55$, $F(.05, 4, 16) = 3.01$, Do not reject
 PW: $F_0 = 0.06$, $F(.05, 2, 16) = 3.63$, Do not reject
 PN: $F_0 = 1.01$, $F(.05, 2, 16) = 3.63$, Do not reject
 NW: $F_0 = 9.58$, $F(.05, 4, 16) = 3.01$, Reject
 W: $F_0 = 59.48$, $F(.05, 2, 16) = 3.63$, Reject
 N: $F_0 = 219.03$, $F(.05, 2, 16) = 3.63$, Reject
 P: $F_0 = 0.05$, $F(.05, 1, 1) = 161.4$, Do not reject

e.

Source	df	MS	F_0
N linear (N_l)	1	2,464.43	389.92*
N quadratic (N_q)	1	304.22	48.13*
W linear (W_l)	1	672.04	106.33*
W quadratic (W_q)	1	79.80	12.63*
$N_l W_l$	1	154.38	24.43*
$N_l W_q$	1	0.46	0.07
$N_q W_l$	1	40.89	6.47*
$N_q W_q$	1	46.35	7.33*
PN_l	1	7.04	1.11
PN_q	1	5.67	0.90
PW_l	1	0.01	0.00
PW_q	1	0.80	0.13
$PN_l W_l$	1	9.77	1.55
$PN_l W_q$	1	0.01	0.00
$PN_q W_l$	1	0.46	0.07
$PN_q W_q$	1	3.64	0.58

*Reject null hypothesis since $F_0 \geq F(.05, 1, 16) = 4.49$

Use a quadratic regression model with N and W for interpretations.

Chapter 15

1. a. Between-subjects design: A completely randomized design with three diet treatments and four subjects randomly assigned to each diet.
  Within-subjects design: Three repeated measures on each subject at 15, 30, and 45 minutes

 b.

Diet	Minutes 15	30	45
1	19.0	35.0	31.5
2	22.0	20.0	11.5
3	26.0	27.3	35.8

 c. $y_{ijk} = \mu + \alpha_i + d_{ik} + \beta_j + (\alpha\beta)_{ij} + e_{ijk}$ $i, j = 1, 2, 3$ $k = 1, 2, 3, 4$
  α_i = fixed diet effect, β_j = fixed time effect, $(\alpha\beta)_{ij}$ = fixed diet × time effect
  d_{ik} = normally distributed independent random experimental error for subjects within diets with mean 0 and variance σ_d^2
  e_{ijk} = normally distributed random experimental error on repeated measures with mean 0 and variance σ_e^2
  The e_{ijk} are assumed to satisfy the Huynh–Feldt condition of equal variance of the difference between all time periods

d.

Source	df	SS	MS
Diet (D)	2	1,020.67	510.33
Error (1)	9	413.33	45.93
Time (T)	2	170.17	85.08
T linear	1	92.04	92.04
T quadratic	1	78.13	78.13
$D \times T$	4	869.67	217.42
$D \times T$ linear	2	631.08	315.54
$D \times T$ quadratic	2	238.58	119.29
Error (2)	18	128.17	7.12

Diet $\times$ Time interaction: $F_0 = 30.5, F(.05, 4, 18) = 2.93$, Reject
Time: $F_0 = 11.9, F(.05, 2, 18) = 3.55$, Reject
Diet: $F_0 = 11.1, F(.05, 2, 9) = 4.26$, Reject
Diet means: 28.5, 17.8, 29.7, standard error $= 1.96$
Time means: 22.3, 27.4, 26.3, standard error $= 0.77$
Cell means: See means in part b., standard error $= 1.33$

e. See sums of squares partitions in analysis of variance table
$D \times T$ linear: $F_0 = 44.3, F(.05, 2, 18) = 3.55$, Reject
$D \times T$ quadratic: $F_0 = 16.8, F(.05, 2, 18) = 3.55$, Reject
T linear: $F_0 = 12.9, F(.05, 1, 18) = 4.41$, Reject
T quadratic: $F_0 = 11.0, F(.05, 1, 18) = 4.41$, Reject

f. $W = 0.41$, chi-square approximation $= 7.12, 2$ d.f., $P = .0285$.

g.

Source	df	SS	MS	F_0
Time (linear)	1	92.04	92.04	12.12*
Diet $\times T$ (linear)	2	631.08	315.54	41.53*
Error	9	68.40	7.60	
Time (quadratic)	1	78.13	78.13	11.76*
Diet $\times T$ (quadratic)	2	238.58	119.29	17.96*
Error	9	59.76	6.64	

*Reject null hypotheses

3. **a** Subjects are random blocks in a randomized complete blocks design and each subject has received all treatments in random order

b. $y_{ij} = \mu + \tau_i + \rho_j + e_{ij} \quad i, j = 1, 2, \ldots, 5$
$\tau_i = $ fixed fabric effect, $\rho_j = $ random subject effect, $e_{ij} = $ random experimental error with mean 0 and variance σ^2, assumed to be independent and normally distributed. Also assume there are no carryover effects of treatments

c.

			Treatment		
Contrast	1	2	3	4	5
(i) 1 vs. 2	1	-1	0	0	0
(ii) 2 vs. 4	0	1	0	-1	0
(iii) 2 vs. 5	0	1	0	0	-1
(iv) 3 vs 4	0	0	1	-1	0

Orthogonal contrast pairs are (i) and (iv) and (iii) and (iv). Sum of coefficient crossproducts of other pairs not equal to zero

d.

Source	df	SS	MS
Subjects	4	17.94	4.49
Treatments	4	316.28	79.07
Error	16	227.08	14.19

Test for treatments: $F_0 = 5.57, F(.05, 4, 16) = 3.01$, Reject

Contrast	SS	F_0
1 vs. 2	246.81	17.39*
2 vs. 4	69.48	4.90*
2 vs. 5	21.32	1.50
3 vs. 4	20.45	1.44

*Reject null hypothesis since $F_0 \geq$ $F(.05, 1, 16) = 4.49$.

5. a. Between-subjects design: Randomized complete blocks design with four varieties randomized to plots in each of three blocks
Within-subjects design: Four levels of water application on the four subplots spaced at equal distances along the length of the variety plot.

b.

		Distance		
Variety	1	2	3	4
1	416.0	447.3	436.3	328.3.
2	570.7	513.7	524.7	468.4
3	358.1	429.2	449.1	326.9
4	455.1	563.1	536.0	338.2

c. $y_{ijk} = \mu + \alpha_i + \rho_k + d_{ik} + \beta_j + (\alpha\beta)_{ij} + e_{ijk}$ $\quad i, j = 1, 2, 3, 4 \quad k = 1, 2, 3$
α_i = fixed variety effect, ρ_k = fixed block effect, β_j = fixed distance effect, $(\alpha\beta)_{ij}$ = fixed variety × distance interaction effect, d_{ik} = whole-plot random error with mean 0 and variance σ_d^2, e_{ijk} = subplot random error with mean 0 and variance σ_e^2. The subplot errors are assumed to satisfy the Huynh–Feldt condition of equal variances for differences between all distances.

d.

Source	df	SS	MS
Blocks	2	27,000	13,500
Varieties (V)	3	128,063	42,688
Error (1)	6	41,902	6,984
Distance (D)	3	119,128	39,709
D linear	1	39,145	39,145
D quadratic	1	76,217	76,217
D cubic	1	3,766	3,766
$V \times D$	9	44,837	4,982
$V \times D$ linear	3	7,499	2,500
$V \times D$ quadratic	3	36,473	12,158
$V \times D$ cubic	3	864	288
Error (2)	24	177,735	7,406

Varieties: $F_0 = 6.11$, $F(.05, 3, 6) = 4.76$, Reject
Distance: $F_0 = 5.36$, $F(.05, 3, 24) = 3.01$, Reject
$V \times D$ interaction: $F_0 = 0.67$, $F(.05, 9, 24) = 2.30$, Do not reject
Variety means: 407.0, 519.4, 390.8, 473.1, standard error = 24.1
Distance means: 450.0, 488.3, 486.5, 365.4, standard error = 24.8
Cell means: See table in part b., standard error = 49.7
e. See sums of squares partitions in analysis of variance table.
$V \times D$ linear: $F_0 = 0.34$, $V \times D$ quadratic: $F_0 = 1.64$, $V \times D$ cubic: $F_0 = 0.04$,
$F(.05, 3, 24) = 3.01$. Do not reject any interaction null hypothesis
D linear: $F_0 = 5.29$, D quadratic: $F_0 = 10.29$, D cubic: $F_0 = 0.51$,
$F(.05, 1, 24) = 4.26$, Reject for linear and quadratic partitions
f. $W = 0.29$, chi-square approximation = 5.82, 5 d.f., $P = .32$
g. G–G epsilon = .5634, adjusted d.f.: (Distance, 1.69), (Variety × Distance, 5.07), (Error, 13.57); Conclusions do not change

h.

Source	df	MS	F_0	$P > F$
Blocks	2	55,834		
D linear	1	39,145	11.4	.0149
$V \times D$ linear	3	2,500	0.7	.5718
Error	6	3,346		
Blocks	2	1,044		
D quadratic	1	76,217	21.7	.0035
$V \times D$ quadratic	3	12,158	3.5	.0916
Error	6	3,516		
Blocks	2	2,138		
D cubic	1	3,766	1.3	.3052
$V \times D$ cubic	3	288	0.11	.9594
Error	6	2,997		

Chapter 16

1. **a.** Yes. Each treatment follows each of the other treatments one time, and each treatment occurs once in each period and once with each subject.

b. $y_{ijk} = \mu + \alpha_i + b_{ij} + \gamma_k + \tau_{d(i,k)} + \lambda_{c(i,k-1)} + e_{ijk}$
$i, k, d, c = 1, 2, 3, 4 \quad j = 1, 2$
α_i = fixed treatment sequence effect, b_{ij} = random subject within sequence effect with mean 0 and variance σ_b^2, γ_k = fixed period effect, $\tau_{d(i,k)}$ = fixed direct effect of the treatment in period k of sequence i, $\lambda_{c(i,k-1)}$ = fixed carryover effect of the treatment administered in period $k - 1$ of sequence i, e_{ijk} = random error for the subplot in period k with mean 0 and variance σ^2. Experimental errors are assumed to be normally distributed and at least satisfy the Huynh–Feldt condition of equal variances of differences between all time periods. Also assume no carryover effects higher than first order.

c.

Source	df	SS	MS
Sequence	3	2.76	0.92
Subjects	4	4.71	1.18
Periods	3	0.05	0.02
Drug (adj.)	3	24.25	8.08
Carryover (adj.)	3	0.06	0.02
Error	15	1.99	0.13

Drug: $F_0 = 62.15$, $F(.05, 3, 15) = 3.29$, Reject
Carryover: $F_0 = 0.15$, $F(.05, 3, 15) = 3.29$, Do not reject

d.

	Least Squares Mean for Treatment					
	A	**B**	**C**	**D**		
Contrast	**1.17**	**1.16**	**3.28**	**1.17**	c	s_c
A vs.B	1	−1	0	0	0.01	0.19
B vs. C	0	1	−1	0	−2.12	0.19
B vs. D	0	1	0	−1	−0.01	0.19

A vs. B: $t_0 = 0.05$, B vs. C: $t_0 = -11.16$, B vs. D: $t_0 = -0.05$,
Bonferroni t: $t(.025, 3, 15) = 2.69$, Reject for B vs. C, Do not reject for A vs. B and B vs. D

3. **a.** Yes. Every treatment follows all treatments including itself one time.

b. Yes. Carryover and direct treatment effects are orthogonal in this design, whereas they are not orthogonal in the four-period design.

c.

Source	df	SS	MS
Steer	3	20.95	6.98
Period	4	11.70	2.93
Roughage	3	114.55	38.18
Carryover	3	3.25	1.08
Error	6	12.50	2.08

d. Roughage: $F_0 = 18.36$, $F(.05, 3, 6) = 4.76$, Reject
Carryover: $F_0 = 0.52$, $F(.05, 3, 6) = 4.76$, Do not Reject

e. $s(\hat{\mu}_i - \hat{\mu}_c) = 0.93$, $d(.05, 3, 6) = 3.10$, Reject if $| \hat{\mu}_i - \hat{\mu}_c | \geq D(3, .05) = 2.88$

Treatment	$\hat{\mu}_i$	$\hat{\mu}_i - \hat{\mu}_{ci}$	95% SCI (L, U)
Cottonseed	72.79	-6.67	$(-9.55, -3.79)$
Bermuda	75.74	-3.88	$(-6.76, -1.00)$
Wheat	77.49	-2.113	$(-5.01, 075)$
Alfalfa	79.62	Control	

The dry matter digestion of cottonseed and Bermuda is less than alfalfa since SCI does not include 0.

5. a. Yes. Each treatment precedes all other treatments three times.

b. $y_{ijk} = \mu + \alpha_i + \gamma_j + \tau_{d(i,j)} + \lambda_{c(i,j-1)} + e_{ij}$ $i = 1, 2, \ldots, 12$ $j, d, c = 1, 2, 3, 4$
α_i = fixed sequence effect, γ_j = fixed period effect, $\tau_{d(i,j)}$ = fixed direct keyboard layout treatment effect in period j and sequence i, $\lambda_{c(i,j-1)}$ = fixed carryover effect of treatment administered in period $(j - 1)$ and sequence i,
e_{ij} = random error with mean 0 and variance σ^2. Experimental errors are assumed to be normally distributed and to at least satisfy the Huynh–Feldt condition of equal variances between all time periods. Assume no carryover effects higher than first order. The normal distribution assumption may not be valid since the response variable is the discrete number of errors committed by subjects.

Source	df	SS	MS
Sequence	11	12.23	1.11
Period	3	0.40	0.13
Layout (adj.)	3	158.73	52.91
Carryover (adj.)	3	1.21	0.40
Error	27	13.58	0.50

Layout: $F_0 = 105.82$, $F(.05, 3, 27) = 2.96$, Reject
Carryover: $F_0 = 0.80$, $F(.05, 3, 27) = 2.96$, Do not reject

Chapter 17

1. a. $y_{ij} = \mu_i + \beta(x_{ij} - \bar{x}_{..}) + e_{ij}$ $i = 1, 2, 3$ $j = 1, 2, \ldots, 6$
μ_i = alloy mean, x_{ij} = weld diameter, β = coefficient for linear regression of weld strength, y, on weld diameter, x, e_{ij} = random experimental error with mean 0 and variance σ^2.
Assume normally distributed errors with homogeneous variance; β is the same for all alloys; and alloy treatments do not affect weld diameter

b.

Source	df	SS	MS
Regression	1	2,838.92	2,838.92
Alloys	2	2,005.97	1,002.99
Error	14	716.16	51.15

Regression: $F_0 = 55.50, F(.05, 1, 14) = 4.50$, Reject, $\widehat{\beta} = 5.505$
Alloys: $F_0 = 19.61, F(.05, 2, 14) = 3.74$, Reject

c.

Alloy	$\bar{y}_{ia}$	Std Err
1	61.73	3.29
2	78.04	3.25
3	50.88	2.92

Average standard error of difference $= 4.63$

d. Plot the following regression lines:
   Alloy 1: $\widehat{y} = 50.42 + 5.505(x - 15.83)$
   Alloy 2: $\widehat{y} = 88.75 + 5.505(x - 19.83)$
   Alloy 3: $\widehat{y} = 51.50 + 5.505(x - 18.00)$

e. $E = 3.69$

f. MS(homogeneity) $= 129.13$, 2 d.f.; $MSE = 38.16$, 12 d.f.
   $F_0 = 3.38, F(.05, 2, 12) = 3.89$, Do not reject

i. coefficient(std error): $\widehat{\beta}_1 = 3.92(1.07), \widehat{\beta}_2 = 7.35(0.96), \widehat{\beta}_3 = 4.19(1.44)$

3. a. $y_{ij} = \mu + \tau_i + \rho_j + \beta(x_{ij} - \bar{x}_{..}) + e_{ij}$ $i = 1, 2, 3$ $j = 1, 2, 3, 4$
   $\tau_i =$ irrigation effect, $\rho_j =$ block effect, $x_{ij} =$ number of plants, $\beta =$ coefficient for linear regression of yield, y, on number of plants, $e_{ij} =$ random experimental error with mean 0 and variance σ^2.
   Assume normally distributed errors with homogeneous variance; β is the same for all irrigations; and irrigation treatments do not affect number of plants

b.

Source	df	SS	MS
Blocks	3	1.50	0.50
Regression	1	1.08	1.08
Irrigation	2	1.52	0.76
Error	5	0.44	0.09

Regression: $F_0 = 12.0, F(.05, 1, 5) = 6.61$, Reject, $\widehat{\beta} = 0.063$
Irrigation: $F_0 = 8.44, F(.05, 2, 5) = 5.79$, Reject

c.

Irrigation	$\bar{y}_{ia}$	Std Err
1	2.91	0.15
2	2.23	0.15
3	2.11	0.15

Average standard error of difference $= 0.21$

d. $E = 2.78$

5. a. $E_{xy} = 1.825, E_{xx} = 0.50, \widehat{\beta} = 3.65 =$ missing value estimate

Source	df	SS	MS
Blocks (unadj.)	3	1.95	
Regression (x)	1	6.66	6.66
Irrigation (adj.)	2	1.38	0.69
Error	5	1.24	0.25

Irrigation: $F_0 = 2.76, F(.05, 2, 5) = 5.79$, Do not reject

Index

661